A Dictionary of

Environment and Conservation

CHRIS PARK

OXFORD

UNIVERSITY PRESS

OXFORD
UNIVERSITY PRESS

Oxford University Press, Great Clarendon Street, Oxford OX2 6DP
Oxford University Press is a department of the University of Oxford.
It furthers the University's objective of excellence in research, scholarship, and education
by publishing worldwide in

Oxford New York
Auckland Cape Town Dar es Salaam Hong Kong Karachi Kuala Lumpur
Madrid Melbourne Mexico City Nairobi New Delhi Shanghai Taipei Toronto

With offices in Argentina Austria Brazil Chile Czech Republic France Greece
Guatemala Hungary Italy Japan Poland Portugal
Singapore South Korea Switzerland Thailand Turkey Ukraine Vietnam

Oxford is a trademark of Oxford University Press in the UK and in certain other countries

Published in the United States by Oxford University Press Inc., New York

British Library Cataloguing in Publication Data

Data available

Library of Congress Cataloging in Publication Data
Data available

Typeset by SPI Publisher Services, Pondicherry, India
Printed in Great Britain by Clays Ltd, St Ives plc

ISBN 0-19-860995-7
ISBN 978-0-19-860995-7

1

Preface

Nature never errs. Things always follow their principles. Though complex, they are never chaotic. Though many, they are not confused.

Wang Pi (Chinese philosopher, AD 226–49)
Quoted by Adelaine Yen Mah (2001) Watching the tree to catch a hare.
Harper Collins.

According to Simon Winchester (The Meaning of Everything, 2003, OUP) the first Oxford English Dictionary was published in 1928. It came in twelve volumes, totalled 15 490 pages, contained 414 825 definitions and had taken a team of writers 70 years to complete. This dictionary was written in less than 3 years, very much on a part-time basis in the gaps between day-job and the-rest-of-life, by one person with a small PC. It contains just under 9000 definitions and around 240 000 words

Wang Pi's advice has informed much of my thinking and decision-making in writing this dictionary, because my aim has been to make clear the principles on which nature operates. Taking on this challenge has involved many choices as I have sought to produce a collection of definitions that is comprehensive but accessible, which embraces words and terms that are in common use throughout much of the English-speaking world (particularly in the UK, North America, and Australasia), and which will not date too quickly. I have also resisted the temptation to include long lists of names (of plants and animals, for example), and tried to balance science and non-science entries. To reinforce Wang Pi's observation that the principles of nature are many but not confused, I have tried to cross-reference entries as much as seems useful, using two conventions—the standard convention *See also*, *Compare* and *Contrast*, and using an asterisk before a term that is defined in its own right (e.g. *niche).

Inevitably, the source material that I have drawn on is very diverse, widely scattered, and rooted in many different disciplines. I have tried hard to weave many different threads together into a coherent tapestry, and (to change the metaphor) to impose one voice on this disparate material, in the rather optimistic hope of turning a cacophony into a symphony! I dread to think quite how lexicographers of old managed to do their work without access to the Internet, notebook-sized computers, and software that allows us to word-process, save back-up copies, and (most importantly!) instantly sort vast amounts of text.

I have adopted a style that I hope works well for the informed lay reader. To this end, whilst a great many technical and specialized terms are included, I have sought to distil what is sometimes complex material down to manageable sound bites and to explain things as clearly as possible without being patronizing.

I owe a great debt of thanks to a number of people who have played a part in the creation of this dictionary. I would never have been brave or stupid enough to think of writing a dictionary (on anything) had not Ruth Langley, Commissioning Editor at OUP, first approached me and suggested this one. Her faith was stronger than mine to start with, but I soon warmed to the idea, and I thank her for the bold approach and for her sustained support and interest during the writing. Jo Walker and Angela Carpenter offered invaluable suggestions for improving the text, for which I am enormously grateful. Bridget Johnson's attention to detail while copy-editing the text was meticulous, and is greatly appreciated. My closest family have sacrificed time with me while I soldiered on with the research and writing, and even when I was able to spend time with them, my mind was often elsewhere. My parents Alex and Margaret have been wonderful; always there in the background, quietly proud and supportive, even when they have little idea what I am doing. Sadly, my mother died while I was writing this dictionary, but her memory lives on through all that I do. My two eldest, Emma and Andrew, have flown the nest but are still very dear to me. My youngest offspring Sam and Elizabeth, both still in the nest, are resigned to having a Dad who seems to do little else but sit silently at the keyboard evening after evening; they tolerate my behaviour with charitable indifference. The true hero of the piece is my wife and travelling companion Penny, who keeps me grounded and brought the sunshine back into my life. I dedicate this dictionary to her, with love.

Chris Park

Lancaster, UK
February 2006

Contents

A Dictionary of Environment and Conservation

Appendices

aa Volcanic *lava rock that is thick and porous and has a rough, jagged surface when it cools. It is a Hawaiian word, pronounced *ah-ah*.

abandoned well Any *well that has not been used for a long time and/or is not properly sealed, and is so poorly maintained that it cannot be used for the purpose for which it was intended.

abatement The elimination or reduction of emissions that create *pollution. Examples include smoke abatement and noise abatement.

abatement debris Waste materials produced by *remediation activities.

Abbey, Edward US author (1927–89) who wrote the book *Desert Solitaire* and is remembered for his strong criticism of public land policies and his passionate advocacy of environmental issues.

abiotic Devoid of life, not *biotic, non-biological. Examples of the abiotic environment include climate, geology, and atmosphere.

ablation 1. The process by which *snow or *ice is lost from a *glacier or *ice cap by melting, *sublimation, and *evaporation, which produces *meltwater.
2. The removal of rock particles by wind action.

Ablation zone The lower part of a glacier, where more ice melts than accumulates over a year. Also known as **ablation area** or **zone of ablation**. *Contrast* ACCUMULATION ZONE.

Aborigines *See* INDIGENOUS PEOPLES.

abrasion The *physical weathering or mechanical weathering of rock or sediment by running water, glaciers, or wind loaded with fine particles, through such processes as scratching, rubbing, grinding, or wearing away by friction.

absolute age Age that is expressed in number of years, rather than age relative to a particular event. In geology absolute age is considered misleading because ages are subject to error. *Contrast* RELATIVE AGE.

absolute chronology A *chronology that determines the age of a feature or event in years. *Contrast* RELATIVE CHRONOLOGY.

absolute dating A range of methods for determining the *absolute age of an object or material, for example using *radiometric dating which measures the *decay product produced by *radioactive decay in minerals, or via *dendrochronology. *Contrast* RELATIVE DATING. *See also* GEOCHRONOLOGY.

absolute humidity A measure of *humidity, or the actual amount of *water vapour in a body of air, usually expressed in grams of water per cubic metre of air $(g\ m^{-3})$. This can also be expressed as *specific humidity or *vapour pressure. *Contrast* RELATIVE HUMIDITY. *See also* DEW POINT, MIXING RATIO.

absolute poverty An extreme state of *poverty, in which the standard of living is below the minimum that is needed for the maintenance of life and *health. *Contrast* RELATIVE POVERTY.

absolute relief The maximum elevation of a particular area above *sea level. *Contrast* RELATIVE RELIEF.

absolute scarcity A condition that exists when there is not enough of a *resource in existence to satisfy existing demand for it. *Contrast* RELATIVE SCARCITY.

absolute zero A temperature of $-273°C$, $-460°F$, the zero point on the

*Kelvin temperature scale, at which atomic and molecular motion stops, at least theoretically.

absorb To take something in (such as the penetration of a solid substance by a liquid, by capillary, osmotic, solvent, or chemical action). In energy terms, to take in energy and then not reflect it.

absorbed dose The amount (*dose) of a chemical substance that is absorbed by, and thus enters the body of, an organism exposed to it. Most commonly expressed as milligrams per kilogram body weight per day (mg kg^{-1}day^{-1}). Also known as **internal dose**.

absorbent The ability of a substance to *absorb, which reflects its *porosity.

absorption 1. The penetration of one substance into or through another, such as the absorption of water into soil, or the uptake of water and nutrients by a cell or organism which *absorbs them. *Contrast* ADSORPTION, DESORPTION, INGESTION.
2. A way of containing an *oil spill, in which oil is absorbed onto special materials from which it can be squeezed out for reuse or disposal.

absorptive capacity The maximum amount of waste material that can be naturally absorbed by the environment on a sustainable basis, without causing environmental damage.

abstraction The action of removing something, such as the abstraction of water from a river, lake or *groundwater, for use in industry. Also known as **extraction**, particularly when applied to a *rock or *mineral.

abundance The number or amount of something. *See also* SPECIES ABUNDANCE, RELATIVE ABUNDANCE.

abyss *See* OCEAN FLOOR.

abyssal hill A hill on the ocean floor, which can range up to several hundred metres in height and several kilometres in diameter.

abyssal plain The flat, deep part of the ocean floor that lies offshore from the

*continental rise, below about 2000 metres below the ocean surface. It usually has a very low slope (less than 1:1000) and is covered by a thick layer of sediment.

abyssal zone An ecological zone within the ocean that includes the *continental rise and the *ocean floor, which accounts for three-quarters of the deep ocean floors. Conditions are cold (temperatures remain steady at about 4°C) and dark in this zone, more than 2000 metres below the ocean surface.

accelerated erosion An increased rate of *erosion above natural levels. This can be natural (for example, caused by hurricane damage) but is often caused directly or indirectly by human activities. It removes soil much faster than soil naturally develops and can lead to serious environmental damage, both where it occurs (because it reduces plant productivity and can lead to *mass movement of soils and sediments) and downslope or downstream where it is deposited (where it can, for example, infill reservoirs and lakes, change river channels and increase the growth of a *delta).

acceptable daily intake The largest daily amount of a substance that a typical person can consume or be exposed to over a lifetime, without producing unwanted health risk, usually expressed in milligrams per kilogram per day (mg kg^{-1} day^{-1}).

acceptable risk The level of *risk that is regarded as acceptable by the community or authorities, below which no specific action by local government is deemed necessary other than making the risk known. *Contrast* AVOIDABLE RISK, UNACCEPTABLE RISK.

accident site The location of an accident. Examples include the failure of a pipe in a factory that leads to the release of hazardous chemicals, or the site of an accidental release of pollutants into a lake or river.

accident type The accident event that causes, or the circumstances that give rise to, injury to people or damage to property. Examples include road traffic accidents, an explosion in a building or a flood that drowns people.

acclimation An adjustment that occurs in an individual organism that allows it to tolerate a change (often in a single factor, such a temperature) in its environment, or to tolerate a new environment. The adjustment is usually gradual and reversible, and it can be physiological or behavioural, or both. Also known as **acclimatization**. *See also* ADAPTATION.

acclimatization *See* ACCLIMATION.

accounting The systematic recording, reporting, and analysis of transactions, particularly financial ones, in order to support effective management and decision-making, by providing the required information. *See also* ENVIRONMENTAL ACCOUNTING, INTERNALIZATION.

accretion A process of growth by accumulation and adhesion. Examples include the build-up or accumulation of *sediment, and the process by which *precipitation particles grow, by the collision of an ice crystal or snowflake with a supercooled liquid droplet that freezes upon impact. On a large scale, the term is also used to describe primitive planetary growth, and to describe the addition of material at the edges of pre-existing continents.

accumulation rate The speed at which something (such as a layer of sediment on the floor of a lake or ocean, or layers of ice on the surface of a glacier) accumulates or builds up. Usually expressed as depth per unit time (for example millimetres per year, mm year^{-1}).

accumulation zone An area where accumulation of material occurs.
   1. The upper part of a glacier, where snow and ice accumulate because annual snowfall exceeds loss by *ablation. Also known as **zone of accumulation**. *Contrast* WASTAGE ZONE.
   2. The lower part of a hillslope, where sediment transported downhill by *mass movement processes accumulates.

ACE *See* US ARMY CORPS OF ENGINEERS.

acid An organic or inorganic compound that contains hydrogen and releases hydrogen ions (H^+) when dissolved in water. Acids also react with metals to release hydrogen, and with bases to form salts. All acids turn litmus paper red and have a *pH of less than 7.0. Many have a sour taste and can cause severe skin burns in humans. *Contrast* ALKALI.

acid aerosol Acidic liquid or solid particles small enough to become airborne. High concentrations can irritate the lungs and have been associated with respiratory diseases like asthma.

acid deposition *See* ACID RAIN.

acid grassland Grassland that develops over nutrient-poor, acidic soils, often as a result of a long period of grazing or burning of woodland. Also known as **grass heath**.

acid loading The total amount of acid material that is added to a particular habitat or area over a given period of time, usually expressed as grams per square metre per year (g m^{-2} year^{-1}).

acid mine drainage The drainage of water from areas that have been mined for coal or other mineral ores (including gold, silver, *copper, iron, *zinc, and *lead). The water has high acidity (low *pH, sometimes less than 2.0) because it has been in contact with material that contains sulphur, producing *sulphuric acid (H_2SO_4) which is strongly corrosive and dissolves metals such as lead, zinc, copper, *arsenic, *selenium, *mercury, and *cadmium into ground water and surface water. Acid mine drainage can poison groundwater and drinking water, and it is harmful to aquatic organisms and habitats.

acid neutralizing capacity *See* BUFFERING CAPACITY

acid precipitation *See* ACID RAIN.

Acid Rain Program A major air pollution control initiative in the USA, established under the *Clean Air Act

ACID RAIN

A form of *air pollution that is created by the transformation within the atmosphere of gaseous pollutants, principally *sulphur dioxide (SO_2) and *nitrogen oxides (NO_x), which are emitted mainly from coal-fired power stations and vehicle exhausts, respectively. These oxides interact with sunlight and other atmospheric contaminants such as *ozone, *photo-oxidants, *hydrocarbons, and *moisture. The chemical chain-reactions turn the *primary pollutants into a variable cocktail of *secondary pollutants. It can occur as *dry deposition and as *wet deposition. Acid rain has a *pH of less than 5.6, and the deposition of acidic material on the ground causes acidification of soils, water bodies, and vegetation. Acidified lakes are noted for large-scale fish deaths and declining fish populations, and acidified forests suffer from *tree dieback and tree death. Acidification affects plant growth by damaging leaves, impairing *photosynthesis, and damaging root hairs. The oxides can remain active for at least seven days, during which time they can be blown thousands of kilometres by wind systems to create *trans-frontier pollution. Swedish scientists estimate that up to nine-tenths of the sulphur deposited on Sweden is blown in from other countries (particularly the UK, France, and Germany). The USA is also accused of exporting acid rain from industrial centres in the north-east, across the Canadian border into the province of Ontario. From the early 1920s onwards many lakes and rivers in southern Scandinavia have displayed signs of acidification, including large-scale fish deaths, declining fish populations, and falling rates of reproduction. But only since the late 1960s have scientists beyond Scandinavia taken the acid rain problem seriously. In 1972 the *Organisation for Economic Co-operation and Development (OECD) launched a major research programme designed to measure air quality at 70 sites, in order to examine the long-range transport of air pollutants. By 1977, OECD studies had shown that the areas receiving acid rain were growing (in size and number), and that acid rain crosses national boundaries, so that some countries (such as Britain) are net exporters whilst others (such as Sweden) are net importers. The first international initiatives designed to deal with acid rain came in 1979 when a *convention was drafted by the United Nations Economic Commission for Europe (ECE). This convention grew out of the *Long-Range Transport of Air Pollutants (LRTAP) project and it aimed to reduce acid rain by reducing and controlling emissions of SO_2 and NO_x using the 'best available control technology economically feasible'. Thirty-one countries signed initially, but the convention was not legally binding and the commitment of some countries quickly evaporated. The Scandinavians were not satisfied with the 1979 convention, which they saw as inadequate to deal with the problem. They proposed that signatories should be responsible for reducing their emissions of SO_2 by 30% between 1980 and 1993. As a result a group of LRTAP countries met in Ottawa in 1984 to form what has since been called the *Thirty Percent Club. By July 1985, 21 of the LRTAP countries had joined the club by promising to cut their emissions by at least 30%. The USA and UK initially refused to join the club and reduce emission levels, insisting that there was not yet enough firm evidence linking SO_2 emissions with observed damage to forests and lakes. The UK had a change of heart and joined in 1986. Tackling acid deposition will require action on a broad front, because there are a number of options for reducing the two main gases (SO_2 and NO_x) that create the problem. Also known as **acid deposition** and **acid precipitation**.

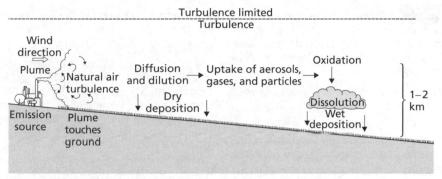

Fig 1 Acid rain cycle

Amendments of 1990, which established a *cap and trade system for reducing emissions of *sulphur dioxide from power plants.

acid rock An *igneous rock that contains more than 60% *silica by weight, most of which is as silicate materials, with the excess of about 10% as free quartz. Acid rock is low in magnesium. Acid rocks tend to be much lighter in colour than *basic rocks, they tend to disintegrate around the mineral *crystals, and they often weather into coarse-textured and rather infertile *soils.

acid shock A sudden increase in the level of *acidity of surface waters (lakes, streams, and rivers) in mid-latitude areas, caused by the melting in spring of snow that has accumulated through the winter, and stored dry fallout of acidic precipitation. An acid shock can cause significant damage to freshwater species and habitats. Also known as **acid surge**.

acid soil Soil with a *pH of less than about 6.6, caused by *acidification. Acidification occurs either naturally beneath *coniferous forest which sheds acidic needles and creates acidic litter, or as a result of air pollution by *acid deposition.

acid surge *See* ACID SHOCK.

acidic A material, liquid, or solid that has a high *acid content. The acidity of solutions (including water in soil) is measured using the *pH scale; acidic means with a pH of less than 7.0.

acidification The process by which *acids are added to a water body or soil, causing a decrease in *buffering capacity and leading to a significant decrease in *pH that may lead to the water body or soil becoming acidic (having a pH less than 7.0). This is a common form of *water pollution, and when pH drops below about 6.5 it damages aquatic ecosystems. For sensitive organisms the chemical *threshold of acid water is about pH 5.5 (zero *alkalinity); below this many aquatic species disappear from the waterbody.

acidity A measure of how acid a material, liquid, or solid is as expressed using the *pH scale. A *pH of less than 7.0 is described as *acidic (the lower the pH value the greater the acidity); a pH of more than 7.0 is described as *alkaline (the higher the pH value the greater the alkalinity).

acquired land General term for land in the USA which is owned by the Government and was obtained through purchase or as a gift or donation, by exchange or to exploit the timber growing there.

acquired trait A *phenotype characteristic of an *organism, that is acquired during the growth and development of an individual, such as the large muscles that an athlete develops as a result of exercise and training. An acquired trait is not genetically based and so it cannot be passed on to the next generation.

acquired value A value (monetary or otherwise) that people attach to something because of its perceived usefulness. *Contrast* INTRINSIC VALUE. *See also* INSTRUMENTAL VALUE.

acrid Bitter or sharp-tasting, usually with an unpleasantly strong or bitter smell.

act A legal document that codifies the result of deliberations by a committee, society, or legislative body. *See also* STATUTE.

actinide Any of a series of *radioactive elements that have an *atomic number of between 89 and 103. Also known as actinoid.

actinoid *See* ACTINIDE.

action level The *threshold level of *contamination, defined in a regulatory programme, that defines when action or a response is formally required. For example, in the US *Superfund programme, the action level makes clear when a particular form of contamination is high enough to warrant action or trigger a response under *SARA and the *NCP. Also in the USA, the *Environmental Protection Agency has defined action levels for levels of indoor *radon, levels of pesticide residues in foodstuffs, and for other environmental factors.

action research Learning by doing. A particular type of applied research that often involves researchers and practitioners working closely together, in which a group of people identifies a problem, does something to resolve it, reflects on how successful their efforts were, and, if not satisfied, tries again. *See also* ADAPTIVE MANAGEMENT.

activated carbon A highly absorbent carbon-rich material, made from burnt wood, that is used to remove odours and toxic substances from liquid or gas emissions by absorbing the solutes onto the activated *carbon. It is used, for example, to remove dissolved organic matter (including *pesticides) and some inorganic solutes (including chlorine) from waste water. Also known as **activated charcoal**. *See also* GRANULAR ACTIVATED CARBON TREATMENT.

activated charcoal *See* ACTIVATED CARBON.

activated sludge A process for treating *sewage and primary effluent water which involves using an active population of *micro-organisms that feed on organic wastes. Raw sewage is mixed with bacteria-laden sludge, and the mixture is then shaken and aerated (oxygen is added) to speed the breakdown of organic matter in *secondary wastewater treatment. The process converts soluble organic matter into solid *biomass. Half of the sewage in Britain is treated like this.

activator The activity of a *catalyst, such as a chemical that is added to a *pesticide to increase its activity.

active fault A geological *fault that is still active rather than *dormant, meaning that significant displacement or movement has occurred on it within about the last 10 000 years.

active ingredient The component in any *pesticide that kills or controls the target pests, and that determines how that pesticide is regulated.

active restoration The use of active measures, such as hand-planting trees and shrubs and removing *exotic plants and animals, in *habitat restoration. *See also* NATURAL PROCESS RESTORATION.

active solar heating A form of *solar heating that uses *solar energy to heat buildings indirectly. *Solar collectors are used to heat water which can then be stored and circulated around a building by pumps and pipes to heat water and space. *Contrast* PASSIVE SOLAR HEATING.

activism *See* ENVIRONMENTAL ACTIVISM, GRASSROOTS ACTIVISM.

Activities Implemented Jointly (AIJ) A pilot programme established by the *United Nations Framework Convention on Climate Change, designed to encourage a cooperative international approach to tackling global warming caused by air pollution. It allowed private bodies and organizations in one country to reduce, *sequester, or avoid reducing *greenhouse gas emissions through funding or running a project in a different country that reduces emissions there. The AIJ programme ended in 2000 and has evolved into *joint implementation under the *Kyoto Protocol.

activity plan A detailed plan for managing a single resource programme, designed to achieve specific objectives, and usually developed only when needed to support more general *land use plan decisions. An example is a habitat management plan for the *conservation of a particular wildlife *habitat.

actual evaporation The amount of *evaporation that takes place from open water. *Contrast* POTENTIAL EVAPORATION.

actual evapotranspiration The rate at which water is lost from vegetation and soil by a combination of *evaporation (from any water surface) and *transpiration (from plants), which is usually lower than *potential evapotranspiration.

actual vapour pressure *See* VAPOUR PRESSURE.

acute In general, extremely serious or severe. Relating to disease, severe and of short duration. *Contrast* CHRONIC.

acute dose A fatal level of exposure to some factor (such as radiation) that is experienced over a short period of time.

acute effect A sudden, rapid onset in any organism of severe symptoms or adverse effects as a result of exposure to some factor (such as high concentrations of particular pollutants), which often disappear after the exposure stops. *Contrast* CHRONIC EFFECT.

acute exposure A single, brief, intense exposure to a hazardous material (such as a toxic substance) or situation that may cause serious injury or death in an organism (including people). Such exposure often lasts less than a day, but can have long-lasting consequences. *Contrast* CHRONIC EXPOSURE.

acute poverty A serious shortage of income or of access to the range of re-sources that usually provide the basic necessities for life for humans, such as food, shelter, sanitation, clean water, medical care, and education.

acute toxicity The short-term effects of a single large exposure to a *toxic substance, which are apparent usually within 24 to 96 hours. *See also* CHRONIC TOXICITY, TOXICITY.

adaptability A measure of the extent to which a *species or *ecosystem is able to adjust to environmental change (such as *global warming). Such *adaptation can be planned or unplanned, and it can be carried out in response to or in anticipation of changes in conditions. Also known as **adaptive capacity**.

adaptation A change in the structure or habit of an *organism that makes it better adjusted to its surroundings. Short-term change is physiological (for example, *acclimation) or behavioural (*phenotypic adaptation), and long-term change is genetic (*genotypic adaptation). Particular genetic adaptations in a population become frequent and dominant if they enhance an individual's ability to survive in the environment, for example as climate changes through time. *See also* EVOLUTION, NATURAL SELECTION.

adaptive capacity *See* ADAPTABILITY.

adaptive management An approach to the management of *natural resources that is based on learning by doing, and on making decisions as part of an on-going process of monitoring, review, and adaptation. A planned course of action is kept under constant review, and is adapted where appropriate as new information becomes available from the monitoring of results, publication of new scientific findings and expert judgements, and changing needs of society. Also known as the **monitor-and-modify approach**. *See also* ACTION RESEARCH.

adaptive radiation Evolutionary divergence through successive generations, which causes the *evolution of two or more distinct *species or *subspecies from a single ancestor, with each of the new species diversifying to become adapted to a different *niche or *environment. This can occur, for example, after a *mass extinction creates some vacant niches or an environmental change gives rise to new niches. For example, Charles *Darwin observed that over many generations finches had adapted to the availability of different food sources on islands in the Galapagos by developing different shapes of beak. *See also* CLADOGENESIS, SPECIATION.

adaptive strategy The means by which *species cope with competition and environmental change over the long term, on an evolutionary time scale. *See also* GENOTYPIC ADAPTATION.

additionality A reduction in or avoidance of *greenhouse gas emissions that exceeds what would have occurred under a *business-as-usual scenario. Under the *Kyoto Protocol, such reductions must be real and measurable.

additive A substance that is added to improve or preserve something. Examples include the chemicals that are added to food and can pose a threat to human health, and the lead that was previously added to gasoline to reduce engine wear in vehicles but which also damaged human health via air pollution.

additive effect The combined effect of two or more changes that is equal to the sum of their individual effects. *Contrast* SYNERGISM.

ADEOS II A *satellite *remote sensing system, launched in 1999, that monitors surface wind speeds and directions over the *oceans.

adhesion 1. The ability to stick to something.
2. Intermolecular attraction between unlike substances where adjacent surfaces cling together.

adiabatic Any change in temperature, pressure, and volume in a parcel of air that is caused by the expansion (cooling) or compression (warming) of the air as it rises or descends in the atmosphere. The parcel of air does not exchange heat with the surrounding atmosphere. *Contrast* DIABATIC.

adiabatic lapse rate The rate at which the temperature of a parcel of air decreases when the air is lifted (or the rate of increase of temperature when air sinks), without any exchange of heat or energy with the surrounding air, by *adiabatic change. When lifted, a parcel of unsaturated air cools at the *dry adiabatic lapse rate; a parcel of saturated air cools at the *saturated adiabatic lapse rate. When it descends, the parcel of air warms at the corresponding rate.

adiabatic process A process that occurs with no exchange of heat or energy between a system (such as a parcel of air) and its environment (the surrounding atmosphere).

adit A more or less horizontal tunnel from a *hillside into an underground *mine, open to the surface at one end, that is used for access or drainage.

adjustment Change or adaptation within a *system that serves to accommodate the factor(s) that are promoting the change and produce a new *equilibrium. A system changes when it is forced to, otherwise it tends to remain stable and unchanging. *See also* HOMEOSTASIS.

administered dose In exposure assessment, the amount (*dose) of a substance that is given to a test subject (a human or animal) to determine *dose–response relationships. Also known as **intake** or **potential dose**.

adsorption The process by which molecules of a gas, liquid, or dissolved substance (in a condensed form) stick to the surface of a solid material, or less frequently on a liquid. This process is used, for example, in advanced wastewater treatment, using *activated carbon to remove organic matter from the water. *Contrast* ABSORPTION.

advanced air emission control device Equipment for controlling air pollution after it has been initially treated (for example by low-energy *scrubbers). Examples include *electrostatic precipitators and high-energy scrubbers.

advanced character *See* DERIVED CHARACTER.

Advanced Very High Resolution Radiometer (AVHRR) The main sensor on US polar orbiting *satellites, which is used for monitoring changing levels of radiation as well as visible and infrared light. It can produce cloud images even at night.

advanced wastewater treatment Any treatment of sewage and wastewater that goes beyond the secondary (biological) stage and includes the removal of nutrients such as *phosphorus and *nitrogen, and a high percentage of *suspended solids. Also known as **tertiary treatment**. *See also* PRIMARY TREATMENT, SECONDARY TREATMENT.

advection The horizontal transfer of properties (such as heat or moisture) or substances (such as air or water pollutants) by the flow of a current of water or air. *See also* FOG, ADVECTION FOG.

advection fog A *fog that forms by *advection when warm air flows over a cold surface (such as a cold ocean current). The air cools from below until air temperature falls to the *dew point, *saturation occurs, and *water vapour condenses to form fog. This type of fog is quite common and can be persistent, lasting longer and dispersing more slowly than *radiation fog.

advection frost *See* FREEZE.

advection inversion A temperature *inversion caused by *advection as warm air passes over a cool surface.

adventitious root A plant root that grows in an unusual location (for example, from a stem). *See also* ROOT CLIMBER.

adverse effect/impact An effect that damages the environment or reduces environmental quality, which can include human health and quality of life to an unacceptable degree.

advisory A term used in the USA to refer to guidance issued by the authorities, usually in a particular area. It usually relates to the incidence of a specific *environmental hazard (such as adverse weather) that is likely to cause significant inconvenience to people but is not serious enough to issue a *warning. *Compare* CAUTION.

Aedes aegypti *See* YELLOW-FEVER MOSQUITO.

aeolian (eolian) Wind-blown or involving wind. For example, *loess is an aeolian *deposit, and a *dune is a *landform produced by aeolian action.

aeolian deposit Fine sediments that are transported and deposited by wind. Examples include *loess, *dunes, desert sand, and some *volcanic ash.

aeolian erosion *See* WIND EROSION.

aeolian soil Soil that has been deposited by wind, usually made up of silt or sand.

aeration To oxygenate or allow air to penetrate something. This process is used to promote biological degradation of *organic matter in water, by exposing it to air. The process may be passive (when waste is simply exposed to air), or active (when a mixing or bubbling device introduces the air into the liquid, generating bubbles and aerosols that lead to the release of the dissolved gases). *See also* DIFFUSED AIR.

aeration tank A chamber used to inject air into water to speed the *aeration process.

aeration zone The subsurface zone between the ground surface and the *water table, where the pores in soil and rock contain both air and water. Also known as **unsaturated zone**, **vadose zone**, or **zone of aeration**.

aerial logging The removal of logs from a timber harvest area by helicopter, which allows access to previously inaccessible sites but causes less environmental damage because fewer access roads need to be built.

aerial photograph A photograph of the Earth's surface that is taken from the air, usually at regular intervals along a *transect. Interpretation of such photographs provides valuable information about such things as *land use, *vegetation, and *landforms. Also known as **air photograph**.

aeroallergen Any of a variety of *allergens (such as pollens, grasses, or dust) that are carried in the air and dispersed by winds.

aerobe An organism that can live and grow only in the presence of oxygen. Also known as an **aerobic organism**. Contrast ANAEROBE.

aerobic A process that occurs, or life that exists, only in the presence of oxygen. Contrast ANAEROBIC.

aerobic biological oxidation Any waste treatment process that uses *aerobes to reduce the pollution load or oxygen demand of organic substance in water.

aerobic composting A method of composting organic waste using *aerobic bacteria. The waste is exposed to air either by turning or by forcing air into it through pipes. This process is reasonably odour-free, and it generates more heat and is faster than *anaerobic composting. Temperatures may become high enough to destroy *pathogens and weed seeds.

aerobic digestion The decomposition of *organic matter by *aerobic organisms in the presence of oxygen, which results in the formation of mineral and simpler organic compounds. Also known as **aerobic treatment**, **sludge processing**.

aerobic organism See AEROBE.

aerobic treatment See AEROBIC DIGESTION.

aerodynamic A body shape which is *streamlined to reduce air resistance and thus allow easy movement through air. Compare HYDRODYNAMIC.

aerodynamics The study of objects in air flow, or of objects moving through air.

aerosol A suspension of tiny liquid droplets or solid particles in a gas which are so small that they do not settle out under the influence of gravity and so remain suspended in the air. Natural aerosols include *fog and minute particles that originate from wind-blown sea salt or dust, volcanic eruptions, and the burning of *vegetation. Manufactured aerosols include fine sprays used in perfumes, insecticides, and antiperspirants. Aerosol particles provide nuclei for the *condensation of water droplets and for the growth of ice crystals, and they also absorb and scatter *solar radiation, through which they influence the *radiation budget of the atmosphere, which in turn influences *climate on the surface of the Earth.

aesthetic Pleasing to the eye or judged to have beauty, particularly as relating to visual appearance, such as a view or *landscape.

aesthetic degradation Any reduction in environmental quality that people interpret as making it less attractive, for example the adverse impact on a natural *landscape of building a large power station.

aesthetic resource The intangible natural qualities of a place that make it appeal to people, such as a sense of solitude, of being in close contact with nature, and of being inspired and emotionally recharged by the experience of being there. Contrast CULTURAL RESOURCE. See also RESOURCE.

aesthetic zoning Land use *zoning which regulates property in order to protect *aesthetic values.

aesthetics The study of beauty, and the basis of judgments about beauty.

aetiology The study of the causes of *diseases.

affected public A North American term for the people who are adversely affected by exposure to a *toxic pollutant in food, water, air, or soil, perhaps by living near or working in a hazardous waste site.

affinity A chemical attraction that causes the atoms of certain elements or compounds to combine with atoms of another element or compound, and then remain in the combined state. The term is also used to describe how likely such a reaction is to occur.

afforestation The process of establishing and growing *forests on bare or cultivated land which has not been forested in recent history. As well as creating new *habitats for wildlife, and new supplies of timber, afforestation provides valuable *carbon sinks which absorb *carbon dioxide and thus help to reduce *global warming. Also known as **forestation** or **reforestation**. *Contrast* DEFORESTATION.

aflatoxin A *toxin that is produced by *moulds in stored crops (particularly in peanuts) and is *carcinogenic.

Afrotropical realm A *biogeographical realm which covers most of Africa and is dominated by *tropical forest, *savanna, and *desert.

afterburner Equipment used in an *incinerator, in which the exhaust gases are passed through a burner to remove smoke and odours. This helps to completely combust unburned or partially burned carbon compounds.

aftershock A lesser *earthquake that follows a stronger one.

age class 1. An age grouping of any organism, including people.
   2. Foresters usually group trees, forests, stands or forest types into 20-year age classes (1–20, 21–40 and so on).

age distribution The proportions of a population that fall in different *age classes, or into different groups such as pre-reproductive (subadult), reproductive, and post-reproductive. Also known as **age structure**.

Age of Reason *See* ENLIGHTENMENT.

age pyramid A diagram that shows the proportions within each *age class of a population.

age ratio The proportion of young individuals to adults in a population. This has an impact on productivity and population growth.

age structure *See* AGE DISTRIBUTION.

agency 1. A form of action or intervention.
   2. An organization or government department that provides a particular service.

agency capture A term used in the USA for the situation in which a regulatory process is 'captured' by those it is supposed to regulate, and then turned to their advantage.

agent Any physical, chemical, or biological entity that can affect (benefit or injure) an *organism, people, or the *environment. Also known as an **environmental agent** or **stressor**.

agent of disease Any factor (such as a *micro-organism, a chemical substance, or a form of *radiation) that is essential for the occurrence of a *disease.

agent orange A toxic *defoliant that was used by the USA in the Vietnam War. Based on a mixture of several *herbicides, it is believed to be a *carcinogen because it includes *dioxin. The name comes from the orange-banded barrels in which the defoliant was marketed.

agglomerate 1. To collect or gather together fine particulates into a larger mass.
   2. A type of volcanic rock.

agglomeration 1. In general, the process of gathering together.
   2. The process by which precipitation particles grow larger, by collision or contact with *cloud particles or other precipitation particles, until they are large enough to fall as *precipitation.

aggradation The accumulation of unconsolidated sediments, which builds up the land surface. *Contrast* DEGRADATION.

AGENDA 21

A comprehensive, long-term plan of action on environmental and development issues that was drawn up at the 1992 *United Nations Conference on Environment and Development (UNCED) and is being implemented globally, nationally, and locally in order to relieve *poverty in developing countries and promote *sustainable development. The main focus of Agenda 21 is the sustainable management of economic, social, and natural capital (including *biodiversity) in order to meet the needs of present and future human generations, with particular emphasis on the needs of the most disadvantaged. Agenda 21 proposals include various ways of saving biodiversity, including: creating a world information resource for biodiversity; protecting biodiversity; offering *indigenous peoples the chance to contribute to biodiversity conservation; making sure that poor countries share equally in the commercial exploitation of their products and experience; protecting and repairing damaged habitats; conserving endangered species; assessing every big project (dams, roads, etc.) for its *environmental impact. The estimated total cost of implementing all of the recommendations in Agenda 21 is nearly $600 000 million (at 1992 prices), and a number of sources of funding were identified. UNCED proposed that $460 000 million (77%) would come from local sources, and the remaining $140 000 million (23%) from international sources including foreign aid and the *World Bank. It was agreed that each country should meet the costs of implementing Agenda 21 from its own resources, but that additional funding would be needed for the *developing countries and this would come mainly from official development assistance, the *Global Environment Facility (GEF) set up by the World Bank, the *United Nations Development Programme (UNDP) and the *United Nations Environment Programme (UNEP), and the International Development Association, the branch of the World Bank that provides interest-free loans to the lowest-income countries. *See also* GLOBAL ENVIRONMENT FACILITY, LOCAL AGENDA 21.

aggregate 1. A mass or cluster of material, such as the particles in a *soil.
2. A mixture of rock and mineral fragments that resemble a rock.

aggregation 1. The process of cementing or binding together of soil particles into an *aggregate.
2. A method by which ice crystals grow around nuclei in the atmosphere, by colliding with other crystals and sticking together.
3. A collection of organisms that live closely together but are not physically connected. *See also* COLONY.

agrarian Relating to *agriculture or farming.

agrarian civilization A society that depends on an *agricultural economy, and which is therefore settled, as opposed to *hunter-gathering.

agribusiness Large-scale, industrialized agriculture controlled by corporations, which includes all of the operations involved in the production, storage, processing, distribution, and wholesale marketing of farm products.

agricultural diversification The process of seeking supplementary or replacement income streams for farms, beyond agricultural activities themselves. Farm-based tourism is an example.

agricultural economy An economic system that is based primarily on crop production.

agricultural intensification The management of increased production in *agriculture by making greater use of *agrochemicals and mechanization. *See also* INTENSIVE AGRICULTURE.

agricultural pollution The liquid and solid wastes produced by any type of agricultural activity. These include runoff and leaching of chemicals (*pesticides and *fertilizers) from fields, *soil erosion and dust from ploughing, animal manure and carcasses, and crop residues. Some such pollution (for example, from large feedlots) is *point source, but much (for example blowing dust, or nutrients from fields) is *non-point source.

Agricultural Revolution A major shift in human society and culture that began about 10 000 years ago, in which people started to raise livestock, cultivate crops and create goods from natural resources. Agriculture replaced a nomadic mode of living, based on hunting and gathering, and it allowed people to adopt a more settled mode of living. *See also* GREEN REVOLUTION, INDUSTRIAL REVOLUTION.

agricultural runoff The surface water that leaves an area of agricultural land use, which is usually enriched with nutrients, sediment, and agricultural chemicals. *See also* EUTROPHICATION.

agricultural sewage Liquid waste material that is produced through the agricultural processes of cultivating soil, producing crops, or raising livestock.

agricultural system *Agriculture viewed as an *open system, including the input and output of materials and energy from farms.

agricultural waste The waste products created by agriculture, including animal *byproducts, *manure, *agricultural sewage, residual materials in liquid or solid form, and the residues from harvesting of grain, vegetables, and fruit.

agricultural zoning A type of *natural resource zoning that only allows *agriculture and other low-density uses.

agriculture The practice of cultivating the soil, growing crops, or raising livestock for human use, including the production of food, feed, fibre, fuel, or other useful products. Also known as **farming**. *See also* TRADITIONAL AGRICULTURE.

Agriculture and Consumer Protection Act (1973) US legislation, some sections of which authorized a forestry incentives programme for non-industrial private landowners for tree planting and timber stand improvement in order to produce marketable timber crops and other values. These sections were repealed and supplanted by the *Cooperative Forestry Assistance Act (1978).

Agriculture Department *See* US DEPARTMENT OF AGRICULTURE.

agrochemicals Synthetic chemicals that are used in agriculture to fertilize soil (*fertilizers), and to control animal pests (*pesticides) and weeds (*herbicides).

agroecological zone A large area of land that has relatively uniform soils and climate, across which agricultural productivity is also relatively uniform.

agroecosystem An agricultural system viewed as an *ecosystem or *open system, where the objective is sustainable management of the interrelationships between system components (including crops, pastures, livestock, other flora and fauna, atmosphere, soils, groundwater, and *drainage networks) and environmentally sensitive management of uncultivated land and *wildlife.

agroforestry A form of *agriculture that is based on the cultivation of *trees with other crops, animals, or both. This is usually done for multiple objectives, including improving the quality of wildlife *habitat, increasing access by humans and wildlife, and growing more woody plant products (including trees, shrubs, palms, or bamboos). *See also* AGROSYLVICULTURAL SYSTEM, SYLVOPASTORAL SYSTEM.

agrometeorology The study of how atmospheric conditions and processes affect agriculture.

agronomy The science of soil management and crop production.

agropastoral system A system of land use based only on crops and livestock, without trees.

agrosilvicultural system An *agroforestry system of land use in which trees and shrubs are grown with herbaceous food crops, pastures, and animals.

A-horizon The upper layer of soil within a *soil profile. This is the fertile topsoil of a mineral soil, that contains much organic material (from the decomposition of litter) and minerals, and is the location of much biological activity (by earthworms and other soil organisms). The organic-rich layer in the A-horizon is termed the *O-horizon. Fine materials (such as iron, aluminium oxides, and silicate clays) are washed out (*leached) from the A-horizon and deposited in the lower horizons, by *eluviation, so this horizon is also sometimes known as the **zone of leaching**. *See also* B-HORIZON, C-HORIZON, SOIL HORIZON.

AIDS Acquired immune deficiency syndrome, an often fatal disease that destroys the human immune system and is spread through direct contact with the bodily fluids of an infected person, especially by sexual contact or via contaminated needles.

AIJ *See* ACTIVITIES IMPLEMENTED JOINTLY

air 1. A mixture of gases that makes up the Earth's *atmosphere. It comprises about 78% *nitrogen, 21% *oxygen, less than 1% of *carbon dioxide and other gases, and varying amounts of *water vapour that humans and other organisms require for breathing.
2. The region above the ground, also known as the *sky or *atmosphere.

air cleaning An approach to the management of indoor air quality based on removing particulates and/or gases from the air, using procedures such as *particulate filtration, *electrostatic precipitation, and gas *sorption.

air conditioning Regulation by mechanical means of the temperature, humidity, cleanliness, and circulation of air within a particular enclosed area, such as a building. *See also* VENTILATION.

air current The movement of air. *See also* WIND.

air curtain A device that forces air bubbles from a perforated pipe in order to create an upward flow of water in a waterbody. This is used for example to contain an oil spill or to stop fish from entering polluted water.

air drainage wind *See* KATABATIC WIND.

air emission Any gas that is released into the air from as a result of human activities, including *ozone, *carbon monoxide, *nitrogen oxides, *nitrogen dioxide, and *sulphur dioxide.

air exchange rate The rate at which outside air replaces indoor air in a space, expressed either as the number of air changes per hour or the rate at which a volume of outside air enters (in cubic metres per minute, $m^3\ min^{-1}$).

air mass A large body of air in which the characteristics (temperature, moisture, and pressure) are relatively uniform at a given height and latitude. Weather changes are most marked at the boundary zone (the *frontal system) between adjacent air masses. Air masses exist at a variety of sizes, from several square kilometres (km^2) to 100 000 km^2. They are usually classified in terms of temperature and moisture content—tropical (T) and equatorial (E) air masses are warm, whereas arctic (A) and polar (P) air masses are cold; continental air masses (c) are normally dry, and maritime air masses (m) are usually wet.

air mass thunderstorm A *thunderstorm produced by local *convection

AIR POLLUTION

The presence in the air of any *air pollutant that reduces *air quality enough to threaten the health and welfare or people, plants, and animals, to adversely affect materials and structures, and/or to interfere with the enjoyment of life and property. Although there are a number of important *natural sources of air pollution (including forest fires and volcanic eruptions), the term is usually applied to substances released into the atmosphere as a result of human activities, which can be either deliberate (such as the continual release of gases from factory chimneys) or accidental (such as the release of material from the damaged *Chernobyl nuclear power station, and the *Bhopal explosion).

within an unstable *air mass. *Contrast* FRONTAL THUNDERSTORM.

air monitoring A programme of sampling and measuring the pollutants present in the atmosphere at a particular place, on a continuous, regular, or intermittent basis. *See also* MONITORING.

air photograph *See* AERIAL PHOTOGRAPH.

air pollutant An unwanted substance that is found in the air in the form of solid particles, liquid droplets, or gases. Some (such as pollen, fog, and dust) come from natural sources, but there are also more than 100 man-made substances including solids, *sulphur compounds, *volatile organic compounds, *nitrogen compounds, *oxygen compounds, *halogen compounds, *radioactive compounds, and odours that cause air pollution.

air pollution control device Equipment that cleans emissions generated by a source (such as an *incinerator, industrial *smokestack, or vehicle exhaust system) by removing pollutants that would otherwise be released to the atmosphere.

air pollution episode A period when the concentration of *air pollutants is unusually high, often as a result of low winds and temperature inversion. Prolonged or severe episodes can cause illness and death amongst people, as happened during the London smogs of the 1950s.

air pressure *See* ATMOSPHERIC PRESSURE.

air quality A general term that describes the condition of the air in a particular place and time, reflecting the degree to which it is pollution-free (particularly in terms of *sulphur dioxide, *nitrogen oxides, *carbon dioxide, ground level *ozone, and airborne particles).

air quality class A designation given to geographical areas of the USA in which *air quality is better than the national standards. Allowable increases in pollutants are specified for each class, with Class II allowed the greatest increase and Class I allowed the least.

air quality control region An area in the USA, designated by the Federal Government, in which communities share a common *air pollution problem. Such a region can sometimes involve more than one state.

air quality criteria Threshold levels of *air pollution and lengths of exposure, above which there may be adverse effects on human health and welfare.

Air Quality Framework Directive A *directive of the *European Community which has a framework for limits for a range of *air pollutants.

Air Quality Index (AQI) A numerical scale to indicate how polluted the air is in a particular place. It was developed by the US *Environmental Protection Agency for measuring pollution levels for the major air pollutants that are regulated under the *Clean Air Act. It replaced the *Pollution Standards Index (PSI) in 1999.

air quality standard The maximum level of air pollutants (particularly *sulphur dioxide, *nitrogen oxides, *carbon dioxide, ground level *ozone, and *airborne particles), defined to protect human health and prescribed by law or regulation, which must not be exceeded during a given time in a defined area.

air sampling The collection and analysis of samples of air to measure *air quality, particularly the concentration of radioactive substances, particulate matter, or chemical pollutants. *See also* AIR MONITORING.

air stripping A system of treating water pollution that involves removing *volatile organic compounds from contaminated *groundwater or surface water, by forcing a stream of air through the water and causing the compounds to evaporate and be separated from the water.

airborne infection A method of transmitting an infectious agent via particles, dust, or nuclei suspended in the air.

airborne particle A *particle that exists in the *atmosphere as a solid or liquid. It can come from natural and/or human sources, and can vary in size from coarse (bigger than 3 *micrometres in diameter) to fine (less than 3 micrometres).

airborne particulate The suspended matter that exists in the *atmosphere as tiny solid particles or liquid droplets. It comes from many sources, including windblown dust, emissions from industrial processes, smoke from the burning of wood and coal, and the exhaust of motor vehicles. The chemical composition of the particulates varies widely, depending on location and the time of year.

airborne pollutant *See* AIR POLLUTION.

airborne release The deliberate or accidental release into the air of any gas or object, including *pollutants.

airshed 1. A geographical area that frequently shares the same air because of topography, meteorology, and/or climate.

2. The catchment area (after *watershed) for air pollutants around known emission sources.

Alar The trade name of a *pesticide (Daminozide) that can be sprayed on apples to regulate their growth, make it easier to harvest them, and improve their colour. It was registered with the US Food and Drug Administration from 1963 to 1989, but was banned as a result of public concern over a controversial study which found that Alar residue could produce *tumours in mice.

ALARA *See* AS LOW AS REASONABLY ACHIEVABLE.

Alaska National Interest Land Conservation Act (ANILCA) (1979) US legislation that offered protection to over 100 million acres of federal lands in Alaska, doubling the size of the country's *national park and *refuge system, and tripling the amount of land designated as *wilderness.

albedo A measure of ability to reflect light (*reflectivity), which is the ratio of the amount of light reflected by a surface to the amount of incident light. Dark surfaces reflect relatively little light so they have a low albedo; for example rock surfaces 12–18%, green grass and forest 8–27%, sand up to 40%. Light surfaces have a high albedo; for example fresh *snow up to 90%; cloud surfaces have an albedo of around 55% on average, but it can be as high as 80% for thick *stratocumulus clouds. The Earth's albedo averages around 30% at the edge of the atmosphere.

alcohol A group of organic chemical compounds composed of *carbon, *hydrogen, and *oxygen that occurs naturally and is commonly obtained by fermentation. It is used in solvents, antifreezes, chemical manufacture, and as a fuel.

aldehyde An organic compound that contains *carbon, *hydrogen, and *oxygen, produced by the oxidation of an *alcohol. Some (such as *formaldehyde) are manufactured, but some are emitted in vehicle exhausts. Aldehydes appear to be *carcinogenic, and formaldehyde was listed as a hazardous air pollutant in the 1990 US *Clean Air Act.

aldrine An *organochlorine *insecticide that was used as a dressing for seeds but led to population decline in some birds of prey (which ate seed-eating birds). Its use has been restricted and controlled since the 1960s.

ALERT *See* AUTOMATED LOCAL EVENT REPORTING IN REAL TIME.

alfalfa A *perennial plant in the pea family that is rich in vitamins, minerals, and protein and is widely cultivated to be made into *hay or used for *forage. *See also* MEADOW.

alfisol A type (*order) of soil that typically develops beneath broadleaved *forest. It has a much thicker *A-horizon than most other soils, because leaves and other litter decompose quickly during the mild, wet winters. The A-horizon often has a white layer, indicating *eluviation of nutrients into the soil. An alfisol also has a relatively thick *B-horizon where clay and plant nutrients accumulate by *illuviation.

algae Simple, aquatic, non-vascular plants (without roots, stems, and leaves), a member of the *Protoctista kingdom. They contain *chlorophyll and live in *colonies, either floating or suspended in sunlit freshwater (*phytoplankton), or attached to structures, rocks, or other submerged surfaces (*periphyton). Algae vary in size from minute single-celled forms to large multicellular seaweeds such as giant kelp (which can often grow to longer than 30 m). Like all aquatic plants, algae *photosynthesize, add oxygen to the water, affect *pH, and are important in the food chain—they provide food for fish and small aquatic animals. Algae grow in proportion to the amount of available nutrients, and excess algal growth (*algal blooms) can reduce water quality and alter water chemistry. *See also* EUTROPHICATION.

algal bloom A sudden rapid growth of *algae (typically blue-green algae or *cyanobacteria), which often develops as a mat on the floor of surface waters (lakes, streams, ponds, reservoirs, and marine waters) due to an increase in nutrients such as nitrogen and phosphorus. It can occur naturally or as a result of pollution. The bloom reduces the availability of oxygen in the water, and can kill fish and other aquatic organisms through oxygen deprivation. *See also* EUTROPHICATION, NUTRIENT ENRICHMENT, RED TIDE.

algal mat Blue-green *cyanobacteria that develop in shallow marine environments as well as in lakes and swamps. The *algae trap sediment and produce alternating *laminations of algal layers and sediment layers.

algicide A substance or chemical that is used specifically to kill or control algae.

alien A non-native species, especially one introduced to an area outside of its historically known natural range through human action (by accident or design) or by other agents. Also known as **exotic** or **introduced species**. *Contrast* NATIVE SPECIES, INDIGENOUS.

alkali A compound that dissolves in water, giving *hydroxide ions. Alkalis neutralize *acids to form a salt and water. Alkalis have a *pH of more than 7.0, turn litmus paper blue, and can cause severe burns to the skin. Common commercial alkalies include sodium carbonate (soda ash), lime, potash, caustic soda, and regular mortar. Alkali is the opposite to acid, and is also known as a *base.

alkalination The process of increasing the pH of a soil to more than 7.0, usually by increasing its content of carbonates of calcium, magnesium, potassium, and, more specifically, sodium.

alkaline Having the qualities of an *alkali, or a solution with an excess of *hydroxide ions (with a *pH greater than 7.0).

alkaline soil A soil that contains a relatively large amount of alkaline salts (usually sodium carbonate), has a *pH greater than 7.0, and restricts the growth of most crop plants.

alkalinity The *buffering capacity of water. This means its capacity to neutralize an acid solution by its content of bicarbonates, carbonates, or hydroxides, adding *carbon to the water and preventing the *pH of the water from becoming too *basic or too *acidic, stabilizing it at a pH of around 7.0. The lower the alkalinity, the less capacity the water has to absorb acids without becoming more acidic. The alkalinity can be raised by adding a *buffer.

all-aged stand A stand of forest in which trees of different ages are found growing together. *See also* EVEN-AGED STAND, UNEVEN-AGED STAND.

allele One of several alternative forms of a *gene which occupy the same relative position on paired *chromosomes, and which control the inheritance of one characteristic.

allelochemical Any substance that affects allelopathic reactions. *See also* ALLELOPATHY.

allelopathy The ability of a plant species to produce substances that inhibit the growth or *germination of other plants.

Allen's rule A 19th century *ecogeographical rule that the warmer the climate the longer the appendages (ears, legs, wings) of warm-blooded animals in comparison with closely related taxa from colder climes. *See also* BERGMANN'S RULE, GOLGER'S RULE.

allergen Any substance that is capable of causing an allergic reaction. These include dust, pollen, certain foods and drugs, fur. *See also* AEROALLERGEN.

allergy Hypersensitivity to a substance (*allergen) that is normally harmless.

alley cropping The planting of crops in strips, with rows of trees or shrubs grown on either side of the produce.

Alliance of Small Island States (AOSIS) A coalition of small, low-lying island countries formed during the Second World Climate Conference in 1990 that includes 35 states from the Atlantic, Caribbean, Indian Ocean, Mediterranean, and Pacific, most of which are members of the *Group of 77, and all of which are particularly vulnerable to *sea level rise. These countries share the threat that *global warming and *climate change pose to their survival, and have called on developed countries to rapidly reduce their emissions of *greenhouse gases.

allocation The process of assigning activities, costs, or facilities to particular people, or of designating certain resources for particular purposes.

allochthonous 1. Not *indigenous, introduced. Can be used to describe flora, fauna, or people.
2. In geology, a deposit that originated elsewhere than its current position. *Contrast* AUTOCHTHONOUS.

allometry The study of the relative growth of a part or aspect of an organism (for example, its shape or length) relative to the growth of the whole organism, either for growth of an individual (ontogenetic allometry) or by comparing related organisms of different sizes (phylogenetic allometry).

allomone An *allelochemical that can be produced by a species and influences the behaviour or growth of another species. *See also* SEMIOCHEMICAL.

allopatric Similar organisms which could cross breed but don't because of geographical separation. *Contrast* SYMPATRIC.

allotment 1. In the UK, a small plot of land rented by local people (usually from the local authority) on which they can cultivate food and flowers.
2. In the USA, the area of land designated for use by a prescribed number of livestock for a prescribed period of time. Also known as a **range allotment**.

allotrope Two or more different physical forms in which a particular element exists. For example, *carbon has two allotropes, diamond and graphite.

allotropy The existence of two or more forms (*allotropes) of the same element.

allowable annual harvest The volume or acreage of timber that can be harvested each year from a particular area of forest, defined by the objectives of management.

allowable cut The amount of wood that may be harvested annually or periodically from a particular area of forest over a stated period of time, defined by the objectives of management.

allowable use The level of use of *rangeland that is sustainable, defined by the objectives of management. To maintain a range in good to excellent condition, use of 40 to 50% of the annual growth is often allowed; less use will be allowed if the aim is to restore overgrazed land.

alloy A material made of two or more *metals.

alluvial Composed of or relating to *alluvium, or caused by or related to river action. Also known as **riverine**. *Compare* COLLUVIAL.

alluvial cone *See* ALLUVIAL FAN.

alluvial deposit *See* ALLUVIUM.

alluvial fan A mass of sediment that builds up at a point on a stream course where there is a sudden decrease in gradient (for example, in a mountain canyon). Adjacent alluvial fans are commonly found in arid and semi-arid environments, and they can coalesce to form a *bajada.

alluvial plain *See* FLOODPLAIN.

alluvial river A river that flows through *alluvium, that it can move and refashion, as opposed to a *bedrock river that flows through or over solid rock.

alluvial soil Soil that has developed on alluvium that has recently been deposited, with as yet little or no modification of the original material by soil-forming processes.

alluvium Loosely compacted sediment that is deposited by a river, which can include particles ranging in size from clay, silt and sand, to gravel. Stones and boulders are often worn by the water into rounded shapes. Also known as an **alluvial deposit**. *Contrast* COLLUVIUM. *See also* ALLUVIAL FAN, ALLUVIAL PLAIN, FLOODPLAIN.

alp 1. A high mountain.
2. Relatively flat area of grazing land in a mountain environment, that is covered by snow during winter but free from snow in the summer.

alpha decay Process of radioactive decay in which the nucleus of an atom emits an *alpha particle.

alpha diversity The diversity of *species within a particular area or *ecosystem, expressed by the number of species (*species richness) present there. A measure of *biodiversity. Also known as **local diversity**. *See also* BETA DIVERSITY, GAMMA DIVERSITY.

alpha particle A *helium nucleus given out by some *radioactive substances during *alpha decay. Alpha particles cannot penetrate very far into materials, even soft tissue, but if they enter the body (by inhalation or through a wound) they can cause great biological damage.

alpha radiation Emission of *alpha particles from a material that is undergoing nuclear transformation. *See also* RADIATION.

alpine Relating to or growing above the *tree line in high mountain areas, higher than about 1500 metres above sea level, where it is too cold for trees to grow. The term is also used to describe things that are typical of high mountain areas, or related to or found in them.

alpine glacier *See* MOUNTAIN GLACIER.

alternative agriculture An approach to farming that reflects traditional practices such as using organic rather than chemical *fertilizers and *pesticides, increased use of *crop rotations, reduced tillage of the soil, greater use of *renewable energy sources and of *intermediate technology, compared with modern high-energy, chemical-dependent *agribusiness. Also known as **alternative farming**. *Contrast* CONVENTIONAL AGRICULTURE. *See also* ECOFARMING, ORGANIC AGRICULTURE.

alternative compliance An approach to *pollution control or the reduction of *environmental risk that is based on setting targets but allowing those responsible to choose amongst a variety of different methods for achieving them, rather than *command and control regulations that define standards and specify how to meet them.

alternative crop Any non-traditional crop that can be grown in an area to diversify *crop rotations (and thus increase soil fertility and productivity) and increase income.

alternative energy Energy that is produced from sources other than *fossil fuels, which includes sources such as compressed *natural gas, *solar, *hydroelectric, or *wind energy.

alternative farming *See* ALTERNATIVE AGRICULTURE.

alternative fuel A non-petroleum (not gasoline or diesel) fuel, in liquid or gas form, that is used to power vehicles and produces less pollution. Examples include *biofuels or alcohol fuels (such as *ethanol and *methanol), mineral fuels, *natural gas, *hydrogen, and fuel cells.

alternative technology *Technology that has been developed or is used to reduce the generation of *hazardous waste, promote *recycling, or develop alternative disposal methods.

altimeter An instrument that indicates the *altitude of an object above a fixed level, usually sea level. Pressure altimeters use an *aneroid barometer with a scale graduated in altitude instead of pressure.

altiplano A high plateau in south-eastern Peru and western Bolivia, at an elevation of about 3500 m.

altithermal 1. In general, any period of high temperature.
2. Often used to refer specifically to a period of time in the mid-*Holocene, between about 4000 and 8000 years ago, when the climate was generally warmer (particularly during summer) than before or since. This period is also known as the **hypsithermal** or **Climatic Optimum**.

altitude Height above *sea level or ground level. Also known as **elevation**.

altitudinal migration A vertical pattern of *migration in which populations that breed in the *alpine or subalpine zones during the summer move to lower levels in winter, to avoid harsh climate and snow cover, and then return to the higher level the following spring. *See also* ALP.

altitudinal zonation The pattern of variation of plant and animal species relative to elevation, in response to vertical differences in climate (particularly temperature and precipitation).

altocumulus (Ac) A category of *clouds which have a *cumulus shape and are found at medium altitude, about 2400–6100 metres (8000–20 000 feet). They are composed mainly of water droplets, and consist of grey-white sheets or soft, rounded patches, which are usually larger and darker than those in *cirrocumulus and smaller than those in *stratocumulus. Altocumulus are associated with weather changes, and often appear in middle latitude *cyclones. They can create a *mackerel sky.

altostratus (As) A type of *cloud that is found at medium altitude, about 2400–6100 metres (8000–20 000 feet), and generally consists of uniform grey sheets or layers that are lighter than *nimbostratus but darker than *cirrostratus. The cloud may be striated, fibrous, or uniform, and it may be composed of ice crystals and well as water droplets. Altostratus often cover the whole sky, and they often arrive ahead of a *frontal system and signal that weather changes are about to occur. *See also* STRATUS.

altruism Unselfish concern for the welfare of others, usually other people. Many animals also display forms of altruistic behaviour.

alum Aluminium potassium sulphate. A white or colourless crystalline compound that is used to dress leather, and as a pigment in dyes. Also known as **potash alum.**

aluminium (Al) A naturally occurring, light, metallic element that is silvergrey in colour, conducts heat and electricity, is easy to weld, and is resistant to corrosion. It is the third most abundant element in the Earth's crust, and is used in alloys with *copper, *zinc, *manganese, and *magnesium. It is a very versatile metal with a wide variety of uses, including utensils, vehicle and airplane bodies, building materials, electrical conductors, explosives and fireworks, abrasives, cosmetics, paints, and even food additives. It is an important ingredient in the Earth's *continental crust (it accounts for 8.13% by mass) and occurs naturally in soil, water, and air. Inhalation of the fine powder can produce serious damage to lungs (pulmonary fibrosis).

aluminium sulphate A compound that is used in sewage treatment and for the purification of drinking water. It causes dirt and other particles to clump together and fall to the bottom of *settling basins. It is also used in the paper industry and as a fire-proofing agent.

Amazonia The Amazon Basin of South America. It covers an area of 5.8 million square kilometres of Peru, Colombia, and Brazil, and is dominated by *grassland, *wetland, *shrubland, lakes, and *tropical forests.

amber The fossilized *resin of conifer trees. It is hard, translucent, yellowish to brownish in colour, and used for the manufacture of ornamental objects including jewellery.

ambient Surrounding environmental conditions, such as pressure, temperature, or humidity.

ambient air The open air or surrounding air.

ambient air quality standard The permissible upper limit for a pollutant in *ambient air, established by the state or federal government, that serves as a target in local air quality improvement or protection programmes. It is based on the maximum acceptable average concentrations of *air pollutants during a specified period of time. The primary standard protects public health, and the secondary standard protects public welfare. *See also* CRITERIA POLLUTANT, HAZARDOUS AIR POLLUTANT, NATIONAL AMBIENT AIR QUALITY STANDARDS, PRIMARY AMBIENT AIR QUALITY STANDARDS, SECONDARY AMBIENT AIR QUALITY STANDARDS.

ambient level The general level of any particular substance or *pollutant in the environment, expressed as an average over a suitably long time and large area.

ambient measurement A measure of the concentration of a particular substance or *pollutant within the immediate vicinity of an organism, which can be related to the maximum amount of possible exposure.

amelioration *See* REMEDIATION.

amenity The useful or desirable features of a place, that provide non-monetary benefits to those who use it but which are not necessary for its use. An amenity can be natural (such as an attractive location or accessible woods or water) or made by people (such as a swimming pool or garden).

amenity horticulture Gardening or cultivation for leisure or *aesthetics rather than for commercial reasons. Also known as **hobby farming.**

amenity planting Planting trees and shrubs for non-commercial purposes, for example to make a landscape more attractive rather than to produce a crop for sale.

amenity resource Any resource that is valued for non-monetary characteristics, such as its beauty or uniqueness. For example, attractive natural scenery can inspire creativity and promote well-being.

amenity value The non-monetary, intangible value of goods or services.

amino acid An organic compound that contains both an amino group (NH_2) and a carboxyl group (COOH). Amino acids are the building blocks of *protein molecules in living things. There are 20 common amino acids that can be combined to form proteins.

amino-acid dating A method of *absolute dating of organic materials that is based on the chemical change that amino acids undergo. This is dependent on both time and temperature.

ammonia (NH_3) A colourless gas with a strong smell. It is a compound of *nitrogen and *hydrogen that is formed naturally when bacteria decompose nitrogen-containing compounds such as manure, and is used to manufacture *fertilizers and nitrogen-based compounds. Ammonia is extremely soluble in water and it reacts with *nitrogen oxides to form ammonium nitrate in soils and streams. *See also* NITROGEN CYCLE.

ammonia fixation The *adsorption of ammonium ions by *clay minerals, which makes them insoluble and non-exchangeable.

ammonification The biochemical process in which *ammonia is produced by *micro-organisms through the *decomposition of organic matter.

ammonium (NH_4) An *ion derived from *ammonia; the primary form in which *nitrogen is applied in *fertilizers.

ammonium nitrate A colourless, crystalline salt that is used in explosives, *fertilizers, and veterinary medicine.

ammonium sulphate A chemical compound that is used as a *fertilizer. It is quite acidic, and will raise the *pH of a soil.

amphibian A *cold-blooded, smooth-skinned *vertebrate animal of the class Amphibia that can live on land or in water but which returns to the water to breed. They usually hatch as aquatic larvae, breathe by means of gills, and metamorphose to an adult form with air-breathing lungs. Examples include *frogs, *toads, and salamanders. *See also* HERPETOFAUNA.

anabatic wind A gentle local wind that blows up a *hillslope when the sloping ground has been warmed by the sun. Also known as an **upslope wind**. *Compare* KATABATIC WIND.

anabolism The process by which a living cell or organism constructs the *proteins, *carbohydrates, and fats which form tissue and store energy. Also known as **biosynthesis** or **constructive metabolism**. *See also* METABOLISM.

anadromous Fish that mature and spend their adult life in the sea but swim upriver to reproduce in freshwater spawning grounds. Examples include salmon and striped bass. *Contrast* CATADROMOUS.

anaemia A condition in which there are low levels of red blood cells or red blood cells that are deficient in haemoglobin. Can be due to iron deficiency or loss of blood and causes tiredness.

anaerobe An organism (such as bacteria) that does not need oxygen to respire. The opposite is an *aerobe.

anaerobic Organisms or processes that do not require oxygen, or which oxygen would damage. Also known as **anoxic**. *Contrast* AEROBIC.

anaerobic biological treatment A process commonly used in the treatment of municipal waste that uses *anaerobic organisms (without air) to reduce the organic matter in wastes. Also known as **anaerobic digestion**.

anaerobic composting Composting based on *anaerobic decomposition. *Contrast* AEROBIC COMPOSTING.

anaerobic decomposition The breakdown of organic matter caused by *anaerobic *micro-organisms in an oxygen-free environment.

anaerobic digestion *See* ANAEROBIC BIOLOGICAL TREATMENT.

ana-front A weather *front in the atmosphere, along which warmer air rises rapidly over a layer of cold air. Ana-fronts are very active, and are associated with uplift, condensation, and pronounced weather changes. *See also* KATA-FRONT.

analogue Similar in structure, appearance, or function but not in origin or development. *See also* CLIMATIC ANALOGUE.

anatomy **1.** The structure of an *organism or one of its parts.
2. The study of those structures.

ancestor Any *organism, *population, or *species from which another organism, population, or species is descended.

ancient forest A *forest that is typically older than about 250 years with large trees, dense canopies, and great diversity of wildlife. In North America this is described as *old growth forest.

ancient woodland Woodland in Britain that originated before AD 1600 and has had a continuous cover of trees since then. Ancient woodland can also be *secondary woodland, which might have previously been cleared for *underwood or timber production.

andesite A fine-grained, *intermediate, volcanic rock characterized by *feldspars which formed by the ejection of *lava in continental areas. Named after the Andes mountains in South America.

andisol An order of deep, texturally-light *volcanic soils that contain *iron and *aluminium compounds.

androgyny Possessing both masculine and feminine characteristics. *See also* HERMAPHRODITE.

anemometer An instrument that is used to measure *wind speed (some can also measure wind direction). There are two main types, the *rotating-cup anemometer and the *pressure-tube anemometer.

aneroid barometer An instrument that is used to measure atmospheric *pressure based on variations in the height or thickness of a sealed metal box from which most of the air has been evacuated. The box has a flexible top that rises or falls in response to changes in atmospheric pressure; variations in the height or thickness of the box are recorded on a pressure scale.

angiosperm Any of the flowering seed plants having their seeds enclosed in an ovary which matures into a fruit. *Contrast* GYMNOSPERM.

angle of incidence The angle at which any ray or wave meets a surface, such as the angle at which the Sun's rays strike the Earth's surface.

angle of repose The steepest angle in which *unconsolidated sediment will lie without sliding down. It is an important measure in determining the stability of a slope and its susceptibility to *mass movement.

angle of slide The slope at which unconsolidated sediment will start to slide, which is slightly greater than the *angle of repose (because of friction).

angstrom (Å) A unit of length equal to 10^{-10} metres that is used to measure wavelengths and intermolecular distances. It has been replaced by the nanometre (nm): 1 Å = 0.1 nm.

angular momentum The tendency of a spinning body or mass of air or water to continue to spin.

angular unconformity An *unconformity in which the *bedding planes of the rocks above and below are not parallel to each other. This occurs, for example, where older underlying beds have undergone alteration and been eroded prior to deposition by younger overlying beds.

ANILCA *See* ALASKA NATIONAL INTEREST LAND CONSERVATION ACT.

animal Any member of the *animal kingdom, which comprises all multicellular *organisms that obtain energy, actively acquire their food and digest it internally, have well-developed nervous systems, have cells organized into tissues, and reproduce sexually. This includes multicellular marine organisms, worms, insects, spiders, crustaceans, fish, amphibians, reptiles, birds, and mammals.

animal community The characteristic assemblage of *animals that is associated with a particular ecosystem.

animal kingdom The taxonomic *kingdom (*Animalia) that comprises all living or extinct *animals. Also known as **zoological kingdom.**

animal rights The belief that *animals have rights similar to those afforded to humans, and that those right need to be respected and protected.

animal waste methane recovery The capture and reuse of *methane produced from the *decomposition of animal waste, before the *manure or *fertilizer is spread on fields. Recovery technologies include the *anaerobic digester, and machines that use the methane to produce energy.

Animalia In *taxonomy, the *animal kingdom. *See also* ANIMAL.

anion A negatively charged *ion. *See also* CATION.

anion exchange capacity A measure of the surface charge of an *anion, expressed in equivalents of exchangeable ions per unit weight of the solid.

anisotropy Having different properties in different directions. For example, variations in hydraulic properties of *groundwater in an *aquifer. *Contrast* ISOTROPY.

Annex A Gases The *greenhouse gases whose emissions are regulated by the *Kyoto Protocol (as defined in Annex A) in order to reduce *global warming and slow down *climate change. The gases are *carbon dioxide, *methane, *nitrous oxide, *hydrofluorocarbons, *perfluorocarbons, and *sulphur hexafluoride.

Annex B Countries The list of industrialized countries and *economies in transition that have their *greenhouse gas emissions capped under the *Kyoto Protocol (as listed in Annex B). The Protocol also defines legally binding obligations on each country to reduce their emissions by set amounts within a defined period of time. For example, most European countries are required to decrease their emissions by 8% (relative to 1990 levels) during the first commitment period from 2008 to 2012, whilst Iceland is allowed to increase its emissions by up to 10% over the same period (because it is at a lower stage of industrial development).

Annex I Countries/Parties Countries that committed themselves specifically to the aim of reducing their emissions of *greenhouse gases to 1990 levels by the year 2000. The list, which appears in Annex I to the *United Nations Framework Convention on Climate Change, includes all the countries in the *Organisation for Economic Cooperation and Development, plus countries designated as *economies in transition. By default other countries were referred to as Non-Annex I countries. When these countries ratified the *Kyoto Protocol they accepted further emission targets for the period 2008–12, and the Annex I list and the *Annex B list are almost interchangeable. Strictly speaking, it is the Annex I countries which can invest in *Joint Implementation (JI) and *Clean Development Mechanism (CDM) projects, as well as host JI projects, and non-Annex I countries which can host CDM projects.

Annex II Countries/Parties The countries listed in Annex II to the *United Nations Framework Convention on Climate Change that have a special obligation to help developing countries with financial and technological resources required to reduce *greenhouse gas emissions and curb *global warming. This includes all 24 original member countries of the *Organisation for Economic Cooperation and Development plus the *European Union.

annual A plant that completes its *life cycle (germination, growth, flowering and fruiting, then death) in a single growing season and so has a *lifespan of 1 year. *See also* BIENNIAL, PERENNIAL PLANT, SEASONAL PLANT.

annual cycle The yearly cycle of changes in the climate system and related environmental systems, caused by variations in the amount of solar radiation that reaches the Earth as it orbits the Sun.

annual growth The amount of new *biomass produced by a plant per year, usually measured as above-ground production.

annual maximum series Flow records that are used in *flood frequency analysis, which include only the highest *discharge each year (the annual maximum). *Contrast* PARTIAL DURATION SERIES.

annual pasture A *pasture consisting of *introduced *forage species that have been planted to grow for only one year or season. *Contrast* PERMANENT PASTURE.

annual ring *See* GROWTH RING.

annular drainage pattern A type of *drainage pattern in a river system. It develops on a dome where concentric outcrops of rocks with different resistance to erosion are exploited by river erosion.

anomaly The difference between a particular measurement (for example of weather elements such as temperature or precipitation) and the mean or expected value.

Anopheles The *genus of mosquito that transmits *malaria.

anoxic Lacking oxygen or oxygen-free. *See also* ANAEROBIC.

ant mosaic The three-dimensional mosaic within a habitat that is created by the non-overlapping territories of dominant ants. This in turn has a significant influence on the diversity and species composition of other organisms, including plants, *vertebrates, and *invertebrates.

antagonism The interaction of two or more substances with opposing effects which partially or wholly cancel each other out, or (for organisms such as *bacteria) where the growth of one inhibits that of the other. *Contrast* SYNERGISM.

Antarctic Circle The area of land and sea that lies south of latitude 66°S, around *Antarctica.

Antarctic circumpolar current A large ocean current that is driven by the wind and flows completely around *Antarctica from west to east.

Antarctic Convergence A sharp change in the physical characteristics of the Atlantic, Indian, and Pacific oceans between latitudes 48° and 60°S. Here, warmer waters moving south from the mid-latitude oceans meet and mix with colder waters moving north from *Antarctica, and there is a steep temperature gradient within the ocean water which effectively isolates life in the *Southern Ocean (i.e. the area within the Antarctic Convergence).

Antarctic ozone hole The seasonal depletion of *ozone over much of *Antarctica. *See also* OZONE HOLE.

Antarctic realm A *biogeographical realm in the southern hemisphere, which contains a variety of *ecosystems from *temperate forest and *grassland in New Zealand to *tundra and *ice sheets in *Antarctica.

Antarctic Treaty An international treaty that is designed to preserve the unique environment of *Antarctica by

a

ANTARCTICA

The fifth largest and by far the coldest of the seven continents. It is centred on the *South Pole, located mostly within the *Antarctic Circle, covered with ice (which in places is more than 2000 metres thick) and surrounded by sea ice, particularly during the winter. The outer limits of Antarctica are defined by the *Antarctic Convergence. It is, in effect, a *cold desert, with an average annual precipitation in the interior of about 50 millimetres. Climate in the interior is dominated by extreme cold and light snowfall, and temperatures are milder and precipitation is much higher (up to about 380 millimetres a year) around the coastal fringe. Over 95% of the land surface is currently covered by ice, and the ice cover is variable in thickness but sometimes thicker than about 2000 metres. This thick ice cover makes Antarctica the highest continent overall, with an average elevation of about 2300 metres above sea level. Wildlife on the ice-covered areas is extremely limited; the most prominent examples are the penguins (particularly Adelie and emperor penguins) which breed on ice and live on ice and in the surrounding oceans. The sea has more abundant life, including six species of seal and large numbers of whales which feed on krill (small, shrimp-like crustaceans that swarm in dense shoals and feed on tiny diatoms). The human population of Antarctica is extremely small, because there is no native population and most residents are short-term scientific visitors. Antarctica is widely regarded as the last great wilderness. Until relatively recently its natural environments have been preserved more by lack of exploitation (because the continent is so remote, inaccessible, and uninviting) than by purposeful action. But this is changing, as geological exploration reveals more details of the large deposits of valuable mineral resources (particularly *coal, *oil, and *natural gas) beneath the cold continent and its surrounding *continental shelf. Pressures on Antarctic marine resources are also mounting, with extensive commercial exploitation of whales and krill. West Antarctica contains enough ice to raise *sea level by between five and six metres, and there is mounting concern about the rapid melting of this ice, which is occurring faster than previously thought, and appears to be associated with, if not directly caused by, *global warming. *See also* ANTARCTIC TREATY.

limiting development there. It was signed in 1959 by the seven countries (Norway, France, Australia, New Zealand, Chile, Great Britain, and Argentina) which claim sovereign rights over the continent, plus twelve other countries, and it dedicated the whole continent to peaceful scientific investigations. All existing territorial claims were suspended when the treaty came into effect in 1961. Since it was signed, international concern has grown over the prospect of renewed mineral prospecting in and around Antarctica, and there have been repeated calls for the continent to be designated a *World Park and protected for ever against development.

In 1991, 24 countries approved a protocol to the treaty (the *Madrid Protocol) that would ban oil and other mineral exploration for at least 50 years.

antecedent moisture The degree of wetness of the soil at the beginning of a period of *runoff, usually expressed as the total inch-depth-equivalent of water stored in the soil. Also known as **antecedent soil water**.

antecedent river A *river network in which an initial pattern, established in the geological past, has been preserved while the area has been uplifted. Rates of river downcutting must have been higher than rates of *uplift for this to occur.

antecedent soil water *See* ANTECEDENT MOISTURE.

antediluvian The period before the Biblical flood (described in Genesis, Chapter 7).

anthophilous Living or growing on flowers, as certain insects do.

anthracite A hard, black *coal that burns slowly with little flame or smoke but gives off intense heat, formed at the later stages of the coal cycle by the folding and hardening of sedimentary strata containing *bituminous coal. This is the most highly metamorphosed form of coal, containing 92 to 98% of fixed *carbon.

anthropic Relating to the period during which humans have existed on Earth.

anthropocentric Human-centred. Based on the belief that only humans have value and thus they have a privileged position in nature, that the environment exists only for the benefit of humans, and that nature has no rights. Also known as **homocentric**. *Contrast* ECOCENTRIC.

anthropocentrism A *worldview that sees humans as the source of all value, since the concept of value itself is a human creation, and that sees nature as of value merely as a means to the ends of human beings. *See also* ECOCENTRISM, HUMAN EXCEPTIONALISM PARADIGM.

anthropogenic Made by humans or resulting from human activities.

anthropoid A member of the group of primates (suborder Anthropoidea) that comprises monkeys, apes, and humans.

Anthropological Reserve *See* NATURAL BIOTIC AREA/ANTHROPOLOGICAL RESERVE.

anthropology The study of human cultural variations, including language, biology, and society. *See also* ETHNOGRAPHY.

anthropomorphism The attribution of human motivation, characteristics, or behaviour to non-human things, which leads to the treatment of animals, gods, inanimate objects, or natural phenomena as if they have human feelings and emotions.

antibiotic A chemical produced or synthesized by living micro-organisms (such as yeast) that destroys or inhibits the growth of other organisms, especially *bacteria. Used to treat or prevent infections by inhibiting the growth of bacteria and *fungi that produce disease.

antibody A special *protein that is produced by the immune system within an organism in response to an invading foreign *antigen (such as *bacteria, dust, or pollen).

anticarcinogen A substance or agent that opposes the action of *carcinogens.

anticline A common type of *fold in rock. It takes the form of a symmetrical fold, arched upward in the middle, in which the rocks at the centre are usually the oldest. *Contrast* SYNCLINE.

anticlinorium A series of small-scale *anticlines and *synclines which may be small enough to be viewed in outcrop.

anticyclone A large area of high *atmospheric pressure around which the winds blow clockwise in the northern hemisphere (and anticlockwise in the southern hemisphere) and which usually results in calm, fine *weather. Also known as a **high.**

antidote A remedy designed to relieve, prevent, or counteract the effects of a poison, usually by eliminating it, neutralizing it, or absorbing it.

anti-environmental group A group that is known for either pushing an agenda contrary to the *environmental movement, or for engaging in efforts to undermine the effectiveness of the environmental movement. Examples from the USA include *think-tanks such as the *Cato Institute, *Heartland Institute, *Heritage Foundation, legal groups such as the *Mountain States Legal Foundation, and *front groups such as the *Climate Council, *Coalition for Vehicle

Choice, *Global Climate Coalition, *National Wetlands Coalition, and *Wise Use Movement.

anti-equitable effect An effect that leads to less *equity and more *inequality.

antigen A substances that the body regards as foreign, which results in the production of *antibodies.

antiquities The collective name for prehistoric and historic artefacts, objects, structures, ruins, sites, and monuments that have some cultural or scientific significance and are considered to be older than 100 years.

Antiquities Act (1906) US legislation that protects ruins or objects of antiquity on *federal lands.

anvil cloud The top part of a *cumulonimbus cloud which spreads horizontally outward in the vicinity of the *tropopause and looks like a blacksmith's anvil.

ANWR See ARCTIC NATIONAL WILDLIFE REFUGE.

AONB See AREA OF OUTSTANDING NATURAL BEAUTY.

AOSIS See ALLIANCE OF SMALL ISLAND STATES.

aphelion The point in the Earth's orbit when it is farthest from the sun (152.5 million kilometres), which occurs on the 3rd or 4th of July each year. The opposite is *perihelion.

aphotic Totally dark. Also known as **disphotic**. Contrast PHOTIC.

apiculture Beekeeping in order to produce honey. Contrast BOMBICULTURE.

apodous Without feet; legless.

applied dose In exposure assessment, the amount (*dose) of a substance that is in contact with the parts of an organism (such as the skin, lung tissue, or gastrointestinal track) through which it can be absorbed.

appropriate technology See INTERMEDIATE TECHNOLOGY.

apterygote A wingless insect of a group that is believed never to have had wings in its evolutionary past. Contrast PTERYGOTE.

AQI See AIR QUALITY INDEX.

aquaculture The cultivation of marine or freshwater food fish or shellfish, such as oysters, clams, salmon, and trout, and the farming of aquatic plants, under controlled conditions to be sold for human consumption. Also known as **fish farming**, **mariculture**, or **pisceculture**.

aquanaut A person who is trained to live in underwater installations and to conduct, assist in, or be a subject of scientific research.

aquarium A tank, pool, or building in which living aquatic animals and plants are kept under controlled conditions for pleasure, study, exhibition, or as a form of *ex situ conservation.

aquasilvicultural system An *agroforestry system that combines trees with fish farming. See also AQUACULTURE.

aquatic Related to, living, or growing in or on water.

aquatic resources The living resources of aquatic habitats, which include fish, invertebrates, and *amphibians.

aqueduct A long channel or raised bridge-like structure built to transport water from a remote source, usually by gravity. See also INTERBASIN TRANSFER.

aqueous Containing water, dissolved in water, or composed mainly of water.

aqueous solubility The extent to which a substance will dissolve in water.

aquiclude An effectively *impermeable layer of rock that confines an *aquifer, preventing the water in it from moving upward or downward into adjacent *strata. Examples include shale and some *igneous rocks. See also PERCHED AQUIFER.

aquifer A body of *permeable and/or *porous *rock that is underlain by *impermeable rock and is saturated with water (*groundwater) or transmits water underground, which can be extracted for use by humans, most commonly by drilling *wells. *See also* AQUICLUDE, AQUITARD, ARTESIAN WELL, CONFINED AQUIFER, PERCHED AQUIFER.

aquifer depletion Withdrawing *groundwater from an *aquifer faster than it is naturally replenished. Also known as **groundwater mining** or **overdraft.**

aquifer mining The withdrawal of *groundwater from an *aquifer, over a period of time, at a rate that exceeds the rate of natural *recharge.

aquitard The less *permeable beds in a rock sequence, that may be permeable enough to transmit water but not permeable enough to allow water to be abstracted from *wells within them.

arable 1. Land that is suitable for cultivation.
2. A farming system in which crops (such as cereals) are raised, rather than livestock.

arachnid Air-breathing *arthropod such as spiders and mites, that has a body made of two segments (except mites) and four pairs of legs.

Aral Sea A very large *freshwater *lake (formerly the world's fourth largest lake) on the border between Kazakhstan and Uzbekistan in Central Asia. Since the 1960s the lake has been drying up and shrinking as a result of a loss of water due to the diversion of *river flow to a major *irrigation project designed to make the area *self-sufficient in cotton, and to increase rice production. The surface area of the Aral Sea has shrunk by nearly half since 1960, and its volume has decreased by two-thirds.

arbitration A formal process for resolving disputes, in which there are strict rules of evidence, cross-examination of witnesses, and a legally-binding decision made by the arbitrator that all parties must obey.

arboreal Relating to or living in or amongst *trees.

arboretum A collection of specimen *trees from which seeds and cuttings can easily be gathered as part of a *living collection.

arboriculture The cultivation and management of *trees. *See also* SILVICULTURE.

arbovirus A *virus that is transmitted by an *arthropod (such as a mosquito or tick), which includes encephalitis, *yellow fever, and dengue fever.

Archaean A period of geological time extending from about 3.9 to 2.5 billion years ago (the start of the *Proterozoic), during which the Earth's crust formed. The earliest part of *Precambrian time. Rocks of this age contain fossils of single-celled organisms, which are the earliest forms of life on Earth.

Archaeological and Historic Preservation Act (1974) US legislation that provides for recovery, protection, and preservation of significant cultural resources that will be irreparably lost or destroyed by alteration of terrain from any federal construction project or federally licensed activity or programme.

archaeological resource All remaining physical evidence of past human occupation, other than historical documents, which can be used to reconstruct life styles of ancient cultural groups. This includes skeletons, sites, monuments, *artefacts, environmental data, and all other relevant information.

archaeological site Any place that contains physical evidence of past human activity, and which provides scientific, cultural, or historical evidence relating to the history of that place.

archaeology The study of past human cultures through the analysis of material remains (as fossil relics, *artefacts, and monuments), which are usually recovered through excavation.

archetype The original form or body plan from which a group of *organisms develops.

Arctic air A very cold and dry *air mass that generally forms north of the Arctic Circle but moves southwards, cooling the areas it passes over.

Arctic front A weather *front that develops at the boundary between the very cold dense arctic *air mass that is associated with the cold arctic *anticyclone, and the warmer *polar air mass.

Arctic haze A persistent thin mist that lies over the Arctic during the winter months, possibly associated with by the long-range transport of *air pollutants that originate in mid-latitude countries.

Arctic high A very cold zone of high *atmospheric pressure that originates over the Arctic Ocean.

Arctic National Wildlife Refuge (ANWR) A *National Wildlife Refuge that covers about 80 000 square kilometres in north-eastern Alaska, in the North Slope region. The area has been protected since 1960, but the refuge was expanded in 1980 under the *Alaska National Interest Lands Conservation Act.

Arctic Ocean The ice-covered waters that surround the *North Pole, which are largely covered with solid ice or with *ice floes and *icebergs.

arcuate In the shape of an arc, bow, or fan.

area mining 1. A type of *strip mining that is practiced on relatively flat land, which involves cutting a trench through the *overburden in order to expose the deposit of *mineral or *ore to be removed. The overburden is placed on unmined land adjacent to the cut, and the mineral or ore is then removed. A second cut is then made parallel to the first, the overburden from which is deposited in the first cut. This process is continued until the mineral or ore is all extracted. Also known as **area strip mining.**
   2. Surface mining, on a large scale, in an area of level or gently rolling ground.

ARCTIC

The area lying above 66.5° North that is dominated by the Arctic Ocean but also includes large land areas in Canada, Russia, Greenland, Scandinavia, Iceland, and Alaska. Some of the land areas, including most of Greenland, are permanently covered in ice and *pack ice is common throughout the Arctic Ocean. Climate in this high latitude zone is extreme, with short cool summers and long cold winters. Temperatures are particularly low in the interior— average mid-winter temperatures on the Greenland *ice cap are around −33°C, for example. Precipitation is generally low (less than 250 millimetres a year on average), and well distributed throughout the zone. Large river and lake systems are rare, because of the low precipitation, but shallow lakes, ponds and marshes are common in areas underlain by *permafrost. Although the Arctic has the appearance of a vast, frozen desert, it does support wildlife. Signs of life are difficult to find during the cold, dark winter months but some species of mammals and birds carry extra insulation (such as fat) to survive the winter. The Arctic appears to wake up in spring. More than 400 species of flowering plant grow in the Arctic, and most of the land which is not covered in ice is *tundra, with a natural vegetation of low creeping *shrubs, grasses, thick growths of *lichens and *mosses, and *herbs and *sedges. The Arctic region is home to a wide variety of birds and fish, and the mammals include polar bear, arctic fox, arctic wolf, walrus, seal, caribou, reindeer, the infamous lemming, and many species of whale.

area of critical environmental concern An area of public land in the USA where management is directed towards the protection of fish and wildlife, important historic, cultural, or scenic values, or other natural systems or processes. *See also* FRAGILE OR HISTORIC LANDS.

Area of Outstanding Natural Beauty (AONB) An area of protected *countryside in the UK, designated under the National Parks and Access to the Countryside Act 1949 on the basis of attractive *landscape, in which development is tightly controlled.

area source Any source of *air pollution that is released over a relatively small area but is not a *point source. Sources include vehicle exhausts, houses and small businesses. Similar to a *nonpoint source. *See also* EMISSION SOURCE.

area strip mining *See* AREA MINING.

area-sensitive species Species that respond badly if the *patch size of their *habitat decreases.

arenaceous rock Medium-grained sandy *clastic *sedimentary rocks, such as *sandstone, with particles ranging in size from 0.06 to 2.0 millimetres.

arene *See* AROMATIC.

arête A long, sharp, knife-edged ridge with a serrated top, formed when *glaciers erode adjacent *cirques on the opposite sides of a mountain ridge. *See also* COL, HORN.

argillaceous rock Fine-grained *clastic *sedimentary rocks, such as *siltstones and *marls, made of *clay and *silt particles smaller than 0.06 millimetres.

argon (Ar) A natural, colourless, odourless, *inert *gas that is the third most abundant constituent of dry air (it comprises 0.93% of the Earth's *atmosphere).

arid Dry, lacking moisture.

arid region An area with a dry climate which cannot support the growth of trees, woody plants, or most crops (without *irrigation), and with a *biome in which only drought-resistant vegetation naturally survives.

aridisol A desert type (*order) of soil that develops in an *arid region, with little or no organic content but significant amounts of calcium carbonate (*gypsum) and other deposited salts. With little if any vegetation growing on the soil surface, there is no surface *litter, no *humus, and no dark organic layer in the *A-horizon. There is a relatively deep *B-horizon where soluble compounds accumulate which are leached down from above. Aridisols are often affected by *salinization.

aridity index A measure of the lack of moisture in a place, based on the level of water deficit. In equation form the index is 100 × water deficit/potential evaporation.

arithmetic growth A pattern of growth that increases at a constant rate over a specified time period, such as 1, 2, 3, 4 or 1, 3, 5, 7. *Contrast* EXPONENTIAL GROWTH, GEOMETRIC GROWTH.

Army Corps of Engineers *See* US ARMY CORPS OF ENGINEERS.

aromatic A major group of unsaturated *hydrocarbons, such as *benzene or *toluene, which have a specific type of benzene ring structure and are added to *gasoline to increase octane. Some are toxic; all have a strong but not unpleasant odour. Also known as **arene**.

arrested development An incomplete ecological *succession that has not reached the natural climax stage, usually as a result of human disturbance or alteration which is reflected in *subclimax and *plagioclimax vegetation.

Arrhenius, Svante A Swedish physicist and chemist (1859–1927) who predicted that the release of *carbon dioxide into the atmosphere would lead to *global warming, and estimated the magnitude of the *greenhouse gas effect.

arroyo A small, deep gully or channel with steep or vertical walls and a flat floor, which is eroded into the surface of a dry *desert by an intermittent stream. The bed is usually dry, and the arroyo contains *streamflow only after rain.

arsenic (As) A *heavy metal element that occurs naturally in the Earth's crust and fossil fuels, which *bioaccumulates and is highly *toxic and *carcinogenic to humans. It is used in the production of glass, enamels, ceramics, oil, cloth, linoleum, electrical semiconductors, pigments, fireworks, pesticides, fungicides, veterinary pharmaceuticals, and wood preservatives.

artefact An object made by humans, usually for a practical purpose, which has been preserved and can be studied to learn about the period it was made and used.

artesian aquifer An *aquifer in which *groundwater is confined between two *impermeable layers (*aquitards), under *hydrostatic pressure, which is significantly greater than *atmospheric pressure and forces the water to flow out of an *artesian well, boreholes or springs, so long as the aquifer is constantly recharged. Also known as **confined aquifer.**

artesian well A *well sunk in an *artesian aquifer, from which *groundwater is naturally forced to the surface by *hydrostatic pressure so the well is free flowing. *See* FLOWING WELL.

arthropod A member of the *phylum Arthropoda, which is by far the largest in the *animal kingdom and includes many different *invertebrates with an *exoskeleton such as *arachnids (spiders, mites) *insects (bees, ants, moths) and *crustaceans (shrimps, crabs).

artificial fertilizer A chemical that is added to soil to increase fertility and enhance crop production.

artificial rain *See* CLOUD SEEDING.

artificial recharge Adding water to an *aquifer to restore *groundwater.

artificial regeneration Establishing a new *forest by planting seedlings or by direct seeding, often after *deforestation. *Contrast* NATURAL REGENERATION.

artificial selection The process by which humans breed animals and cultivate crops to ensure that future generations have specific desirable characteristics, by exploiting the process of *natural selection. Breeders select the most desirable variants in a plant or animal population, and then selectively breed them with other desirable individuals. Examples include particular plants that are resistant to disease, high yielding, or attractive in appearance, and particular breeds of cattle which produce more meat or milk. Domesticated animals such as dogs, cats, and farm animals are the product of selective breeding.

artificial wetland An engineered *wetland, often constructed to treat *sewage or other organic wastes.

As Low As Reasonably Achievable (ALARA) An approach to the control of radiation in order to protect public health and the environment, based on keeping radiation emissions and exposures to levels set as far below regulatory limits as is reasonably possible.

asbestos A group of naturally occurring fibrous minerals that can be spun and woven. They are strong, do not corrode or burn, and good at insulating heat and electricity, and are used for insulation, construction and brake linings. Asbestos minerals separate into fibres that can cause air and water pollution, and cause cancer or *asbestosis when inhaled by humans. In the US their use in construction and manufacturing is now banned or severely restricted.

asbestos abatement Measures used to control the release of dangerous fibres from materials that contain asbestos, or remove them completely. These include removal, enclosure, and encasement.

asbestosis A lung disease in humans, associated with chronic exposure to asbestos and inhalation of asbestos fibres, that results from scarring of the lung tissues by the fibres. The disease makes breathing progressively more difficult and can be fatal.

asexual reproduction A mode of reproduction, common among lower animals, micro-organisms, and plants, in which offspring are created by a single parent so they inherit only the *genes of that parent.

ash 1. The mineral content of a product that remains after complete combustion, which consists mainly of minerals in oxidized form. *See also* FLY ASH.
  2. Volcanic dust that erupts from a *volcano, and either flows out (as a *pyroclastic flow) or forms a cloud.

ash cone A steep-sided *volcano composed of fine *volcanic ash erupted from the volcano.

ask *See* OFFER.

ASOS *See* AUTOMATED SURFACE OBSERVING SYSTEM.

aspect The direction in which a slope or feature faces, usually expressed in terms of points on a compass. For example, an aspect of 90° is due east, 180° is due south and 315° is north-west. *See also* AZIMUTH.

asphyxia Suffocation dues to inhalation of toxic gases.

asphyxiant A vapour or gas that can cause unconsciousness or death due to a lack of oxygen (suffocation). Chemical asphyxiants like *carbon monoxide (CO) reduce the ability of the blood to carry oxygen; those like *cyanide interfere with the way the body uses oxygen.

assay Analysis of a specific chemical, microbe, or biological response. *See also* BIOASSAY, LIMIT OF DETECTION.

assemblage A group of plants and/or animals that is indicative of a particular environment. *See also* COMMUNITY.

asset Anything owned by an individual or group that has a value, whether financial or otherwise.

assigned amount The total amount of *greenhouse gas that each country is allowed to emit during the first *Commitment Period of the *Kyoto Protocol, taking into account certified *emissions reductions acquired via the *Clean Development Mechanism. Also known as **emissions budget.**

assimilation The process by which organisms take in and convert nutrient substances for growth, reproduction, or repair.

assimilative capacity The ability of a body of water to naturally purify itself of *pollutants.

association Two or more species living in the same place at the same time. *See also* COMMUNITY.

Association for Biodiversity Information *See* NATURESERVE.

association, index of A measure of the likely occurrence of particular two species together, in *association.

asteroid A small, rocky or metallic star-like body that orbits the Sun in the asteroid belt in between Mars and Jupiter. There are probably more than 100 000 asteroids in space, which might be remnants of a former planet or planets that disintegrated, or may be part of the original matter of the solar system that never became a planet. The risk of an asteroid impact on Earth is measured using the *Torino scale. *See also* COMET.

asthenosphere The weak upper part of the *mantle within the *Earth's interior, just below the *lithosphere.

astronomy The study of the universe and celestial bodies.

asynchronous Not coinciding in time. *Contrast* SYNCHRONOUS.

at risk species Species whose survival is not guaranteed. These are often classified as *endangered, *threatened, or *species of concern. *See also* IMPERILLED SPECIES.

a

Atlantic Basin The *Atlantic Ocean north of the equator, the Caribbean Sea, and the Gulf of Mexico.

Atlantic Ocean The Earth's second largest *ocean, which covers about a fifth of its surface, and is bounded by North and South America to the west and by Europe and Africa to the east. It is relatively shallow, and is the warmest and most *saline of the major oceans.

Atlantic Period *See* CLIMATIC OPTIMUM.

atmospheric boundary layer *See* BOUNDARY LAYER.

atmospheric deposition The deposition of solids, liquids, or gaseous materials from the air onto land or oceans. *See also* ACID RAIN.

atmospheric pressure The force exerted by the weight of the overlying *atmosphere. *Pressure is measured with a *barometer and expressed in *bars or millibars (mb), and pressure readings are normally adjusted to sea level equivalents in order to eliminate the effects of altitude. Mean atmospheric pressure on Earth is 1013.25 mb. High pressure cells within the atmosphere, which generally produce stable weather conditions, have pressures up to about 1060 mb; low pressure cells, which can result in storms, have pressures down to about 940 mb. Also known as **air pressure** or **barometric pressure.**

atmospheric transport The movement of *air pollutants from one region to another via the *atmosphere, which may be for hundreds or even thousands of kilometres and can cross national boundaries to create international or *trans-frontier pollution.

atmospheric window A region of the *electromagnetic spectrum, from 8 to 12 micrometres, where the atmosphere is effectively transparent to *longwave radiation.

ATMOSPHERE

1. The mixture of *gases that surrounds the *Earth, which includes *nitrogen (78.08%), *oxygen (20.94%) and *carbon dioxide (0.035%). It is a major *environmental system, which interacts with the *lithosphere, the *hydrosphere, and the *biosphere. It contains a variable amount of fine solid material, held up in suspension by air currents, wind systems, and convection currents, which comes from a variety of sources and in a variety of forms, including *dust (for example from volcanoes and desert sandstorms), *pollen and mould spores (from natural vegetation), *smoke (from natural forest fires, and from air pollution), and salt spray (from the oceans). There is no definite outer limit to the atmosphere, it simply becomes less dense with increasing height above the ground until it eventually becomes part of the solar atmosphere (at an altitude of about 80 000 kilometres) and merges into space where near-vacuum conditions exist. *Weather is mostly confined to the lowest 16 km of the atmosphere, although air movements in the lowest 30 km affect weather patterns and changes. The atmosphere is composed of a number of vertical layers or zones, each of which is quite clearly defined and has distinctive physical properties (particularly temperature and pressure). Abrupt changes in temperature and pressure mark the boundaries between successive zones. It is convenient to think in terms of three main layers within the atmosphere—the lower atmosphere (*troposphere, which contains three-quarters of the mass of the atmosphere), the middle atmosphere (*stratosphere), and the upper atmosphere (*mesosphere, *thermosphere and beyond into the *exosphere).
2. A unit of pressure equal to *atmospheric pressure at *sea level.

atoll A low-lying island composed of *coral reef, which is usually circular or horseshoe-shaped, with a *lagoon in the middle.

atom The smallest part of an *element that can exist. An atom is composed of a nucleus containing protons and neutrons surrounded by *motile electrons. *See also* ELECTRON, PROTON, NEUTRON, SUBATOMIC.

atomic energy Energy that is released in a nuclear reaction, either by *fission or *fusion.

atomic number (Z) The number of protons in the *nucleus of an *atom, which defines and identifies the element, indicating its place in the *periodic table of the elements. The atomic number is also equal to the number of electrons in the atom.

atomic pile North American term for a *nuclear reactor.

atomic weight *See* RELATIVE ATOMIC MASS.

attainment area A geographical area in which levels of a *criteria air pollutant meet the *national ambient air quality standard for that pollutant. An area may have an acceptable level for one criteria air pollutant but not for others, so it could be both an attainment and a non-attainment area at the same time. *See* NON-ATTAINMENT AREA.

attenuation Weakening, such as the dilution of concentration of a compound or agent, or a reduction in the size and energy of a signal (such as a *seismic wave).

attractant A chemical or agent that attracts insects or other pests by stimulating their sense of smell.

attributable risk The difference between the *incidence rate (of a disease or effect) in groups who have been exposed to an environmental risk and that in non-exposed groups.

attrition The wearing or grinding down of a substance by friction.

auction *See* EMISSIONS AUCTIONING.

audit An inspection or systematic examination of a process or activity, often carried out by an independent or external specialist, designed to ensure compliance with requirements. *See also* ENVIRONMENTAL AUDIT.

Audubon, John James A US wildlife artist (1785–1851) who published the book *Birds of America* (a collection of 435 life-size prints). His name is synonymous with birds and bird conservation worldwide, although he played no role in the organization that bears his name (the *Audubon Society).

Audubon Society A non-profit, US based environmental conservation organization whose mission is to conserve and restore natural *ecosystems, focusing on birds, other wildlife, and their habitats for the benefit of people and biodiversity.

auger An instrument that is used for drilling holes, mainly for extracting samples of soil or peat.

auger mining A type of surface *mining that involves drilling large, closely spaced holes into a deposit, where the slope is too steep for *contour strip mining or where there is a lot of *overburden relative to the workable deposit.

aurora Spectacular lights and colours (usually green, red, or yellow) in the sky that can sometimes be seen within the *ionosphere, particularly in *polar regions. These are the aurora borealis (northern lights) in the *northern hemisphere and the aurora australis (southern lights) in the *southern hemisphere. They are created when the high speed *solar wind interacts with the upper atmosphere.

austral Relating to high southern latitudes. *See also* BOREAL

Australian realm A *biogeographical realm which is largely *desert, surrounded by *tropical forest and *savanna.

autecology The study of individual *organisms or *species, with particular reference to their interaction with their environment. *See also* GENECOLOGY. *Contrast* SYNECOLOGY.

autochthonous Formed *in situ*, in its present location. **1.** A rock whose main constituents were formed in that position rather than being transported from elsewhere. Organic deposits that turn into *coal are of this type.
2. Organic matter produced within the given *habitat, *community, or *ecosystem. *See also* ENDOGENOUS. *Contrast* ALLOCHTHONOUS.

Automated Local Event Reporting in Real Time (ALERT) A network of automatic rain gauges in the USA that transmit via VHF radio link when precipitation occurs. Some sites are also equipped with other sensors to report temperature, wind, pressure, river stage, or tide level.

Automated Surface Observing System (ASOS) A climate monitoring system that automatically collects and stores continuous information from a place or area about factors such as sky conditions, *temperature and *dew point, *wind speed and direction, and *atmospheric pressure.

automated weather station An unmanned station that has a range of sensors to measure weather elements such as *temperature, *wind, and *pressure and that transmits these readings for use by meteorologists in *weather forecasting.

autotroph An *organism that makes its own food by synthesizing organic matter from inorganic substances, unlike a *heterotrophic organism which derives food from *organic matter. Higher plants and algae use *carbon dioxide as a source of *carbon and obtain their energy from the Sun, via *photosynthesis, and are called *photoautotrophs. Bacteria and other autotrophs oxidize inorganic substances such as *sulphur, *hydrogen, ammonium, and nitrate salts, and are called *chemoautotrophs. Autotrophs are also known as **primary producers.**

autumn The *season of the year, between *summer and *winter, when temperatures at mid-latitudes decrease as the Sun approaches the winter solstice. Astronomically this is the period between the *autumnal equinox and the *winter solstice, which covers the months of September, October, and November in the northern hemisphere, and March, April, and May in the southern hemisphere. This is the season when leaves fall from deciduous trees, hence it is known in North America as **fall.**

autumnal equinox The *equinox at which the Sun approaches the southern hemisphere and passes directly over the equator, which occurs around 23 September each year. *See also* VERNAL EQUINOX.

available element *See* AVAILABLE NUTRIENT.

available nutrient The elements in the soil solution that can readily be taken up by plant roots, usually only a fraction of the total amount of a nutrient that is present in the soil. Also known as **available element.**

available water The water in soil that can be taken up by plant roots. *Contrast* UNAVAILABLE WATER.

avalanche A sudden, rapid *mass movement process like a *landslide, but involving a large mass of snow, ice, and rock crashing down a mountainside under its own weight. Most avalanches occur in spring when the snow starts to melt, although some are triggered by *earthquakes.

average life expectancy The number of years that an average person can expect to live, in a particular country or area.

avermectin A naturally occurring *toxic *antibiotic that is produced by the bacterium *Streptomyces avermitilis* and is often used to control mites.

avian Having to do with *birds.

aviary A building or large enclosure in which *birds are kept.

avicide A *pesticide which can be used to kill *birds.

aviculture The rearing of *birds in captivity.

avifauna The *birds of a particular region, *habitat or period of time.

avoidable risk A *risk which it is not necessary to take because the individual or public goals can be achieved by other means, at the same or less total cost, without taking the risk. Contrast ACCEPTABLE RISK, UNACCEPTABLE RISK.

avoidance Emissions of *greenhouse gases that have not occurred, either because of improvements in energy efficiency (thus less energy is used) or because of a switch in energy sources (from high to low or no emission).

avoidance cost The actual or estimated cost of preventing environmental damage by adopting alternative production and consumption processes, or by reducing or abstaining from particular activities.

avoided emissions Emissions of *greenhouse gases that would have been made under a *business-as-usual scenario, but have been avoided through the implementation of an *emissions reduction project.

azimuth The angle along the Earth's horizon made with *magnetic north, which is measured clockwise. See also ASPECT.

azoic Devoid of life, or before life began.

azonal soil A poorly developed soil with young and evolving *horizons. Examples include *alluvium and *loess. Through time these types of soils are likely to mature and develop into a *zonal soil.

Bacillariophyta *Diatoms. Also known as golden-brown algae.

***Bacillus thuringiensis* (Bt)** A naturally occurring *bacterium that produces a *protein which kills caterpillars and some moths and butterflies. It is a *biopesticide that is used extensively by the *microbial *pesticide industry.

backcountry A general term used in North America for all parts of *wildlands in which there are no permanent, improved, or maintained access roads or working facilities (such as lumber mills, ski resorts, or settlements with permanent residents). Any current uses of the area only have primitive facilities, such as cabins, base camps, or undeveloped campgrounds.

backcountry recreation A simple or primitive recreation experience in a roadless *backcountry area, where travel is mostly by horse, foot trails, or canoe.

backfill The process of filling in an excavation, and the material that is used in doing so.

backfit *See* RETROFIT.

backflow A reverse flow condition in a *potable water supply system, caused by a difference in water pressures, which makes water flow back into the distribution pipes from a unintended source.

background The part of a scene or *landscape that is furthest away from the viewer. *See also* FOREGROUND, MIDDLE-GROUND.

background concentration *See* BACK-GROUND LEVEL.

background extinction The normal rate of *extinction of species, which reflects changes in local environmental conditions, in the absence of human influences. *See also* MASS EXTINCTION.

background level The *ambient or typical level of a substance that occurs in an environmental medium (air, water, or soil) in a particular place or area, through natural processes or from natural sources. This is often used as a *baseline against which to measure changes that result from human activities, including *pollution control.

background radiation Low-intensity *radiation in the natural environment that originates from cosmic rays and from the naturally *radioactive elements of the Earth. *See also* RADON.

backing An anti-clockwise change in *wind direction that might, for example, be associated with cold air *advection.

backwash The return flow of water down a beach after a wave has broken.

backwater A small, generally shallow body of stagnant water that is attached to the main channel but has little or no current of its own.

backwoods A remote, undeveloped area where few people live.

backyard composting The composting of organic food waste and organic garden waste in one's yard, using bacteria and fungi to decompose it into a humus-like product. It is regarded as a *source reduction form of waste management.

BACM *See* BEST AVAILABLE CONTROL MEASURES.

BACT *See* BEST AVAILABLE CONTROL TECHNOLOGY.

bacteria A diverse group of single-celled microscopic *organisms (singular 'bacterium') that lack *chlorophyll and break down organic matter, making its components available for reuse by other organisms. Some bacteria help to control pollution by consuming or breaking down organic matter in *sewage or by similarly acting on oil spills or other water pollutants. Bacteria in soil, water, or air can also cause infectious disease in humans (for example diphtheria, tetanus, and typhoid fever), and in animals and plants.

bacterial agent A live *pathogenic *organism that can cause disease, illness, or death.

bactericidal That which destroys bacteria, such as *antibiotics and *disinfectants.

bacteriophages A group of *viruses whose hosts are *bacteria.

badland Steep, barren land, usually broken by narrow channels and sharp crest ridges that result from rapid erosion, often of *unconsolidated sediments. Most common in dry areas.

bag limit The maximum number of animals that a hunter may legally harvest during one hunting season. *See also* GAME BAG, GAME SPECIES.

baghouse filter A fabric filter device that is used to remove *particulate pollutants from air.

bajada (bahada) A gentle slope of *unconsolidated material where several *alluvial fans overlap at the foot of a mountain.

baling The compacting of solid waste into blocks, in order to reduce volume.

ballast water Water that is carried in the lower hold of ships to make them heavier, which causes them to float lower in the water and thus be less likely to roll. The water is discharged from the ship when it enters harbour.

banded iron ore formation A *sedimentary rock composed of alternating layers of *chert and ferric iron oxides (haematite and limonite).

bank erosion The wearing back of a river bank, usually during a *flood.

bankfull The height (*stage) of water in a river channel that equals the level of the surrounding *floodplain, so that

*flooding would occur if the water level increased further. *See also* FLOOD STAGE.

Bankhead-Jones Farm Tenant Act (1937) US legislation that authorized federal acquisition of eroded and exhausted farm lands, which were ultimately designated *National Grassland.

bankside On the side of a river or stream.

banner cloud A stationary layer *cloud that extends downward from an isolated mountain peak, often on an otherwise cloud-free day, and is produced by rising air downwind of the mountain peaks.

BAP *See* BIODIVERSITY ACTION PLAN.

bar 1. A shoal of coarse-grained *sediment (sand, gravel, and pebbles) that has been deposited either on the bed of a *river (*see also* POINT BAR) or along a *coast, offshore from the *beach (*see also* LONGSHORE DRIFT).
   2. A measure of *atmospheric pressure. One bar is the force required to lift a column of mercury up a distance of 750.1 millimetres in a glass tube at 0°C at 45° latitude. Scientists usually measure pressure in *millibars (mb); one bar = 1000 mb. *See also* KILOBAR.

barchan A crescent-shaped *dune form common in *hot deserts with strong prevailing winds. The convex side faces upwind and the concave side faces downwind. The horns of the barchan point downwind, and the dunes migrate continuously downwind across the desert floor.

barium (Ba) A soft, silver-grey alkaline-earth metal that is naturally abundant in nature and is found in plant and animal tissue. It has lots of uses, including in various alloys, paints, soap, paper, rubber, ceramics, glass, and insecticides. Barium can *bioaccumulate in the human skeleton and it can cause muscular problems if ingested.

bark The tough protective outer covering of the branches and roots of *trees and other woody plants.

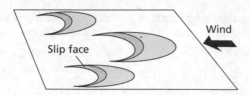

Fig 2 Barchan dune

bark beetle An insect that bores through the *bark of forest trees to eat the inner bark and lay its eggs. Many forest trees are killed by bark beetles.

barograph A *barometer that provides a continuous record of variations in *atmospheric pressure through time.

barometer An instrument that measures *atmospheric pressure, which is used in making *weather forecasts. The two most common types are the *aneroid barometer and the *mercury barometer.

barometric gradient See PRESSURE GRADIENT.

barometric pressure See ATMOSPHERIC PRESSURE.

barothermograph An instrument that measures *atmospheric pressure and temperature, for use in making *weather forecasts.

barrage A dam across a river or estuary.

barrel, petroleum A unit of volume that is equal to 42 US gallons.

barren Devoid of life, incapable of sustaining life.

barren land Land that is uninhabited wilderness, often in a hostile climate, and offers no prospect of being developed or used for agriculture or other economic activities. See also FORESTLAND.

barrier hedge A hedge that is grown to help prevent *runoff and *soil erosion from a field. See also CONTOUR HEDGE, HEDGE.

barrier island A long, low, narrow, sandy *island that forms offshore from a *coastline, parallel to the *coast. Such features help to protect the mainland during hurricanes, tidal waves, and other maritime hazards.

barrier reef A *coral reef that is separated from a mainland or island shore by a lagoon too deep for coral to grow in.

basal ice The ice at the bottom of a glacier, which is in direct contact with (and both affects and is affected by) the sediment and rock beneath the glacier.

basal lamina A layer of *proteins and glycoproteins that surrounds tissues in a body.

basal sediment The sediment (*till) that is transported at, or deposited from, the bottom of a *glacier.

basal sliding/slip The process by which the bottom of a *glacier slides directly over the bedrock below, often lubricated by *meltwater.

basalt A fine-grained, dark-coloured, *basic, *extrusive *igneous rock. It is the most common type of volcanic rock in the Earth's *crust, and is rich in *iron and *magnesium and has a relatively low *silica content. Under suitable climatic conditions basalt can weather into deep productive *soils.

base A metal *oxide, *hydroxide, or compound (such as *ammonia) that gives off hydrogen ions in aqueous solution. See also ALKALI.

base exchange capacity See CATION EXCHANGE CAPACITY.

base level The level below which a river cannot erode. This is usually *sea level, but local base levels are provided by resistant layers of rock or the level of a lake or *reservoir. *See also* LONG PROFILE, WATERFALL.

base map A map that shows important basic information, on which can be superimposed more specialized information relevant to a particular study or purpose.

base metal A non-precious metal, such as *copper, *lead, or *zinc, which is inferior in value to *gold and *silver but still has commercial uses and value. Base metals easily corrode, *oxidize, and tarnish in air, moisture, or heat.

base neutral acid compound *See* SEMI-VOLATILE ORGANIC COMPOUND.

base pair The pair of complementary, nitrogen-rich molecules within a strand of *DNA which are held together by weak chemical bonds. The bonds between the base pairs hold the two strands of DNA held together in the shape of a *double helix.

base year In environmental regulation the year for which a national inventory is to be taken, or the baseline year against which changes are measured. For *Annex I countries, for example, the base year is 1990. Under the *Kyoto Protocol, the base year for *hydrofluorocarbons, *perfluorocarbons, and *sulphur hexafluoride is 1995.

baseflow The portion of *streamflow that comes from the seepage of *groundwater and *throughflow into the channel, rather than from *direct runoff or *precipitation.

Basel Convention on the Control of Transboundary Movements of Hazardous Wastes and their Disposal (1992) An international *convention that seeks to minimize the generation of *hazardous wastes, ensure that such wastes are disposed of in the countries that generate them, and control the import, export, and movement of hazardous waste.

baseline emission The emissions that would occur in a *business-as-usual scenario, without any change or intervention. Estimates of what these emissions would be are needed in order to determine the effectiveness of *emissions reduction programmes.

baseline information Information collected at the start of a study or programme which gives an initial or known value as a standard against which later measurements can be compared.

baseline study A description and analysis of the existing conditions and trends in a location or area where some form of change is proposed or likely, such as a change in *land use, or the likelihood of air or water pollution. This provides a reference point from which effects of proposed actions might be predicted, or against which observed effects can be compared.

basement The oldest rocks in a given area. Usually a complex of *metamorphic and *igneous rocks underlying *sedimentary formations.

base-rich Containing an abundance of *basic (less *acidic) materials.

basic *See* ALKALINE.

basic needs The basic items and services that a person needs to ensure a reasonable standard of living.

basic rock Any of a number of *igneous rocks that are rich in the heavy compounds of *iron, *magnesium, and *calcium. They tend to be darker than *acidic rocks and many of them are weathered by *chemical weathering which breaks down the *crystal structure into the main *mineral constituents. Basic rocks produce more fertile and productive *soils than acidic rocks because *nutrients are more freely available for plants. *See also* ACID, INTERMEDIATE, ULTRABASIC.

basin **1.** An area of land that drains into a lake or river. *See also* DRAINAGE BASIN.
   2. A circular depression of rock *strata which dips towards the centre.
   3. The site where a large thickness of *sediments accumulates by deposition.

basin and range *Topography dominated by numerous mountain ranges separated by broad valleys (basins), found in the west and south-west of the USA.

bastion A large rock *outcrop which projects out from the side of a valley or from an ice field.

BAT *See* BEST AVAILABLE TECHNOLOGY.

Batesian mimicry A form of *mimicry in which one non-poisonous *species (the Batesian mimic) evolves to resemble the coloration, body shape, or behaviour of another species that is protected from *predators by some defensive adaptation (such as a venomous sting or a bad taste). This adaptation gives the mimic species a much better chance of survival. *Contrast* MÜLLERIAN MIMICRY.

bathing water quality The state of purity or pollution of water that people have access to for swimming, usually along a coastline.

batholith The largest and most common type of *discordant igneous *intrusion, most often composed of granite, similar in form to but much larger than a *laccolith. They vary in size between 100 and several thousand square kilometres. Batholiths often form the core of major mountain ranges, such as parts of the Sierra Nevada and Rocky Mountains in the USA.

bathyal zone The upper boundary of the *continental shelf, comprising the *continental slope and *continental rise

bathymetry Measurement of the depths of large bodies of water (particularly seas and oceans) in order to determine the bottom *topography.

bathypelagic zone The deep *aphotic zone of a waterbody (including lakes and the sea) into which no light penetrates, so no *photosynthetic organisms are found here. *Contrast* PHOTIC ZONE.

BATNEEC *See* BEST AVAILABLE TECHNIQUES NOT ENTAILING EXCESSIVE COSTS.

bauxite A rock composed mainly of *hydroxides of aluminium, formed by *weathering of aluminate silicate rocks in tropical areas with good drainage. Bauxite is the major ore of aluminium, from which the metal is extracted.

bay A large curved inlet of the sea; an *estuary.

BCIS *See* BIODIVERSITY CONSERVATION INFORMATION SYSTEM.

BDAT *See* BEST DEMONSTRATED AVAILABLE TECHNOLOGY.

beach A strip of land that borders the sea. The upper and lower limits of a beach are usually defined by the *high water mark and the *low water mark, respectively.

beach dune A sand *dune on a *beach, formed by deposition of fine particles of sediment that are carried by the wind, in the same way that dunes develop in hot *deserts.

beach nourishment *See* BEACH REPLENISHMENT.

beach replenishment The deliberate addition of sand to beaches where erosion is a major problem, in order to stabilize losses and restore badly eroded (and sometimes unsightly) beaches. Also known as **beach nourishment.** *Contrast* SAND MINING.

beach seining *See* FISH TRAPPING.

Beaufort scale A scale that is widely used to describe wind speed. It was defined in 1806 by Admiral Sir Francis Beaufort (1774–1857), and uses the effect that wind has on certain familiar objects such as trees and the sea surface. The scale is given in Appendix 3.

becquerel (Bq) The *SI unit for measuring *radioactivity. 1 Bq represents the activity of a *radionuclide which decays on average at one spontaneous nuclear transition per second. This unit replaced the *curie and *picocurie.

bed A layer of deposited *sediment within a *sedimentary rock. Also known as a **stratum** (plural strata).

bedding plane The junction between the layers in a *sedimentary rock (such as sandstone), which shows the original surface onto which *sediment was deposited. See also DIP.

bedform Patterns that are produced on the surface of sediment by the flow of air or water, such as *ripples in water or *dunes in a desert.

bedload The coarse portion of the *sediment load in a river that is too large or too heavy to move in the body of water as *suspended load, but instead rests on the river bed and is pushed or rolled along by the flow of water.

bedrock A general term for the solid, unweathered *rock that lies beneath *soils or *deposits and is often exposed at the ground surface. See also OUTCROP.

bedrock river A river that flows directly on *bedrock, rather than through *alluvium.

benchmark A reference point or standard for comparison. For example, a site for measuring natural processes or features, where there has been no human impact, or a set of observations that is used to establish standards by which to compare the effectiveness of alternative policies or strategies. See also REFERENCE SITE, REFERENCE YEAR.

benefit The degree to which effects are judged desirable, through giving advantage, profit, or gain.

benefit–cost analysis See COST–BENEFIT ANALYSIS.

benefit value An estimate or projection of the expected positive results or outputs of a proposed activity, project, or programme, expressed in either monetary or non-monetary terms. See also COST–BENEFIT ANALYSIS.

Benguela current The eastern boundary *current of the South Atlantic subtropical *gyre, which flows north from the *Southern Ocean that surrounds *Antarctica and moves cold water along the west coast of Africa.

benign A mild, non-lethal illness or disease, which is not dangerous to health. Contrast MALIGNANT.

benthic Of or relating to the *benthos.

benthic region/zone The bottom layer of a body of water.

benthic–pelagic coupling The cycling of nutrients in a body of water between the water itself and the sediments on the bottom.

benthos An organism that lives on or in the bottom of a body of water such as a river, lake, or sea. Unlike *plankton and *nekton, benthic organisms are not free-floating but are attached to or lie on the sediment. Nearly 98% of all marine animals are benthic organisms. There are two type of benthic organisms—*epifauna and *infauna. See also MACROBENTHOS, MEIOBENTHOS, MICROBENTHOS. Contrast PELAGIC.

bentonite A type of clay, derived from weathered *volcanic ash, that is highly plastic and expands when wet, and is used to seal *landfills, *lagoons, and *wells. See also MONTMORILLONITE.

benzene A colourless, highly flammable liquid *hydrocarbon present in coal tar and petroleum, from which it is distilled for use as a solvent and in making dyes and drugs. See also AROMATIC.

bequest motive See EXISTENCE VALUE.

bequest value The value that people place on knowing that future generations will have the option to enjoy a particular environmental *asset, which is usually measured by *willingness to pay.

Bergmann's rule A 19th century *ecogeographical rule that the colder the climate (or higher the latitude), the larger the body size of a *warm-blooded animal when compared with close relatives in warmer regions. See also ALLEN'S RULE, GOLGER'S RULE.

bergschrund A deep gap (*crevasse) that forms between the ice and headwall (back wall) in a *cirque glacier as the ice is pulled away from the headwall and moves downslope.

Berlin Mandate A proposal under the *United Nations Framework Convention on Climate Change that was adopted in 1995 and was designed to make reductions of *greenhouse gas emissions mandatory. Under the Framework Convention, developed countries pledged to take measures designed to return their greenhouse gas emissions to 1990 levels by the year 2000. The Berlin Mandate established a process that would enable countries to take appropriate action for the period beyond 2000, including a strengthening of *developed country commitments, through the adoption of the *Kyoto Protocol in 1997.

berm 1. A narrow embankment (often made from earth, bales of hay, or timber framing) along a slope, which is often used as *dike or *dam.
2. A strip of large rocks placed at the bottom of a spoil pile to help hold the material in position.
3. A small mound of earth piled up on the outer edge of a mountain or secondary road.
4. An embankment at the rear of a *beach, above the *high tide mark.

Berne Convention An international nature conservation treaty, signed by 40 nations in 1979, which came into force in 1982. Its full title is the Convention on the Conservation of European Wildlife and Natural Habitats, and it seeks to conserve wild flora and fauna and their natural habitats.

beryllium (Be) A greyish-white, brittle, toxic, metallic substance that occurs naturally in certain rocks, soils, and volcanic dust. It is one of the lightest of all metals, resists oxidation in air, and is nonmagnetic. It is used in nuclear reactors, radio and television tubes, fluorescent tubes, and powders, is discharged by machine shops, ceramic and propellant plants, and foundries, and also enters the environment through the fly ash from combustion of coal and fuel oil.

It can cause severe skin problems and is hazardous to human health if inhaled.

Best Available Control Measures (BACM) An approach to *pollution control in the USA that is based on adopting the most effective methods of controlling emissions of pollutants from sources such as roadway dust, soot and ash from woodstoves, and open burning of timber, grasslands, or rubbish.

Best Available Control Technology (BACT) An approach to *pollution control in the USA that is based on adopting, for any specific source, the technology that is currently available and produces the greatest reduction in emissions of air pollutants, taking into account energy, environmental, economic, and other costs. In the USA, major sources are required to use BACT unless they can show that it is not feasible for energy, environmental, or economic reasons.

Best Available Techniques Not Entailing Excessive Costs (BATNEEC) An approach to *pollution control in the UK that is based on adopting the most effective techniques for an operation at the appropriate scale which are commercially available and where the benefits gained are more than the costs of obtaining them

Best Available Technology (BAT) An approach to *pollution control in the USA that is based on adopting the best technology, treatment techniques, or other means which are available, taking account of cost.

Best Demonstrated Available Technology (BDAT) An approach to *pollution control in the USA that is based on adopting the most effective commercially available means of treating specific types of *hazardous waste.

Best Management Practice (BMP) An approach to *pollution control in the USA that is based on adopting methods that have been determined to be the most effective, practical means of preventing or reducing *water pollution from non-point sources.

Best Practicable Environmental Option (BPEO) An approach to *pollution control in the UK that is based on seeking to establish the option which causes the least damage to the environment at an acceptable cost, taking into account the total pollution from a process and the technical possibilities for dealing with it.

Best Practical Control Technology (BPCT) An approach to *pollution control that is based on adopting the best technology for pollution control available at reasonable cost and operable under normal conditions.

best-current-data approach An approach to decision-making in *resource management that uses current data collected through new or existing sampling programmes. Managers analyse the data using the latest techniques, assess their management options, and then choose the best option to implement.

beta decay The process of *radioactive decay in which a *neutron is converted to a *proton (emitting an *electron and an antineutrino) or a proton is converted to a neutron (emitting a positron and a neutrino). The result is that an unstable atomic nucleus becomes stable by altering its proton number.

beta diversity The difference in diversity of *species between two or more *ecosystems in an area, expressed as the total number of species that are unique to each of the ecosystems being compared. A measure of *biodiversity. Also known as **species turnover**. *See also* ALPHA DIVERSITY, GAMMA DIVERSITY.

beta particle A high-energy *electron that is emitted from a nucleus during *beta decay. It has a single negative electric charge, is halted by a thin sheet of metal, and cause skin burns.

Bhopal The location in India of the world's worst industrial accident in terms of loss of life. A major explosion

at the Union Carbide *pesticide factory there, in December 1984, released nearly 40 tonnes of *methyl isocyanide which blew over the residential area nearby. Nearly 200 000 people were exposed to the poison, and within ten years more than 4000 had died as a result. Many more have suffered long-term ill-health and disability because of exposure to the *toxic cloud.

B-horizon The intermediate layer in a *soil profile, beneath the *A-horizon and above the *C-horizon. The B-horizon is dominated by the deposition (*illuviation) of material *leached from the A-horizon, so it is a zone of accumulation and enrichment. It contains little organic material, and is usually more compacted than the A-horizon because *clay washed down from the A-horizon fills many of the voids between particles. *See also* SOIL HORIZON.

bicarbonate A sodium salt ($NaHCO_3$) of *carbonic acid that can act as a *buffer and resists changes in the *pH of soil and waterbodies.

bid The price that a prospective buyer is willing to pay in *greenhouse gas *emissions trading.

biennial A plant that completes its life cycle in two years, producing leaves in the first year, blooming and producing seed in the second year, and then dying. *Contrast* ANNUAL, PERENNIAL.

big game The collective term for large mammals, such as deer, elk, bear, bison, and antelope, that are hunted for sport. *See also* GAME.

bilateral trade/transaction A direct trade between two parties, on a one-to-one basis, with no intermediary exchange or third party involvement.

bilharzia A disease in humans caused by an infestation of or an infection caused by a *parasite of the genus *Schistosoma*, that is common in the tropics and South East Asia. The symptoms depend on the part of the body infected. Also known as **schistosomiasis**.

bill A piece of legislation introduced by government that is intended to become law.

billow cloud A distinctive type of *cloud in the form of broad parallel bands oriented at right angles to the wind, with distinct clear areas between the bands.

binding target An agreed or mandatory environmental standard (such as a reduction in emissions of *greenhouse gases) that a country must meet in the future, usually within a defined period of time.

bio- Referring to living organisms.

bioaccumulant A substance that increases in concentration through time in living organisms, as they take in contaminated air, water, or food, because the substance is used or excreted only very slowly. *See also* BIOACCUMULATION.

bioaccumulation The tendency for a *pollutant (particularly a *toxic substance, such as *mercury, *PCBs, and some *pesticides) to accumulate in the tissues of plants or animals because it is absorbed faster than the organism can break it down through *metabolism. The organism absorbs the pollutant either through exposure to it or through digestion (via food), and organisms towards the top of the *food chain can have high levels of a particular pollutant through consuming other organisms further down the food chain. Also known as **bioamplification, biological amplification, bioconcentration, biological concentration, biomagnification** and **biological magnification.**

bioamplification *See* BIOACCUMULATION.

bioarchaeology The study of plant and animal remains collected from *archaeological sites, in order to reconstruct the *environment in which they grew.

bioassay A laboratory test using living organisms (tissues, cells, live animals, or humans), designed to measure the effects of various substances on them.

bioassessment *See* BIOLOGICAL ASSESSMENT.

bioaugmentation The deliberate addition of *microbes to soil or groundwater in order to enhance *biodegradation or the *bioremediation of organic contaminants.

bioavailability The degree to which chemicals (such as *contaminants) can be absorbed by organisms.

bioavailable Available for biological uptake.

biobarrier An obstruction (such as a filled trench or a *membrane) that inhibits living tissue. Used, for example, to control plant root growth in contaminated soils.

biocatalyst An enzyme that activates or speeds up a biochemical reaction in a *bioprocess. *See also* BIOTECHNOLOGY, INDUSTRIAL BIOCATALYST.

biocentric *See* ECOCENTRIC.

biocentrism *See* ECOCENTRISM.

biochemical Chemicals that are present in living organisms, and reactions associated with them.

biochemical decay The breakdown of pollutants in water through the action of *bacteria.

biochemical oxygen demand (BOD) *See* BIOLOGICAL OXYGEN DEMAND.

biochemical precipitate A *sedimentary rock, such as *limestone, that has formed from elements extracted from seawater by living organisms.

biochemicals Chemicals (such as hormones, pheromones, and enzymes) that are found naturally occurring in living organisms and are synthesized by them. Biochemicals can be very effective *pesticides by disrupting the mating pattern of insects, regulating growth, or acting as repellents.

biochemistry The chemistry of living organisms, especially the chemical components, their processes, and reactions.

biocide A chemical agent that kills a wide range of organisms.

bioclastic rock A *sedimentary rock made up of broken fragments of organic skeletal material.

biocoenosis See ECOSYSTEM.

bioconcentration See BIOACCUMULATION.

biocontrol See BIOLOGICAL CONTROL.

bioconversion The conversion of a compound from one form to another by the actions of organisms or enzymes. Also known as **biotransformation.**

biocriteria Biological measures of the health of an environment, such as the incidence of a particular disease in a particular species.

biodegradable Capable of being decomposed rapidly by natural biological processes. Most organic waste such as food, wood, paper, wool, and cotton is biodegradable.

biodiesel A *renewable, *biodegradable, *alternative fuel or fuel additive for diesel engines, produced from organic material such as soybean or sunflower seed oil.

Biodiversity Action Network (Bionet) An international network of *non-governmental organizations working to strengthen *biodiversity policy and law.

biodiversity action plan (BAP) A national plan, programme or strategy for the *conservation and *sustainable use of biological diversity, drawn up in response to the *United Nations Framework Convention on Biological Diversity.

Biodiversity Conservation Information System (BCIS) An international initiative that seeks to support environmentally sound decision-making and actions affecting the status of *biodiversity and *landscapes at the local, national, regional, and global levels through cooperative provision of data, information, advice, and related services.

Biodiversity Convention See UNITED NATIONS CONVENTION ON BIOLOGICAL DIVERSITY.

Biodiversity Forum A web-based forum funded by the US National Science Foundation, that allows people to express and exchange views on topics related to *biodiversity loss and conservation.

biodiversity hotspot A biogeographical region define by *Conservation International (CI) that is both a significant reservoir of *biodiversity and threa-

---

BIODIVERSITY

Short for biological diversity. A measure of variation (the number of different varieties) amongst living things. The word is most commonly used to describe 'species diversity'—the number and relative abundance of different species, within a particular area (*local biodiversity) or within the world (*global biodiversity), which is not the same as *species richness (the count or number of *species). A number of ways of measuring biodiversity are in use, the most common of which is the *Shannon–Wiener index. The term biodiversity is also sometimes used to describe the number of genetic strains (differences) within species, and the number of different *ecosystems in an area. The most common expression of biodiversity is the number of different species within a particular area (local biodiversity), a specific habitat (*habitat biodiversity), or the world (global biodiversity). The *United Nations Framework Convention on Biological Diversity, agreed by the *United Nations Conference on Environment and Development in 1992, is a framework for international action in protecting biodiversity. See also ALPHA DIVERSITY, BETA DIVERSITY, GAMMA DIVERSITY.

tened with destruction. The CI list includes 25 hotspots that cover 1.4% of the land area of the Earth but support nearly 60% of the world's plant, bird, mammal, reptile, and amphibian species. The CI hotspots are larger than the *ecoregions identified by the *Worldwide Fund for Nature in their *Global 200 initiative, and they only cover terrestrial areas whereas the latter also include *freshwater and *marine ecoregions.

biodiversity prospecting The search for new products among genes found in wild organisms, which may be of potential commercial value.

bioenergy Energy that is made available by the combustion of materials derived from biological sources.

bioengineering The application to biological science of engineering principles or equipment.

bioerosion Erosion or decay caused by living things.

bioethics The study of the ethical implications of biological research and applications, including *biomedicine and *biotechnology.

biofilm A layer of micro-organisms that is attached to a surface such as the rocks in natural streams.

biofuel An *alternative fuel that is produced from biological materials including crops (especially trees) and animal wastes. Examples include *ethanol, *methanol, and *biodiesel.

biogas A mixture of *methane and *carbon dioxide that is generated by the *anaerobic fermentation of organic matter such as plant residues and animal *manure, and can be used as a *renewable fuel or a *fertilizer.

biogenic Material that results from the activity of living things.

biogenic source Biological sources, such as plants and animals, that emit *air pollutants such as *volatile organic compounds.

biogenically reworked zone The upper zone within a *sediment that is actively burrowed by *benthic organisms.

biogeochemistry The study of the exchange of materials between living (*biotic) and non-living (*abiotic) components of the *biosphere.

biogeocoenosis *See* ECOSYSTEM.

biogeographical Relating to the geographical distribution of plants and animals.

biogeographical province/realm One of the eight geographical regions from which particular assemblages of plants and animals evolved and dispersed. *See also* AFROTROPICAL REALM, ANTARCTIC REALM, AUSTRALIAN REALM, INDOMALAYAN REALM, NEARCTIC REALM, NEOTROPICAL REALM, OCEANIAN REALM, PALAEARCTIC REALM.

biogeography The study of the geographical distribution of organisms, past and present. *See also* ISLAND BIOGEOGRAPHY.

biogeophysical feedback A *feedback mechanism that connects the biological and geophysical parts of the climate system.

biohazard The health risk to humans or animals that is posed by the possible release of a *pathogen into the environment.

biohydrology The study of the interactions between the *water cycle and plants and animals.

bioindicator An organism (plant or animal species) that has a known sensitivity to particular types of stress (such as pollution) at relatively low levels, so its and health can be used as an *indicator of environmental quality. Also known as **environmental indicator.** *See also* INDICATOR.

bioinformatics The use of computers and statistical methods in the classification, storage, retrieval, and analysis of

BIOGEOCHEMICAL CYCLE

One of the large-scale long-term environmental cycles that circulates elements (such as *carbon, *nitrogen, *oxygen, *hydrogen, *calcium, and *sulphur) between the *biotic and *abiotic components of the *environment (including the *atmosphere, *soil, *water cycle, and *ecosystems) by living organisms, geological processes, or chemical reactions. Biogeochemical cycles function at the global scale, and *nutrient cycles are those biogeochemical cycles that involve the elements necessary for life. Flow through the environment is cyclic because of the finite supply and the relatively constant form of the individual elements. Individual cycles can be identified for each of the elements, but they all have in common a basic two-part structure involving an *inorganic component (comprising the abiotic or non-living parts of the environment, with sedimentary and atmospheric phases), and an *organic component (comprising plants and animals, living and dead, and their physical and chemical interactions). Natural biogeochemical cycles are being disrupted by a range of human activities, including changes in *land use and the burning of *fossil fuels. The nitrogen, carbon, and sulphur cycles are particularly important to the functioning of the biosphere, and they are also closely linked to the climate system. Despite recent advances in our understanding of how these great cycles work (partly a result of the *IGBP studies), many uncertainties and gaps remain. Many of the more pressing global and regional environmental problems we face today—such as *greenhouse gases, *eutrophication, *acid rain, and *ozone depletion—are closely related to these cycles. Also known as **mineral cycle.**

biological information, particularly relating to *genomes. Also known as **computational biology.** *See also* INFORMATICS.

biolimiting Any factor that determines or restricts the growth of a particular *life form.

biological additive A microbiological *culture, *enzyme, or nutrient additive that is deliberately introduced into an oil discharge to promote *biodegradation in order to reduce the effects of the discharge.

biological amplification *See* BIOACCUMULATION.

biological assessment An evaluation of the biological condition of a population or habitat. Also known as **bioassessment.**

biological classification *See* CLASSIFICATION.

biological concentration *See* BIOACCUMALATION.

biological contamination The presence of infectious agents (such as viruses,

bacteria, fungi, and mammal and bird antigens) in an environment. *See also* CONTAMINATION.

biological control The use of natural predators, pathogens, or competitors to regulate *pest populations. Examples include introduced or naturally occurring predators such as wasps, or hormones that inhibit the reproduction of pests.

biological diversity *See* BIODIVERSITY.

biological half-life *See* HALF-LIFE.

biological integrity The ability of an *ecosystem to support and maintain a balanced, adaptive *community of *organisms that has a species composition, diversity, and functional organization comparable to that of natural *habitats within the same region. *See also* INDEX OF BIOLOGICAL INTEGRITY.

biological invasion Processes by which species become established in *ecosystems to which they are not native.

The invading species are often weeds, pests, or disease-causing organisms.

biological magnification See BIOACCUMULATION.

biological monitoring Measurement of the levels of particular chemicals that are present in biological materials (such as blood or urine) in order to determine whether chemical exposure has occurred. Also known as **biomonitoring**.

biological oxidation The decomposition of complex organic materials by bacteria and micro-organisms, as occurs in the self-purification of water bodies and in *activated sludge *wastewater treatment.

biological oxygen demand (BOD) An indirect measure of the concentration of biologically degradable material that is present in organic wastes, usually based on the amount of oxygen that is consumed in five days by biological processes breaking down the organic waste. Also known as **biochemical oxygen demand**. See also CHEMICAL OXYGEN DEMAND.

biological pesticide A substance that is biological in origin, such as a *virus, *bacteria, *pheromone, or natural plant compound, used as a *pesticide. Also known as a **biopesticide**. See BACILLUS THURINGIENSIS.

biological productivity The amount of organic matter, *carbon, or energy content that is accumulated in a given area over a given period of time. Usually expressed in terms of weight per unit area per unit time (grams per metre squared per year, g m^{-2} $year^{-1}$).

biological resource See BIOTIC RESOURCE.

biological simplification The reduction of *biodiversity that results from altering the environment in ways that favour certain *species over others, either directly (through management) or indirectly (for example, through pollution).

biological species concept The most commonly used definition of *species, as a group of natural populations that interbreed between themselves but not with other such groups. This explains why members of a particular species resemble one another and differ from other species. Compare CLADISTIC SPECIES CONCEPT, ECOLOGICAL CONCEPT, RECOGNITION SPECIES CONCEPT.

biological stressor An organism that finds itself, by accident or design, in a habitat to which it does not naturally belong. Examples include the fungus causing *Dutch elm disease and certain types of algae and bacteria.

biological survey The collection, processing, and analysis of a representative sample of a *community of plants and animals in order to determine the structural and/or functional characteristics of that community. Also known as **biosurvey**.

biological treatment A *waste treatment technology that uses bacteria to consume waste material.

biological warfare The intentional use of micro-organisms or toxins derived from living organisms to cause death or disease in humans, animals, or plants. See also BIOTERRORISM.

biological weapon Any biological substance (such as a deadly *virus or *bacterium) that can be used to kill or injure people, animals, and plants. A type of *weapon of mass destruction.

biologically unique species A *species that is the only living representative of an entire *genus or *family.

biologicals A collective term for preparations (such as vaccines and cultures) that are made from living organisms and their products, and are used in diagnosing, immunizing, or treating humans or animals, or in related research. Also known as **biologics**.

biologics See BIOLOGICALS.

biology The study of living organisms.

bioluminescence The production of light without heat by living organisms.

The process occurs in many *bacteria and *protists, as well as certain animals and fungi. Also known as **phosphorescence**.

biomagnetism 1. The *magnetic field that is created by a living organism.
2. The effect of an external magnetic field on living organisms.

biomagnification *See* BIOACCUMULATION.

biomarker A chemical compound that is produced by an organism which can be used as an indicator of the presence or health of that organism.

biomass 1. The total weight of living matter in an area, including plants, animals, and insects. The term is sometimes used to refer specifically to organisms of one type.
2. Living material (such as wood and vegetation) that is grown or produced for use as fuel.

biomass burning The burning of organic matter for energy production, forest clearing, and agricultural purposes, which releases *carbon dioxide and other *greenhouse gases into the air.

biomass energy Energy that is produced from burning organic waste, such as sawmill wood waste or crop waste.

biomass fuel Organic material produced by plants, animals, or micro-organisms that can be burned directly as a heat source or converted into a gaseous or liquid fuel. Examples include wood and forest residues, animal manure and waste, grains, crops, and aquatic plants.

biomass generation plant A plant that produces electricity from *biomass fuel.

biomass harvesting A forest harvest method in which whole trees are chipped and used as fuel.

biome A large naturally occurring regional *ecosystem that contains *communities of plants and animals that are adapted to the conditions in which they occur. Biomes are strongly influenced by climate, and their distributions often coincide with climate regions, although many other factors also influence the distribution of plants and animals. Biomes are characterized by a dominant vegetation, and defined by the *species within them. Examples include *desert, *tundra, *grassland, *savanna, *woodland, *coniferous forest, *temperate deciduous forest, and *tropical rain forest. Species composition can vary from place to place within a biome because of local differences in soils, drainage, *topography, *microclimate, and other factors. It can also vary within and between continents for a given type of biome, reflecting *adaptation to the environment and local *speciation in widely separated populations.

biomedicine The study of the ability of organisms to cope with environmental stress, and the application of basic sciences to problems in clinical medicine.

biometric A measurable physical characteristic or personal behavioural trait that can be used to recognize or confirm the identity of an individual.

biometrics 1. The science of measuring and statistically analysing biological data. Also known as **biometry**. *See also* BIOSTATISTICS.
2. The use of biological properties (such as fingerprints, retina scans, or voice recognition) to identify individuals.

biometry *See* BIOMETRICS.

biomonitor A *species that is sensitive to changes in the environment, such as changes in pollution levels, and shows measurable responses to them.

biomonitoring *See* BIOLOGICAL MONITORING.

BIONET *See* BIODIVERSITY ACTION NETWORK.

biopesticide *See* BIOLOGICAL PESTICIDE.

biopharmaceuticals *Proteins produced by living organisms that have medical or diagnostic uses.

biopile A mound of soil built to allow *aerobic *bioremediation by *aeration.

biopiracy The unauthorized and uncompensated collection of biological resources (indigenous plants and animals) by individuals or companies who then use or patent them for their own benefit. Illegal *bioprospecting or biological theft. Also known as **ecopiracy.**

bioprocess A process in which living cells, or components of them (such as enzymes), are used to produce a desired product. *See also* BIOCATALYST, BIOTECHNOLOGY.

bioprospecting The search for economically valuable new genetic and biochemical resources from nature that may serve as sources for natural products. *See also* BIOPIRACY.

bioreactor A large container in which *micro-organisms are grown for the production of biologically useful materials, such as *enzymes and insulin.

bioreclamation *See* BIOREMEDIATION.

bioregion A territory defined by ecological systems (such as *drainage basins or *ecosystems), rather than by political or administrative units. An area of relatively homogeneous ecological characteristics, or a specific assemblage of ecological communities. Similar to a *biome but smaller and with more specific characteristics. *See also* BIOREGIONAL PLANNING, ECOREGION.

bioregional planning Planning based on seeking to preserve the integrity of a *bioregion rather than a political or administrative unit.

bioremediation The use of living organisms to clean up *oil spills or remove pollutants from soil, groundwater, or wastewater. Short for biological *remediation. Also known as **bioreclamation.** *See also* LAND FARMING.

bioreserve An area of natural *ecosystems and high *biodiversity that has been designated and protected for *nature conservation purposes. *See also* BIOSPHERE RESERVE.

biosafety Safety from exposure to infectious *agents, particularly those created by *biotechnology.

biosecurity Biological security, particularly protection against *bioterrorism and the use of *biological weapons.

biosolids *See* SEWAGE SLUDGE.

biosphere The part of the Earth in which living organisms exist and interact, which includes the *atmosphere, *hydrosphere, and *lithosphere; the sum of all *ecosystems. The biosphere extends from less than 11 kilometres below *sea level, to the *tropopause which is less that 17 kilometres above sea level. This gives it a maximum thickness of 38 kilometres—roughly 0.5% of the radius of the Earth (6371 kilometres). The concept of the biosphere as the Earth's integrated living and life-supporting system was first proposed in the 1920s, but only in recent decades has it been widely adopted and used. Originally, the concept was applied just to the Earth's surface where plants and animals obviously make their home. But it has more recently been extended by the *Gaia hypothesis to include parts of the atmosphere and subsurface geology that were previously thought of as *abiotic. Also known as **ecosphere.**

Biosphere II A man-made, closed *ecological system that was constructed in the Arizona desert between 1987 and 1989, in which experiments were conducted to study how *ecosystems operate and to evaluate whether people could live and work in a closed *biosphere (for example during colonization of other planets). It involved artificially recreating a range of habitats under a geodesic glass dome, using nearly 4000 natural species representing *tropical rainforest, *saltmarsh, *desert, *coral reef, *savanna, and intensive agriculture. Eight humans were part of the experiment, and they lived on a self-sufficient basis inside the sealed dome for two years, effectively isolated from the outside world. The dome was self-contained and self-sufficient, except for some electricity fed in from outside.

Biosphere Reserve A large, protected area of natural habitat, established by *UNESCO under the *Man and the Biosphere Programme. Designed to create a representative network of the world's *ecosystems where research and monitoring activities are conducted, with the participation of local communities, to protect and preserve healthy natural systems threatened by development. Each reserve contains a *core area (dedicated to preserving *biodiversity with no human interference) and a *buffer zone. *See also* BIORESERVE.

biostatistics The use of statistics to analyse biological data. *See also* BIOMETRICS.

biostimulation The addition of nutrients to a population of micro-organisms in order to stimulate growth and activity during *bioremediation or *biotreatment.

biostratigraphy The organization of layers of *sedimentary rock into units, based on their *fossil content, for the purposes of *dating and *correlation. *Contrast* LITHOSTRATIGRAPHY.

biosurvey *See* BIOLOGICAL SURVEY.

biosynthesis The production of a chemical substance by a living organism. *See also* ANABOLISM.

biota All of the living organisms (including animals, plants, fungi, and micro-organisms) that are found in a particular area.

biotechnology Any technology that uses living organisms (or parts of organisms) to make or modify products, improve plants and animals, or develop micro-organisms for specific use. An example is the use by industry of *recombinant DNA. *See also* BIOCATALYST, GENETIC ENGINEERING, INDUSTRIAL BIOTECHNOLOGY.

bioterrorism The unlawful use of biological agents, such as a deadly *virus or *bacteria, to kill or harm people, ani-

mals, or plants. *See also* BIOLOGICAL WARFARE, BIOSECURITY, TERRORISM.

biotic Living, containing life. *Contrast* ABIOTIC.

biotic association *See* BIOTIC COMMUNITY.

biotic community A natural assemblage of plants and animals that live in the same environment and are mutually sustaining and interdependent. Also known as **biotic association**. *See also* BIOME.

biotic potential The maximum reproductive rate of an organism, given unlimited resources and ideal environmental conditions.

biotic resource *See* BIOLOGICAL RESOURCE.

biotope *See* ECOTOPE.

biotrade Trade in biological resources within and between countries. *See also* BIOPIRACY, CITES.

biotransformation *See* BIOCONVERSION.

biotreatment The process of reducing pollution in waste streams, such as industrial wastewater, by biological treatment (particularly the use of micro-organisms).

biotroph An organism that derives nutrients from the living tissues of another organism (its *host).

bioturbation The mixing of a *sediment by the burrowing, feeding, or other activities of living organisms.

biotype A group of genetically identical individuals.

bioweapon *See* BIOLOGICAL WEAPON.

bipedalism The ability to walk upright on two legs, as in humans.

bird A *warm-blooded *vertebrate animal that lays eggs and has two legs, feathers, and wings. Most birds can fly.

birth cohort A group of individuals who are born during a specific period of time.

b

birth control Any method that is used to reduce births, including celibacy, delayed marriage, contraception, and sterilization. *See also* FAMILY PLANNING.

birth rate The number of live births per year per 1000 population. Also known as **natality.**

bitter lake A general term for a saltwater lake.

bitumen A black, sticky, *hydrocarbon substance that is obtained naturally or from *petroleum, used for surfacing roads.

bituminous coal A soft, black *coal that burns with a smoky yellow flame, derived from the *lithification of *lignite. It can be turned into *anthracite by folding and hardening.

bivalve A mollusc that has two shells hinged together, for example oysters, clams, scallops, mussels, and other shellfish.

bivoltine Having two broods and generations in a year or season. *See also* VOLTINISM. *Contrast* MULTIVOLTINE, UNIVOLTINE.

black body A mass or body that absorbs and emits all wavelengths of electromagnetic *radiation, and at constant temperatures will radiate as much heat as it absorbs to remain in a *steady state. *See also* EMISSIVITY.

black body temperature The temperature that the surface of a body (such as a *planet, like the *Earth) would be if it were not warmed by its own *atmosphere. It can be calculated using the *Stefan–Boltzmann equation. The black body temperature of the Earth is −23°C, but the actual surface temperature is about 15°C. The difference (38°C) is the amount by which the planet is warmed by the absorption of *radiation within its atmosphere, by the natural *greenhouse effect.

black box In systems terms, an unknown and often unknowable mechanism, process, or system which is judged solely by observing its inputs and outputs.

black earth A dark-coloured *chernozem soil with a high *humus content, which is often deep and very fertile.

black ice Thin, new ice that forms when rain falls on surfaces that are below freezing. It is common on roads during autumn and early winter. It appears black because it is transparent and can create hazardous driving conditions because it often cannot be seen.

black lignite *See* SUB-BITUMINOUS COAL.

black shale A type of *sedimentary rock that is deposited in conditions with very little oxygen. It is generally dark-coloured and has a high *organic content.

black smoker A vent in a geologically active region of the ocean floor from which superheated water laden with minerals (sulphide precipitates) flows out into the ocean, looking like black smoke.

black water Wastewater that contains animal, human, or food waste. *Contrast* GRAY WATER, WHITE WATER.

blade The green leaf of a grass or cereal plant.

blanket bog A *peatland formed in areas of high rainfall and humidity that covers large areas of flat and gently sloping ground.

bleaching *See* CORAL BLEACHING.

blizzard A severe storm characterized by blowing snow, low temperatures, strong winds (56 kilometres per hour or higher), and reduced visibility (less than 0.40 kilometres). It usually lasts for three hours or longer.

BLM *See* US BUREAU OF LAND MANAGEMENT.

block fault A type of geological faulting in which the topography is divided by *faults of different heights and orientations.

bloom *See* ALGAL BLOOM.

blowout **1.** A depression in the surface of sand or dry soil that is caused by wind erosion.
2. A sudden release of oil or gas which can be disasterous.

blue moon **1.** The second full moon in a calendar month.
2. A sky condition caused by the presence of large quantities of suspended particles in the atmosphere which selectively remove the longer visible wavelengths more than the blue or green wavelengths.

blue-green algae *See* CYANOBACTERIA.

Blueprint for Survival A book published by *The Ecologist* in 1971 as a manifesto for radical changes in *lifestyle and patterns of *economic development, which had a significant impact on attitudes towards the *environment, and on the development of modern *environmentalism.

bluff A high steep bank or cliff.

BMP *See* BEST MANAGEMENT PRACTICE.

boardwalk A walkway made of wooden planks that is used in fragile sites which are heavily used by people, such as on sand or across boggy ground.

BOD *See* BIOCHEMICAL OXYGEN DEMAND, BIOLOGICAL OXYGEN DEMAND.

body burden The total amount of a chemical that is stored in the body of an organism at a given time, particularly a potential *toxin (such as *lead) to which the organism has been exposed.

bog A poorly drained area of shrubby *peat dominated by specialized acid-tolerant vegetation including shrubs, sedges, and peat moss. It is a spongy *wetland habitat with a high *water table, and has a high content of organic remains. A bogs is similar to, but more acidic than, a *fen. Also known as a **peat bog.**

boiling water reactor (BWR) A type of *light-water nuclear reactor in which water is boiled in the *core, producing steam that drives a turbine to generate electricity. *Contrast* PRESSURIZED WATER REACTOR.

bole **1.** The main trunk of a tree.
2. A reddish clay.

boll weevil A small, greyish beetle (*Anthonomus grandis*) that is common throughout the south-eastern USA. It has destructive larvae that hatch in and damage cotton bolls.

bolson In arid regions, a basin that is filled with *alluvium and intermittent *playa lakes, and has no outlet.

bomb calorimeter A sealed apparatus that is used to measure the heat produced by combustion of a particular material.

bombiculture The use of bees as pollinators of specific crops such as hothouse tomatoes and other fruit. *Contrast* APICULTURE.

bond The force that holds together two atoms in a molecule or crystal.

bonding **1.** The physical and chemical processes that join substances together (such as individual *crystals in rock, to make *minerals).
2. The making of close relationships between individual of the same species.

Bonn Convention An international treaty, formally called the Convention on the Conservation of Migratory Species of Wild Animals, which came into force in 1983. Its aim is the to protect those species of wild animals that migrate across or outside national boundaries, and the species covered include marine mammals, sea turtles, and sea birds.

boom **1.** A floating device that is used to contain oil on a body of water.
2. Equipment used to apply pesticides from a tractor or other vehicle.

bora A cold *katabatic wind that blows downslope from former Yugoslavia into the coastal plain of the Adriatic Sea, usually in winter and associated with heavy *precipitation.

Bordeaux mixture A *fungicide that contains copper sulphate, lime, and water and is used as a spray on grapevines to fight fungal diseases.

bore A tidal wave that surges upstream in an *estuary, and occurs during high *tides (particularly during *spring tides). Also known as a **tidal bore.**

boreal Relating to cool or cold temperate regions at high latitudes in the northern hemisphere. *See also* AUSTRAL.

boreal climate *See* SUBPOLAR CLIMATE.

boreal forest *See* CONIFEROUS FOREST.

borehole A hole drilled into the ground either for subsurface exploration (for example in the search for reserves of *oil or *gas) or for extraction (for example the pumping *groundwater from an *aquifer).

Borlaug, Norman American scientist (born 1914), considered by some to be the 'father of modern agriculture'. Winner of the Nobel Peace Prize in 1970 for his contribution to the *Green Revolution because of his efforts in the 1960s to introduce crossbred seeds into agricultural production in Pakistan and India, which saved over a billion people from starvation.

boron (B) A *trace element that occurs naturally as boric acids in soil solutions and in the ore borax. It is a *micronutrient that is important for plant growth, essential for bone development in animals, and can be *toxic at high concentrations in humans.

boss *See* STOCK.

botanical Relating to plants and plant life.

botanical garden A facility in which trees and shrubs are cultivated for exhibition, or preserved for *conservation or *reintroduction, as part of a *living collection. *See also* ZOO.

botanical insecticide *See* BOTANICAL PESTICIDE.

botanical medicine The use of plants or plant extracts for medicinal purposes.

botanical pesticide An *organic *pesticide that is derived from another plant, for example *pyrethrin or *rotenone.

botany The study of plants and their structures and functions. Also known as phytology.

bottle bank A place where glass bottles can be deposited for *recycling.

bottom ash The heavy residue from burning coal in a boiler which falls to the bottom of the boiler and is removed mechanically. It may include *toxic compounds.

bottom berg An *iceberg that originates from near the base of a *glacier. They are usually black (from trapped rock material) or dark blue (made from old, coarse, bubble-free ice), and are heavy so they sit low in the water.

bottom fish *See* GROUNDFISH.

bottom land A North American term for low-lying land along a waterway.

bottom water The dense, cold water at the deepest depths of the ocean, near the ocean floor, which has different properties (such as temperature, *salinity, and *dissolved oxygen content) from the overlying deep water.

boulder A large particle of *sediment, which has a *particle size greater than 256 millimetres.

boulder clay A glacial deposit, consisting of unsorted *boulders and smaller particles in a matrix of *clay, that is laid down beneath a *glacier or *ice sheet. Also known as **drift, glacial diamicton, glacial drift,** and **till.**

boundary The edge between two adjacent things. **1.** In ecology, the edge between adjacent habitats (*ecotone).
2. In a system, the defining limit across which matter and energy can move (in an *open system) or within which matter and energy are confined (in a *closed system).

3. The outer edge of a political or organizational unit, including a national boundary or frontier.

boundary layer 1. The layer of fluid that is in contact with a surface. *Contrast* MAINSTREAM FLOW.
2. The layer of air closest to the ground the ground (up to about 100 metres), which is affected by exchanges of heat, moisture, or momentum with the surface. Also called the **atmospheric boundary layer.**

boundary water A river or lake that is part of the *boundary between two or more countries or provinces that have rights to the water.

Boundary Water Treaty A treaty that was signed in 1909 providing the principles and mechanisms to help resolve disputes and to prevent future ones relating to water quantity and water quality along the *boundary between Canada and the USA.

bovine Relating to cattle.

Bovine Spongiform Encephalopathy (BSE) Mad cow disease. A chronic, progressive and ultimately fatal disease that affects the brains and central nervous systems of adult cattle. Humans have been known to develop new variant *CJD as a result of eating the meat of affected animals.

BP The number of years before the present time.

BPCT *See* BEST PRACTICAL CONTROL TECHNOLOGY.

BPEO *See* BEST PRACTICABLE ENVIRONMENTAL OPTION.

Bq *See* BECQUEREL.

BR *See* US BUREAU OF RECLAMATION.

brachiopod A group of *benthic *marine animals that have hinged half shells and a soft body, and attach themselves to the sea floor with a stalk.

bracken A large coarse fern, often several feet high, with a widespread distribution. Common on *moorland.

brackish Mixed fresh water and seawater, as occurs in a river *estuary. Also known as **briny.**

Braess's paradox In traffic planning, the tendency for traffic flows (and thus congestion) to increase when new links are added to a transport network such as a highway or motorway.

braided river A river that carries coarse sediment as *bedload, and flows through multiple channels with islands in between them.

bramble Any prickly shrub or bush, such as a raspberry.

Brandt Commission *See* INDEPENDENT COMMISSION ON INTERNATIONAL DEVELOPMENT.

breaker zone In the *sea, the zone within which waves approaching the *shore start to break, normally in water depths of between 5 and 10 metres. *See also* SURF ZONE.

breccia A *clastic *sedimentary rock composed of angular pieces of broken rock cemented together by a fine matrix. *Contrast* VOLCANIC BRECCIA.

breed 1. To produce more of a species; animals have babies and plants produce seeds or spores.
2. A group of animals or plants that are related by descent from common ancestors and look similar. One species can have numerous breeds, which can look very different. Breeds are often domesticated or cultivated by humans.

breed at risk Any *breed that is in danger of becoming *extinct if the factors that are causing a decline in its numbers are not removed or controlled.

breed not at risk Any *breed in which the population size is increasing, and there are more than 1000 breeding females and more than 20 males.

breeder nuclear fission reactor *See* BREEDER REACTOR.

breeder reactor A *nuclear reactor that produces more *fuel than it consumes, by bombarding *isotopes of *uranium and *thorium with high-energy *neutrons. This converts inert (unreactive) atoms to ones that can be split (*fission), which sustains the nuclear *chain reaction but can be dangerous because the chain reaction is difficult to control. Also known as **breeder nuclear fission reactor**, or **fast breeder**.

breeding 1. The sexual activity of conceiving and bearing offspring.
2. The production of animals or plants by *inbreeding, *outbreeding, or *hybridization.

breeding dispersal The movement between successive breeding sites of individuals that have reproduced. *See also* DISPERSAL.

breeding ground A place where animals breed.

breeding line The genetic history of an individual or species of plant or animal, which breeders need to know in order to *breed new *varieties.

breeding parasitism *See* BROOD PARASITISM.

breeding population A group of individuals (plants or animals) that tends to reproduce among itself, and much less frequently with individuals from other members of the same species. These separated subpopulations are important sources of migrants, and their genetic variability can prove critical to the survival of a species as a whole.

breeding rate The number of new individuals that are born in a given period of time.

breeding stock 1. Individuals (usually animals) that are kept specifically for *breeding purposes.
2. The group of individuals within a given population which is capable of breeding.

breeze Light *wind, or a local air movement such as a *sea breeze or *land breeze.

brine A salt solution. Water that is saturated with or contains a large amount of salts, especially of *sodium chloride.

brine mud Waste material that comes from well-drilling or mining and is composed of mineral salts or other inorganic compounds.

briny *See* BRACKISH.

British thermal unit (Btu) A unit of measurement for *heat. It is the amount of energy that is required to raise the temperature of 1 lb of water by one degree from 62–63°F (about the amount of energy released when a match tip burns). One Btu is equal to 252 *calories or 1055 joules.

broadcast application The process of spreading a chemical (such as a *pesticide) over an area.

broadcast burn A controlled fire that is started deliberately and is allowed to burn a designated area. Such fires are used to remove organic debris after a clear-cut *deforestation project, and as part of ongoing vegetation management to reduce wildfire hazards, improve *forage for wildlife and livestock, and encourage successful *regeneration of trees.

broadcast seeding Scattering seed on the surface of the *soil. *Contrast* DRILL SEEDING.

broadcaster Equipment that is used to scatter (broadcast) seeds.

broadleaved Any tree with broad leaves and not leaves like pine-needles. This includes both *evergreens (such as holly) and *deciduous trees (such as oak). They belong to the *angiosperm group of plants.

broad–sclerophyll forest A North American *chaparall type of forest vegetation with a closed canopy of trees that are adapted to drought by having small, broad, waxy-coated leaves, which is found in dry areas like California (for example, on the western foothills of the Sierra Nevada).

broken clouds *Clouds that cover between six-tenths and nine-tenths of the sky.

broker In *greenhouse gas *emissions trading, a broker acts as an intermediary between a buyer and a seller, for which they usually charge a commission.

bromine (Br) A *halogen chemical that works as a disinfectant, and is often used as an alternative to *chlorine to kill bacteria and algae in the water of pools and spas. It is one of the most common chemicals found in seawater. It exists as a volatile liquid at room temperature and is also used for petrol additives. Liquid bromine is harmful to human tissue, and the vapour irritates the eyes and throat.

Bronze Age The *prehistoric period of human culture in Europe from about 2000 BC to about 1000 BC, during which bronze (an alloy of *copper and tin) was the main material used for making tools and weapons. It followed the *Stone Age and ended with the start of the *Iron Age.

brood The young of certain animals, particularly young birds and fowl, that are hatched and cared for one at one time.

brood cover Low vegetation (such as grasses or herbs) that offers protection for ground nesting birds to raise their young.

brood parasitism A form of *parasitism in which a bird of one species (such as a cuckoo) lays an egg in the nest of a bird of a different species, and the latter then looks after the young. Also known as **breeding parasitism, nest parasitism.**

Brower, David US environmental campaigner and activist (1912–2000) who founded many environmentalist organizations including the Sierra Club Foundation, the John Muir Institute for Environmental Studies, *Friends of the Earth, the League of Conservation Voters, Earth Island Institute, North Cascades Conservation Council, and

Fate of the Earth Conferences. He was nominated for the Nobel Peace Prize three times, served as the first Executive Director of the *Sierra Club between 1952 and 1969, and served on its board three times between 1941 and 2000.

brown earth soil A type of soil that develops under *deciduous forest on calcium-rich parent material, has a high *base status, and lacks a well-developed *illuvial horizon.

brownfield An abandoned industrial or commercial site where expansion or redevelopment is compromised by the possibility that it may be contaminated with hazardous substances from operations on the site. After *remediation it may be reclassified as a *greenfield site. *See also* INFILL.

browse 1. Leaves, young shoots, herbs, shrubs, trees, and other vegetation that serve as food for livestock and wildlife.
2. The act of eating such food. *See also* GRAZE.

browse line The height limit on trees and tall shrubs to which livestock and big game browse. Also known as **grazing line.**

Brundtland Commission *See* WORLD COMMISSION ON ENVIRONMENT AND DEVELOPMENT.

brush 1. A type of woody vegetation, comprising shrubs and small trees. Also known as **undergrowth** or **underwood.**
2. Material such as twigs that are cut from undergrowth.

brush control The deliberate removal or reduction of *brush in order to reduce *wildfire fuel, or to allow the growth of plant species that are preferred for growing timber or *forage.

brush management The management of stands of *brush using mechanical, chemical, or biological methods, or by *prescribed burning.

brushing In *silviculture, the removal from beneath the trees of *brush and *weed species, which compete with seedlings for sunlight, water, and soil nutrients.

bryophyte A plant of the division Bryophyta (a moss, liverwort, or hornwort) that grows in damp places. *Contrast* VASCULAR PLANT.

BSE *See* BOVINE SPONGIFORM ENCEPHALOPATHY.

Btu *See* BRITISH THERMAL UNIT.

bubble A concept used in *air pollution control, based on the idea that reductions anywhere within a specific area should count towards a common reduction goal. *See also* EMISSIONS BUBBLE, EU BUBBLE.

bubble approach In *pollution control, an approach that allows polluters to discharge more pollutants at one source, if an equivalent reduction occurs at other sources within the *bubble. Also known as **bubble policy**. *See also* EMISSIONS BUBBLE, EMISSIONS TRADING.

bubble policy *See* BUBBLE APPROACH.

bucolic Typical of a *rural or *pastoral life.

Budget Period *See* COMMITMENT PERIOD.

buffer **1.** A solution that works to keep *pH levels fairly constant when acids or bases are added, effectively neutralizing them.
    2. *See* BUFFER STRIP.

buffer solution A solution that contains a *buffer and is thus able to oppose changes in *pH when small quantities of acid or base are added to it.

buffer species A non-game or undesirable species of animal that provides food for *predators and thus reduces the loss of *game or other desirable species.

buffer strip An area of land or water, usually around or beside a sensitive wildlife habitat (such as a *wetland), that contains undisturbed vegetation and is designed to minimize sharp changes in habitat, inhibit soil erosion,

or prevent disturbance from surrounding land uses. Also known as **buffer**, **buffer zone** or **filter strip**.

buffer zone *See* BUFFER STRIP.

buffering agent Anything that turns an acidic (*pH less than 7.0) or alkaline (pH greater than 7.0) solution neutral (pH 7.0). *See also* BUFFER SOLUTION.

buffering capacity The ability of a medium (such as soil or water) to resist rapid changes in *pH. Also known as **alkalinity** or **acid neutralizing capacity**.

building code Local regulations that control the design, construction, and materials used in construction, usually based on health and safety standards.

building-related illness A discrete, identifiable disease or illness (such as legionnaire's disease) that can be linked to a specific pollutant or source within a building. *Contrast* SICK BUILDING SYNDROME. *See also* LEGIONELLA.

bulk density The mass or weight of unconsolidated material, such as soil, usually expressed in grams per millilitre (g mL^{-1}). Compaction (for example by recreational use, grazing animals, or logging equipment) increases bulk density.

bulky waste Large items of waste materials (such as appliances, furniture, large vehicle parts, trees, and stumps) that cannot be handled by normal procedures for processing *municipal solid waste.

bunchgrass A North American *grassland vegetation in which different types of native *perennial grass grow in tufts or clumps rather than forming a sod or mat. Dominant grasses include little bluestem (*Schizachyrium scoparium*) and buffalo grass (*Buchloe dactyloides*), which provide good *forage for grazing animals (such as cattle and buffalo). Common throughout the western states.

bund An artificial barrier on a slope, made, for example, from soil or agricultural waste, designed to reduce runoff and soil erosion.

buoyancy An object's ability to float, for example in the air or in water, because the medium it is in has a greater density.

Bureau of Land Management *See* US BUREAU OF LAND MANAGEMENT.

Bureau of Reclamation *See* US BUREAU OF RECLAMATION.

Burgess shale A layer of *Cambrian rocks in British Columbia, Canada. It contains many unique *fossils of a variety of soft-bodied animals that have not been found elsewhere, and scientists are uncertain whether or not these are the ancestors of animals that are alive today.

burial ground 1. A disposal site for unwanted radioactive materials, where earth or water are used as a shield.
2. A piece of land where dead bodies are buried. Both are also known as a **graveyard**.

burn 1. A land management technique that involves setting fire to vegetation to clear it, encourage new growth and restore soil fertility. *See also* PRESCRIBED BURN.
2. An area of land that has recently been burned. *See also* FIRE.

bush 1. A low woody *perennial plant that has no distinct trunk but usually has several major branches.
2. An area of dense vegetation dominated by stunted trees or bushes.

bush encroachment The unplanned conversion of vegetation dominated by grassland to one dominated by woody species, often as a result of *overgrazing or invasion by *aliens.

bush fallow The natural vegetation that grows on land that is left uncultivated for some time, and includes small trees, shrubs, grasses, sedges, and herbaceous plants. *See also* ENRICHED FALLOW, FALLOW, SHIFTING CULTIVATION.

bush land An open area with tall, scattered *bushes covering at least 40% of the area.

business-as-usual scenario A scenario for future patterns of activity which assumes that there will be no significant change in people's attitudes and priorities, or no major changes in technology, economics, or policies, so that normal circumstances can be expected to continue unchanged.

butane (C_4H_{10}) A gaseous *hydrocarbon that is extracted from *natural gas or refinery gas, and is used as household fuel, a propellant, and a refrigerant.

butte A narrow flat-topped hill of resistant rock with very steep sides, found in an arid area where there is little vegetation. It is formed by the erosion of horizontal *strata where remnants of a resistant layer protect the softer rocks underneath.

butterfly effect The phenomenon by which a small change at one place in a complex system can have large effects elsewhere. For example, through *teleconnection a butterfly that flaps its wings in Rio de Janeiro might ultimate cause the weather in New York to change.

butyric fermentation The *decomposition of various types of organic matter by chemical reaction. One of the many forms of *fermentation involved in *putrefaction.

buy-back centre A facility where individuals or groups take *recyclable material in return for payment.

Buys Ballot's law A law named after a Dutch meteorologist Buys Ballot (1817–90), who in 1857 defined the relationship between the *pressure gradient and the *Coriolis force. The simple rule of thumb is: in the northern hemisphere, if you stand with the wind blowing into your back, low pressure is always to your left and high pressure is always to your right; in the southern hemisphere the pattern is reversed, and low pressure is on your right and high pressure on your left.

BWR *See* BOILING WATER REACTOR.

bycatch Fish and other marine life that is caught incidentally while fishing for something else. Unwanted bycatch is normally thrown back into the sea as a waste product, even if it is commercially valuable. Also known as **incidental catch** or **incidental take**. *See also* LONGLINE, NON-TARGET SPECIES.

bypass A route or road that passes around a *city or other congested area.

byproduct Any material, other than the principal product, that is generated as a consequence of an industrial or biological process, and for which there is often no commercial market.

C:N ratio *See* CARBON: NITROGEN RATIO.

cable logging A method of *logging in which timber is dragged by cable from where it is cut to a collecting point. *See also* GROUND-LEAD LOGGING, HIGH-LEAD LOGGING.

cadmium (Cd) A naturally occurring, soft, bluish trace *heavy metal that accumulates in the environment. Sources include smelter fumes and dust, some incineration products, *phosphate fertilizer, municipal wastewater, and sludge discharges. It is also an industrial *byproduct of the manufacture of *zinc, *copper, and *lead. It corrodes galvanized pipes, and can be extremely toxic to humans and aquatic life. *See also* ITAI-ITAI DISEASE.

caesium (Cs) A silver-white, soft *ductile element, part of the *alkali metal group. Used in *photoelectric cells. The naturally occurring *isotope is ^{133}Cs but there are 15 other radioactive isotopes.

CAFE *See* CORPORATE AVERAGE FUEL ECONOMY.

CAFO *See* CONCENTRATED ANIMAL FEEDING OPERATION.

Cainozoic *See* CENOZOIC.

Cairo Conference *See* INTERNATIONAL CONFERENCE ON POPULATION AND DEVELOPMENT.

calcareous 1. A substance that is composed of, or contains, *calcium carbonate, which typically causes an *alkaline condition (*pH greater than 7.0).
2. A species that accumulates *calcium carbonate in its tissues.

calcicole A plant that is adapted to growing on *limestone or on soils that are derived from limestone and have a high *calcium content.

calcification 1. A *soil forming process in which *calcium carbonate accumulates in the lower *horizons. This occurs mainly under grass or *xerophytic shrub in subhumid, semi-arid, and arid climates and produces neutral or *basic soils.
2. Replacement of the hard body parts of an organism by *calcium carbonate (for example in the form of a shell).

calcite A common rock-forming *mineral of *calcium carbonate. It is crystalline, stable, one of the most common minerals, and a main ingredient of *limestone.

calcium (Ca) An *alkaline metal that occurs naturally in the *environment and accounts for 3.63% of the Earth's *crust. It readily forms salts with various *metals and *halogens, and is an *essential element (a *minor element) for animals and humans.

calcium carbonate ($CaCO_3$) A *chemical precipitate which is formed from the skeletons of corals and the shells of molluscs, and occurs as *calcite, *chalk, and *limestone. It dissolves in water and so affects *salinity and *alkalinity, and at high concentrations it can make water hard. It is usually white, colourless, or light shades of grey, yellow, and blue, and is used in polishes and the manufacture of lime and cement.

calcrete *See* CALICHE.

caldera A large, steep-sided, circular depression formed by the collapse of a volcanic *crater. The caldera is usually much larger than the original *vent of the volcano and has a relatively flat floor.

caliche A layer of *calcium carbonate that has been deposited on or near the surface of a stony soil in an *arid or *semi-arid area, by the *evaporation of moisture in the soil. Also known as **calcrete**.

California grassland A North American native *grassland vegetation common throughout California, which was originally a mixture of short and tall bunch grasses. Much of the habitat has been lost through development and conversion, and in the areas that remain many of the native *perennials have been replaced by introduced *annuals.

calm Undisturbed air in which there is no apparent motion. The lowest point on the *Beaufort scale.

calorie A measure of energy. One calorie is equal to the total amount of heat that is required to raise the temperature of one gram of water by 1°C.

calving The process of forming *icebergs, in which pieces of a *glacier or *ice sheet break off when the ice meets the ocean.

cambium A single layer of cells between the woody part (*xylem) and the bark (*phloem) of a tree, where the growth takes place. These are the growth rings that can be counted to determine the age of the tree. *See also* DENDROCHRONOLOGY.

Cambrian The first geological period of the *Palaeozoic era, ranging from 570 to 510 million years ago, during which marine life (particularly invertebrates) evolved and spread widely. The oldest *fossils of organisms with mineralized skeletons date from the Cambrian. *See also* BURGESS SHALE.

Cambrian Explosion The common name for the huge increase in the number and diversity of life forms in the Earth's oceans that started during the *Cambrian period.

camouflage The disguise that a species of animal develops (through *natural selection) that enables individuals to hide from predators by blending in with their surroundings, usually through the colour and pattern of their skin or fur. *See also* COUNTERSHADING, CRYPTIC COLORATION.

Campbell–Stokes sunshine recorder An instrument for measuring how long the Sun shines at a particular place, based on a glass sphere which focuses the Sun's rays to burn a hole through a graduated card. It is the most common sunshine recorder in use outside the United States.

canal An artificial waterway that has been built for navigation, water supply, land irrigation, or drainage.

canalization Engineering works that straighten and/or deepen a natural channel to improve navigability, or that artificially force a river to flow along a particular course (often within a concrete channel) to speed the flow and/or reduce flooding.

cancer A group of more than 120 different diseases in which abnormal cells divide and grow rapidly and uncontrollably, producing tumours that can be malignant and often invade surrounding cell tissue. *See also* CARCINOMA.

candidate species Species of plants and animals that are considered (by the *US Fish and Wildlife Service or the National Marine Fisheries Service) to be candidates for listing (and thus protecting) under the *Endangered Species Act.

canopy The roof-like cover formed by the leafy upper branches and *crowns of the tallest trees in a *forest. In tropical forests, the canopy may be 50 metres or more above the ground. *See also* OVERSTOREY.

canopy closure The progressive reduction of space between the *crowns of trees in a *forest as they spread laterally, increasing *canopy cover.

cantharophily *Pollination by beetles.

canyon A large, narrow valley or *gorge with deep sides that is cut by a river.

CAP *See* COMMON AGRICULTURAL POLICY.

cap A layer of *clay or other impermeable material that is spread over the top of a closed *landfill or *hazardous waste site in order to seal it, and thus prevent the *infiltration of rainwater and minimize the *leaching of *pollutants into adjacent soil and rock. Also known as **cover material.**

cap and trade An approach to *air pollution control, introduced in the *Kyoto Protocol, that is based on the trading of emissions allowances between countries, within a total allowance that is strictly limited or capped. See also EMISSIONS ALLOWANCE, EMISSIONS TRADING.

capability, land The suitability of a particular area of land for a given type and intensity of use, without permanent damage, which reflects factors such as site conditions (particularly climate, soils, geology), broader socioeconomic factors (including political, social, and economic constraints), and land management practices.

capacity The total amount of *sediment that a *river can transport, which is determined by both discharge and particle size.

capillary action The process by which water moves through very small spaces between particles in *soil. It is caused by attraction (capillary forces) between the liquid and the particles. This process allows water to be drawn upwards in the soil.

capillary fringe In *rock or *soil, the zone of porous material just above the *water table that remains more or less saturated with water, because of *capillary action.

capillary water Water that is held within a *soil by *surface tension in and on soil particles, and in the pore spaces between particles. It can move in any direction within the soil, by *capillary action. See also FIELD CAPACITY.

capital Any form of wealth, resources, or knowledge that is available for use in the production of more wealth.

capital cost The total investment that is needed to complete a project and make it commercially operable, such as

the cost of building a new factory or engineering scheme.

capitalism An economic system that is based on private enterprise, in which the means of production are owned privately and resources are exchanged through market processes. Also known as **capitalist economy.** Contrast MARXISM, SOCIALISM. See also MARKET ECONOMY.

capitalist economy See CAPITALISM.

captive breeding The breeding and raising of *endangered species of animals in captivity (for example, in a *zoo), where they can be protected and cared for, as part of a *living collection, before being released into the wild in an attempt to prevent their extinction. See also SELECTIVE BREEDING, REINTRODUCTION.

captive conservation Keeping small numbers of animals in protective isolation, to ensure that they breed successfully before being released.

capture–recapture method A procedure for estimating the size of a biological population, based on the distinctive marking and subsequent recapture or sighting of individuals. Also known as **mark–recapture.**

carapace The hard case or shell that covers the body of animals such as turtles or crabs. Contrast PLASTRON.

carbamate A class of *insecticides derived from carbamic acid, which is less *toxic and more *biodegradable than many *organophosphate insecticides.

carbohydrate An *organic chemical compound that contains only *carbon, *hydrogen, and *oxygen and supplies energy to organisms. Examples include *starch, *glucose, and *cellulose.

carbon (C) A non-metallic *element that occurs in many *inorganic compounds and all *organic compounds, and is one of the most widely distributed elements on Earth. It occurs in combination with other elements in all plants and animals, and is a basic building block of life. Carbon occurs in four basic forms in nature—in the pure form as graphite and diamonds; as *calcium carbonate in *carbonaceous rocks such as *chalk and *limestone; as *carbon dioxide in the atmosphere; and as *hydrocarbons in *fossil fuels.

carbon-14 (^{14}C) A naturally occurring *carbon *isotope that is produced in the atmosphere by cosmic ray bombardment and has a half-life of 5700 years. It is useful for the *absolute dating of samples up to 40 000 years old. *See also* RADIOCARBON DATING.

carbon absorber An *air pollution control device that uses *activated carbon to absorb *volatile organic compounds from a stream of gas, which are later recovered from the carbon.

carbon adsorption A water treatment system that removes *contaminants from groundwater or surface water by *adsorption, by forcing the water through tanks that contain *activated carbon, which attracts the contaminants.

carbon assimilation The process by which plants take in (assimilate) *carbon dioxide.

carbon budget The balance of the exchanges (incomes and losses) of carbon

between different reservoirs of *carbon in the *carbon cycle, which shows whether particular reservoirs (such as oceans or forests) are operating as *sources or *sinks for *carbon dioxide.

carbon credit In *greenhouse gas *emissions trading, this is a credit granted to a particular country that counts towards their emission targets as agreed in the *Kyoto Protocol.

carbon dating A method of dating *organic material that is applicable to material of up to at least 40 000 years old. It uses the known rate of disintegration of the *carbon-14 atom as a basis for determining the age of a sample, based on the principle that the ratio of $^{14}C/^{12}C$ is directly related to the age of the sample. Also known as **radiocarbon dating**.

carbon dioxide (CO_2) A naturally occurring, heavy, colourless, odourless gas, which is a part of the *ambient air, accounting for 0.035% of the Earth's *atmosphere, by volume. It is produced by *respiration, decomposition of organic substances, and the combustion of *fossil fuels and *biomass (particularly forest fires). It dissolves in water to form *carbonic acid, is the principal *greenhouse gas (it accounts for 80% of all green-

CARBON CYCLE

The natural *biogeochemical cycle through which *carbon circulates through the *biosphere, *atmosphere, and *hydrosphere. This involves *abiotic and *biotic parts of interconnected cycles on land, sea, and the atmosphere. In the atmosphere carbon is found as *carbon dioxide (CO_2), which plants use in the process of *photosynthesis to manufacture *carbohydrates. Plants can then be consumed by *herbivores and the carbon is passed along *food chains, as these animals are then eaten by *carnivores. Some of the carbon is returned to the atmosphere when these animals respire or when they die and decay through the respiration of decomposing bacteria. If plants are not eaten the carbon contained in them may become fossilized as fuels such as oil and coal, or it may return to the atmosphere as CO_2 by the decay of the plant matter. Plants may also return carbon to the atmosphere by the process of *respiration. The carbon cycle is also being disrupted by changes in land use, particularly the removal of natural vegetation. Natural *sinks of atmospheric CO_2 are decreasing as a direct result of vegetation clearance and changing land use. Levels of carbon dioxide are building up in the atmosphere as a result, thus contributing to the *greenhouse effect.

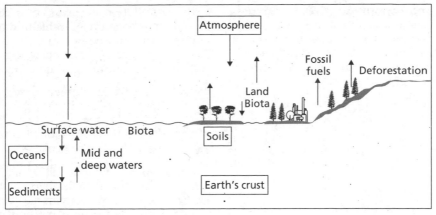

Fig 3 Carbon cycle

house gas emissions), and persists in the air for more than 100 years. About 20% comes from natural sources, the rest from human activities (particularly from the burning of fossil fuels to run vehicles, heat buildings, and power factories). Roughly half of the CO_2 that is released is soon absorbed by the oceans or by increased plant *photosynthesis, but the other half remains in the atmosphere for many decades. Concentrations of carbon dioxide in the atmosphere have risen by about a quarter since the *Industrial Revolution. Analysis of *ice cores from *Antarctica shows that natural background levels of atmospheric CO_2 are in the order of 200 parts per million by volume (ppmv) during *glacial periods, rising to around 280 ppmv during warmer *interglacials. In 1958 the atmospheric concentration was measured at 315 ppmv, and by the early 1990s it had risen to more than 350 ppmv. By 1997 the concentration stood at more than 360 ppmv, the highest level in 160 000 years. It is predicted to continue rising at a rate of 0.4% a year, and to rise above 600 ppmv over coming decades. *Industrialized countries produce much more than their fair share of CO_2; for example the USA houses 5% of the world's population but produces 25% of the CO_2. *Per capita emissions of CO_2 are much lower amongst *developing countries, largely because per capita energy use is much lower.

carbon dioxide equivalent *See* GLOBAL WARMING POTENTIAL.

carbon dioxide fertilization The increased plant growth or rise in *net primary production in natural or agricultural systems that is or would be promoted by an increase in the concentration of *carbon dioxide in the atmosphere.

carbon fixation A chemical process by which *carbon from the atmosphere is assimilated into *organic compounds and made available to *organisms through *food chains. *See also* PHOTOSYNTHESIS.

carbon flux The rate of movement of *carbon between different pools or reservoirs in the *carbon cycle.

carbon isotope There are three naturally occurring *isotopes of carbon: ^{12}C or carbon-12 (which accounts for 98.9 of all the carbon), ^{13}C or carbon-13 (about 1.1%) and ^{14}C or carbon-14 (negligible amount). Changes through time in the absolute and relative amounts of the carbon isotopes can be measured in *ice cores and lake sediments, and these provide *indicators or *proxy measures of long-term climate change.

carbon management An approach to limiting the emission of *carbon dioxide that is based on storing it (for example in *carbon sinks), or using it in ways that prevent its release into the air.

carbon monoxide (CO) A colourless, odourless, non-irritating but highly *toxic gas that is produced by the incomplete combustion of *hydrocarbon fuels, incineration of *biomass or solid waste, or partially *anaerobic *decomposition of organic material. Much comes from vehicle exhausts. It is a *criteria pollutant which affects human health by interfering with the ability of blood to carry oxygen to cells, tissues, and organs in the body, and exposure to high levels is harmful to people, but particularly to those with diseases of the heart, lung, and circulatory system.

carbon: nitrogen ratio (C:N ratio) The ratio between the amount of *carbon and *nitrogen in a particular soil or organic material, which affects *fertility and suitability for *composting.

carbon offset *See* OFFSET.

carbon pool The total amount of *carbon that is stored, cycling, or otherwise available for use in *biogeochemical cycles.

carbon reservoir Any of the locations within the *carbon cycle at which carbon compounds are stored, including the atmosphere, oceans, vegetation and soils, and reservoirs of fossil fuels.

carbon sequestration The long-term natural storage of *carbon in forests, soils, oceans, or underground in oil and gas reservoirs and coal seams. There have been initiatives to reduce the amount of carbon that is in the atmosphere by increasing the number and size of *carbon sinks, in order to slow down *global warming.

carbon sink A *geochemical reservoir (*sink) that can absorb or *sequester *carbon which has been released from another part of the *carbon cycle, thus removing it from the flowing part of the cycle for a period of time. Examples include *soils, *peat, large *forests (which store organic carbon), and ocean sediments (which store *calcium carbonate). One of the *flexibility measures introduced in the *Kyoto Protocol for dealing with *air pollution by *greenhouse gases is to increase carbon sinks by *afforestation or other means.

carbon source Any part or reservoir of the *carbon cycle that releases carbon to some other part of the cycle. Examples include the burning of *fossil fuels, *decomposition of *organic waste, and burning of trees or other vegetation.

carbon stock The *carbon that is naturally stored in vegetation, decomposing organic matter, soils, and wood products.

carbon tax A tax or surcharge on the sale of *fossil fuels (oil, coal, and gas) that varies according to the carbon content of each fuel, and is designed to discourage the use of fossil fuels and reduce emissions of *carbon dioxide.

carbon tetrachloride (Cl$_4$) A colourless, non-flammable, *toxic liquid compound that was once widely used as an industrial raw material, as a solvent, and in the production of *CFCs. It is highly *carcinogenic, and its use has been banned or strictly regulated in most countries since 1995, under the *Montreal Protocol. Also known as **tetrachloromethane.**

carbonaceous A rock, sediment, or aerosol that contains *carbon.

carbonate A mineral that is mostly found in limestones and dolomites (such as *calcite). The term is also used to describe sedimentary rocks that are largely composed of carbonate minerals.

carbonate hardness The hardness in water that is caused by *bicarbonates and *carbonates of *calcium and *magnesium. *See also* HARD WATER.

carbonation A common type of *chemical weathering of rock by weak *carbonic acid that is created from *carbon dioxide dissolved in rainwater. This can be very significant in areas of *limestone and *dolomite.

carbon-based resource The amount of recoverable *fossil fuel and *biomass that can be used to produce *fuel.

carbonic acid (H₂CO₃) The weak *acid that is formed when *carbon dioxide (CO₂) is dissolved in water.

Carboniferous A period of geological time during the *Palaeozoic era, dating from about 360 to 290 million years ago, during which extensive *coal measures were deposited.

carcinogen Any substance that produces *cancer, by causing changes in the *DNA of cells. *See also* MUTAGEN, TERATOGEN.

carcinogenic *Cancer causing.

carcinoma A type of *cancer that occurs in particular types of tissue.

cardenolide A *toxic material that certain plants are able to produce as a defence against their own predators.

Caring for the Earth A strategy for *sustainable development that was published in 1991 by the *IUCN, jointly with the *United Nations Environment Programme and the *World Wide Fund for Nature.

carnivore A flesh-eating animal (*predator or *carrion eater).

carnivorous Meat-eating.

Carolina parakeet A species of parrot (*Conuropsis carolinensis*) that was the only parrot species *native to the eastern USA. It lived in old forests along rivers, and was found from the Ohio Valley to the Gulf of Mexico. By the early 1890s it had become *extinct as a result of shooting by farmers (who saw it as a *pest), over-collecting (for its colourful feathers), and destruction of its native *habitat.

car-pooling Sharing a car to a destination in order to reduce fuel use, pollution, and travel costs.

carr Wet *woodland that grows on soils with permanently high water levels and is usually dominated by alder or willow trees.

carrier A person or animal that carries a specific infectious *agent and is a potential source of infection. *See also* DISEASE VECTOR.

carrion The decaying flesh of a dead animal that is used as food by animals that *scavenge.

carrying capacity The maximum number of individuals (people or animals) that can be supported by a particular *ecosystem (or area) on a *sustainable basis without degrading it. Global carrying capacity refers to the total number of people which the Earth and its environmental systems can support on a sustainable basis. There are no clear-cut answers to the question of what the Earth's carrying capacity really is; published values range between about three billion and 14 billion people. Experts expect there to be natural limits, but there is no consensus over precisely what those limits are likely to be. Also known as **ecological limit** or **environmental capacity.** *See also* MALTHUS, TRAGEDY OF THE COMMONS.

Carson, Rachel A scientist and writer whose book **Silent Spring* (1962) looks at the effects of *insecticides and *pesticides on populations of songbird throughout the USA. *See also* BIOACCUMULATION.

Cartagena Protocol on Biosafety An international agreement arising from the *United Nations Framework Convention on Biological Diversity, that came into force in January 2000, and is designed to protect *biodiversity from the potential risks posed by *living modified organisms that result from *biotechnology. *See also* BIOSAFETY.

cartel A consortium of producers of a single product who agree to limit production to keep the price of the product high.

cartography The construction and study of maps.

cascade 1. A small waterfall or series of small waterfalls.
2. A sudden downpour.

cash crop Any crop that is grown for sale in a market, to agents, or directly from the farm. *See also* SUBSISTENCE FARMING.

casing The pipe that provides the walls of oil, gas, or water wells.

cask A thick-walled container, usually made from lead, that is used to transport *radioactive material. Also known as a coffin.

casualty Any person suffering physical and/or psychological damage that kills, injures, or leads to material loss.

CAT See CLEAR AIR TURBULENCE.

catabolism The process by which a living cell or organism breaks down complex *organic molecules to smaller ones, producing energy and waste matter. Also known as **destructive metabolism**. See also METABOLISM.

catadromous Fish (such as eels) that spend most of their lives in freshwater rivers and lakes, but migrate downstream to the sea in order to breed. Contrast ANADROMOUS.

catalysis Increasing the rate of a chemical or biochemical reaction by the addition of a *catalyst.

catalyst An *inorganic substance that increases the rate of a chemical or biochemical reaction without being consumed or chemically changed in the process.

catalytic converter A device fitted to the exhaust systems of motor vehicles which uses a catalytic agent (tiny quantities of metals such as platinum) to stimulate chemical reactions that remove pollutants from the exhaust gases. The process reduces emissions of noxious gases such as *nitrogen oxides, unburnt *hydrocarbons, and *carbon monoxide and replaces them with less harmful emissions, either by *oxidizing them into *carbon dioxide and water or reducing them to *nitrogen. See also END-OF-THE-PIPE.

catalytic incinerator A pollution control device that *oxidizes *volatile organic compounds by using a *catalyst to promote the combustion process.

catastrophe A sudden, severe event that causes unusually large-scale losses (including deaths, injuries, damage to property, and financial loss), beyond normal expectation. Examples include extensive damage cause by flooding, hurricanes, or large fires. Also known as **disaster**. See also EMERGENCY.

catastrophism An interpretation of features of the Earth's surface (including rocks, fossils, and landforms) as the product of sudden, cataclysmic upheavals, particularly caused by major floods (*neptunianism) or volcanic activity (*vulcanism). This view was popular during the 19th century, but it was challenged and eventually replaced by *uniformitarianism.

catch basin A sedimentation area that is designed to remove pollutants from *runoff before it is discharged into a stream or pond.

catchment See DRAINAGE BASIN.

catena 1. A sequence of soils in a landscape that are of similar age and have developed from similar parent material in the same climate, but which have different properties because of variations in relief and in drainage. Also known as **toposequence**.
2. A series of volcanic *craters.

cation An ion that has a positive charge. See also ANION.

cation adsorption capacity See CATION EXCHANGE CAPACITY.

cation exchange The chemical exchange of *cations that takes place between plant roots and the water, minerals, and organic matter within a soil.

cation exchange capacity (CEC) A measure of the ability of the *colloids in a soil to attract and retain *cations, which is usually expressed in terms of milliequivalents per 100 grams of soil (mEq 100 g^{-1}). This capacity is highest in soils that contain a lot of *humus and *clay.

Cato Institute A US non-profit *anti-environmental *think-tank based in Washington, DC that seeks to promote traditional American principles of limited government, individual liberty, free markets, and peace within public policy debates.

cauliflory The growth of flowers on the large branches and trunk of a tree.

caustic Burning or corrosive.

caustic soda Sodium hydroxide. A strongly *alkaline substance that is used in the manufacture of soap, paper, *aluminium, and a range of sodium compounds.

caution A warning against danger. *See also* ADVISORY.

cave A hollow chamber in rock, which is accessible from the surface, is large enough for a person to enter, and has usually been formed by natural processes.

cavernicolous Living in caves.

cavity Any hollow or hole, such as a hole in a tree, that is regularly used by species of wildlife (particularly birds) for nesting, roosting, and reproduction.

cavity tree *See* DEN TREE.

CBNRM *See* COMMUNITY BASED NATURAL RESOURCE MANAGEMENT.

CDC *See* CONSERVATION DATA CENTER.

CDM *See* CLEAN DEVELOPMENT MECHANISM.

CEC *See* CATION EXCHANGE CAPACITY.

ceiling In the *atmosphere, the height of the lowest layer of *clouds when the sky is broken or overcast.

celerity Speed, swiftness of motion.

celestial Of or relating to the sky or outer space.

celestial equator The projection onto the sky of the Earth's *equator, which is 90° from each of the Earth's *celestial poles.

celestial pole The points in the sky that are above the Earth's North and South poles.

cell The smallest unit of living matter that is able to grow and reproduce independently, of which all living organisms are composed. Cells contain *DNA (information store), ribosomes (protein synthesis), and mechanisms for converting energy into usable forms. There are two main types of cell, *prokaryotic and *eukaryotic. *See also* CYTOPLASM, GENOME, NUCLEUS.

cell division The process by which two identical *cells are formed from one.

cell fusion A technique of joining two *cells from different *species to create one hybrid cell, in order to combine some of the genetic characteristics of each original.

cell membrane The outer *membrane of a *cell, which separates it from the surrounding environment. The term is used particularly to refer to the *plasma membrane.

cell wall A tough *permeable layer that surrounds the *cells of a plant.

cellular respiration The process by which a *cell breaks down *organic compounds (such as sugar) in order to release the energy it needs to perform work. This process may be *aerobic or *anaerobic, depending on the availability of *oxygen.

cellulose ($C_6H_{10}O_5$) A colourless, solid *carbohydrate *polymer that is made from the simple sugar *glucose, and is the most abundant compound on Earth that is manufactured by living things.

Celsius (°C) A temperature scale that is used internationally and was devised by the Swedish astronomer Anders Celsius (1701–44). In it water freezes at 0°C and boils at 100°C. The equation for converting temperature from *Fahrenheit to Celsius is °C = 5/9 (°F – 32), so that 100°C is equal to 212°F. Previously known as **Centigrade.**

cement 1. To bind or join together.
2. Any material that hardens to act as an adhesive.
3. A building material made of *limestone and *clay, which is mixed with water and sand or gravel to make concrete and mortar.

cemented The grains of a *sedimentary rock that have been bonded together by substances such as *calcium carbonate, *silica, oxides of iron and aluminium, or *humus.

Cenozoic The current geological era, after the *Mesozoic era, from about 65 million years ago to the present, during which the modern continents formed and modern animals and plants evolved. The Cenozoic is subdivided into the *Tertiary and *Quaternary periods.

census 1. A count of all of the individuals in a specified area over a specified time interval or at a particular point in time.
2. The act or process of counting all individuals within a specified area and estimating population density or a total population for that area.

Centigrade *See* CELSIUS.

centrally planned economy An economic system (like those in most communist countries, such as China) in which investment and production are coordinated by a central government body rather than by market forces.

centre of diversity A geographical area that has high levels of genetic or species diversity. *See also* BIODIVERSITY.

centre of endemism A geographical area that has a relatively high number of locally endemic species. *See also* ENDEMIC.

centre of origin A geographical area in which a *taxon originated, which may or may not correspond with the *centre of diversity for that taxon.

centrifugal force The force that causes a body moving in a curve to move away from the centre of the curve. *Contrast* CENTRIPETAL FORCE.

centripetal drainage pattern A *drainage pattern that is the opposite of a *radial drainage pattern, in which the rivers drain in towards the centre of a basin, like the spokes of a wheel.

centripetal force The force that causes a moving object to follow a curved path rather than continue on a straight line. The force is directed towards the cenrtre of the curve. *Contrast* CENTRIFUGAL FORCE.

century A period of a hundred years.

CEQ *See* COUNCIL ON ENVIRONMENTAL QUALITY.

CERCLA *See* COMPREHENSIVE ENVIRONMENTAL RESPONSE COMPENSATION, AND LIABILITY ACT.

CERCLIS *See* COMPREHENSIVE ENVIRONMENTAL RESPONSE, COMPENSATION, AND LIABILITY INFORMATION SYSTEM.

cereal A plant in the grass family Gramineae, which produces seeds (*grain) that are edible. Examples include wheat, oats, barley, rye, rice, and maize.

CERFA *See* COMMUNITY ENVIRONMENTAL RESPONSE FACILITATION ACT.

certification In the context of pollution control, this is an important part of the *Clean Development Mechanism agreed by the *Kyoto Protocol. It is the process by which an accredited body gives written assurance of the emission reductions that have been achieved, which can be confirmed by independent third parties, and which then allows the emissions reduction to become a separate commodity for *emissions trading.

certified emissions reduction A unit of *greenhouse gas reductions that has been achieved and has been subjected to *certification under the *Clean Development Mechanism agreed by the *Kyoto Protocol. *See also* CREDIT, EMISSIONS REDUCTION UNIT.

cesspit/cesspool A covered hole or pit in which untreated *sewage is stored.

cetacean Any member of the order of marine *mammals that includes whales, dolphins, and porpoises.

cetane ($C_{15}H_{34}$) A colourless, liquid *hydrocarbon that is found in *petroleum and is used as an indicator of fuel efficiency for diesel oils.

CFCs *See* CHLOROFLUOROCARBONS.

chain reaction 1. A general term for any self-sustaining process.

2. A nuclear reaction in which the fission of nuclei produces *neutrons that then split other nuclei and the process becomes self-sustaining.

chalk A soft, white *organic sedimentary rock, the purest form of *limestone, which is made of very pure *calcium carbonate derived from the shells and skeletons of marine micro-organisms.

chalk grassland A type of *grassland that grows in thin *calcareous soils over *chalk bedrock.

channel 1. A waterway, such as a *stream, *river, or *canal, that contains moving water (continuously or periodically) and has a definite bed and banks. Also known as a **watercourse**.
2. A narrow seaway, such as the English Channel.

channel bar A ridge of coarse *gravel that is deposited on the bed of rivers with steep slopes, where *flow velocity decreases, for example where a steep *tributary stream flows into the main river.

channel improvement An engineering scheme that is designed to stabilize a river channel to increase flow speeds, reduce *bank erosion, and/or protect against flooding. This might involve decreasing channel roughness (for example by lining the channel with smooth artificial bed and banks), widening and/or deepening the channel (by *dredging), or shortening the channel (by straightening it and cutting off *meander bends). Also known as **channelization**.

channel pattern The shape of a river *channel when viewed from above or on a map. There are three main patterns, which are straight, *meandering, and *braided.

channel storage The volume of water that is temporarily stored in a river *channel and its *floodplain while it flows towards an outlet (such as a *lake, *reservoir, or *estuary).

channelization *See* CHANNEL IMPROVEMENT.

chaos The breakdown of predictability, or a state of disorder in a complex system in which patterns exist but they are complicated, not truly *random, and difficult to discover.

chaos theory The study of complex systems that exhibit discontinuous change, with a focus on irregular and complex behaviour which is only partially predictable.

chaparral An *arid *biome in California that is dominated by *broadleaved, *evergreen, drought-resistant *shrub, associated with the climate of hot dry summers and mild wet winters.

character Any recognizable trait, feature, or property of an organism.

character displacement An ecological pattern in which two *species having overlapping *niche requirements differ more when they live together than when they live apart. The difference is usually morphological (such as head size) and is caused by competition and adaptation to local conditions (such as the size of available *prey).

character type An area of land that has relatively uniform vegetation, geology, hydrology, and topography.

characteristic Any one of the four categories that are used to define *hazardous waste in North America: *corrosivity, *ignitability, *reactivity, and *toxicity.

characteristic diversity The natural pattern of distribution and abundance of *populations, *species, and *habitats in the absence of human impact.

characteristic species *Species that are special to or especially abundant in a particular situation or *biotope, which can usually be easily identified.

characterization The process of determining the scale and nature of a release of a *pollutant, which helps to determine appropriate ways of dealing with it and cleaning it up.

charcoal A very dark grey porous form of *carbon that is made by incomplete burning of wood or bone.

charge An excess or deficiency of *electrons in an *atom.

chattermark See STRIATION.

chelate The combination of *nutrients in a form that is easy for plants to absorb, chemically bound to a carbon-based substance.

chelation The process of chemically binding a metal to another substance, which is used to increase *absorption of mineral *nutrients (such as *calcium and *magnesium), to remove toxic substances (such as *lead and *arsenic), and to prevent the precipitation of metals (*copper).

chemical Any *element, *chemical compound, or mixture of elements and/or compounds.

chemical agent A *chemical that is used for some purpose, particularly to reduce or remove *pollutants from water.

chemical bond The force that holds atoms together in a molecule or crystal.

chemical compound The combination of two or more *elements in defined amounts.

chemical control Using *pesticides and *herbicides to control *pests and unwanted plant species.

chemical decomposition See CHEMICAL WEATHERING.

chemical element See ELEMENT.

chemical emergency An *emergency created by an accidental release or spill of *hazardous chemicals that threatens the safety of workers, residents, the environment, or property.

chemical energy Energy that is stored in *chemical compounds and released as a result of *chemical reactions.

chemical equilibrium A stable chemical reaction or series of reactions in which the concentrations of all substances remain constant through time.

chemical fixation See SOLIDIFICATION.

chemical of concern (COC) In *risk assessment, particularly in the USA, these are particular *chemicals that are identified for evaluation, to determine whether they pose a risk to people or the environment.

chemical of potential concern (COPC) In *risk assessment, particularly in the USA, these are *chemicals identified as possibly posing a risk to people and the environment, which are further investigated.

chemical oxygen demand (COD) A measure of the amount of *oxygen that is required to oxidize all of the *organic and *inorganic compounds in a particular *waterbody. *Biological oxygen demand only measures the oxygen used in breaking down the biodegradable material. See also BIOCHEMICAL OXYGEN DEMAND.

chemical partitioning The separation of a *chemical into different media or states. Many metals, for example, are more likely to partition to *sediments than remain in *groundwater.

chemical pollution The introduction of chemical contaminants into air, water, or soil.

chemical precipitate See EVAPORITE.

chemical reaction A change in one or more chemical elements or compounds which results in a new compound.

chemical stress The result of a *chemical reaction between two or more materials, such as corrosive materials attacking a metal.

chemical stressor Any *chemical that is released into the environment through human activities (such as industrial waste, vehicle emissions, and pesticides) that can damage or kill plants and animals.

chemical symbol A shorthand way of representing an element, using a one- or two-letter symbol. For example, sodium chloride is represented in chemical symbols as NaCl (Na is sodium and Cl is chlorine).

chemical synthesis The production of complex chemical compounds from simpler ones, for commercial or research purposes.

chemical treatment 1. The use of *chemicals or chemical processes to treat waste materials.
  2. The use of chemicals to control unwanted vegetation. *See also* HERBICIDE.

chemical warfare The use of *chemical agents to kill, injure, or incapacitate the enemy during a conflict.

chemical weapon Any chemical substance that can be used to kill or injure people, animals, and plants. A *weapon of mass destruction.

chemical weathering The breaking down of rocks and minerals as a result of *chemical reactions with the surrounding air and water. The most common processes are *oxidation, *carbonation, and *hydrolysis, and decomposition is usually fastest in hot, humid climates. *Contrast* MECHANICAL WEATHERING. *See also* WEATHERING.

chemistry The study of the properties, composition, and structure of matter.

chemoautotroph An organism that uses *inorganic *carbon (*carbon dioxide or *carbonates) as a source of energy for the synthesis of organic matter. *See also* AUTOTROPH, PHOTOAUTOTROPH.

chemoreception The ability to sense changes in the concentration of chemicals in the environment.

chemoreceptor A sense organ that responds to chemical stimuli.

chemosynthesis The process in which chemical energy is used to make *organic compounds from *inorganic compounds, such as the *oxidation of *ammonia to nitrite by *nitrifying bacteria. *Compare* PHOTOSYNTHESIS.

chemotroph An *autotrophic *organism (such as certain *bacteria and *blue-green algae) that obtains energy from *inorganic substances by *oxidation.

chernozem A dark-coloured *soil that is common under *grassland in cool, temperate climates. It is a productive, well-developed soil with a thick *A-horizon which is rich in *organic matter and exchangeable *calcium, and a zone of *calcium carbonate accumulation beneath it. Also known as **dark earth.**

CHERNOBYL

A *nuclear power plant in the Ukraine (former Soviet Union) that was the site of the world's worst nuclear power accident in 1986. One of the four *nuclear reactors at the plant was badly damaged by a chemical explosion after it seriously overheated and went out of control. Highly *radioactive and dangerous *fission products poured out uncontrollably into the atmosphere over a period of ten days while the struggle was underway to plug the leak. This material created an invisible radiation cloud which blew over much of Western Europe within days, and many areas were contaminated by dry *fallout (of particles and dust) and when rain washed the material back to the ground. More than 1000 people in the area were injured, 31 were killed and 135 000 people were evacuated from surrounding towns and villages. Up to 100 million people across Europe received some exposure to low levels of radiation (particularly *iodine-131 and *caesium-137) from the accident. The whole world is believed to have received a *radiation dose equivalent to all previous years' fallout from *nuclear weapons, and some individual doses in Britain were equivalent to one year of natural background radiation. In the area around Chernobyl there has since been a dramatic rise in the incidence of thyroid diseases, *anaemia, *cancer and symptoms of *radiation sickness (including fatigue, loss of vision, and appetite) amongst humans, and some major birth defects among livestock.

chert A hard, fine grained siliceous *sedimentary rock composed of interlocking *quartz crystals, that varies in colour between white, pink, brown, and grey, and is made from the skeletons of microscopic marine organisms.

chestnut blight A disease of American chestnut trees caused by a *fungus (*Cryphonectria parasitica*) that was accidentally introduced to the USA around 1900–8, either in imported chestnut lumber or in imported chestnut trees. By 1940 the disease had made the American chestnut virtually *extinct.

chestnut soil A type of *soil that has a relatively thick, dark brown *A-horizon over a lighter coloured *B-horizon.

Chihuahuan Desert A North American hot *desert, which is located mostly in Mexico but extends into southern New Mexico and Western Texas.

chimney A vertical *flue or *stack.

china clay A fine, pure *clay which is non-plastic and takes the form of a white powdery material made by the chemical decomposition of *feldspar in *granite. Also known as **kaolin**.

China Syndrome Non-scientific expression for the meltdown of a nuclear reactor, based on the science-fiction idea that an immensely hot damaged reactor could vaporize (melt) its way through the centre of the Earth from the USA to China.

chinook A warm, dry *adiabatic wind that blows down the eastern slopes of the Rocky Mountains in North America. *See also* FOEHN, SANTA ANNA.

Chipko Movement A local *environmental movement that began in India in the early 1980s, when a group of village women engaged in *direct action to save the forest on which their livelihoods depended by literally hugging the trees in order to prevent *deforestation. *See also* ECO-ACTIVISM.

chitin A long chain of *carbohydrates (a polysaccharide) that is used for structural support in *invertebrates, for example in insect *cuticle.

chitin inhibitor An *insecticide that works by stopping the formation of *chitin, so that when the insect starts to moult it dies.

chlorides (Cl⁻) **1.** The *anions that are formed when the element *chlorine picks up an electron.
2. The salts of hydrochloric acid that contain chloride ions.

chlorinated hydrocarbon See ORGANOCHLORINE.

chlorinated solvent An organic solvent that contains *chlorine atoms. Such solvents are used in aerosol spray containers, highway paint, and dry cleaning fluids.

chlorination **1.** The addition of *chlorine to drinking water, *sewage, or industrial waste in order to disinfect it or to *oxidize unwanted compounds within it.
2. The addition of *chlorine atoms to compounds by chemical reactions.

chlorine (Cl) A highly reactive *halogen gas that is added to drinking water to kill *bacteria and *algae, and used as a raw material for products such as plastics, pharmaceuticals, and *pesticides. It is a very *toxic *biocide, poisonous to fish and invertebrates, persistent in the environment, and important in the destruction of *ozone.

chlorocarbon A compound (such as *carbon tetrachloride) that contains *carbon and *chlorine and destroys *ozone.

chlorofluorocarbons (CFCs) A group of odourless, inert, synthetic, non-toxic, and easily liquefied chemical compounds consisting of *chlorine, *fluorine, and *carbon. They are very long-lasting (50 to 200 years) and very efficient absorbers of *infrared radiation. Until the early 1990s CFCs were widely used as propellants in aerosol cans, as refrigerants in refrigerators and air conditioners, and in the manufacture of foam boxes for take-away food cartons. CFCs are important *greenhouse gases that are broken down by strong *ultraviolet light in the *stratosphere and release chlorine atoms that then deplete *ozone. Since the early 1990s most industrial countries (including the USA and UK) agreed under the *Montreal Protocol to phase out production of CFCs and a range of other ozone-depleting chemicals by the year 2000. There are a number of important species of CFC, including CFC-11 and CFC-12 which remain active in the atmosphere for 50 to 100 years, the concentration of which is increasing at a rate of more than 5% a year, CFC-113, the atmospheric concentration of which is increasing at a rate of about 10% a year, and CFC-22, which remains active in the atmosphere for about 15 years, the concentration of which is increasing at a rate of about 11% a year.

chloronicotinyl A systemic form of *insecticide that works by disrupting the central nervous system of an insect.

chlorophyll A green pigment that exists in plants and transforms light energy into chemical energy in *photosynthesis.

chloroplast A structure that is found in some cells of plants and contains *chlorophyll.

chlorosis A discoloration of plant leaves from green to pale green or yellow, where production of *chlorophyll is prevented. It is caused by disease, lack of essential nutrients, or exposure to various air pollutants.

CHM See CLEARING-HOUSE MECHANISM.

cholera An acute infectious *disease caused by a *bacterium that is transmitted in drinking water contaminated by faeces. It causes severe vomiting and diarrhoea, leading to dehydration and death if left untreated.

C-horizon The mineral layer beneath the *B-horizon within a *soil profile, which is relatively unaffected by biological activity and is composed of only partially weathered material. See also SOIL HORIZON, A-HORIZON.

choropleth map A map showing discrete areas, such as counties, soil units, and vegetation types. Uniform conditions are assured within each map unit.

chromatography An analytical technique that is used for the chemical separation of mixtures and substances.

chromium (Cr) A naturally occurring, hard, brittle, greyish, non-toxic *heavy metal, that is resistant to corrosion and tarnishing. It enters surface waters in wastewater from electroplating operations, leather tanning industries, and textile manufacturing, and is harmful to aquatic organisms. One of eleven *pollutants of concern, it is corrosive to skin and *carcinogenic in humans.

chromosome A package of *genes in the nucleus of a *cell, composed of *DNA and *proteins, which contains the genetic information for that cell. Humans normally have 46 chromosomes in 23 pairs, and children get half of their chromosomes from each parent.

chronic Long-lasting or taking a long time to manifest itself. Contrast ACUTE.

chronic effect An adverse effect on a human or animal (for example, of exposure to a *toxin) that can result from either one *acute exposure or a period of continuous, low-level exposure.

chronic exposure Continuous or repeated exposure of a human or animal to a hazardous substance over a long period of time, usually at least seven years. Contrast ACUTE EXPOSURE.

chronic respiratory disease A persistent or long-lasting intermittent disease of the respiratory tract.

chronic toxicity The effects of repeated or long-term exposure to a *toxic substance, which can often become apparent only after many years. Contrast ACUTE TOXICITY.

chronology The order of events according to time. See also ABSOLUTE CHRONOLOGY, RELATIVE CHRONOLOGY.

chronosequence A sequence of *soils that changes gradually from one to the other through time.

chronostratigraphy The organization of layers of *rock or *sediment into units, based on age or time.

chrysalis The *pupa of a butterfly or moth, which is not covered in silk. *Contrast* COCOON.

CI *See* CONSERVATION INTERNATIONAL.

cinder cone A steep-sided conical hill that is built of coarse *ash material throw out of a volcano by escaping gases.

circadian rhythm Biological activity that occurs at 24-hour intervals, related to sleep cycles. *See also* DIURNAL.

circle of influence The circular outer edge of a depression in the surface of a *water table that is caused by pumping water from a well. *See also* CONE OF INFLUENCE, CONE OF DEPRESSION.

circle of life The idea that all livings things need other living things to exist and prosper, and so they are all linked together.

circumboreal Found throughout the high latitude forests of the northern hemisphere, in North America and Eurasia.

circumpolar Found in an area around either the *North Pole or the *South Pole.

cirque Bowl-shaped depression in an upland area, that has been eroded by glacial *quarrying processes at the head of valleys in which *glaciers form. There are many regional names for a cirque, including *corrie (Scotland) and *cwm (Wales).

cirque glacier A small *glacier within a *cirque.

cirriform High-altitude ice clouds that appear very thin and wispy.

cirrocumulus (Cc) A thin, *cirrus *cloud characterized by thin, white patches that contain ice crystals and look like cotton wool. These clouds are high, stretching between about 6000 and 12 000 metres (20 000 to 40 000 feet). They often mark the leading edge of a *frontal system and indicate that major weather changes are likely.

cirrostratus (Cs) A thin, white, *cirrus *cloud that appears in layers and looks like a flat sheet. Sometimes it covers the whole sky and is so thin it can hardly be seen. These clouds are high, stretching between about 6000 and 12 000 metres (20 000 to 40 000 feet), and they contain ice crystals which can refract moonlight and shine like a halo. They are often associated with weather changes.

cirrus (Ci) High, thin, wispy, feather-like *clouds, usually between about 6000 and 12 000 metres (20 000 to 40 000 feet), that are composed of ice crystals which diffuse sunlight or moonlight. Cirrus clouds are associated with fair weather, but thick cirrus clouds bring rapid weather changes.

cistern A small tank that is used to store water for a home or farm, and is often used to store rain water.

CITES *See* CONVENTION ON INTERNATIONAL TRADE IN ENDANGERED SPECIES.

CITES species *Species that are listed under the 1975 *Convention on International Trade in Endangered Species, which cannot be commercially traded as live specimens or wildlife products because they are *endangered or threatened with *extinction.

city A large, densely populated *urban settlement, larger than a *town, which can include two or more independent administrative districts within it and usually has *suburbs.

city planning *See* URBAN PLANNING.

CJD *See* CREUTZFELDT–JAKOB DISEASE.

clade A group of *species which are all descended from a single common ancestor.

cladistic species concept The definition of *species as a lineage of populations between two *speciation events (*phylogenetic branch points), based on the branch points not on how much change has occurred between them. *See also* BIOLOGICAL SPECIES CONCEPT, ECOLOGICAL SPECIES CONCEPT, PHENETIC SPECIES CONCEPT, RECOGNITION SPECIES CONCEPT.

cladistics A method of classifying *organisms into groups (*taxa) that is based on order of evolutionary branching rather than on present similarties and differences. Also known as **phylogenetic systematics**.

cladogenesis The development of a new *clade by the splitting of a single lineage into two distinct lineages which are *taxonomically different. *See also* ADAPTIVE RADIATION, SPECIATION.

cladogram A branching tree diagram, based on *cladistics, that shows the apparent sequence of branching by which individual *clades have come into being and charts the emergence of a particular *taxon or taxa. It shows the sequence but ignores the time-scale of evolutionary divergence.

clarification The clearing action that happens during water treatment when solids settle out. It can be speeded up by *centrifugal action and chemically induced *coagulation.

clarifier A tank, used in water treatment, in which solids settle to the bottom and are removed as sludge. Also known as **settling basin** and **sedimentation basin**.

clarity The degree of transparency of a liquid such as water.

Clarke-McNary Act (1924) US legislation that authorized technical and financial assistance to states for forest fire control and for production and distribution of forest tree seedlings. Some sections of the act were repealed by the *Cooperative Forestry Assistance Act (1978).

class 1. In *taxonomy, a group of related and similar orders. It is the category above an *order and below a *phylum, so each phylum comprises more than one class, and each class comprises more than one order.
2. A person's ranking in a social hierarchy, largely based on access to wealth and influence.

class action suit A North American term for a lawsuit in which one or more parties file a complaint seeking damages for injury or loss on behalf of themselves and all other people that have an identical interest in the alleged wrong doing

Class I area A geographical area designated by the US *Clean Air Act where only a small amount of deterioration, or an increase in deterioration, of air quality is permitted. Examples include national parks, wilderness areas, monuments, and other areas of special national and cultural significance.

Class I Substance One of several groups of chemicals that have a *global warming potential of 0.2 or higher, including *chlorofluorocarbons and *halons.

Class II Substance A substance that has a *global warming potential of less than 0.2, including *hydrofluorocarbons.

classical economics An economic theory based on the principles that both individuals and society prosper most with a minimum of political intervention, and that the allocation of a scarce resource is best decided through competition for supply and demand of goods and services in the marketplace. This is the foundation of modern, western economic theory and the basis for *capitalism. *Compare* ECOLOGICAL ECONOMICS, ENVIRONMENTAL ECONOMICS. *See also* ECONOMICS.

classification A means of arranging species on the basis of degree of evolutionary relatedness, in which members of a particular group share a more recent common ancestor with one another than with the members of any other group. This classification is based on order of evolutionary branching rather than on present similarities and differences. The most common system of *taxonomy for classifying organisms was developed by

*Linnaeus, who divided the plant kingdom into several *divisions, and the animal kingdom into *phyla (singular phylum). Divisions and phyla are, in turn, divided into *classes, which are further subdivided into *orders. Orders are divided into *families, which are subdivided into *genera (singular genus), and genera into *species. Thus the normal classification scheme is hierarchical, going down from the kingdom (highest level) to the species (lowest level): kingdom → division/phylum → class → order → family → genus → species. Also known as **phylogenetic systematics**. *See also* CLADISTICS, EVOLUTIONARY CLASSIFICATION, PHENETIC CLASSIFICATION, TAXONOMY.

clast An individual mineral grain, constituent, or fragment of broken down rock.

clastic Often refers to fragmentary sedimentary rocks that are made up of fragments of pre-existing rocks (*clasts).

clay A fine-grained mineral *soil or *sediment with a *particle size of less than 0.002 mm (greater than 9.0 on the *phi (ϕ) scale).

clay pan A highly compacted *clay soil horizon. *See also* HARDPAN.

clay soil Any soil material that contains more than 40% *clay, less than 45% *sand, and less than 40% *silt.

Clean Air Act (1963) US legislation that gave the *Environmental Protection Agency authority to set standards for air quality and to control the *emission of *air pollutants from industries, power plants, and cars, in order to protect public health. The act was passed in 1963 and revised in 1970 and 1990. *See also* ACID RAIN PROGRAM, PREVENTION OF SIGNIFICANT DETERIORATION, NATIONAL EMISSIONS STANDARDS FOR HAZARDOUS AIR POLLUTANTS.

Clean Air Act Amendments (1970) US legislation that built on the *Clean Air Act (1963) and expanded the federal role in setting and enforcing ambient *air quality standards, including regulating land management practices to achieve and maintain such standards. It set *emissions standards for stationary sources (such as factories and power plants), motor vehicles (automobiles and trucks), *criteria pollutants (including *lead, *ozone, *carbon dioxide, *sulphur dioxide, *nitrogen oxides, and *particulate matter), and *toxic pollutants.

Clean Air Act Amendments (1977) US legislation that amended the *Clean Air Act (1963) by establishing the national goal of preventing any future reduction of visibility of Class I federal areas (which include all International Parks, all National Wilderness Areas that exceed 5000 acres, all National Memorial Parks that exceed 5000 acres, and all National Parks that exceed 6000 acres) from *air pollution caused by humans.

Clean Air Act Amendments (1990) US legislation that builds on the *Clean Air Act Amendments (1977) in an attempt to reduce *smog and *air pollution, particularly by dealing with problems such as *acid rain and *ozone depletion. The law allows the *Environment Protection Agency to set limits on how much of a particular pollutant can be in the air anywhere in the USA, but it allows individual states to have stronger pollution controls, and requires them to develop state implementation plans which detail how they aim to achieve the national targets.

Clean Development Mechanism (CDM) One of the *flexibility mechanisms defined in the *Kyoto Protocol for dealing with *air pollution by *greenhouse gases. This involves a project between a developed country and a developing county in which the former is helped to comply with its *emissions reduction commitments in return for providing the latter with financing and technology to help with sustainable development. *See also* EMISSIONS TRADING, BUBBLE.

clean fuel *Alternative fuels such as *gasohol, *natural gas, and *liquefied petroleum gas, which emit less pollution than conventional fuels such as *gasoline and *diesel.

Clean Water Act (1972) US legislation that regulates water pollution. It gave the *Environment Protection Agency the right to set and enforce uniform national standards of water quality in order to restore and maintain the chemical, physical, and biological integrity of US surface waters.

cleanup The treatment or *remediation of an area of material that has been contaminated with a hazardous substance. *See also* CORRECTIVE ACTION, ENVIRONMENTAL RESTORATION.

clear A sky condition of less than one-tenth cloud cover.

clear air turbulence (CAT) The erratic movement of air masses without any visual signs (such as clouds), which is caused when bodies of air moving at widely different speeds meet. It can cause violent shaking of aircraft.

Clear Skies Initiative (2002) An initiative from US President George Bush that is designed to create a mandatory programme that would dramatically reduce emissions of *sulphur dioxide (SO_2), *nitrogen oxides (NO_x), and *mercury from *power plants, by setting a national limit on each *pollutant.

clear well A reservoir for the storage of previously treated *potable water, before it is distributed to users.

clearance A patch of land that has been cleared of vegetation in order to prevent the spread of *wildfires.

clearcut 1. A forest management technique in which all or most of the trees are felled and removed at one time.

2. A previously forested area after all or most of the trees have been removed by *clearcutting.

clearcutting The practice of felling and removing most or all of the trees in a stand of *forest at the same time. *See also* DEFORESTATION.

clearing 1. A relatively small treeless area in the middle of a *wood or *forest.

2. The process of disposing of undergrowth and debris left after trees have been felled and trimmed, usually by burning it.

clearing index *See* VENTILATION INDEX.

clearing-house A service or organization associated with an exchange between a large number of traders, where trades are confirmed, matched, and settled.

Clearing-House Mechanism (CHM) An international network of agencies that work together to implement the *United Nations Framework Convention on Biological Diversity by compiling and exchanging information on biodiversity around the world, and by promoting technical and scientific cooperation at all levels.

cleavage The ability of a *mineral or rock to split (cleave) along planes of weakness.

cleavage plane The smooth flat surface along which *minerals or rocks tend to break (*cleavage).

cleptobiosis An association between *species that is found in some social organisms, in which one species steals food from the stores of another species but does not live or nest close to it. Also known as **lestobiosis.**

cliff A very steep, high face of rock, sediment, or soil, which might be vertical or even overhanging.

CLIMAP *See* CLIMATE, LONG RANGE INVESTIGATION, MAPPING AND PREDICTION PROJECT.

climate The long-term average *weather conditions of a place, in terms of *precipitation, *temperature, *humidity, sunshine, and *wind velocity and phenomena such as *fog, *frost, and *hail storms. These are determined by factors that are fixed through time, such as latitude, position relative to oceans or continents, and altitude. Climate changes from place to place, but much more slowly than weather because *climate zones are usually quite large. It also changes through time, again much more slowly than weather, over decades rather than days.

climate change Any natural or induced change in *climate, either globally or in a particular area. Examples include the natural climate change that has caused *ice ages in the past, and *global warming that is now being caused by rising concentrations of *greenhouse gases in the *atmosphere. By AD 1700 most of the standard *meteorological instruments that are used to measure *temperature, *precipitation, and other climate variables had been invented (they have since been greatly improved), although good quality records going back centuries are available for only a few sites. Direct measurements of *weather have been collected regularly only since about the 1850s, and many parts of the world (particularly in the tropics and southern hemisphere) only have records covering much shorter periods available. Long-term changes in climate can be reconstructed using historical evidence, and geological and biological *proxy indicators which include analysis of *ice cores, oxygen *isotope analysis, *pollen analysis, and *dendroclimatology.

Climate Change Convention See UNITED NATIONS FRAMEWORK CONVENTION ON CLIMATE CHANGE.

Climate Change Levy A tax on the use of energy in industry, commerce, and the public sector in the UK, which is designed to encourage these sectors to make more efficient use of energy and to assist the UK to meet its reduced *greenhouse gas emissions commitments under the *Kyoto Protocol. There are different rates of levy for electricity generated from gas, coal, and liquefied petroleum gas. The levy came into effect on 1 April 2001 and is a key part of the UK Government's overall climate change programme. It is expected to lead to reductions in *carbon dioxide emissions of at least 2.5 million tonnes of carbon a year by 2010.

Climate Change Protocol A series of international meetings held in Bonn, Germany between 1995 and 1997 to decide how best to address the problems associated with *greenhouse gas emissions and *global warming. It led to the *Kyoto Protocol.

Climate Council A US non-profit *anti-environmental *front group, backed by self-interested funders including the coal industry, railways, and utilities, which has strongly opposed the *Kyoto Protocol and actively worked behind the scenes against most US government efforts to address *climate change.

climate lag The delay or time it takes for any factor that promotes *climate change (such as an increase or decrease in *greenhouse gas emissions) to produce a change that can be measured. Delay is often caused by *adjustment and *feedback in complex *environmental systems.

climate model A simulation of how the *climate system works, usually using large and complex computer programs that are based on mathematical equations derived from knowledge of the physics that governs the Earth–atmosphere system. See also GENERAL CIRCULATION MODEL.

climate modification Changes of the climate that can occur as a result of either natural processes (such as *volcanic eruptions, *El Niño–Southern Oscillation events, or *sunspot activity) or human activities (such as the effects of *pollution and *deforestation). See also CLIMATE CHANGE.

climate sensitivity A measure of how rapidly and strongly the Earth's climate is likely to react to increases or decreases in radiation (*radiative forcing), such as those caused by the greenhouse effect, and reach a new *equilibrium state. See also SENSITIVITY.

climate system The five interacting environmental systems (*atmosphere, *hydrosphere, *cryosphere, *lithosphere, and *biosphere) that are responsible for the climate and its variations.

climate zone An area with generally consistent climatic characteristics. From the equator towards the poles there are the *tropical, *mid-latitude, *subpolar, and *arctic climates.

Climate, Long Range Investigation, Mapping and Prediction project (CLIMAP) An international research project during the 1970s and 1980s that reconstructed the Earth's climate over the last million years and produced maps of sea-surface temperature at various periods, based on analysis of *proxy data from ocean sediment cores.

climatic analogue A climatic event or situation from the past that displays similar characteristics to one occurring in the present. Comparing the two is often helpful in projecting how the present event might develop in the future. *See also* ANALOGUE.

climatic anomaly The deviation or departure of a particular climatic variable from the norm, as defined over a specified period of time.

climatic climax *See* CLIMAX COMMUNITY.

Climatic Optimum The warm, moist climatic phase during the *Holocene, between about 7500 and 5200 years ago, when average global temperatures were 1–2°C warmer than they are today, glaciers and ice sheets receded greatly from their post-glacial maximum, and the *meltwater raised sea level by about three metres. Also known as the **Atlantic Period.** *See also* ALTITHERMAL, HYPSITHERMAL.

climatology The study of *climate, climate processes, and *climate change.

climax community The natural vegetation that exists in an area at the end of a *succession. This develops over a long enough period of time without major environmental change or human interference, for example *woodland in a *temperate area. Climax is indicated by a *community of vegetation that is stable, self-maintaining, and self-reproducing and in *equilibrium with its environment. Its species composition is stable through time, so long as its environment remains unchanged, because

the growth of other species is inhibited by competition for food, energy, *nutrients, and space. It would normally take a long time to re-establish a climax community if it is destroyed or seriously damaged. Also known as **climatic climax** or **climax vegetation.** *See also* BIOME, FIRE-CLIMAX COMMUNITY.

climax forest A forest *community that represents the final stage (*climax community) of natural forest *succession for its *environment.

climax vegetation *See* CLIMAX COMMUNITY.

climber A vine or climbing plant that grows up surrounding structures such as the branches of trees and shrubs for support and in order to raise its foliage and flowers above the ground.

cline A gradual change in a plant population within a given species over a geographical area. It is genetically based and shows *adaptation to an *environmental gradient.

clique A group of species that has a similar diet but is different from other such groups.

clod A compact mass of *soil produced by disturbance such as digging or ploughing.

clone A genetically identical copy of an *organism, derived from creating or reproducing a group of *cells or *DNA molecules from a single parent or ancestor. Most *plants, *fungi, *algae, and many other organisms naturally reproduce by making clones of themselves.

cloning The process of making genetically identical copies (*clones) of a *gene or *organism. Cloning is an essential element in *genetic engineering and *biotechnology. Growing cuttings from a plant is a traditional form of cloning.

close management measure Any technique that is used to aid the *reintroduction and establishment of viable populations of a species into its natural *habitat, such as artificial nesting boxes.

closed area An area that is closed to people, where specified activities are temporarily restricted, or where resource extraction (including harvesting) is temporarily or permanently banned, except by special permit for specific purposes.

closed canopy A *woodland or *forest in which tree *crowns spread over at least 20% of the ground, often touch one another, and shade out light the forest floor. *See also* CROWN CLOSURE.

closed forest A forest with a *closed canopy.

closed season A specified period of time each year during which particular *game species (such as certain fish) cannot be legally caught.

closed system A *system that can exchange energy but not matter with its surroundings. The global *water cycle is a closed system because there is a fixed amount of material (water) within the system, and the circulation of water is maintained by energy from the Sun (which is imported across the system boundary). *Contrast* OPEN SYSTEM.

closed-loop recycling The reuse of *wastewater for purposes other than drinking in an enclosed process.

closure 1. Any structure that is designed to close off the opening of a container and prevent loss of its contents.
2. In North America, the process of preparing a *landfill site for closure, which involves covering it with soil and vegetation such as grass.

cloud A mass of water droplets and ice crystals in the air, formed by *condensation of *water vapour around nuclei such as dust, salt, and soil particles as a mass of air rises and cools. There are three main types of cloud: *stratus, *cumulus, and *cirrus.

cloud base The lowest level in the atmosphere that contains *cloud droplets and at which clouds can be seen. The bottom of a cloud.

cloud condensation nuclei Aerosols (including sea salts and byproducts of combustion) on which water vapour condenses to form *cloud droplets.

cloud cover The extent of *cloud in the sky at a given place and time, which is usually measured in eighths. A coverage of four-eighths means that half the sky is covered by cloud.

cloud deck The top of a *cloud layer, usually viewed from an aircraft.

cloud droplet A particle of liquid water that is created by the *condensation of *water vapour onto a *cloud condensation nucleus. Cloud droplets are usually smaller than *raindrops.

cloud forest A high mountain forest in which the temperature remains uniformly cool and fog or mist keep vegetation permanently damp. It is covered in cloud for most of the day.

cloud seeding The deliberate introduction of artificial substances (usually silver iodide or *dry ice) into a *cloud in order to change the way it develops or to increase *precipitation, as a form of *weather modification. Also known as artificial rain.

cloud street A row of *clouds that are aligned parallel to the wind.

cloud type Clouds are usually classified by height into low, middle, and high. Low clouds, which form from ground level up to about 2000 metres, include *stratus, *cumulus, *cumulonimbus, *nimbostratus, and *stratocumulus. Middle clouds, which form between 2500 and 6000 metres, include *altostratus and *altocumulus. High clouds, which form above 6000 metres and up to about 10 600 metres, include *cirrocumulus, *cirrostratus, and *cirrus.

cloudburst A sudden, heavy shower of rain, that is usually intense but short-lasting.

cloudy The state of the sky when nine-tenths or more is covered by *clouds.

Club of Rome A non-profit, non-governmental, global *think-tank that brings together scientists, business people, and bureaucrats who share an interest in tackling what the Club calls the *world problematique. One of its most famous reports is *Limits to Growth*.

cluster zoning A type of *open space zoning that allows a developer to reduce the minimum residential lot size below that required in the *zoning ordinance, if the land so gained is preserved as permanent open space for the community.

clutch A number of birds that are hatched at the same time.

CNG See COMPRESSED NATURAL GAS.

CO See CARBON MONOXIDE.

co-adaptation The evolution, by *natural selection, of characteristics in two or more *species that are to their mutual advantage.

coagulant Any chemical that helps *colloidal particles to group together irreversibly to form larger masses. For example, chemicals such as *lime, *alum, and iron salts are often added to *wastewater to help settle out impurities.

coagulation The process in which *colloidal particles group together irreversibly to form larger clusters.

coal A *carbonaceous deposit that forms from the remains of *fossil plants. The plants are deposited initially as *peat, which (after compaction and burial) undergoes physical and chemical changes and is progressively turned into coal, to produce the *coal series. Coal is the dirtiest of the *fossil fuels when burned, and it contributes to the atmosphere a significant amount of *particles and *aerosols, *greenhouse gases, and *sulphur dioxide. There are much bigger reserves of coal and *lignite in the world than there are of *oil and *natural gas. Roughly half (48%) of the world's known coal reserves are in the former Soviet Union, Eastern Europe, and China, and there are also sizeable reserves in Western Europe (9%), Africa (6%), North America (26%), and Australia and Asia (9%). Known reserves are thought to be large enough to meet projected world demand

for up to the next 500 years, but the life expectancy will fall sharply if large amounts of coal are used to make synthetic liquid fuels as a substitute for oil. See also ANTHRACITE, BITUMINOUS COAL.

coal cleaning technology The physical or chemical treatment of *coal before it is burned, in order to remove some of its *sulphur content and thus help to reduce *sulphur dioxide emissions (to reduce *acid rain). See also COAL WASHING.

coal gasification The conversion, by heating and partial combustion, of *coal into gases such as *methane and *carbon monoxide. It is an efficient fuel that produces less *air pollution than burning unmodified coal. Coal gasification was a major energy source in the 19th and early 20th centuries but production declined after the availability of *natural gas increased in the 1970s. *Coal tar was a byproduct.

coal liquefaction The conversion of coal into a liquid *hydrocarbon fuel such as synthetic gasoline or methanol.

coal measures A *stratigraphic unit that usually contains coal, occurring in the upper *Carboniferous in Europe (the Pennsylvanian in North America).

coal series The different types of *coal, ranging from *peat, through *bitumen and *lignite, to *anthracite. Successive coals in the series contains less moisture, fewer *volatiles, and more *carbon.

coal tar A thick black liquid *hydrocarbon that is distilled from *coal and contains organic chemicals. See also COAL GASIFICATION.

coal washing A *coal cleaning technology that involves crushing coal and washing soluble sulphur compounds out of it with water or other solvents.

Coalition for Vehicle Choice A US non-profit *anti-environmental *front group, based in Washington, DC, that was 'created to preserve the freedom of Americans to choose motor vehicles that meet their needs and their freedom to travel'. It claims a membership of more than 40 000 state and local organizations and individuals, including vehicle and tyre manufacturers.

coarse filter approach An approach to the *conservation of *biodiversity that involves maintaining a diversity of structures within stands and a diversity of *ecosystems across the landscape, in order to meet most of the habitat requirements of most of the native species. *See also* FINE FILTER APPROACH.

coarse woody debris Bits of wood, including rotting logs and stumps, that provide *habitat for plants, animals, and insects, and *nutrients for soil development.

coarse-grained 1. A *rock that has large (five millimetres to three centimetres) grains or crystals.
2. Ecological processes or factors that occur in large patches relative to the activity patterns of an *organism. *See also* FINE-GRAINED.

coast The zone where the land meets the sea, which is the boundary between *terrestrial and *marine *environmental systems.

coastal defence Any activity that is designed to protect a coastal area from erosion and/or flooding. Examples include engineering schemes such as sea walls and embankments. Also known as **coastal protection.**

coastal flooding The flooding of coastal areas by seawater, usually caused by unusually high *tides and/or *storm surges.

coastal pelagic fish Fish (such as mackerel, anchovies, and sardines) that live in the sea at or near the surface of the water but remain relatively close to the coast.

coastal plain A low-lying, relatively flat area of land beside the sea.

coastal processes The action of natural forces on the *shore and the nearshore sea bed.

coastal protection *See* COASTAL DEFENCE.

coastal reef A *coral reef that is located close to the *coast. *Contrast* FRINGING REEF.

coastal upwelling A seasonal process that occurs along many coasts, especially on the western sides of continents, that involves the rising to the surface of cold water from the deep ocean. This brings *nutrients (particularly *nitrates and *phosphates) to the surface, which filters down through the *food chain.

coastal wetland A *wetland habitat (such as a *tidal marsh, *lagoon, *tidal flat, or *mangrove swamp) that is found along a coastline and is covered with saltwater from the ocean for all or part of the year.

coastal zone The shallow part of the ocean that extends from the high-tide mark on land to the edge of the *continental shelf, and in which *marine and *terrestrial *environmental systems interact with one another. *Contrast* OPEN SEA.

Coastal Zone Management Act (1972) (CZMA) US legislation designed to encourage 30 seacoast and Great Lakes states to develop coastal management programmes. The Act established a method for the federal government to approve states' plans and funding for approved plans, and it requires federal actions to be consistent with the states' plans.

coastline The line that separates land from sea. Also known as **shoreline.**

cobalt (Co) A magnetic metallic element that is soluble in water. It is one of eight *trace elements that plants and animals require, and an important *catalyst in *nitrogen fixation.

cobble A rock or particle of sediment that has a *particle size between 64 and 256 millimetres (–6.0 to –8.0 on the *phi (ϕ) scale).

COC *See* CHEMICAL OF CONCERN.

cocoon A covering that protects the eggs and larvae of *invertebrates, such as the spinning of a silk sheath by larvae of moths, in which pupae develop. *Contrast* CHRYSALIS.

COD *See* CHEMICAL OXYGEN DEMAND.

codominant Trees or shrubs with *crowns that receive full light from above but relatively little from the sides, and whose crowns usually form the general level of the canopy.

co-evolution Evolution in two or more *species that interact, in which the adaptive changes of each species influence those of the other.

coffin A thick-walled container, usually made of lead, that is used for the safe transport of *radioactive materials. Also known as a **cask**.

cogeneration The generation of two different and useful forms of energy (usually electricity and steam or hot water) at the same time, from a single fuel source. Also known as **combined heat and power**.

cohesion The force of attraction which holds like molecules together.

cohort A group of individuals of similar age.

cohort study *See* PROSPECTIVE STUDY.

coke A hard, dry *carbon substance that is produced by heating *coal to a very high temperature in the absence of air.

col 1. A notch or natural pass cut through a mountain ridge, usually by *glacial erosion, such as an *arête between two back-to-back *cirques.
2. A region of low *atmospheric pressure that is found between two high-pressure and two low-pressure systems.

Colburn, Theo Senior scientist and director of the Wildlife and Contaminants project at the *World Wildlife Fund, pioneer of low-dose toxicity research, and co-author of the book *Our Stolen Future*. Regarded by some as a latter-day Rachel *Carson.

cold climate A climate that has temperatures of 10° to 20°C for up to four months a year, and is cooler over the rest

of the year, as found in places like the *Arctic and *Antarctica. The two *biomes associated with cold climates are the *tundra and the *taiga.

cold desert A *desert *biome that is hot or warm in the summer and cold in the winter, such as the Gobi Desert. *Contrast* HOT DESERT.

cold front A frontal boundary between two *air masses, where the cooler, denser mass advances, pushing under and replacing the warmer one. Often associated with steady *precipitation followed by *showers. *Contrast* WARM FRONT. *See also* FRONT, FRONTAL LIFTING.

cold occlusion *See* OCCLUDED FRONT.

cold pole The place that has the lowest annual mean temperature in its *hemisphere.

cold sector The region of cold air in a *depression that is in contact with the surface.

cold-blooded An animal whose body temperature changes with the external environment, and is not internally regulated. Also known as an **ectotherm**. *Contrast* WARM-BLOODED.

cold-water fishery A lake or stream that supports species of fish (such as salmon and trout) which are intolerant of sustained water temperatures higher than about 21°C in the summer.

coliform bacteria *Bacteria such as *E. coli* and *Salmonella* that live in the gastrointestinal tracts of vertebrates and indicate the presence of *faeces in water, exposure to which causes diseases such as *cholera.

coliform index A measure of the purity of water based on a count of how many faecal bacteria are present within a given volume of it.

coliforms *See* COLIFORM BACTERIA.

colliery A coal mine and associated coal processing facilities.

collision theory A theory of chemical reactivity. Collision theory assumes that, in order for reaction to occur, molecules must collide with at least some minimum energy and with proper orientation.

collision zone The zone along which adjacent convergent *crustal plates push into each other. *See also* PLATE TECTONICS.

colloid A suspension in a liquid of tiny particles, that neither dissolve nor settle.

colluvial Composed of or relating to *colluvium. *Compare* ALLUVIAL.

colluvial deposit *See* COLLUVIUM.

colluvium Loosely compacted sediment that has moved downhill and accumulated on the lower slopes and at the bottom of a *hill, as a result of *weathering, *erosion, and *mass movement processes. Also known as **colluvial deposit**. *Compare* ALLUVIUM.

colonial Of or relating to a *colony.

colonization 1. The establishment of a new *colony.
2. The arrival and establishment of plants and animals on a new area of land. *See also* ISLAND BIOGEOGRAPHY, SUCCESSION.

colony 1. A group of organisms of the same *species that live together, such as *algae, seabirds, or people. *See also* AGGREGATION.
2. A geographical area that is politically controlled by a distant country.

co-management An approach to the management of natural resources which is based on the sharing of authority, responsibility, and benefits on a cooperative basis, either informally or legally, between different stakeholders, such as local government and local communities.

combined heat and power *See* COGENERATION.

combined sewer A sewer system that carries both *sewage and storm-water *runoff.

combustion Burning (rapid *oxidation) of fuel, which releases energy in the form of heat and light, and also releases pollutants (such as *sulphur dioxide, *nitrogen oxides, and *particulates).

combustion chamber A place where *combustion takes place, for example in an *incinerator or an *internal combustion engine. *See also* FURNACE.

combustion nucleus A *condensation nucleus that is formed as a result of *combustion.

comet A small astronomical object similar to an *asteroid but made largely of ice.

command and control An approach to *environmental regulation that relies on the ability of the state to regulate (*control) the behaviour of individuals, companies, and other groups by setting and imposing controls, rather than by *market-based mechanisms. Control is usually exercised through *legislation, introduction of *regulations that require specific action, and setting targets and defining how to meet them, sometimes including what technologies to adopt (for example in order to regulate emissions of *pollutants). *Contrast* ALTERNATIVE COMPLIANCE.

commensal A harmless *parasite.

commensalism A form of *symbiosis between two different species in which one benefits and the other neither benefits nor is harmed. An example is *epiphytes such as orchids, which use other plants for support.

commensurable Values or benefits that can be measured by a common standard (for example in dollars or cubic metres) and objectively compared. *Contrast* INCOMMENSURABLE.

commercial forestland *See* TIMBERLAND.

commercial reactor A *nuclear reactor that is used for commercial purposes to generate electricity, rather than for military purposes or scientific research.

commercial sector Non-manufacturing parts of the *economy, which includes leisure and recreation, health, and education. *See also* INDUSTRIAL SECTOR, RESIDENTIAL SECTOR, TRANSPORTATION SECTOR.

commercial thinning In *forest management, cutting and selling trees that are large enough to be sold as products (for example as poles or fence posts), which thins the forest and helps to improve the health and rate of growth of the crop trees that remain.

commercial waste All *solid waste that is generated by businesses (which includes stores, markets, office buildings, restaurants, shopping centres, and theatres).

commercial waste management facility A facility that accepts waste (for treatment, storage, disposal, or transfer) from a variety of sources, unlike a private facility which usually manages waste that it generates itself.

commercial whaling The hunting of *whales in order to sell whale products.

commercial/industrial land A North American category of *land use that covers land used primarily for buying, selling, and processing goods and services. This includes sites for stores, factories, shopping centres, and industrial parks.

commingled recyclables Mixed waste materials that can be *recycled and are collected together.

comminution 1. The shredding or pulverizing of waste in order to reduce its size, as used in *solid waste management and *wastewater treatment.
2. The extraction of valuable *minerals from their *ores by crushing and grinding.

commission 1. The granting of authority to undertake certain functions (for example, to open and operate a *nuclear plant that has been built). *Contrast* DE-COMMISSION.
2. A special group that has been set up and authorized to consider some particular matter, such as the *Independent Commission on International Development and the *Commission on Sustainable Development.

Commission on Sustainable Development (CSD) *See* UNITED NATIONS COMMISSION ON SUSTAINABLE DEVELOPMENT.

commitment The *environmental change that will inevitably occur at some time in the future as a result of human activities that have already taken place because of lags (such as *climate lags) in *environmental systems, or that are likely to occur in a *business-as-usual scenario. Also known as **environmental commitment**. *See also* EQUILIBRIUM WARMING COMMITMENT.

Commitment Period The period of time, defined in the *Kyoto Protocol, during which a country's average level of emissions of *greenhouse gases must be kept within the agreed emission targets. The first Commitment Period covers 2008–12. Also known as **Budget Period** or **Compliance Period**.

commodity Any good that is traded (bought and sold).

Common Agricultural Policy (CAP) A set of regulations and practices, adopted by members states of the *European Union, designed to provide a common, unified policy framework for *agriculture. The overall aim is to increase farm productivity, stabilize markets, ensure a fair standard of living for farmers, guarantee regular supplies, and ensure reasonable prices for consumers. *See also* SET-ASIDE.

common ancestor A species from which two or more different species have evolved.

common law An unwritten body of *law that is based on general custom and usage, and which is recognized and enforced by the courts.

Common Market The original (pre-1994) name for the *European Union.

common name The familiar, non-scientific name for a species of plant or animal, which is widely used. This can vary within and between countries, so there may be many common names for a particular species. *Compare* LATIN NAME.

common property resource *See* COMMON RESOURCE.

common property resource management The management of a particular local *common resource (such as a forest or pasture) by a well-defined group of resource users who have the authority to regulate its use by members and outsiders.

common resource A resource (such as the air, migratory birds, fish in the oceans, and wilderness areas) which is not privately owned. Rights of use are communally shared, so it is freely available, at least in theory, to anybody who wants to use it. *See also* TRAGEDY OF THE COMMONS.

common variety mineral A natural material such as sand, stone, gravel, and clay that has no market value as a metal or for energy production, which is often extracted in large amounts and used by the construction industry.

Commoner, Barry US cellular biologist (born 1917), author of *Science and Survival*, proposer of the four *'laws of ecology', and an influential writer who helped to initiate the modern *environmental movement.

commons An area of open land that is available for common use, such as shared grazing of animals on common *pasture. *See also* COMMON RESOURCE, TRAGEDY OF THE COMMONS.

communal resource management system Resources that are managed by a community for long-term sustainability.

community A group of plant and animal populations that live together in a given area, are adapted to local environmental conditions, and interact with each other. Also known as **ecological community**. *See also* CLIMAX COMMUNITY.

community based conservation A bottom-up (*grassroots based) approach to conservation, usually within the context of *ecosystem management, that is based on two broad concepts—that people who participate in decision-making will be more inclined to implement agreed outcomes, and that people are quite capable of deciding for themselves what the most appropriate solutions should be, provided they are given sufficient information and support. *See also* COMMUNITY BASED NATURAL RESOURCE MANAGEMENT.

community based natural resource management (CBNRM) An approach to the management of *natural resources that is based on engagement with local communities, as a means of focusing attention on natural resource problems or opportunities that require action at community level or that involve the management of shared resources (such as land, forest, water resources, fisheries, and wildlife). *See also* COMMUNITY BASED CONSERVATION.

community based planning An approach to *urban planning which engages citizens and local communities as stakeholders in the development of comprehensive neighbourhood plans.

community ecology The study of the interactions between the populations of organisms within a particular *ecosystem.

Community Environmental Response Facilitation Act (CERFA) A 1992 law in the USA that requires the Federal Government to identify land that it owns that is not contaminated with *hazardous substances and that can be made available for public use with a minimum of *cleanup.

community forestry An approach to *forestry management that includes local people and takes their needs into account. It includes the establishment of village woodlots, *farm forestry, tree planting in private fields, and joint management of forests by communities and governments. *See also* SOCIAL FORESTRY.

community involvement An inclusive and democratic approach to planning and management that allows all *stakeholders to have their concerns heard and taken seriously, by giving them access to reliable and up-to-date information and allowing them to exercise their right to have their say and to protest against decisions they are unhappy with.

community right-to-know Making information about environmental threats (particularly toxic pollution) readily available to anyone who would like access to it.

Community-Right-To-Know Act (EP-CRA) (1986) US legislation on community safety, which was designated to help local communities protect public health, safety, and the environment from chemical hazards. It is based on local emergency planning committees for each district which include representation by fire fighters, health officials, government and media representatives, community groups, industrial facilities, and emergency managers.

community services land A North American category of *land use that covers land used primarily for schools, hospitals, churches, libraries, sewerage and water treatment plants, sanitary land fills, public parking areas, and other community service facilities.

compaction Decrease in volume caused by *compression.

comparative risk assessment A comparison of the *risks associated with two or more *hazards, judged by experts using a common scale.

compartment A subdivision of a site or area, for example a particular part of a *forest or *nature reserve, that is recognized in management plans.

compatible Able to exist together harmoniously. *Contrast* INCOMPATIBLE.

compatible use Different uses of land or other resources which are harmonious and can exist together in the same area, so that one does not inhibit or adversely affect another. *Contrast* INCOMPATIBLE USE. *See also* COMPLEMENTARY USES.

compensation level/point The point within a *water column at which plant *photosynthesis just equals plant *res-

piration. It defines the lower boundary of the *euphotic zone.

compensatory restoration *See* MITIGATION.

competence The ability of water, ice, or air to carry particles of a given size. The competence value is defined by the *viscosity of the medium, so ice has the highest competence and wind the lowest.

competent authority A government agency that is responsible for regulating complex things such as *biotechnology, *biosafety, and *intellectual property rights.

competition Interaction between two or more organisms in the same space, where each requires the same resource (such as food, water, or space) which has a limited supply and where the presence of each is in some way harmful to the other(s). Competition can occur within (*intraspecific competition) and between species (*interspecific competition) and it is an important factor in *adaptation and *evolution by *natural selection.

competitive consumerism A competitive form of *consumerism, driven by the quest for unlimited control of *resources for personal use.

competitive exclusion principle The local extinction of one species by another in the same area, as a result of *competition to occupy the same *niche and to use the same *resources in the same *habitat. Also known as **Gause's principle.**

complementary uses Different uses of land or other resources, each of which benefits from the other(s) where they exist together in the same area.

complex terrain Any highly variable land surface (*terrain), particularly mountains.

compliance coal Any *coal that meets *sulphur dioxide emission standards for air quality without the need for *flue gas desulphurization. Also known as **low sulphur coal.**

compliance monitoring Monitoring that is done to ensure that statutory requirements (such as limits on the amount of a particular *pollutant that can be emitted) are met.

Compliance Period *See* COMMITMENT PERIOD.

compliance schedule An agreement that has been negotiated between a polluter and a government agency, which details dates and procedures by which *emissions will be reduced in order to comply with a regulation.

component An identifiable, functional part of a *system that is defined for a particular reason. Within an *ecosystem, for example, this could be a *species, a *habitat, an individual organism, or even part of an individual, depending on the purpose of the study.

composition The relative proportion of space or *biomass that is accounted for by each *species or system *component in a given area.

compost Completely or partly decayed *organic matter that is dark, odourless, and rich in nutrients. It results from the controlled biological decomposition of solid organic waste materials under *aerobic conditions by *bacteria, *fungi, and other micro-organisms, and is used to fertilize the soil and increase its *humus content.

compound A substance that is made from two or more *elements in fixed proportions.

Comprehensive Environmental Response, Compensation, and Liability Act (CERCLA) (1980) US legislation that provides for the *cleanup of *hazardous substances released to the environment, usually after the substances were disposed of improperly, by giving the *Environmental Protection Agency the authority to clean up abandoned hazardous waste sites. Also known as the **Superfund**.

Comprehensive Environmental Response, Compensation, and Liability Information System (CERCLIS) A database and management system operated by the *Environmental Protection Agency that contains the official inventory of *CERCLA sites in the USA.

comprehensive plan An official plan, developed and adopted by local government, that contains a vision of how the local economy is expected to change and a set of plans and policies to guide *land use decisions. Also known as **general plan** and **master plan**.

Comprehensive Soil Classification System A system for classifying *soils that was developed by the US Department of Agriculture and is now widely used around the world. All soils are divided into 11 *soil orders, which in turn are subdivided into 47 suborders, 185 great soil groups, many subgroups, and a great many soil families and series. *See also* ALFISOL, ANDISOL, ARIDISOL, ENTISOL, HISTOSOL, INCEPTISOL, LATOSOL, OXISOL, ULTISOL, VERTISOL.

compressed natural gas (CNG) *Natural gas stored in a high-pressure container. It provides a clean alternative fuel for motor vehicles because it emits little *hydrocarbon or *ozone-producing material, but it does emit a significant quantity of *nitrogen oxides.

compression The process of becoming smaller or closer together, which increases density. Geological compression causes *folding in rocks.

computational biology *See* BIOINFORMATICS.

computer model A *model of a complex *environmental system that is based on known mathematical relationships between system components and uses a computer to do the complicated modelling, usually in order to predict future states or patterns. Examples include *Daisyworld, *general circulation models, and *Limits to Growth*.

concentrated animal feeding operation (CAFO) An agricultural business in which animals are raised in confined situations and fed an unnatural diet, rather than allowing them to roam and graze freely.

concentrated recreation *Recreation activities associated with developed recreation sites, such as campgrounds, picnic grounds, ski areas, fishing ramps, scenic viewpoints, and interpretive sites. *Contrast* DISPERSED RECREATION, INTENSIVE RECREATION.

concentration The amount of a particular substance that is contained or dissolved in a given amount of another substance or medium (such as air, water, or soil).

concession A defined area of *land that is licensed or leased to a company for a given period of time, for exploration and development of *natural resources (such as *logging) under specified terms and conditions.

concretion A small, hard, local mass of mineral matter (such as *calcite, *gypsum, *iron oxide, or aluminium oxide), often rounded, which is found in *sedimentary rock and often contains a *fossil nucleus. *See also* MANGANESE NODULE.

condensate The liquid that is produced when a vapour is cooled or *compressed, such as the liquid *hydrocarbons that are condensed from *natural gas and oil wells.

condensation The process by which energy is released from a vapour or gas when it changes into a liquid. For example, *water vapour in the air can collect as droplets on a cold surface or in the atmosphere (around *condensation nuclei) to give rise to *precipitation. *Contrast* EVAPORATION.

condensation level *See* LIFTING CONDENSATION LEVEL.

condensation nuclei Minute particles in the air which are derived from *air pollution or from natural processes (such as pollen, salt from sea spray, and dust from volcanic eruptions), around which moisture can collect via the *condensation of *water vapour, to produce *precipitation. Also known as **hygroscopic nuclei.**

condensation point *See* DEW POINT.

condensation trail *See* CONTRAIL.

condition indicator *See* ECOLOGICAL INDICATOR.

conditional instability Stable unsaturated air that will result in instability if *condensation occurs. Also known as **conditionally unstable.**

conditionality A state of being conditional, or qualified by reservations. The granting of development assistant funds to a country might depend on it signing a climate change policy agreement, for example.

conditionally unstable *See* CONDITIONAL INSTABILITY.

conditioned air Air that has been heated, cooled, humidified, or dehumidified to maintain comfort inside a building, vehicle, or other space. Also known as **tempered air.**

conditioned response A response that is learned or altered by conditioning.

conductance *See* CONDUCTIVITY.

conduction The transfer of *heat within a substance or between substances that are in direct physical contact. *See also* ELECTRICAL CONDUCTION.

conductivity A measure of the ability of a sample of water to carry an electrical current, which reflects concentration of ionized substances (*dissolved solids) in the water. Also known as **conductance, electrical conductivity.** *See also* THERMAL CONDUCTIVITY.

conductor Any material that readily transmits (conducts) heat or electricity.

cone of depression A cone-shaped depression in the surface of a *water table, in the vicinity of a *well, that is caused by pumping. *See also* DRAW DOWN.

cone of influence The cone-shaped depression produced in a *water table by the pumping of water from a *well. The surface area included in the cone is known as the *circle of influence of the well.

confined aquifer *See* ARTESIAN AQUIFER.

confinement The act of confining or restraining, such as the techniques that are used to confine a release of hazardous material (such as a spill or leak) to a limited area, and prevent damage elsewhere.

confining unit A layer of rock that is less permeable than those above or below it, and which prevents or restricts the vertical movement of water and pressure.

confluence 1. A place where things merge or flow together (such as where a tributary flows into a river).
2. The process of *convergence.

conformable contact A boundary between two layers of rock that does not indicate a major change in sedimentary conditions.

congener A member of the same *species or *genus.

conglomerate A coarse-grained *clastic *sedimentary rock composed of large rounded clasts (*gravels, *cobbles, *boulders) which are cemented together in a matrix of fine materials.

Congressionally Reserved Area An area of land in the USA that has been reserved by an act of Congress for a specific purpose, such as a *wilderness area or a *wildlife refuge.

conifer A tree or shrub that has needles and produces seeds in cones, for example pine, spruce, or fir. There are 570 species of conifer in the class Gymnospermae; most are *evergreen, but a few (such as larch) are *deciduous. They usually grow faster and develop less dense wood (*softwood) than other trees, and are commercially important as a source of timber for the papermaking, building, and furniture industries.

coniferous forest A forest dominated by *conifer trees, with very few other tree species. It grows in thin, acidic soils that have few nutrients, in climates with long, cold winters, and short, occasionally warm, wet summers. It is the largest terrestrial *biome on Earth, and grows in a broad band at high latitudes across much of North America, Europe, and Asia, where it forms the *climax vegetation, and extends to the southern border of the arctic *tundra. Also knows as **taiga** or **boreal forest.**

conk *See* FRUITING BODY.

connectedness The state of being connected. Used by ecologists to describe the networks of *corridors that link isolated patches of *habitat in an area. *See also* FRAGMENTATION, HABITAT CONNECTIVITY.

conservancy 1. An official term for *conservation of the environment and natural resources.
2. In the USA, a commission that has jurisdiction over fisheries and navigation in a port or river.

conservation assessment An evaluation of the natural *ecosystems and *biomes of a region, with a special focus on the need and potential for *nature conservation and protecting *biodiversity.

conservation biology The multidisciplinary science that deals with *biodiversity.

Conservation Data Center (CDC) In North America, an organization or government programme that compiles, maintains, and disseminates information about *biodiversity in the area it is responsible for. Also known as the **Natural Heritage Program.**

conservation easement An *easement that restricts a landowner to use that particular land for functions that are compatible with long-term *conservation objectives (such as protecting wildlife habitat, agricultural lands, natural areas, scenic views, historic structures, or open spaces). It is a means by which the development rights to that land are secured (by the government), without buying the land itself.

conservation farming Farming practices that are designed to protect *wildlife and *habitats, for example by

CONSERVATION

The planned protection, maintenance, management, sustainable use, and restoration of natural resources and the environment, in order to secure their long-term survival. Conservation recognizes that natural communities of plants and animals are not static, and it involves preventing any development that would alter or destroy natural habitat but not interfering unduly with ecological changes that occur naturally. It differs from *preservation because of the emphasis on positive management, not simply preventing environmental change. Since the 1960s the word conservation has been used more widely to embrace the rational use of all types of natural resources (*biotic and *abiotic). This is a core belief of modern *ecology and central tenet of *sustainable development. There are many arguments in favour of conservation, including economic or utilitarian arguments (plant and animal resources provide the material basis for human life, and the genetic variation in plant and animal species provides the requisite materials for sustaining and improving farm production, forestry, animal husbandry, and fisheries), moral or ethical arguments (we owe it to future generations to leave an environmental heritage or estate for them; all species have a right to live), aesthetic arguments (wildlife is attractive and interesting, and it makes the world pleasurable, e.g. watching birds and butterflies, hearing birdsong, seeing wild flowers and natural forest), and environmental arguments (natural wildlife is important to *environmental systems such as *biogeochemical cycles, on which all life on Earth depends). *See also* EX SITU CONSERVATION, IN SITU CONSERVATION.

minimizing *soil erosion and the release of *pollutants.

Conservation International (CI) A non-profit, US-based international organization whose mission is to conserve global *biodiversity and demonstrate that human societies are able to live harmoniously with *nature. It has headquarters in Washington, DC, and works in more than 40 countries on four continents. *See also* BIODIVERSITY HOTSPOT.

Conservation Reserve Program A long-range federal programme in the US, in which farmers voluntarily contract to take *cropland out of production for 10 to 15 years and to devote it to nature conservation uses.

conservation tillage The cultivation (*tillage) of soil with minimum disturbance (no ploughing), which leaves plant residues on the soil surface, controls soil erosion, and conserves soil moisture.

conservation zoning *Land use *zoning which limits the development of un-suitable areas such as steep slopes, scenic areas, and spaces with other natural values.

conservative margin *See* CONSERVATIVE PLATE BOUNDARY.

conservative plate boundary A junction in the Earth's *crust between two *crustal plates that move past each other in opposite directions along a *transform fault, where crust is being neither created nor destroyed. Also known as **conservative margin**. *Contrast* CONSTRUCTIVE PLATE BOUNDARY.

conserve To save, use sustainably, avoid waste, and preserve for the future. For example, protecting a natural resource (such as *water or an *endangered species) through wise management and use.

consilience The view, put forward in the 1990s by biologist Edward O. Wilson, that everything in the world can be understood and explained through a small number of natural laws or principles, which apply across the sciences and the humanities, but which are based on physics and evolve according to the laws of evolution.

consolidated Tightly packed or composed of particles that are not easily separated. *See also* UNCONSOLIDATED.

consolidation Compression or compaction.

conspecific Belonging to the same *species.

conspicuous consumption Buying expensive services and products that are not really needed, in order to impress others.

constructive margin *See* CONSTRUCTIVE PLATE BOUNDARY.

constructive metabolism *See* ANABOLISM.

constructive plate boundary A divergent boundary between two *lithospheric plates that are moving apart, where new crustal rocks are being formed. For example, at *mid-ocean ridges associated with *sea-floor spreading. Also known as **constructive margin.** *See also* PLATE TECTONICS, DESTRUCTIVE PLATE BOUNDARY. *Contrast* CONSERVATIVE PLATE BOUNDARY.

consumer An *organism (such as an *animal or a *parasitic plant) that obtains energy and nutrients by feeding on other organisms or their remains. There are several different kinds of consumers including *carnivores, *herbivores, and *detritivores. *Contrast* PRODUCER. *See also* HETEROTROPH.

consumerism A term used to describe the effects of equating personal happiness with purchasing material possessions and consumption.

consumption 1. The process of using *resources to satisfy human wants or needs.
2. In water supply, the fraction of the water that is not available for use by humans.

consumptive use The use of *resources in ways that reduce supply. Example include *mining and *grazing, or *hunting, *fishing, and *logging in a *forest. *Contrast* NON-CONSUMPTIVE USE.

consumptive wildlife Any *game and fur-bearing species of *wildlife that are harvested for sport, food, fur, study, or commerce.

contact herbicide A *herbicide that only kills the part of the plant it is sprayed on, not the root systems. *Contrast* SYSTEMIC HERBICIDE.

contact metamorphism The *metamorphism of rock that is caused mainly by heat and pressure from pockets of *magma, which bakes and hardens the surrounding pre-existing rocks. *Contrast* REGIONAL METAMORPHISM.

contact pesticide A *pesticide that kills *pests when it touches them, instead of when they eat it.

contagious Any disease that is easily spread between individuals by direct contact.

contained use Any activity involving organisms that limits their contact with people and the environment, by the use of a physical, chemical, or biological barrier.

containment Any method of enclosing or containing *hazardous substances in a structure, in order to prevent the release of *contaminants into the *environment. Examples include the shielding of a *nuclear reactor within a thick concrete structure and the containment of an *oil spill using a floating *boom.

contaminant Any substance that pollutes or *contaminates another. In particular any physical, chemical, biological, or radiological substance that causes an impurity in the environment (air, water, and soil) and/or which may be harmful to human health.

contaminant level A measure of how much of a *contaminant is present.

contaminate To pollute or make impure or unclean, either by contact or by mixture.

contaminated land Land which has had its quality and usefulness reduced by the presence of one or more *contaminants in the *soil and/or *groundwater.

contaminated site Any site or land that contains harmful contaminants (particularly in its soil and water) because of previous *land use.

contamination 1. The act of polluting or contaminating.
2. An increase in the background concentration of a chemical, micro-organism, or radionuclide. Also known as **environmental contamination.**

contiguous Very close or connected without a break, sharing a common boundary or edge. For example, the contiguous states of the USA do not include Alaska and Hawaii.

contiguous zone A zone of the *high sea around each coastal state, that was established by the *Convention on the Territorial Sea and Contiguous Zone, and extends seaward up to 12 miles from the state's *territorial sea. A state has the authority within this zone to exercise the control necessary to prevent violations of regulations within its territories and territorial sea.

continent One of the seven major land areas on the Earth, which are North America, South America, Europe, Africa, Asia, Australia, and Antarctica.

continental Originating in an inland area, away from the coast.

continental air mass A dry *air mass that develops over a *continent or large land mass.

continental climate A climate that is found in the interior of continents, having low humidity, low rainfall, and large diurnal variations in temperature compared with *marine climates.

continental crust That part of the Earth's *crust that forms the continents and the *continental shelves. It is composed of rocks that are rich in *silica, and varies in thickness from 35 to 60 kilo-

metres. *Contrast* OCEANIC CRUST. *See also* CONTINENTAL PLATE, SIMA.

continental divide The line of high ground that separates *rivers that flow toward opposite sides of a *continent, usually into different *oceans.

continental drift A geological theory that was used to explain the relative positions and shapes of *continents, and the formation of *mountains, *folds, *faults, *earthquakes, and *volcanoes, as caused by the floating of continents across the Earth's *crust on the *mantle, like logs floating on water. The lighter *continental crust (*sial) floats on the more plastic *ocean crust (*sima) below, and geologists believe that the continents more or less drifted from their original locations. The evidence suggests that all of the continents were originally joined together as one huge landmass, called *Pangaea, probably between 200 and 250 million years ago. The rest of the Earth was covered by ocean (*Panthalassa). The theory of continental drift is based on the idea that Pangaea broke up and moved apart to form the continents we have today. The northern part of Pangaea (called *Laurasia) broke up to create North America, Greenland, Europe, and Asia, and the southern part of Pangaea (called *Gondwanaland) provided the continents of South America, Africa, Australia, and Antarctica. Between Laurasia and Gondwanaland, in this original continent, lay the *Tethys Sea (of which the Mediterranean is a surviving remnant). This pattern of evolution of the continents helps to explain the remarkably close fit not only of the continental margins but also of old mountain belts and shield areas of the Earth's crust. *See also* PLATE TECTONICS, SEA-FLOOR SPREADING.

continental glaciation The formation of an *ice sheet over a large area of a *continent during a glacial period.

continental glacier *See* ICE SHEET.

continental margin The area between the *shoreline and the beginning of the *ocean floor, which includes the *continental shelf, *continental rise, and *continental slope.

continental plate A thick section of the Earth's *continental crust that is composed mainly of granite, floats on the *asthenosphere, and moves over the surface of the Earth.

continental polar air mass A dry, cold *air mass that develops over *continental areas at high latitudes in the *northern hemisphere, and is very cold in winter and mild in summer.

continental rise The part of the *ocean floor that extends from the *continental slope down to the *abyssal plain, with a gentle slope and a smooth bed.

continental shelf The gently sloping part of the *ocean floor that extends from the *shore of a continent down to the *continental slope, usually to a depth of about 200 metres.

continental slope The relatively steep-sloping part of the *ocean floor that extends from the *continental shelf to the *continental rise or *ocean trench.

continental tropical air mass A warm, dry *air mass that develops over subtropical *deserts and *continental areas at low latitudes. *Contrast* TROPICAL MARITIME AIR MASS.

continentality In climatology, the degree to which the climate of a place is influenced by a neighbouring land mass.

contingency plan A plan for action that is prepared in anticipation of an incident (such as a release of toxic chemicals, a fire, or an *earthquake), that clarifies responsibilities, resource requirements, and appropriate action. *See also* DISASTER PLAN.

contingent valuation method In *economic evaluation, a method of valuing *non-market uses of a *natural resource (such as clean air, wildlife, or an attractive view) by directly asking people how much they would be willing to pay for it under certain circumstances, or

what sort of compensation they would regard as necessary if it were not available. *See also* NON-MARKET VALUES.

continuous cropping The growing of *crops each year in succession, without a period of *fallow, which allows *soil fertility to recover.

continuous discharge A permitted release of *pollutants into the *environment that occurs without interruption, except for infrequent shutdowns for things like maintenance.

continuous monitoring Monitoring of flows or processes that continues without interruption, normally using equipment that is fitted with a recording device.

continuous stocking Allowing grazing livestock unrestricted and uninterrupted access to a given area of land over the whole period when grazing is allowed.

contour *See* CONTOUR LINE.

contour cropping *See* CONTOUR FARMING.

contour farming Planting a *crop and ploughing in horizontal rows that follow the *contour lines of a hill, which minimizes *soil erosion. Also known as **contour cropping, contour ploughing,** and **contour tillage.**

contour hedge A type of *barrier hedge that is grown along contours to help prevent *runoff and *soul erosion from a field.

contour line A line drawn on a map that connects points of the same elevation. Also known as a **contour.**

contour map A *topographic map that shows relief by means of *contour lines.

contour mining A type of *strip mining in hilly areas where the *mineral is exposed at the surface at roughly the same height along the *hillside. It involves removing the *overburden from the mineral seam, starting at the *outcrop and proceeding around the hillside, so the cut appears as a contour line. Also known as **contour strip mining.**

contour ploughing See CONTOUR FARMING.

contour strip mining See CONTOUR MINING.

contour stripping A form of *strip mining used in areas of steep topography, where the mineral seam outcrops at roughly the same elevation along a hillside.

contour tillage See CONTOUR FARMING.

contrail The condensation trail from a jet aircraft.

control 1. To manage or regulate. See also CONTROL MEASURE, COMMAND AND CONTROL.
2. In a scientific experiment, a standard or baseline against which other conditions can be compared.

control measure An activity that is undertaken to ensure that a standard is maintained, or a risk or hazard is eliminated or reduced to an acceptable level. See also CONTROL TECHNOLOGY.

control rod A rod made of *cadmium or *boron (which absorb *neutrons) that is used to control the power in a *nuclear reactor. Large numbers of rods are lowered into or lifted from the reactor *core in order to control the number of neutrons that cause a *chain reaction.

control technology A *control measure that uses special equipment, for example in order to reduce air pollution. See also BEST AVAILABLE CONTROL TECHNOLOGY, BEST AVAILABLE CONTROL MEASURES, MAXIMUM ACHIEVABLE CONTROL TECHNOLOGY.

controlled burn See PRESCRIBED BURN.

conurbation An extensive *urban settlement that is formed when two or more *cities, which were originally separate, grow together to form a continuous metropolitan region or *megalopolis.

convection The vertical circulation of air, water, or molten rock within the Earth which is driven by density differences and results in heat transfer.

convection column A rising column of gases, smoke, ash, particulates, and other debris that is produced by a fire.

convection current A current that transfers material (such as air, water, or molten rock within the Earth) due to differences in density that result from differences in temperature. See also SEA-FLOOR SPREADING, TURNOVER.

convection rain A form of *rain that occurs when heated air rises and cools, so that air temperature drops to the *dew point and *condensation occurs.

convective instability Instability that is caused by the rising of very dry air over warm, moist air below. Also known as **potential instability**.

convention 1. An international agreement that is usually legally binding between the states who signed up to it. Also known as **international convention**.
2. A large formal assembly.

Convention Concerning the Protection of the World Cultural and Natural Heritage (1972) An international *convention that was adopted by *UNESCO in 1972, which aims to encourage the identification, protection, and preservation of cultural and natural heritage. It recognizes that nature and culture are complementary and that cultural identity is strongly related to the natural environment in which it develops, and provides for the protection of those cultural and natural 'properties' that are regarded as being of greatest value to humanity. It is not intended to protect all properties of great interest, importance, or value, but to protect a select list of the most outstanding of these from an international viewpoint. Cultural heritage refers to monuments, groups of buildings, and sites with historical, aesthetic, archaeological, scientific, ethnological, or anthropological value. Natural heritage covers outstanding physical, biological, and geological formations, habitats of *threatened species, and areas with scientific, conservation, or aesthetic value. The level of *biodiversity within a given site is a key indicator of its importance as a natural property. Also known as the **World Heritage Convention.**

Convention for the Prevention of Marine Pollution by Dumping from Ships and Aircraft (1972) A regional component of the *Convention on the Prevention of Marine Pollution by Dumping of Wastes and Other Matter, which applies to states bordering the northeast Atlantic Ocean, from Iceland to Spain. Also know as the **Oslo Convention.**

Convention on International Trade in Endangered Species (CITES) An international agreement, launched under the auspices of the *IUCN in 1975, designed to regulate trade in *endangered wildlife and thus help to conserve those *species. It has been signed by 132 nations, and protects nearly 90 species of plants and 400 species of animals which are listed as 'threatened with extinction'.

Convention on the Conservation of Migratory Species of Wild Animals *See* BONN CONVENTION.

Convention on the Prevention of Marine Pollution by Dumping of Wastes and Other Matter (1972) An international *convention that prohibits the dumping at sea of certain *hazardous materials, requires a prior special permit for the dumping of a number of other identified materials, and a prior general permit for other wastes or matter. Also known as the **London Convention.**

Convention on the Territorial Sea and Contiguous Zone A United Nations convention that came into force in 1958 and defines the *territorial sea and the *contiguous zone for each coastal state.

conventional agriculture * Farming practices that involve the use of chemical *fertilizers, *pesticides, and *machinery. *Contrast* ALTERNATIVE AGRICULTURE.

conventional pollutant *See* CRITERIA POLLUTANT.

convergence Coming together or *confluence. *Contrast* DIVERGENCE.

1. The process by which a similar character evolves independently in two species. *See* also CONVERGENT EVOLUTION.

2. The flowing together of air masses or ocean currents.

3. The coming together of plates in *plate tectonics.

convergent evolution The *evolution of two or more *species from different places but under similar environmental conditions, so that they come to closely resemble one another. Also known as **convergence.** *Contrast* DIVERGENT EVOLUTION.

convergent margin *See* CONVERGENT PLATE BOUNDARY.

convergent plate boundary A junction between two *lithospheric plates on the Earth's *crust that are moving towards each other, which forces one plate down beneath the other into the *mantle (where it melts), by the process of *subduction. Also known as **convergent margin.**

conversion The change of mass into energy, for example converting a fuel (such as gasoline, coal, or *biomass) into heat or electricity.

conveyance loss The loss of water from a pipe or *canal that is caused by leakage, seepage, *evaporation, or *evapotranspiration.

convivium A population within a *species which is isolated geographically and different from others within the same species, usually a *subspecies or *ecotype.

coolant A liquid or gas that is used to cool a machine, including the liquid or gas used to transfer heat from the *core of a *nuclear reactor to the steam generators or directly to the turbines.

cooling tower A structure that helps to remove heat from water that has been used as a *coolant (for example in an electric power generating plant) by direct contact (*evaporation) between the warm water and cooler atmospheric air.

Cooperative Forest Management Act (1950) US legislation that authorized technical and financial assistance to states to enable them to provide technical assistance to private forest landowners and processors.

Cooperative Forestry Assistance Act (1978) US legislation that brought together authority from nine cooperative assistance programmes in forestry, expanded some of them, and authorized consolidated programmes to participating states. It built on the *Clarke-McNary Act (1924), the *Forest Pest Control Act (1947) and the *Agriculture and Consumer Protection Act (1973).

coordinate 1. To bring order and organization to. *See also* COORDINATION.
2. A number that identifies a location relative to an axis, on a graph or map.

Co-ordinated Information on the Environment in the European Community *See also* CORINE PROJECT.

coordination The planned collaboration of individuals and organizations, and their resources, in order to achieve a common goal.

COPC *See* CHEMICAL OF POTENTIAL CONCERN.

copepod A member of a large group of species of tiny shrimp-like *crustaceans (*zooplankton).

copper (Cu) A natural metallic *trace element that occurs naturally in rock, soil, water, sediment, plants, and animals. It is a good conductor of heat and electricity and is used in making brass, copper alloys, and electrical conductors, in agriculture (to treat plant diseases), for water treatment, and as a preservative for wood, leather, and fabrics. In high concentrations copper can be *toxic to plants.

copper acetoarsenite An *arsenic-based *insecticide that is *toxic by ingestion. It is also used as a wood preservative and has an emerald green colour. Also known as **Paris Green.**

coppice A traditional form of *woodland management in which multiple stems (poles) are allowed to grow up from the base of a felled *tree. The poles are then cut every few years to provide fuel and wood for making tool handles, fencing, and charcoal. *Contrast* HIGH FOREST. *See also* POLLARD.

coprolite Fossilized excrement or *faeces.

coprophage An organism (such as a dung beetle) that feeds on *faeces.

copse A small group of *trees.

coral Small *colonial invertebrates (*polyps) that live in shallow saltwater seas found in the *coastal zones of warm tropical and subtropical oceans. The coral can be white, red, or black in colour. When they die, their vacant protective skeletons (also called coral) form layers on which other corals build, and this creates a *coral reef.

coral bleaching The whitening of a *coral colony, indicative of environmental stress. This occurs when the coral *polyp expels symbiotic algal cells (*zooxanthellae), which contain photosynthetic pigments, from its body. The *coral reef appears bleached because most reef-building corals have white *calcium carbonate skeletons, and these show through the tissue of the corals.

coral budding The way in which a *coral colony expands in size, by forming an offshoot.

coral reef A *reef made by colonies of stony *coral invertebrates (*polyps), together with algal and mineral components, which through time have been *consolidated into *limestone. Coral reefs are found only in shallow regions of tropical oceans. *See also* ATOLL, COASTAL REEF, FRINGING REEF, LAGOON.

corbiculum *See* POLLEN BASKET.

cordwood Small-diameter and/or low-quality wood that is suitable for pulp, woodchips, or firewood, but not for *sawlogs.

core 1. The central part of the *Earth's interior, which is composed mainly of *nickel and *iron, divided into a relatively small inner core and a much thicker outer core, and surrounded by the *mantle. The inner core has a radius of about 1255 kilometres (about 20% of the total radius of the Earth). It is solid, and made up mainly of iron molecules with some nickel and possibly some *silicon and *sulphur. This material is extremely dense (more than 13 times the density of surface water), extremely hot (temperatures are estimated at between 4500 and 5500°C, almost as hot as the surface of the Sun), and under extremely high pressure (estimated at 3.5 million times atmospheric pressure). These extreme conditions mean that the iron material in the inner core behaves rather differently than it would at the Earth's surface. The outer core is about 2220 kilometres thick. This layer is also made up mostly of *iron and *nickel, but this material is *molten. With increasing distance from the inner core temperature drops from around 4500°C to less than 2000°C, pressure drops by half, and the density of material also decreases.
 2. *See* REACTOR CORE.
 3. *See* CORE AREA.

core area The inner area of a *biosphere reserve which is legally protected and where only the minimum amount of human activity is allowed, to enable plants and animals to thrive without disturbance from people. Often surrounded by a *buffer zone. Also known as a **core**.

CORINE project A project (Co-ordinated Information on the Environment in the European Community) that is designed to build a common environmental database for the *European Union, using *geographical information systems.

Coriolis effect The tendency for air above the *Earth, and for ocean currents, to appear to be deflected to the right (in the northern hemisphere) or the left (in the south) because of the rotation of the Earth. *See also* GEOSTROPHIC WIND.

cornucopian A view that natural resources are unlimited and perpetual economic growth is not only possible but essential. *See also* OPTIMIST, ENVIRONMENTAL.

corona A series of coloured rings in the sky that surround the Sun or Moon, caused by the diffraction of light by small water droplets.

Corporate Average Fuel Economy (CAFE) (1975) A US approach to reducing energy consumption by increasing the fuel economy of cars and light trucks, based on calculating the sales weighted average *fuel economy, expressed in miles per gallon (mpg), of a manufacturer's fleet of passenger cars or light trucks manufactured for sale in the USA for any given model year.

corporate social responsibility Awareness, acceptance, and management of the implications and effects of all corporate decision-making, taking particular account of community investment, human rights, and employee relations, environmental practices, and ethical conduct.

corrective action The adoption of solutions that result in the reduction or elimination of an identified problem. For example, the *cleanup of *hazardous waste contamination at particular sites.

correlation 1. A measure of the association between two or more variables.
 2. The matching of sedimentary deposits (for example on a lake bed or in a rock) that are of similar age but in different places.

corridor A strip of natural *habitat that connects two or more larger areas of natural habitat (or *nature reserves) surrounded by developed land, which allows the *migration of *organisms from one place to another. Also known as **buffer, buffer strip, buffer zone, greenway,** or **migration corridor**.

corrie *See* CIRQUE.

corrosion A process in which a metal is attacked in a chemical reaction. *See also* RUSTING.

corrosive A liquid or solid that destroys metals and other materials or burns the skin.

corrosivity The ability to dissolve or break down certain substances, particularly metals. This is one of the four *characteristics that is used to define *hazardous waste in North America.

cosmic radiation A variety of high-energy particles (mainly *protons (92%) and *alpha particles (6%)) that bombard the Earth from outer space.

cosmology The study of the origin and nature of the *universe.

cosmopolitan Widely distributed across the globe, as opposed to local or regional.

cost The quantity of resources that are required in order to achieve a desired end, usually expressed in monetary terms.

cost effective Achieving specified objectives under given conditions, for relatively little expenditure or at equal or lower cost than current practice.

cost recovery A legal process in the USA that allows the government to recover the cost of *cleanup at hazardous (*Superfund) waste sites from those who cause the contamination. See also COST SHARING.

cost sharing Sharing the costs of a project between different people or agencies. For example, the government might share the cost of a *cleanup operation with those who caused the pollution. See also COST RECOVERY.

cost–benefit analysis A formal quantitative assessment of the short-term and long-term social and financial costs (losses) and benefits (gains) that arise from an economic decision, for example about investing in a major project which may have social as well as economic outcomes. Financial values are assigned to each cost and benefit, and the decision whether or not to proceed with the project is made based on the comparison of anticipated costs and benefits. The technique is widely used but has some inherent difficulties, including how to assign monetary values to intangible things (such as *aesthetics), and how to take into account future conditions (*discounting) and externalities (such as the generation of *pollution). This approach assumes that everything has a monetary value, and that this value is more important than any ecological, health, or environmental issues involved which have not been costed. It cannot include political judgements about what is acceptable at a particular point in time, or moral judgements about what is right and wrong and who should decide these issues, and it does not distinguish between who may or should benefit from a proposed scheme, and who may or should pay the costs. See also FULL COST PRICING.

coulee A long, deep, winding channel formed by water erosion (probably by *glacial meltwater at the end of the *Pleistocene), which is often dry or carries intermittent flow, and might have an *underfit stream flowing through it.

Council of Ministers The Council of the *European Union which directly represents the Member Governments, and is the principal decision-making body that acts on proposals from the *European Commission.

Council on Environmental Quality (CEQ) An advisory council to the President of the USA that was established by the 1969 *National Environmental Policy Act and reviews federal programmes for their effect on the environment, conducts environmental studies, and advises the President on environmental matters.

countershading A form of *camouflage in some birds and aquatic organisms, that are dark coloured on top but light coloured underneath.

countries with economies in transition Those Central and East European countries and former republics of the Soviet Union that are in transition to a *market economy.

country 1. Short for *countryside.
2. The land occupied by a nation.

country breeze A light *breeze that blows into a city from the surrounding countryside, and is most obvious on clear nights when the *urban heat island is strongest.

country sports *See* FIELD SPORT.

countryside The land and scenery of a *rural area.

Countryside Agency A statutory agency that was established in the UK in 1999 to conserve and enhance the countryside, promote social equity and economic opportunity for the people who live there, and to help everyone, wherever they live, to enjoy it.

Countryside and Rights of Way Act (2000) UK legislation that contains new provisions for access in the countryside and for protection of *Sites of Special Scientific Interest.

covalent bond The linkage of two *atoms that share a pair of *electrons. *See also* IONIC BOND.

cover 1. The percentage of ground that is covered by living or organic material. *See also* GROUND COVER.
2. Any feature that conceals wildlife or fish and allows them to escape from predators, rest, or feed. Also known as **covert, escape cover,** or **shelter.**

cover crop Plants, such as rye, *alfalfa, or clover, that are planted in between the main crop in order to maintain a plant cover on the land after harvest, and thus reduce erosion and *leaching.

cover forage ratio The ratio between hiding *cover and *forage area for wildlife, in a given area of land.

cover material *See* CAP.

cover type The dominant vegetation in an area.

covert A covering that helps to conceal, such as *cover for wildlife (usually *game).

covert release A release of a biological agent that is unannounced and causes illness. *Contrast* OVERT RELEASE.

covey A small flock of *game birds such as grouse or partridge.

cowboy economy General term to describe an economy that behaves as if *natural resources are infinite in supply and nature can absorb all *wastes.

CPM *See* CRITICAL PATH METHOD.

cracking An oil-refining process that breaks large *hydrocarbon molecules into smaller ones.

cradle-to-grave system A procedure used in North America in which *hazardous materials are identified and tracked as they are produced, treated, transported, and disposed of by a series of permanent, linkable, descriptive documents (manifests). Also known as **manifest system.**

crag A steep rocky cliff.

crag-and-tail *Landform created by glacial *scour and (like a *rôche moutonnée) formed around resistant rock underneath moving ice, which has a rocky, angular upstream (crag) side and a gently sloping downstream (tail) side.

crater The bowl-shaped *vent at the top of a *volcano.

craton A very large, ancient, stable core area within the *continental crust, made of highly deformed *metamorphic rocks, which are effectively the roots of the continents. Also known as **shield.**

created opening An opening in a forest that is created by the application of *even-aged *silviculture practices. *See also* CLEARCUT.

creationism The doctrine that all of the universe (including each species of organism) was created separately, in much its present form, by a supreme being or deity.

Creative Act (1891) US legislation that authorized the President of the USA to set aside public lands as public reservations.

creature A living organism.

credit An amount by which a particular country has actually reduced its pollution emissions, particularly of *greenhouse gases, beyond the agreed or required amount. *See also* CERTIFIED EMISSIONS REDUCTION, EMISSIONS REDUCTION UNIT.

creek A small natural *stream, smaller than a *river.

creep The extremely slow continuous movement of *unconsolidated soil or rock debris down a slope in response to gravity. A very slow form of *mass movement. Also known as **soil creep**.

creeper Any plant (such as ivy) that grows by sending out a shoot that grows along the ground, rooting all along its length. The term is also applied to a tight-clinging vine.

crest The high point of a *wave.

Cretaceous The final geological period of the *Mesozoic era that began 145 million years ago and ended 65 million years ago, during which *deciduous trees evolved and grew over a wide area and shallow seas submerged much of the surface of the Earth. Dinosaurs became *extinct after this period, in a *mass extinction.

Creutzfeldt-Jakob disease A rare brain disease in humans, which usually affects in middle age, is usually fatal, is characterized by progressive dementia and gradual loss of muscle control, and is caused by a *prion protein. *See also* BOVINE SPONGIFORM ENCEPHALOPATHY.

crevasse A deep crack in the ice of a *glacier or *ice sheet that is caused by ice movement.

crisis management 1. Management of a crisis: measures that are taken to identify, acquire, and plan the use of resources to anticipate, prevent, and/or resolve a threat (for example to public safety arising from terrorism).
  2. Management in a crisis: decision-making based on a seat-of-the-pants response to the immediate issue, with little if any consideration of longer-term effects.

criteria pollutant A group of common *air pollutants (*sulphur dioxide, *carbon monoxide, *particulates, *hydrocarbons, *nitrogen oxides, *photochemical oxidants, and *lead) that pose the greatest threat to human health and are regulated in the USA by the *Environmental Protection Agency. Also known as **conventional pollutant**. *See also* DESIGNATED POLLUTANT, HAZARDOUS AIR POLLUTANT.

critical factor The environmental factor that is closest to a *tolerance limit for a given species at a given time. *See also* LIMITING FACTOR.

critical habitat An area that has been designated (usually by government) as critical for the survival and recovery of *threatened or *endangered species.

critical load The maximum amount of a particular *pollutant that an *environment or *ecosystem can tolerate without suffering long-term damage.

critical path method (CPM) In *project management, a technique that is used to plan and control the activities in a project, based on identifying the series of successive activities which take up most time (the 'critical path'). This determines the total lead time of the project. Also known as **programme evaluation and review technique**.

critical wildlife habitat A *habitat that is vital to the health and survival of one or more *species, because it provides such features as nesting sites, food sources, and *breeding grounds.

critically endangered A *species which, according to the *IUCN *Red List, faces an extremely high risk of *extinction in the wild in the immediate future. *See also* EXTINCT, ENDANGERED, VULNERABLE SPECIES.

croft Scottish name for a small *farm, generally run on a *subsistence basis.

crop 1. A plant (such as cereals, vegetables, or fruit plants) that is cultivated and harvested for use by people or livestock. *See also* COMPANION CROP.
  2. *See* CROP YIELD.

crop dusting The application of *pesticides to plants by a low-flying plane.

crop productivity A measure of efficiency, that is, output (production) per unit of input over time; for example grams of biomass per square metre per day, or crop yield expressed in tonnes per hectare per season. Can also be expressed in terms of labour or financial inputs, solar energy inputs, and so forth.

crop residue The portion of a plant or crop that is left in the field after *harvest.

crop rooting zone The depth of soil in which crop roots are found.

crop rotation The practice of planting an area with different crops from year to year, in order to maintain soil fertility and organic matter content, and reduce *soil erosion. Most rotations include *legume-type crops to rebuild stores of *nitrogen in the soil. See also ROTATION PASTURE.

crop yield The amount of a crop that is harvested from a given area of land (such as a field) in a single *growing season, usually expressed as yield per unit area (for example, tonnes per hectare). Also known as **crop**. See also HARVEST.

cropland Land that is used for the production of cultivated crops for harvest, either alone or in rotation with grasses and *legumes.

cropping pattern The layout of crops and fallow in a given area of land, and how these vary through the year.

cropping regime The way in which crops are grown. See also CROP ROTATION.

cropping season See GROWING SEASON.

cropping system A subsystem of the farming system, or a unit of land use, that comprises the soils and plants, water, nutrients, labour, and other inputs that are used to produce food, feed, fuel, and fibre.

crossbreeding The production (*breeding) of a *hybrid plant or animal by deliberately mixing different *species, *breeds, or *varieties. Also known as **crossing**.

cross-contamination 1. The transfer of micro-organisms from one place or food to another, usually by direct contact.
2. The movement of underground contaminants from one level or area to another.

crossing See CROSSBREEDING.

cross-pollination The fertilization of a flower by pollen from a different plant of the same species, which creates an offspring with a different genetic makeup from either of the parent plants. See also CROSSBREEDING.

cross-section A view or representation of the interior of an object, viewed along a plane.

crown The top or highest part of a tree and the *canopy in a forest, made of branches and *foliage. See also OVERSTOREY.

crown class A classification of individual *trees in a *forest, based on dominance relative to adjacent trees, which reflects position in the *canopy and amount of sunlight received. The four classes, in descending order of crown height and size, are *dominant, *codominant, *intermediate, and *suppressed.

crown closure The growth of the crowns of trees in a *forest so that they touch and effectively block out sunlight from the area below. See also CLOSED CANOPY.

crown density The amount of *foliage in the crown of a tree.

crown fire An intense forest fire that burns the top of living trees as well as the plants below it.

crown height The distance from the ground to the base of the crown in a tree.

crude birth rate The number of live births in a given year, per thousand individuals in that *population.

crude death rate The number of individuals who die in a given year, per thousand individuals in that *population. Also known as **crude mortality rate**.

crude mortality rate *See* CRUDE DEATH RATE.

crude oil *See* PETROLEUM.

crust The outer layer of the *Earth, which varies in thickness between 6 and 48 kilometres, and floats on and surrounds the *mantle. It comprises *oceanic crust and *continental crust, and the thick continental crust and much thinner oceanic crust are separated from the mantle below by the *Mohorovicic discontinuity. The crust is made largely of *oxygen (47%) and *silicon (28%), with much smaller amounts of *aluminium (8%), *iron (5%), *calcium (4%), *sodium (3%), *potassium (3%), and *magnesium (2%). It is effectively the Earth's outer skin, occupying much less than 1% of the volume of the planet. All of the Earth's *landforms (mountains, plains, and plateaux) are contained within it, along with the oceans and seas. Whilst the crust appears to be solid, it is subject to repeated movement (including bending, folding, and breaking) associated with the movement of material on the mantle below.

crustacean An aquatic *invertebrate *arthropod of the class Crustacea that lives in freshwater or seawater, has a segmented body, paired jointed limbs, and a hard shell. Examples include shrimp, crabs, lobsters, and crayfish.

crustal deformation The bending up or down of rocks in the Earth's *crust that is caused by pressure, either from within the Earth (for example by mountain building) or from on top of it (for example by *crustal subsidence or *crustal rebound). Deformation includes *faulting, *folding, *shearing, *compression, and extension caused by *tectonic forces.

crustal plate Large sections or blocks of the Earth's *crust, which are curved to the spherical shape of the Earth and fit together like a jigsaw. There are seven major plates (the North American,

South American, African, Eurasian, Indo-Australian, Pacific, and Antarctic plates), and at least twelve minor plates. These plates float on the *asthenosphere, allowing them to move relative to one another, driven by *sea-floor spreading, giving rise to *plate tectonics. Also known as **plate, tectonic plate**.

crustal rebound The raising of part of the Earth's *crust after the melting of heavy continental *ice sheets.

crustal subsidence The depression of part of the Earth's *crust that is caused by the immense weight of continental *ice sheets.

cryogenic storage *See* CRYOPRESERVATION.

cryogenics 1. The branch of physics that studies the production and effects of very low temperatures.
2. The preservation of living organisms in a dormant state by freezing, drying, or both.

cryopreservation The *in vitro preservation of organic material (such as embryos, sperm, or eggs) by freezing, usually in liquid nitrogen. Also known as **cryogenic storage**.

cryosphere The frozen part of the Earth's surface, which includes the polar *ice caps, continental *ice sheets, mountain *glaciers, sea ice, snow cover, lake and river ice, and *permafrost.

cryptic Behaviour or coloration that helps to conceal an animal. *See also* CRYPTIC COLORATION.

cryptic coloration A form of *camouflage, in which many animals develop hereditary colouring and markings which match and conceal them in their usual surroundings, in order to protect themselves against *predators. *See also* CRYPTIC.

cryptogam Non-vascular plants such as fungi, mosses, lichens, and liverworts.

cryptosporidium A single-celled micro-organism that is commonly found in lakes and rivers, is highly resistant to disinfection, and causes acute diarrhoea, abdominal pain, vomiting, and fever in humans.

crystal A homogeneous regularly shaped solid with flat surfaces (faces) and specific angles between the faces. The crystal form varies from one substance to another, reflecting the atomic, molecular, or ionic structure of the crystal.

CSD *See* COMMISSION ON SUSTAINABLE DEVELOPMENT.

cuesta *See* ESCARPMENT.

cull 1. An organism or other object that is rejected or set aside because it is of inferior quality.
2. The process of killing designated animals in a particular place, as part of a management plan, in order to reduce the number of individuals within the population.

culling *See* ROGUING.

cultivar A domesticated variety of plant that has been cultivated by *selective breeding and is not normally found in wild populations. Also known as **cultivated species**.

cultivate 1. To prepare and use land for growing crops.
2. To adapt a wild plant to a new environment. *See also* DOMESTICATE.

cultivated species *See* CULTIVAR, DOMESTICATED SPECIES.

cultural eutrophication *Eutrophication caused by human activities.

cultural extinction The accelerated *extinction of *species that is caused by human activities such as the over-hunting of animals, over-collection of plants, introduction of non-native species, removal of habitat, and induced environmental change (for example, related to *pollution).

cultural landscape Natural *landscape that has been modified (both by accident and by design) by humans.

cultural resource The physical remains (including *artefacts, objects, and structures) of past human activities. *Contrast* AESTHETIC RESOURCE. *See also* RESOURCE.

culture 1. The growth of cells or an organism in a prepared medium, under laboratory conditions.
2. A set of beliefs, attitudes, and rules for behaviour that are commonly held in a particular society.

culvert A drain or pipe that carries surface water under a built structure such as a road or railway.

cumulative effect Effects that result from separate, individual actions that collectively become significant over time.

cumulative exposure The sum of exposures that an *organism has to a *pollutant over a period of time.

cumuliform Having the appearance or character of *cumulus clouds.

cumulonimbus (Cb) A tall *cumulus *cloud (sometimes reaching an altitude of 9000 metres (30 000 feet)), formed in turbulent, rising air. They are usually grey or dark grey. Their tops may be smooth or striated or be flattened into an anvil shape, and they often bring heavy *rain or *hail (formed in the strong vertical uplift) and are sometimes associated with *thunderstorms.

cumulus (Cu) A type of *cloud that is associated with rising air currents, usually below about 2400 metres (8000 feet). In Latin cumulus means heap. They are dense, well-defined mounds with dark bases and white rounded upper regions. Relatively thin cumulus clouds, which are common on warm summer afternoons in middle latitudes, indicate fair weather and relatively stable conditions.

cup anemometer An instrument used to monitor wind speed, via the rotation of cups by the movement of air.

curie An obsolete measure of radioactivity, equal to 3.7×10^{10} (37 000 000 000) nuclear disintegrations per second. Replaced by the *becquerel.

current 1. A horizontal movement of water, for example along a stream or through an ocean.
2. The movement or flow of electricity.
3. Occurring in or belonging to the present time.

cursorial Adapted for running.

cuticle 1. A waxy film made of *cutin that covers the external surface of the stems and leaves of plants, which helps to prevent water loss.
2. The *exoskeleton made of *chitin that is secreted by *invertebrates.

cutin A waxy substance that, together with *cellulose, forms the outer layer of the skin (*cuticle) of many plants.

cut-off *See* OX-BOW LAKE.

cutting A part (stem, leaf, or root) that is removed from a plant and is capable of developing into a new plant through rooting or grafting.

Cuyahoga River A river in northeast Ohio, USA, into which industries around Cleveland dumped oily wastes, and which has literally caught fire on numerous occasions. The most serious fire occurred on 23 June 1969, and that event triggered a series of important pollution control initiatives including the *Clean Water Act (1972) and the *Great Lakes water quality agreement, and the creation of federal and state *environmental protection agencies.

cwm *See* CIRQUE.

cyanides A group of inorganic salts that contain the cyanide ion (CN⁻). They are highly poisonous and are used in ore processing to extract *gold and *silver from crushed rock.

cyanobacteria *Bacteria, also known as blue green algae, that contain *chloro-phyll and can *photosynthesize. They can form large coloured mats on the surface of lakes and rivers. One species causes a *red tide.

cybernetic A *system that continuously changes in response to *feedback.

cycle of erosion A hypothetical model of how *landscapes evolve through a progressive sequence of stages, starting from an uplifted surface dissected by some rivers, through a youthful landscape with steep valley sides and steep long profiles, to a mature landscape dominated by low hills and low river slopes, to the gentle undulating or flat landscape of old age. Also known as **geomorphic cycle**.

cyclic/cyclical Recurring in cycles.

cyclogenesis The development of low-pressure systems (*depressions) in the atmosphere. *See also* FRONTOGENESIS.

cyclone 1. The name for a *hurricane in the Indian Ocean.
2. *See* DEPRESSION.
3. *See* TORNADO.

cyclonic circulation The movement of air around a *depression. In the northern hemisphere the flow is anti-clockwise, and in the southern hemisphere it is clockwise.

cyclonic rain *Rain that occurs in a *depression, when warm air rises above cold air and cools *adiabatically, causing the *condensation of *water vapour and formation of *raindrops. Also known as **frontal rain**.

cytoplasm The contents of a *cell, other than the *nucleus. *See also* PROTO-PLASM.

CZMA *See* COASTAL ZONE MANAGEMENT ACT.

dairy Relating to the production of milk or milk products, such as butter and cheese.

dairy farm A farm where *dairy products are produced.

Daisyworld A *computer model developed by James *Lovelock to illustrate his *Gaia hypothesis on any planet that supports life, based on how black and white daisies adapt to changes in *radiation and temperature.

DALR See DRY ADIABATIC LAPSE RATE.

dam An artificial barrier constructed across a river or valley, usually for flood control, *irrigation, and/or power generation. See also RESERVOIR.

damage assessment Evaluation of the magnitude, physical extent, and types of damage that a biological or physical *resource experiences as a result of an event such as a *natural hazard.

Daminozide See ALAR.

danger The condition of being susceptible to harm or injury, for example as a result of exposure to a *natural hazard.

dark earth See CHERNOZEM.

Darwin Day An international celebration of science and humanity that is held on 12 February each year, to celebrate the birth of Charles *Darwin in 1809.

Darwin Declaration An international agreement reached in Darwin, Australia in 1988, in response to the *United Nations Framework Convention on Biological Diversity, that countries should work together to improve knowledge and understanding of the world's biological diversity.

Darwin, Charles The 19th-century British naturalist (1809–82) who is widely considered to be the father of the theory of *evolution. His landmark work, *On the Origin of Species*, published in 1859, supported the idea of evolution by means of *natural selection.

Darwinism The theory put forward by Charles *Darwin that *species originate and develop by evolution from simpler species, and that evolution is driven mainly by *natural selection.

data A collection of facts (usually created by measurement) from which conclusions may be drawn. See also INFORMATION.

data mining The processing of large amounts of data in order to extract new kinds of useful information from it, based on patterns and relationships. See also INFORMATICS.

dating See ABSOLUTE DATING, RELATIVE DATING.

datum A reference (such as mean *sea level) against which to measure heights or depths.

daughter product See DECAY PRODUCT.

dawn The first light of day, in the eastern sky before sunrise. Also known as daybreak. See also DUSK.

day 1. A 24-hour period of time that begins at midnight and ends the following midnight, during which the *Earth makes a complete rotation on its axis.
2. The time between *sunrise and *sunset, when it is light outside. *Contrast* NIGHT.

day length The number of hours between *sunrise and *sunset.

daybreak See DAWN.

daylight Light that is received from the Sun and the sky, during the daytime.

daylighting 1. Using sunlight to provide supplementary lighting for the interior of a building.

2. The act of cutting back vegetation bordering a road, in order to encourage the growth of new herbs and shrubs.

DDT A colourless *insecticide (dichlorodiphenyltrichloroethane), which is persistent and can collect in the fatty tissues of certain animals. It was used during the Second World War to control the spread of typhus and malaria by insects, and was widely used as an agricultural *pesticide during the 1950s and 1960s. It has been banned in the US since 1973 as a result of growing public concern because of its persistence in the environment and accumulation in the *food chain. It particularly damaged the peregrine falcon, whose numbers declined rapidly because DDT was causing the birds' egg shells to become thin and the eggs to break before hatching.

death rate The number of individuals in a population who die in a given year, usually expressed as a ratio (number of deaths per year per 1000 individuals). Also known as **fatality rate, mortality**.

debris Dead organic material (leaves, twigs, etc.) and sediment.

debris flow A type of overland or submarine *mass movement involving a downslope flow of a saturated mass of soil and rock debris, with more than half of the material being particles larger than *sand size. Slower than a *debris slide. Also known as a **flow**.

debris slide A rapid downslope movement of an *unconsolidated mass of mud, sediment, and rock. Faster than a *debris flow. Also known as a **slide**.

debt-for-nature swap An emerging approach to the *conservation of *wildlife, which involves conservation organizations (usually from developed countries) acquiring part of the international debt of a particular developing country at an agreed discount price. The organization then redeems the debt in local currency, and uses it to fund conservation activities such as the setting up and running of *nature reserves and *national parks. Supporters of such schemes claim that there are no losers and few financial risks in this innovative form of international financial transaction, and for this reason it is an ideal way of raising the financial resources required for major conservation initiatives. The debt-for-nature swap mechanism also helps to break down some of the persistent barriers that inhibit collaborative policies in a world of sovereign states—including lack of cooperation, different agendas and different objectives, short-term perspectives, and inward-looking and territorially defined concerns.

decade A period of ten years.

Decade of the Environment The 1970s, when a large number of the most important US policies towards the environment were passed because of greater public engagement, broader support for environmental policies in Congress, pro-environmental leadership from Republican and Democratic presidents, and favourable interpretations of statutory laws by key US courts (particularly the Supreme Court).

decant To pour of draw off the upper layer of liquid after the heavier material has settled. For example, drawing off water after a solid pollutant has settled out, as a means of cleaning up polluted water.

decay To graduate decrease or decompose. **1.** The breakdown of *organic matter by *micro-organisms.
2. The decrease of a radioactive substance because of nuclear emission of alpha or beta particles, or gamma rays. Each *radionuclide has a specific *half-life and *decay constant.

decay constant The rate of *radioactive decay, which is characteristic of the given nuclide. The decay constant is the probability that an atom of the *radionuclide will decay within a stated time (it is measured in units of reciprocal time).

decay product An element (which may itself be radioactive) that is formed by the *radioactive decay of another element, which may be stable or may itself be radioactive. Also known as **daughter product**.

dechlorination The partial or complete removal of chlorine from a substance, for example by using *activated carbon and *sulphur dioxide.

decibel (dB) A unit that measures the intensity or loudness of sound.

deciduous A tree or plant that sheds its leaves at the end of the growing season (autumn) each year. Examples include ash, beech, hickory, maple, and oak. Also known as **hardwood**. *Contrast* CONIFEROUS, EVERGREEN.

deciduous forest A *temperate or *tropical forest in which the *deciduous trees shed their leaves during either cold or dry seasons. It is relatively fertile (because of the abundant leaf litter), plant biodiversity is high, and typical trees include oak, hickory, maple, ash, and beech. Found in areas with moderate rainfall and marked seasons.

decision-making The process of acting upon the best information available in order to determine the most appropriate course of action.

declination 1. A downward slope or bend.
2. Angular distance north or south of the *celestial equator.
3. The angle between *magnetic north and geographical north.

decommission To permanently shut down a facility (such as a *nuclear reactor or *power plant) or withdraw it from service or active use *Contrast* COMMISSION.

decommissioning waste The waste materials that are produced when a facility (such as a nuclear *power plant) is *decommissioned.

decomposer *See* DETRITIVORE.

decomposition 1. The breakdown of *organic matter into smaller particles by organisms such as bacteria and fungi, releasing energy, simple organic material (such as sugars and proteins), and inorganic compounds (*nutrients) back into the environment where they become available to *plants.
2. The *chemical weathering of *rock.

decontamination Removal of harmful substances (such as chemicals, bacteria, or radioactive material) from exposed individuals, rooms, and furnishings in buildings or from the exterior environment.

deduction *See* DEDUCTIVE REASONING.

deductive reasoning Reasoning from the general to the particular, for example by developing a hypothesis based on theory and then testing it from an examination of facts. Also known as **deduction**. *Compare* INDUCTIVE REASONING.

deep ecology A *worldview or set of beliefs that calls for a major shift in human attitudes, values, and behaviour that rejects *anthropocentrism and directs personal action to protect *nature and improve the *environment. *Contrast* SHALLOW ECOLOGY. *See also* ECOCENTRIC, RADICAL ECOLOGY.

deep ocean Areas of the *oceans that are seaward of the *continental shelf.

deep-scattering layer A well-defined layer in the *ocean that reflects *sonar and indicates the presence of fish, squid, or other large marine organisms.

deepwater habitat Any area of open water that is deeper than two metres or covers the deepest emerging vegetation.

deep-well injection Disposal of waste fluid (particularly *hazardous waste) by pumping it into a deep well, where it is stored underground in permeable rock surrounded by impermeable layers.

deflation The removal of *clay and dust from dry soil by wind. *See also* DESERT PAVEMENT.

deflocculate The breakdown of clusters of soil particles. The term is usually applied to *clay soils.

defluoridation The removal of excess *fluoride in drinking water in order to protect teeth from damage or staining.

defoliant A chemical herbicide that causes the leaves to fall from trees and growing plants. *See also* AGENT ORANGE.

defoliate To remove the leaves from plants, trees, or shrubs.

deforestation The permanent clearance of a *forest, usually rapidly by cutting or burning over a large area, without replanting or natural *regeneration. Also known as **forest clearance**. *Contrast* AFFORESTATION.

deformation A change in the original shape of a material, such as a layer of *sedimentary rock that has been affected by *folding. *See also* CRUSTAL DEFORMATION.

DEFRA In the UK, the Department for Environment, Food and Rural Affairs, formerly *MAFF.

degasification The removal of dissolved gases from water in order to purify it.

deglaciation The melting (*ablation) and receding of *ice sheets and *glaciers, which uncovers the land beneath. *Contrast* GLACIATION.

degradable Able to be broken down or separated into a simpler form, using chemical, physical, or biological means. *See also* BIODEGRADABLE.

degradation 1. The chemical or biological breakdown of a complex compound into simpler compounds.
2. General lowering of land surfaces by *erosion. *Contrast* AGGRADATION.

DEIS *See* DRAFT ENVIRONMENTAL IMPACT STATEMENT.

delist The process of changing the designated status of a particular resource, such as removing a species of animal or plant from the *IUCN *Red List, or removing a *Superfund waste site from the *National Priorities List in the USA.

dell A small wooded hollow.

Delphi method A method of long-term forecasting or *decision-making based on expert judgement. It involves a number of stages: experts are interviewed anonymously and separately, in order to gain their views on a particular issue or problem; the results are combined and fed back to the experts as a group; each expert is then interviewed again, in the light of peer group opinion. In theory, the process continues until consensus emerges within the group.

delta A fan-shaped *alluvial deposit at the mouth of a *river, built up by the deposition of successive layers of fine-grained *sediment is caused by a reduction in velocity of the current so that the sediment drops out.

deluge A heavy downpour of *rain, or a heavy *flood.

demand The ability and desire to buy goods and services, or the quantity of a good or service that is needed in order to meet the requirements of the user. *See also* MARKET EQUILIBRIUM, SUPPLY.

demand management An approach to the allocation of scarce *resources that is based on minimizing wastage, restricting supply, and educating people to use less of the resource and use it more carefully. Also known as **demand-side management**.

demand schedule The relationship between price and quantity demanded of a good or service, which shows how much would be bought or consumed at various prices at a particular point in time.

demand species *Native and desired *non-native species of plants and animals that have a high social, cultural, or economic value.

demand-side management *See* DEMAND MANAGEMENT.

demersal Fish and *aquatic animals that live at or near the bottom of a *sea or *lake.

demersal fish *See* GROUNDFISH.

demineralization The removal of dissolved minerals and mineral salts from a liquid, particularly water.

demographic Relating to *population *statistics, changes, and trends based on measures of *fertility, *mortality, and *migration.

demographic balancing equation
An equation that is used to calculate *population changes from one year to the next in a given area, based on number of *births, *deaths, and *migrations. The general form of the equation is a *mass balance equation, in which end population = starting population ± natural increase ± net migration, where natural increase = births − deaths, and net migration = immigrants − emigrants. At the global scale there is no net migration (unless the Earth is invaded by aliens from another planet, or people permanently leave the Earth to set up home in space), so the balancing equation is then simply end population = starting population ± natural increase.

demographic force Any of the three factors (*fertility, *mortality, and *migration) that determine the level, composition, and distribution of population in a particular place or area. *See also* DEMOGRAPHIC BALANCING EQUATION, DEMOGRAPHIC TRANSITION MODEL.

demographic transition The growth pattern that many human *populations go through as countries develop, undergo *industrialization and *urbanization, and improve their *standard of living. Birth and death rates converge and *zero population growth is achieved.

demographic transition model A four-stage model that shows how *birth rates and *death rates change through time in a particular country or area, giving rise to phases of population stability and change. During stage one the death rate is usually very high because of poor health and difficult living conditions, *life expectancy at birth is less than 30 years, but birth rates are also high and so *population growth rate is zero or close to zero. Stage two begins when the death rate begins to fall (usually because of improved living conditions, better food supplies, and better health practices such as immunization). The birth rate remains high, and may even increase because women are healthier. A high population growth rate is triggered by the excess of births over deaths. It takes time, sometimes generations, for social attitudes (such as the high value attached to having children) to change, so during this phase birth rates remain high. In stage three the birth rate declines (because of better education, better family planning, more career options for women, and reduced *infant mortality), and it eventually catches up with the death rate. Population growth remains relatively high during the early part of this stage, but it falls close to zero towards the end of the stage. By stage four the birth rate and the death rate have converged, and they oscillate around a relatively low level. The population growth rate is once again zero or close to zero during this phase, in which birth rates and death rates are once again similar but now both low. By the year 2000 most *developed countries had completed all four stages, whilst most *developing countries lagged behind at stage two or early in stage three.

demographics *Statistics that describe the characteristics of a *population, such as age, sex, race, family size, income, and location of residence.

demography The statistical study of the characteristics of *populations, including distribution, age structure and composition, and patterns of *fertility, *mortality, and *migration.

den A rain-proof, weather-tight cavity in a tree or a space dug by animals among rocks or in soil. *See also* DEN TREE.

den tree A tree with cavities in which birds, mammals, or insects (such as bees) can nest. Also known as **cavity tree**.

dendritic A type of *drainage pattern that develops as a random network and has branches like a tree.

dendroarchaeology The use of *growth rings in trees to date when timber was felled, transported, processed, and used for construction.

dendrochronology A dating technique based on the analysis of *growth rings in trees. The number of rings indicates age, and the width of individual rings indicates climatic conditions at the time of growth and is used in *dendroclimatology. *See also* TREE RING.

dendroclimatology A technique used to reconstruct past *climate, based on analysis of *growth rings in trees, as *proxy climate indicators. Ring width reflects the climatic conditions of the growing season, and rings can be counted to give ages. *See also* TREE RING.

dendrogram A branching, tree-like diagram, which usually has a single source, and shows relationships produced by *classification. *See also* CLADOGRAM.

dendrology The study of the identification, habits, and distribution of *trees. *See also* SILVICULTURE.

denitrification The biological process in which *nitrates are reduced to *nitrogen gas, by *denitrifying bacteria, in soil under *anaerobic conditions. *Contrast* NITRIFICATION.

denitrifying bacteria Free-living *bacteria in soil that convert *nitrates to gaseous *nitrogen and *nitrous oxide.

dense fog A *cloud with its base on the ground surface, that reduces *visibility to 0.5 kilometres or less.

dense non-aqueous phase liquid (DNAPL) A liquid that does not dissolve in water, and which is also denser than water so it sinks. Many *chlorinated solvents are DNAPLS.

density 1. The mass of a substance per unit volume (kilograms per cubic metre, $kg\,m^{-3}$).
  2. The number of *individuals of a defined group that are present in a given area at a particular time. *See also* FREQUENCY, POPULATION DENSITY.

density current A *current in water that flows because of differences in *density. For example density currents in the *ocean are caused by differences in *temperature, *salinity, and *turbidity.

density stratification The creation of layers in a water body due to differences in *density, which may be caused by differences in *temperature and in concentrations of *dissolved solids and *suspended solids.

density-dependent The regulation of the size of a population by mechanisms that are controlled by *population density (such as the availability of *resources, or the incidence of *contagious disease), and that become more effective as density increases. *Contrast* DENSITY-INDEPENDENT.

density-independent A factor that influences the individuals in a population in ways that do not vary with *population density. *Contrast* DENSITY-DEPENDENT.

denudation The wearing down of the land surface by natural geological processes, which involves *weathering, *mass movement, and *erosion.

denudation chronology The long-term history of *landscape development by *denudation. *See also* CYCLE OF EROSION, EROSION SURFACE.

denuded Laid bare, stripped of all *vegetation cover, often as a result of a disturbance such as a *landslide.

deoxygenation The consumption of *oxygen in water by aquatic organisms, as they decompose (*oxidize) organic materials.

deoxyribonucleic acid *See* DNA.

dependency In *conservation assessment, the reliance of a *species, *community, or ecological process on a particular location (such as a *feeding ground or a *migration corridor) or structure (such as a *coniferous forest) for survival.

dependency ratio In a given human population, the number of economically dependent, non-working members (who are younger than 15 or older than 65) compared with the number of productive, working members (aged 15 to 64).

dependency theory A theory that the relationships between advanced *capitalist societies (the wealthy *First·World) and undeveloped, poor *Third World countries has always been exploitative. Development of the former has resulted in underdevelopment of the latter; the latter are kept dependent on the former by the capitalist system, and as a result the former continue to exploit the latter for their own gain.

depletable resource *See* NON-RENEW-ABLE RESOURCE.

depletion Decline in availability that results from using *abiotic (*non-renewable) resources, or from using *biotic (*renewable) resources faster than they can be renewed.

depletion time The time it takes to use up a *non-renewable resource.

deposit 1. An accumulation of *sediment, minerals, and precipitated substances. *See also* ALLUVIUM, COLLUVIUM, BOULDER CLAY.
  2. A facility where things can be stored for safekeeping.

deposit feeder *See* DETRITIVORE.

deposition 1. The accumulation or laying down of *sediment.
  2. The conversion of *water vapour directly into *ice, without first becoming a liquid. In chemistry this is known as sublimation.

deposition nuclei Tiny particles in the atmosphere on which an ice crystal can grow by the process of *deposition. Also known as **ice nuclei**.

depositional environment Any part of the *environment in which *sediment is deposited, such as a *lake, *coast, or *river.

depression A cyclonic circulation of winds towards the low pressure at the centre, where *convergence occurs. It can produce violent, damaging *thunderstorms with winds of force 12 or above on the *Beaufort scale. Depending on where it occurs a depression is also known as a *low, *tropical cyclone (or just *cyclone), *hurricane or *typhoon.

depression storage *See* DETENTION STORAGE.

deranged drainage pattern A *drainage system having no obvious pattern. Sometimes found in areas of low relief, low slope, and large sediment loads, and typical of recently glaciated areas.

derived character In *cladistics, a feature that is shared among members of smaller groups or *clades and is believed to have evolved at a later date than primitive features. Also known as **advanced character**.

derived demand The *demand for a particular *resource (such as bricks) that is created by the desire to satisfy the demand for some other resource (such as houses).

dermal exposure Contact between a gas, liquid, or solid (such as a *pollutant) and the skin.

dermal toxicity The ability of a *pesticide or *toxic chemical to poison an organism (including people and animals) by contact with the skin.

DES *See* DIETHYLSTILBESTROL.

desalination The removal of *salt from water, usually by *distillation, *freezing, *electrodialysis, or *reverse osmosis.

desalting The process of removing of *salt from *crude oil.

descendant A person who is descended from a particular ancestor or race, as an *offspring.

desert A dry *biome that is common in the *subtropics, and contains plants and animals that are adapted to extreme water shortage. It is created by an *arid climate in which *evaporation exceeds *precipitation, and precipitation is infrequent, unpredictable, and less than 25 centimetres of rain a year. *See also* TROPICAL DESERT, TEMPERATE DESERT, COLD DESERT, SAND DESERT.

desert climate Different climates produce *tropical deserts, *temperate deserts, and *cold deserts, but they have in common wide *diurnal variations in *temperature, low *relative humidity, limited *precipitation, more or less continuous sunshine throughout the year, little *cloud cover and clear skies, and occasional intense local convectional showers. Middle-latitude deserts are cooler than the low-latitude deserts and they receive low and unreliable rainfall. Most are located in the dry *rain shadows *leeward of topographic barriers, or in the interior of large land masses far from *maritime sources of moisture. Such deserts have much less extreme climates than low-latitude deserts and are much less inhospitable to people.

desert crust A hard surface layer that develops in a *desert climate, caused by the *evaporation of water from the ground which leaves *calcium carbonate, *gypsum, or other cementing materials as a *precipitate. *Contrast* DESERT PAVEMENT.

desert grassland A North American *grassland vegetation that grows in semi-arid climates on *bajadas and in valley bottoms where clay-rich soils prevent water from draining. It is often dominated by bunch grasses or (where disturbed) various shrubs.

desert lake basin A type of hot *desert in the form of a depression, which is often salty and receives drainage from inflowing streams.

desert pavement Gravel and stones left after wind and water erosion have removed finer material from a desert floor. *Contrast* DESERT CRUST.

desert province A region of the Earth in which deserts are found. There are five main desert provinces: the Sahara, southern Africa, South America, North America, and Australia.

Desert Solitaire A book by Edward *Abbey who regarded the *wilderness as a place where people can separate themselves from society but from which they need to return to society refreshed and ready to deal with the problems of society. Regarded by some as a latter-day *Walden.*

desert varnish A dark, hard, shiny coating of *iron and *manganese oxides that is found on *rocks in *arid regions.

DESERTIFICATION

The spread of desert-like conditions in *semi-arid regions due to human activities (including *overgrazing, *deforestation, or *soil erosion), prolonged *drought, and/or climatic change (*global warming), which cause long-term changes in the soil, climate, and plants and animals of an area. The term was originally used to mean the spread of *desert conditions over adjacent areas as a result of climatic factors and/or use of resources by humans, but it is now used more broadly to describe a range of processes occurring in many arid and semi-arid areas. An estimated 135 million people are directly affected by desertification, mainly in Africa, the Indian subcontinent and South America. But the problems are not confined to the particular places where desert expansion and soil degradation are concentrated. On a global scale the major impacts include stress on food producing capacity, reduction in *biodiversity, and modification of climate. In 1977 the *United Nations Conference on Desertification (UNCOD) concluded that desertification was then one of the most serious environmental problems in the world. UNCOD data suggest that about 6000 square kilometres of land are turned into desert each year, and that the productive capacity of a further 210 000 square kilometres is ruined each year, making it unprofitable for farmers to use. More recent data suggest that over 3.1 million square kilometres of the world's rangelands (80% of the total), 3.35 million square kilometres of rain-fed cropland (60% of the total), and 0.4 million square kilometres of irrigated *drylands (30% of the total) are threatened by moderate to severe desertification.

desert wind A hot, dry, dusty wind that blows outwards from a low air pressure cell over a *desert, *desiccating surrounding areas and withering plants. Examples include the *sirocco and the *harmattan.

desiccant Any material (such as a chemical) that absorbs moisture.

desiccation The process of drying out as a result of the removal of water.

design capacity The maximum volume of flow that an engineering structure is designed to handle, such as the number of vehicles using a road per day or the average daily flow of material into a water treatment plant.

designated pollutant An *air pollutant (such as *acid mist or *fluorides) which is not defined in the US *Clean Air Act as either a *criteria pollutant or a *hazardous air pollutant, but for which *emission standards exist for new sources.

designated use Uses for water that are identified in state water quality standards in the USA for which water quality must be protected. These include *hydroelectric power generation, navigation, public water supply, fisheries, and recreation (such as swimming, boating, and fishing).

designer bug A common term for *microbes that are developed through *biotechnology and can degrade specific *toxic chemicals in toxic waste dumps or in *groundwater.

desired condition The future condition of a particular resource (such as a forest or grassland) that is expected to result if the goals and objectives of resource management are fully achieved.

desired plant community The plant *community that best meets the agreed objectives for a particular site, as defined in a *land use or management plan.

desorption The release of atoms, molecules, or ions that are attached to solid surfaces. *Contrast* ABSORPTION.

destratification The mixing of water within a *lake or *reservoir in order to reduce or remove separate layers (for example of *temperature or *aquatic organisms). *Contrast* STRATIFICATION.

destructive margin *See* DESTRUCTIVE PLATE BOUNDARY.

destructive metabolism *See* CATABOLISM.

destructive plate boundary A convergent boundary between adjacent *crustal plates that are moving towards each other, where crust is being destroyed by *subduction. Usually associated with *volcanoes and *earthquake activity. Also known as **destructive margin**. *See also* PLATE TECTONICS, CONSTRUCTIVE PLATE BOUNDARY.

desulphurization The reduction or removal of the *sulphur content of *fuel in order to reduce *air pollution. Also known as **fuel desulphurization**.

detention storage The temporary storage of surface water in low areas such as puddles, *bogs, *ponds, and *wetlands, from which it evaporates or flows overland towards a stream channel. Also known as **depression storage**.

detention time The period of time over which a particular material is stored temporarily in a particular way, such as the storage of water in a reservoir before being used or in a *settling basin.

detergent A synthetic cleansing agent that, when added to water, helps to remove oil and dirt. *See also* SURFACTANT.

deterministic A process or event that is predictable because it is governed by definite rules of system behaviour which produce the same results each time. *Contrast* RANDOM, STOCHASTIC.

detoxify To remove a poison or render safe through removal of a poison.

detrital carbonate Crystalline *calcium carbonate that is found in ocean sediments and is usually derived from the *weathering of carbonate rock.

detrital sediment A sediment (such as *detrital carbonate) that is produced by the mechanical breakdown of rock by *weathering and *erosion.

detritivore An animal (such as an earthworm, maggot, or woodlouse) that feeds on dead and decaying *organic matter from plants and animals. Also known as **decomposer, deposit feeder, detritus feeder**. See also DECOMPOSITION.

detritus 1. *Organic debris from dead *organisms, which is often an important source of *nutrients in a *food web.
2. Loose, *unconsolidated particles or sediments that are formed by the *weathering of rocks.

detritus feeder See DETRITIVORE.

deuterium (D) An *isotope of *hydrogen that contains one *proton and one *neutron in its *nucleus.

developed country A country that has evolved through the *demographic transition and is technologically advanced, capital-intensive, highly urbanized, and wealthy. These are usually industrialized countries such as European countries, Canada, the USA, Australia, Japan, New Zealand, and Russia. Compare DEVELOPING COUNTRY, MORE DEVELOPED COUNTRY, TRANSITIONAL NATION. See also ANNEX I COUNTRIES, ANNEX B COUNTRIES.

developed land Any land that has been built on or substantially altered to benefit people, including development of settlements, factories, and roads.

developed recreation Any form of recreation requiring facilities that, in turn, result in concentrated use of the area. Examples include skiing (that requires ski lifts, car parks, buildings, and roads) and campgrounds (that require roads, picnic tables, and toilet facilities). Contrast DISPERSED RECREATION, INFORMAL OUTDOOR RECREATION.

developing country A low-income country with an economy that is largely based on agriculture, which may be going through the *demographic transition, is often in the process of industrialization, and usually has few resources to spare to solve its own socio-economic and envir-

onmental problems. Most developing countries are in the southern hemisphere. Also known as a **less developed country** or a **Third World** country. Compare DEVELOPED COUNTRY, TRANSITIONAL NATION.

development A process in which something changes by degrees to a different stage, particularly a more advanced stage. **1.** The process by which a living organism is produced and matures from a single cell.
2. Conversion of land into *developed land.
3. Conversion of a *developing country/nation into a *developed country/nation.

development control The statutory process *land use planning system in the UK, through which a local authority determines whether a proposal for a large-scale building development should be granted planning permission, and whether any particular conditions or restrictions should apply.

development rights The rights to develop land, as distinguished from ownership of it. Compare CONSERVATION EASEMENT.

development well A *well that is drilled with the intent of producing *oil or *gas from a reservoir that is known to be productive.

Devensian The most recent *continental glaciation, known as the Devensian in Britain and the Wisconsinan in North America, which began about 70 000 years ago, reached a peak about 18 000 years ago, and ended about 10 000 years ago. *Ice sheets covered much of North America and Europe/Asia and sea surface temperatures were about 2–2.5°C below present ones.

Devonian A period of geological time in the *Palaeozoic era, dating from about 409 to 360 million years ago, during which *fish were dominant and the first *trees, *insects, and *vertebrates appeared.

dew Small drops of moisture that form on cool surfaces at night, caused by the *condensation of *water vapour from the air.

dew point The temperature at which *water vapour in air will condense on a cool surface and form drops (*dew). Also known as **condensation point** or **isobaric saturation point**.

dewater 1. To remove some of the water in a *sludge or *slurry, in order to dry the sludge so that it can be handled and disposed.
   2. To drain water from a tank or trench.
   3. To extract groundwater by wells, *electro-osmosis, *sumps and drains, or by exclusion.

D-horizon The unweathered *bedrock beneath the soil in a *soil profile. Also known as the **R-horizon** (rock). *See also* SOIL HORIZON.

diabatic The direct transfer of heat energy. Examples include *condensation, *evaporation, *solar radiation, and the release of *latent heat. *Contrast* ADIABATIC.

diagenesis The processes that take place in a sediment after *deposition. These include *compaction, cementation (*lithification), *dissolution, and recrystallization.

diamicton A general term for unsorted, unstratified rock debris that is composed of a wide range of particle sizes and could be formed by a variety of different processes.

diamond A crystalline *allotrope of *carbon that is very hard (the hardest known substance) and durable. It is valued as a precious gem when it is cut and polished.

diapause A period of dormancy in insects and other *invertebrates, during which growth or development is suspended. Roughly equivalent to *hibernation.

diastrophism The deformation of the Earth's *crust by *tectonic forces, which gives rise to continents, *mountains, and major geological features such as *folds and *faults, through the processes of *warping, *folding, and *faulting.

diatom Microscopic single-celled *algae that grow in both fresh and salt-water. They produce *silica skeletons that are often preserved in lake or marine sediments after death and are studied in *palaeolimnology. There are more than 11 500 different species of diatoms, and they are a common form of marine *phytoplankton and a primary food source for aquatic animals, especially *filter-feeding shellfish. The proper name for diatom is Bacillariophyta.

diatom ooze A fine muddy deep sea *sediment that is formed mainly from the hard *silica remains of *diatoms.

diatomaceous earth *See* diatomite

diatomite A light, easily crumbled, *silica-rich material that is formed mainly from *diatom remains and is used to filter out solid waste in *wastewater treatment plants. Also known as **diatomaceous earth.**

dichlorodiphenyltrichloroethane *See* DDT.

dicotyledon A flowering plant of the *angiosperm group, which has a seed with two seed leaves (cotyledons).

dieback *See* POPULATION CRASH.

diesel engine *See* INTERNAL COMBUSTION ENGINE.

diesel oil A form of *petroleum that is used to fuel *internal combustion engines. *See* CLEAN FUEL.

diet The usual food and drink that an organism consumes.

diethylstilbestrol (DES) A synthetic oestrogen drug that was once widely prescribed to prevent miscarriages, and was used in feed for livestock and poultry.

differentiation 1. The process by which cells grow to become more specialized.
   2. In tackling climate change under the *Kyoto Protocol, the setting of different *emissions reduction targets and timetables for individual countries, to allow for differing national circumstances.

diffraction The bending of light around objects (such as cloud and fog droplets) which produces fringes of light and dark or coloured bands.

diffuse radiation *Radiation within the *atmosphere that is scattered by tiny particles and travels in all directions, rather than travelling in a beam.

diffused air A method of *aeration that forces *oxygen into *sewage by pumping air through perforated pipes inside a *holding tank.

diffusion The random movement of atoms, molecules, or ions from areas of higher to lower *solute concentration, which tends to distribute them more uniformly.

digester An enclosed *composting system that includes a device to mix and aerate the waste materials in order to speed *decomposition by *aerobic bacteria.

digestion The biochemical *decomposition of *organic matter into mineral compounds and simple organic compounds that can be absorbed into the body of an *organism or excreted from it. *Bacteria decompose *sewage in this way.

dike *See* DYKE.

diluent A fluid that is used to *dilute a substance.

dilute To become weaker in strength or affect, for example thinning out a solution or making it less concentrated by adding a liquid, usually water.

dilution The process of reducing the concentration of a *solute in solution, usually by mixing it with more *solvent.

dilution ratio The ratio between the concentrations of a *solute in a solution before and after *dilution.

dimethyl sulphide (DMS) A molecular gas that is produced by *phytoplankton in the surface waters of the *ocean and escapes into the *atmosphere. It might play a significant role in future *climate change because oxidation of DMS creates a *sulphate-based *aerosol that acts as *cloud condensation nuclei, which might create more cloud cover and thus affect the Earth's heat balance.

dimictic A *freshwater lake or reservoir that has two mixing periods (*turnovers) a year, usually in spring and autumn. *See also* MIXING CYCLE.

diminishing returns A situation in any *open system (such as a company or the economy) in which an increase in the input of a particular resource, with other inputs fixed, results in smaller and smaller returns.

dinoflagellate Microscopic, single-celled marine organisms which can be classified as *plants or *animals because some contain *chlorophyll. They provide a major food source of *plankton at the bottom of the marine *food chain. *See also* RED TIDE.

dinosaur An ancient terrestrial *reptile that stood upright on its legs, did not fly or swim, and could be either carnivorous or herbivorous. Dinosaurs became *extinct at the end of the *Cretaceous Period, 65 million years ago. *See also* K–T BOUNDARY.

dioxin A group of about 75 different *chlorinated hydrocarbon compounds that are created in the production of *pesticides, have no industrial use, are persistent and very *toxic and *carcinogenic, and are hazardous to human health. Exposure to dioxin affects humans in various ways, including disfiguring skin complaints, birth defects, miscarriages, and *cancer. Dioxins are regularly released into the environment in fairly small amounts when chlorinated materials (such as treated wood, plastics, and some specially treated fuels) are burned. Recent European Union directives have significantly decreased dioxin emissions from large incinerators. More dramatic and concentrated releases occur from time to time, particularly in chemical accidents. The 1976 accident at a chemical plant in *Seveso, Italy, spread dioxin over a wide area and contamination only came to light slowly as animals died and people showed signs of dioxin exposure.

dip The angle between a geological surface in *rock (such as a *bedding plane or a *fault) and the horizontal. *See also* INCLINATION.

diploid Having a full set of genetic material consisting of paired *chromosomes, one from each parent.

dip-slope fault A geological *fault that involves vertical movement up and down the *dip of the *fault line.

diptera A large order of *insects with two wings, which includes flies, gnats, midges, and mosquitoes.

direct action Protective action by a particular group of people, that is based on non-violent, passive resistance (for example, using protest marches, picketing a site, or blocking roads) which forces opponents to defend their position. *See also* CHIPKO MOVEMENT, ECO-ACTIVISM.

direct circulation Any form of air circulation (such as a *land breeze) that is driven mainly by warm air rising and cold air sinking.

direct effect An effect that occurs at the same time and place as the initial cause, without any other cause or factor being involved. Also known as **primary effect**. *Contrast* INDIRECT EFFECT.

direct filtration A method of treating *wastewater that involves adding chemicals but not using *sedimentation.

direct insolation *See* DIRECT RADIATION.

direct radiation *Radiation within the *atmosphere that reaches the ground surface directly from the Sun, without being deflected or scattered. *See also* DIFFUSE RADIATION.

direct runoff Water that flows through or over the ground directly into streams, rivers, and lakes.

direct use value Economic values that are associated with the direct use of a natural resource.

directional selection Preferential selection that leads to a consistent change in how a particular character of a population changes through time, such as selection for larger eggs. *Compare* NATURAL SELECTION.

directive An instruction that defines what is to be done or achieved, without necessarily specifying how. *See also* REGULATION.

disaster A *hazard event (natural or induced) that seriously disrupts the normal functions of society and causes widespread human, material, or environmental losses which exceed the ability of the affected society to cope using only its own resources. *See also* CATASTROPHE.

disaster management A comprehensive approach to reducing the adverse impacts of particular *disasters (natural or otherwise) that brings together in a *disaster plan all of the actions that need to be taken before, during, immediately after, and well after the disaster event. These include *mitigation, *preparedness, *emergency response, *recovery, *rehabilitation, and *reconstruction. Also known as **emergency management**.

disaster plan A formal record of agreed roles, responsibilities, strategies, systems, and arrangements for managing a particular type of disaster in a particular place. Also known as **contingency plan**.

disaster risk management An approach to *disaster management that is based on reducing the likelihood of situations or events that leads to disasters, and increasing the capacity to deal with them effectively should they arise.

disaster vulnerability A measure of the ability of a community to cope with the effects of a severe *disaster and to recover afterwards.

discharge 1. The rate of flow of surface water in a *stream or *canal, or of groundwater from an *artesian well, a ditch, or a *spring. Usually expressed in cubic metres per second (m^3 s^{-1}) (cumecs) or litres per second ($L s^{-1}$). 2. Any substance (a gas, liquid or solid, or a *micro-organism) that is emitted, deposited, or allowed to escape from any activity.

3. To remove the electrical energy from something.

discharge consent Approval to *discharge an agreed amount of a substance into the *environment, such as a *pollutant released into the air or into a stream.

disclimax community See EQUILIBRIUM COMMUNITY.

discontinuity A major interruption or break in continuity. In geology, a sudden change in the physical properties of rocks within the Earth.

discordant Not in agreement. **1.** In geology, a body (such as a *dike or *vein) that cuts across primary rock structures (such as bedding).
2. In ecology, a set of individuals that display different traits.

discount The difference between present and future value.

discount rate A rate that is used to convert future costs or benefits to their present value, based on assumptions about how to *discount or reduce the value of each in the future. See also COST–BENEFIT ANALYSIS.

disease Any disorder in the normal function of an *organism, which can be genetic but is often caused by *pathogenic *bacteria, *fungi, or *viruses.

disease agent A physical, chemical, or biological factor that causes disease. See also PATHOGEN.

disease host The individual who contracts a particular *disease.

disease reservoir Any source that harbours disease-causing organisms and could thus cause an outbreak of that *disease. See also DISEASE VECTOR.

disease transmission chain The routes by which a particular *disease is transmitted and spread, including direct contact, indirect contact, via droplets, orally, and by *vectors.

disease vector An organism that transmits a *disease or a *parasite without necessarily being affected itself. See also CARRIER, VECTOR.

disinfect To kill all *pathogenic *micro-organisms.

disinfectant A chemical or physical process that kills *pathogenic organisms in water, air, or on surfaces. Examples are *chlorine and *ozone that are used to disinfect water supplies, wells, swimming pools, and effluent from *sewage treatment.

disinfection See STERILIZATION.

disjunct *Populations that are fragmented or separated from one another by an area where the species is not present at all.

disking See HARROWING.

dispersal **1.** The movement of *organisms away from the places where they were born (*natal dispersion) or where they breed (*breeding dispersal).
2. An approach to containing *oil spills, in which chemical *detergents are used to break up the slick.

dispersant A chemical agent that is used to break up concentrations of organic material, such as spilled oil.

dispersed recreation Forms of recreation (such as hunting, fishing, backpacking or scenic driving) that are usually low-density and take place without facilities, not in a developed recreation site. Contrast DEVELOPED RECREATION, INFORMAL RECREATION, CONCENTRATED RECREATION.

dispersed settlement Scattered individual homesteads in the *countryside. A dispersed form of *rural settlement.

dispersion Scattering or spreading widely. See also DISPERSAL.

dispersion model A computer model that is used to predict how *pollutants disperse across an area, based on local environmental factors. See also POINT SOURCE.

disphotic zone *See* APHOTIC.

disposable Designed to be thrown away after use.

disposables Common term for consumer products and packaging that are used once or a few times and then thrown away.

disposal 1. To get rid of something.
2. The final placement or destruction of toxic, radioactive, or other wastes. *See also* WASTE DISPOSAL.

disposal charge A fee or tax that is paid by a producer of waste materials, which reflects the *social cost of the pollution. Also known as **throughput tax**.

disposal facility Equipment and land that is used to receive waste and dispose of it, including *landfills, *surface impoundments, *land farming, *deep-well injection, *ocean dumping, or *incineration.

disposal well A *well that is used for the disposal of waste into rocks beneath the ground surface.

disruptive coloration A form of *camouflage in which colours and patterns disrupt the body shape and outline of an individual.

dissolution The dissolving of a substance in a solvent.

dissolved load The fraction of a river's load that is carried in solution. It is made up for the most part of ions of *chloride, *sulphate, *bicarbonates, *sodium, and *calcium. It is estimated by measuring *electrical conductivity and provides an indicator of *salinity or *hardness. Excessive amounts make water unfit to drink or use in industrial processes. Also known as **dissolved solids, solution load, total dissolved solids**.

dissolved oxygen The amount of oxygen that is dissolved in a given volume of water at a given temperature and *atmospheric pressure, which is usually expressed in milligrams per litre(mg L^{-1}), parts per million (ppm), or per cent of saturation. It is a widely used *indicator of *water quality, and adequate levels are necessary for fish and other aquatic life. *Secondary wastewater treatment and *advanced wastewater treatment are usually designed to ensure adequate level of dissolved oxygen in waterbodies that receive *wastewater. *See also* BIOLOGICAL OXYGEN DEMAND.

dissolved solids *See* DISSOLVED LOAD.

distillation A process that is used to separate the components of a liquid mixture, by boiling the liquid and then condensing the *vapour. It is used to purify water by removing *inorganic contaminants, and in *desalination to remove salt from water.

distribution area *See* RANGE.

disturbance Any activity or event (such as a forest fire or change in land use) that disrupts natural environmental systems and processes.

diurnal 1. Active during daylight hours.
2. Recurring every day or having a daily cycle. *See also* CIRCADIAN RHYTHM.

divergence Movement or flow in different directions, such as winds or ocean currents. *Contrast* CONVERGENCE.

divergent evolution The evolution of two or more unique *species from one ancestral species, through the separate evolution of isolated *populations. *Contrast* CONVERGENT EVOLUTION.

divergent margin *See* DIVERGENT PLATE BOUNDARY.

divergent plate boundary A boundary between adjacent *lithospheric *plates that are being pulled or pushed apart, and where new lithosphere is created, such as at *mid-oceanic ridges. Also known as **divergent margin, constructive margin** or **constructive plate boundary**.

diversification The process of increasing *diversity or variety.

diversion A change in direction. Planned diversions are used, for example, to direct part of a stream flow into a *water supply system and to control the spread of *hazardous material in order to contain risk.

diversion ditch An artificial channel that intercepts and transports water from one place to another.

diversion rate The percentage of waste materials that are diverted from traditional disposal (such as *landfill or *incineration) to be *recycled, *composted, or *reused.

Diversitas An international partnership of intergovernmental and *non-governmental organizations that was formed in 1991 to promote and facilitate scientific research on *biodiversity.

diversity Variety, the presence of a wide range of variation in the qualities or attributes of the thing under discussion. See also BIODIVERSITY.

divide See DRAINAGE DIVIDE.

diving A method of marine *fishing in which some species (such as octopus, sea urchins, and sea cucumbers) are hand-picked by divers, brought to the surface, placed in boats, and taken ashore for processing.

division One of the major subgroups within the *plant kingdom, comprising a number of similar *classes of plants. See also BIOLOGICAL CLASSIFICATION.

DMS See DIMETHYL SULPHIDE.

DNA Deoxyribonucleic acid, the genetic material that is found in all living organisms. It is the long molecule composed of *carbon, *nitrogen, and *phosphorus that is in the *nucleus of every *cell, controls inheritance, and directs the development and functioning of all cells. The strand or string of DNA has a *double helix structure so that the genetic instructions (*gene code) of the *cell are passed on when the cell divides (in one of the split pairs of the helix, which then builds its own matching helix), to produce a perfect copy. See also CHROMOSOME, GENE, RECOMBINANT DNA.

DNA bank The storage of *DNA, for example to assist future medical research, allow individuals to trace the pattern of *diseases in families, or create an archive of the genetic material of different organisms.

DNA fingerprinting Techniques that are used to detect the unique patterns in *DNA, which operate like a genetic barcode and can be used to indicate the presence of a *gene associated with a given trait. This has applications in forensic science, and in selecting breeding stock with favourable characteristics (such as disease resistance or rapid growth).

DNA replication The duplication of a *DNA molecule by itself, in order to create new strands of DNA.

DNAPL See DENSE NON-AQUEOUS PHASE LIQUID.

Dobson unit (DU) A unit that is used to measure the amount of *ozone in the *atmosphere, based on thickness. 100 DU of ozone would form a layer 1 millimetre thick.

Dodo Common name for the Mauritius dodo (*Raphus cucullatus*), a flightless bird that lived on the island of Mauritius and has been extinct since about 1660 because of over-killing. It was about 1 metre high, lived on fruit, nested on the ground, and was captured and eaten by visiting sailors.

DOI See US DEPARTMENT OF THE INTERIOR.

doldrums A climate region near the *equator that has low *atmospheric pressure and light, shifting winds. Also known as **equatorial low**. See also INTERTROPICAL CONVERGENCE ZONE.

doline A closed depression in an area of *karst, that is created by the *solution and *subsidence of *limestone.

dolomite 1.A common rock-forming *mineral, composed of calcium–magnesium carbonate, that is crystalline and forms extensive beds as a compact *limestone. See also KARST.
2. A term also often used to refer to the *sedimentary rock dolostone, but strictly speaking it applies only to those dolostones which contain more than 90% dolomite.

dome 1. The high, stable central part of an *ice sheet, with low rates of accumulation and slow ice movement.

2. A round *anticline from which the ground slopes downward in all directions.

domestic sewage Organic waste and wastewater that is produced by a household and carried from houses to treatment works by sewers.

domestic use Water that is used for *household purposes such as washing, food preparation, and bathing.

domestic waste *See* HOUSEHOLD WASTE.

domesticate To *breed a *species for specific characteristics that humans value, or the species that is so bred. *See also* CULTIVATE.

domesticated species A species in which the evolutionary process has been deliberately influenced by humans to meet their needs. Also known as **cultivated species**.

dominance The extent to which a particular species or organism exerts the most influence in a *community because of its size, abundance, or cover.

dominant plant/species/tree The tallest individuals or species in a stand of vegetation (such as a forest), which receive the greatest amount of sunlight.

dominant social paradigm (DSP) The view that humans are superior to other all other *species, the Earth provides unlimited *resources for humans, and that progress is an inherent part of human history. *Contrast* NEW ENVIRONMENTAL PARADIGM.

dominant use An approach to *resource management in which land and water are the main type of land use, and others are treated as of secondary importance.

dominant use management An approach to *resource management that is based on the idea that although a given area of land may be capable of many uses, it will provide for one use better than any other, so the land is managed for the single purpose of maximizing that use to the exclusion of other uses where conflicts exist.

do-no-harm principle *See* PRECAUTIONARY PRINCIPLE.

donor site The site from which individuals of an *introduced species are taken in order to be introduced to another place (the *receptor site).

Doppler radar A type of weather radar that determines whether air (and thus weather) is moving toward or away from the radar.

dormant Temporarily inactive, such as a volcano that is no longer active but not known to be extinct. *See also* QUIESCENCE, TORPOR.

dosage *See* DOSE RATE.

dose The quantity of a chemical that is given to an organism or to which it is exposed, or the amount of ionizing radiation that someone or something has been exposed to. *See also* ABSORBED DOSE, ADMINISTERED DOSE, APPLIED DOSE, HUMAN EQUIVALENT DOSE, LD50, POTENTIAL DOSE.

dose effect *See* DOSE RESPONSE.

dose equivalent A measure of the amount of ionizing *radiation that an individual person receives, which is based on the type of radiation, the amount of the body exposed, and the risk of exposure. Expressed in *rem.

dose rate A measure of the dose that is received per unit of time, expressed for example in milligrams per day (mg day $^{-1}$). Also known as **dosage**.

dose response An indication of how an individual, or a population, is likely to be affected by a change in the amount, intensity, or duration of an exposure. Also known as **dose effect**.

dose-response assessment An estimation of the potency of a chemical.

dose–response relationship The quantitative relationship between the amount of exposure to a substance and the extent of *toxic injury or *disease that is produced in response to that exposure.

dosimeter An instrument that is used for measuring and registering the *radiation dose a person has received or that something has absorbed.

double helix The twisted-ladder shape that two linear strands of *nucleotides assume when they are bonded together via their *base pairs to form *DNA.

doubling time The time it takes, usually expressed in years, for a population to double in size. A population growing at 5% a year would double in 14 years, whilst a population growing at 1% a year would double in 70 years. Doubling time can be estimated by dividing the annual percentage population growth rate (r) into 70, i.e. doubling time = 70/r. The importance of doubling time as an indicator of rate of change is that it shows very clearly how quickly population grows as the growth rate changes by even a relatively small amount. This is because the change involves *geometric growth rather than *arithmetic growth, in the same way that money accumulates in a bank savings account with compound interest. Population cannot double if the *growth rate is zero (there is no growth) or negative (population is declining, not expanding).

downburst A severe, localized downward push of air from a thunderstorm or shower. Also known as **downdraft**. *See also* MACROBURST, MICROBURST.

downdraft *See* DOWNBURST.

downpour Heavy *rainfall.

downstream In the same direction as the current or stream. *Contrast* UPSTREAM.

downwarping The *warping or bending downwards of the Earth's *crust.

downwelling The process by which surface water in the sea increases in density and sinks, usually along a *coastline or where water masses converge in the oceans. *Contrast* UPWELLING.

downwind In the direction that the wind is blowing, with the wind. *Contrast* UPWIND.

draft environmental impact statement (DEIS) The draft version of an *environmental impact statement that is made available to the public and other agencies for review and comment.

dragline An excavating machine that uses a bucket that is operated and suspended by means of cables, which allows the bucket to be dragged toward the machine for loading.

drain An open channel (*drainage ditch) or pipe that is used to move water such as surface runoff or wastewater from a factory *downstream.

drainage 1. A method of improving land quality by removing excess water from the soil, which decreases *waterlogging. *See also* SUBSURFACE DRAINAGE, SURFACE DRAINAGE. **2.** The removal of water from land by streams and rivers, as part of the local *water cycle.

drainage area The area of land that is drained by a particular river within a *drainage basin, defined by the *drainage divide.

drainage basin The area of land that is drained by a *river and its *tributaries. The boundary of a basin is defined by the *drainage divide, and the area within the divide is the *drainage area. Also known as **basin, catchment, water catchment, watershed, river basin**.

drainage density The length of river channel per unit of *drainage area measured in kilometres per square kilometre ($km\ km^{-2}$). It provides a measure of the concentration of streams in an area and thus a measure of the amount of dissection of the land surface. *Contrast* STREAM FREQUENCY.

drainage ditch A ditch or open *channel that is used to carry excess water or sewage downstream. *See also* SURFACE DRAINAGE.

drainage divide A ridge of land that separates adjacent *drainage networks. Also known as **divide, hydrographic divide, interfluve, watershed**.

drainage network The network of *river channels and *tributaries that forms within a *drainage basin. Also known as **river network**.

drainage pattern The pattern displayed by a *river network, as seen from the air or on a map, which is usually controlled by *geology. The most common ones are *dendritic,

*trellis, *rectangular, *annular, *radial, *centripetal, *parallel, and *deranged.

drainage system A network formed by a main *river and its *tributaries.

drainage tile *See* FIELD TILE.

drainage wind *See* KATABATIC WIND.

drainfield 1. An area of land that is used as a filter for *wastewater.
2. The part of a *septic system where the wastewater is released into the *soil.

draw down To decrease water level, such as in a storage tank or *reservoir, or in groundwater around a well, by the

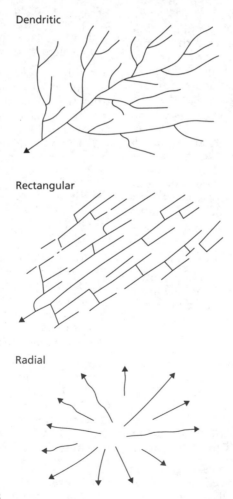

Dendritic

Rectangular

Radial

Fig 4 Drainage patterns

*extraction of water. *See also* CONE OF DEPRESSION.

dredge A method of marine fishing that uses a bag dragged behind a vessel in order to scrape the bottom, usually to catch shellfish.

dredgeate The material that is excavated during *dredging.

dredging The process of excavating, creating, or altering a water body such as a *river, *lake, or *estuary, by scooping or sucking up sediment from the bed in order to deepen it.

drift 1. Sprayed or dusted material that does not settle on the target area but blows further away.
2. Sediment created by glacial activity, such as *boulder clay or till.
3. Material that is transported by wind and deposited in heap, such as snowdrift.
4. A horizontal tunnel in an underground *mine.
5. *See* GENETIC DRIFT.

drift current A broad and slow-moving type of *ocean current.

drift net A fishing net that can be several kilometres long and is suspended vertically from floats at a specific depth in the water. It is left to drift freely with the current to catch fish.

driftwood Wood that is floating on the sea or has been washed ashore.

drill seeding Sowing seed in rows in the soil. *Contrast* BROADCAST SEEDING.

drinking water *See* POTABLE WATER.

drinking water supply Any source of raw or treated water that is, or may be, used by a public water system, or as *drinking water by one or more individuals.

drip irrigation A form of *irrigation that uses a perforated pipe to deliver water one drop at a time directly to the soil around each plant, in order to eliminate *runoff and *soil erosion. Also known as **trickle irrigation**.

drizzle Very light *rain composed of very fine droplets formed by *stratus clouds.

drop-off A method of collecting *recyclable materials in which individuals take them to a designated collection site.

DROUGHT

A prolonged, continuous but temporary period of very dry *weather compared with the long-term average for that place. A drought is different from a dry climate, which is usually associated with a region that is normally, or at least seasonally, dry. A drought is also drier and lasts much longer than a dry spell, which is usually defined as more than 14 days with no significant *precipitation causing a relatively short period of moisture deficiency. Droughts often last for months or even years. Direct impacts of drought include decreased biological productivity (and thus lower yields) in cropland, rangeland and forest land; increased fire hazard; reduced water levels; increased mortality rates amongst livestock and wildlife; and damage to fish and wildlife habitats. There are four main types of droughts: meteorological drought (usually defined by rainfall deficit or the degree of dryness compared with the average amount), hydrological drought (usually defined by river flow deficit), agricultural drought (usually defined by soil moisture deficit), and famine drought (usually defined by food deficit). Drought severity is usually gauged by a number of factors, the most important of which are the degree of moisture deficiency (the greater the deficiency, the worse the drought), the duration of the deficiency (the longer it lasts, the worse the drought), the size of the area affected (the bigger the area, the worse the drought), and the number of people affected (the greater the number, the worse the drought). *See also* DUST BOWL.

drumlin A long, elliptical hill usually composed of glacial *till (but occasionally composed of solid rock), streamlined in the direction of ice movement, that was deposited beneath an advancing *glacier. Usually found in a cluster (*drumlin field).

drumlin field A group of *drumlins which are aligned more or less parallel to one another (in the direction of ice movement). Also known as a **drumlin swarm**.

drumlin swarm *See* DRUMLIN FIELD.

dry adiabatic lapse rate (DALR) The rate (9.8°C per kilometre) at which a rising parcel of dry air cools, by *adiabatic expansion, and at which a sinking parcel of dry air warms. *Contrast* SATURATED ADIABATIC LAPSE RATE. *See also* LAPSE RATE.

dry alkali injection A method of controlling *emission of *air pollutants by spraying dry sodium bicarbonate into *flue gas in order to absorb and neutralize *acidic *sulphur compounds.

dry deposition The *fallout of gases, aerosols, and fine particles with no associated *precipitation. This usually occurs relatively close to the emission sources of the gases. An important process in the formation of *acid deposition.

dry farming A method of *farming in *arid and *semi-arid areas without using *irrigation, which relies on treating the land in ways that conserve moisture (including using a *mulch).

dry ice Frozen *carbon dioxide that is used in *cloud seeding.

dry prairie A type of *prairie that occurs on slopes and well-drained uplands, and is often interspersed with areas of *mesic prairie in valleys.

dry rock geothermal energy A method of extracting *geothermal heat from the Earth based on pumping water through hot rocks.

dry valley A valley that has no river flowing through it. It would have been formed under a climate that is very different from today's (for example by *meltwater at the end of the *Pleistocene) and indicates previous erosion by water.

dry wash The dry bed of an *intermittent stream.

dryland An area with an *arid climate, with low precipitation and rainfall mainly in localized, brief, high-intensity storms.

drystone wall A stone wall that is built without mortar.

dry-summer subtropical climate *See* MEDITERRANEAN CLIMATE.

DSP *See* DOMINANT SOCIAL PARADIGM.

DU *See* DOBSON UNIT.

Ducks Unlimited A US-based non-profit conservation organization which is committed to conserving, restoring, and managing the country's *wetlands, *waterfowl, and *wildlife.

ductile The physical property of being able to sustain large *plastic *deformations without fracture, as occurs in metals that are capable of being hammered thin without breaking, or drawn into a thin shape such as a wire. *See also* MALLEABLE.

duff A North American term for *organic material (such as dead grass, leaves, conifer needles, and other plant parts) that is in various stages of *decomposition on the floor of a *forest.

dulosis Enslavement, usually of worker ants by a rival species.

dump An uncontrolled area where *solid wastes have been left on or in the ground, sometimes illegally, without environmental controls. *See also* LANDFILL.

dune 1. A *bedform in an *alluvial channel that forms at higher *flow velocity than a *ripple and which gradually moves downstream along the bed of the channel, much more slowly than the water flows.
2. A mound of sand that is built up by prevailing winds, in a *desert or on a sandy *beach. Also known as **sand dune**.

dune form The shape assumed by a sand *dune, which depends on wind speed and direction, and the availability of sand. Examples include crescent-shaped dunes (*barchans), *seif dunes, and star-shaped dunes.

dung Animal manure. *See also* FARM-YARD MANURE.

dusk The period of declining daylight between sunset and dark. Also known as **twilight**. *See also* DAWN.

dust General term for fine *particles that are light enough to be suspended in *air.

dust devil A strong miniature *whirlwind that develops best on clear, dry, hot afternoons and throws up dust, litter, and leaves into the air.

dust storm A strong wind that blows fine dust across the surface of a *desert, which can significantly reduce *visibility and adversely affect communications and agriculture. *Contrast* SAND STORM. *See also* HABOOB.

dust veil Fine particles of *ash, dust, and *sulphur dioxide that are thrown up into the *stratosphere by a volcanic eruption, and which can serve as *condensation nuclei for the formation of sulphate *aerosols.

dust whirl *See* DUST DEVIL.

Dutch elm disease A disease of elm trees that is caused by a *fungus (*Ceratocystis ulmi*) which is spread by a bark-beetle. The disease is usually fatal, was first introduced to the UK in the 1930s, and since the 1960s has devastated the elm population of the UK.

duty of care The obligation to avoid negligence, particularly to take reasonable care not to cause physical, economic, or emotional loss or harm to others.

DUST BOWL

An area in the *Great Plains region of the USA that experienced prolonged *drought and *soil erosion in the late 1920s and 1930s, partly caused by dry weather but aggravated by over-intensive farming practices. The area—which includes parts of Kansas, Oklahoma, Texas, New Mexico, and Colorado—has a long history of persistent droughts and had a natural vegetation of grass which protected the fine-grained soils from *wind erosion. Between about 1885 and 1915 many people moved into the area to settle, many new farms were established, and large areas were given over to wheat and crops and to cattle grazing. The *marginal land was unsuited to intensive agricultural use and serious soil erosion began in the early 1930s when a period of severe droughts affected the area. Up to 10 centimetres of valuable *topsoil (including *organic matter, *clay, and *silt) was blown from some places, removing the very resource on which farming was dependent, and it drifted against fences and buildings in other areas. Thousands of families migrated westward, abandoning their farms, and up to a third of the families who stayed behind survived only with the aid of government relief. Farm incomes were seriously depressed during the drought in the 1930s and they also suffered badly during a drought in the 1950s. From 1942 onwards concerted efforts were made by federal and state governments to develop programmes for soil conservation and for rehabilitation of the Dust Bowl region. These included seeding large areas with grass, introducing 3-year *crop rotations, and use of *contour ploughing, *terracing, strip planting, and shelter belts. Whilst these measures have protected much of the remaining soils in the region, droughts in the 1950s, 1960s, and late 1970s caused further wind erosion and reinforced the need for more sustainable use of resources in this marginal area.

dyke 1. An embankment made of earth and rock, for example to prevent a spill from spreading or to prevent flooding.

2. A steeply inclined layer of *intrusive *igneous rock, formed from cooling *magma, that is relatively resistant to *erosion and is *discordant and cuts across existing layers.

dynamic equilibrium A form of *equilibrium in which short-term changes are superimposed on a background state which is itself changing, which allows open systems to remain stable over long periods of time.

dynamics The study of the relationships between motion and the forces that affect motion.

dysgenic The creation of low-quality offspring, for example by harvesting the best individuals and leaving poorer ones to reproduce. *Contrast* EUGENIC.

dystrophic A shallow, acidic lake that contains a lot of *organic matter, but has a low *pH, few available nutrients and high oxygen demand. It is almost *eutrophic, and supports many plants but few fish.

E. coli A species of *bacteria (*Escherichia coli*) of the coliform type that are found in the intestine and indicate *faecal contamination of water. *See also* COLIFORM BACTERIA.

early action The technical term for taking steps to reduce *air pollution by *greenhouse gases before the start of the *Kyoto *Commitment Period, by reducing *emissions, investing in *Clean Development Mechanism projects, *Joint Implementation, or *emissions trading.

early forest succession The *community of plants and animals that develops first after the removal or destruction of vegetation in an area. An early stage in *succession. *Contrast* LATE FOREST SUCCESSION.

earlywood The light-coloured wood cells within the annual *growth ring of a tree that are formed at the beginning of the *growing season. Also known as **springwood**. *Contrast* LATEWOOD.

Earth Charter An international initiative that was formed in 1994 in response to the *World Commission on Environment and Development, in order to 'establish a sound ethical foundation for the emerging global society and to help build a sustainable world based on respect for nature, universal human rights, economic justice, and a culture of peace'.

Earth Council An international *non-governmental organization that was created in September 1992 to promote and advance the implementation of the agreements reached at the *United Nations Conference on Environment and Development.

Earth Day An event that was established in 1970, is supported by the United Nations, and is held every year on 22 April in many countries in order to raise environmental awareness.

Earth Liberation Front (ELF) A North American radical, underground environmental movement that has no leadership, membership, or official spokesperson but is a loose coalition of individuals and groups engaged in radical *environmental activism.

EARTH

Our home planet, the fifth largest of the nine planets within the *solar system. It rotates around its axis (once every 24 hours), orbits around the *Sun 150 million kilometres away (once a year), and the *Moon orbits around it 385 000 kilometres away (once every 27 days 7 hours and 43 minutes). These rhythmic motions create night and day and the seasons of the year. Although we can't feel it, the Earth is moving through space at great speed. Planetary movement is far from random and chaotic, and the order and pattern within it are vital for life on Earth. The Earth moves within the solar system, and the solar system itself is moving through the vastness of space. By planetary standards the Earth is relatively small; it has a surface area of 510 million square kilometres and a mean radius of 6371 kilometres. The Earth is almost but not entirely spherical; it is an oblate spheroid. The circumference around the equator is 40 077 kilometres, compared with 40 009 kilometres around the poles.

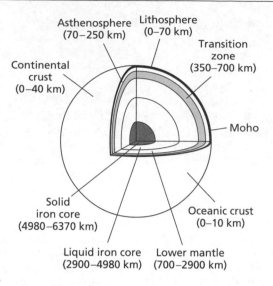

Asthenosphere Lithosphere
(70–250 km) (0–70 km)

Transition
zone
(350–700 km)

Continental
crust
(0–40 km)

Moho

Solid
iron core
(4980–6370 km)

Oceanic crust
(0–10 km)

Liquid iron core   Lower mantle
(2900–4980 km)   (700–2900 km)

Fig 5 Earth's interior

Earth Observation Satellite (EOS)
A *remote sensing programme developed by NASA in 1998, as part of *Mission to Planet Earth, which uses a series of *satellites to monitor long-term global *environmental change.

Earth Resources Technology Satellite (ERTS) A *satellite that was launched by NASA in 1972 and used for *remote sensing of land resources on Earth. Now called *Landsat.

Earth Summit See UNITED NATIONS CONFERENCE ON ENVIRONMENT AND DEVELOPMENT.

EarthFirst! A radical activist movement within the broader *environmental movement, based on local or regional groups who take *direct action to try to prevent activities (such as highway construction or airport developments), which would cause environmental damage and *ecosystem changes. See also ENVIRONMENTAL ACTIVISM, MONKEY-WRENCHING.

earthflow A rapid *mass movement process in which *soil and other loose *sediment moves down a slope, often caused by water saturation from *rainfall.

earthquake Shaking and vibration of the surface of the Earth that is caused by the release of stress accumulated along a *fault or by *volcanic activity, usually associated with *plate tectonics. As *seismic energy accumulates in rocks subjected to *strain, they bend slowly until they can no longer withstand the strain. They then split apart along a fault. The sudden failure and displacement produces an earthquake which might vary in intensity from barely detectable trembling to violent and damaging shaking of the ground. As the stressed area ruptures, this relieves the strain and causes a sudden ground movement on one or both sides of the fault. *Aftershocks occur as the fault settles back, and as slower-moving seismic waves affect the area. Stress builds up through time, and the greater the accumulated stress released, the larger the earthquake. The *focus is the centre of the earthquake underground and the *epicentre is found on the ground surface directly above the focus. Earthquakes often trigger other geological hazards too, including *landslides and *liquefaction. Submarine earthquakes can cause *tsunamis. See also MERCALLI SCALE, RICHTER SCALE.

earthquake intensity See MERCALLI SCALE.

earthquake magnitude *See* RICHTER SCALE.

Earthwatch Popular name for a programme (United Nations System-wide Earthwatch) that was established in 1973 as part of the *United Nations Environment Programme, through which UN agencies work together on global environmental issues by exchanging and sharing environmental data and information.

earthworks The disturbance of land by moving, loosening, depositing, shaping, compacting, and stabilizing *soil and *rock.

easement A North American term for a voluntary binding agreement that allows the right to use someone else's land for a specific purpose, such as for a right-of-way or for a utility company to build a power line. *See also* CONSERVATION EASEMENT.

east The compass point that is at 90° from north.

easterlies *Winds that blow from the east, for example in equatorial regions (trade wind region) and polar regions (polar casterlies).

ebb current *The tidal current that flows away from the shore, or down a *tidal river or *estuary.

ebb tide The part of the *tide cycle between high water and the following low water, during which the water level is falling. Also known as **falling tide**.

EC *See* EUROPEAN COMMUNITY.

eccentricity A measure of how much an ellipse or planetary orbit departs from a perfect circle, which has an eccentricity of zero.

ecdysis Shedding of the skin by *arthropods and *reptiles.

ecdysone The moulting hormone of insects and *arthropods, which causes them to *moult.

ecesis The ability of a plant or animal to become established in a new *habitat. *See also* SUCCESSION.

echo-sounder An instrument that measures the depth of a body of water by emitting pulses of sound into it and measuring the time it takes them to return.

ecliptic The plane of the *Earth's *orbit about the *Sun, which defines the apparent annual movement of the Sun among the stars.

eclosion The emergence of an insect *larva from the egg or an adult insect from the *pupa.

eco- A prefix derived from the Greek word for house.

eco-activism *Direct action designed to raise awareness of, lobby against, or stop particular activities that damage the environment, often at a particular site (such as a planned new road site). *See also* CHIPKO MOVEMENT, ECODEFENCE, ECO-EXTREMISM.

ecocatastrophe An ecological or environmental *catastrophe, or a situation or event that causes major ecological or environmental change.

ecocentric Nature-centred. Based on the belief that all living organisms are equally important, that the environment exists for the benefit of all of nature and not just people, and that nature has rights. Also known as **biocentric**. *Contrast* ANTHROPOCENTRIC, TECHNOCENTRIC.

ecocentrism A *worldview that sees all of nature as having inherent value, and is centred on nature rather than on humans. Also known as **biocentrism**. *See* ANTHROPOCENTRISM.

ecocide A planned effort to eliminate all or part of an *ecosystem.

ecocline A gradient of gradual and continuous geographical change in the environmental conditions of an *ecosystem or *community.

ecodefence The use of non-violence, civil disobedience, and *direct action to prevent damage or destruction of ecological resources. A form of *eco-activism.

ecodevelopment An *environment-friendly approach to *sustainable development that seeks to include social balance, ecological balance, economic efficiency, respect of cultural identity, and harmonious regional development.

ecodoom pessimist See PESSIMIST, ENVIRONMENTAL.

eco-efficiency A management process that is designed to reduce the *environmental impact of producing goods and services, for example by increasing mineral recovery, using fewer inputs (such as energy and water), *recycling more, and reducing *emissions.

eco-extremism Extreme and often aggressive forms of *eco-activism, including *ecotage, *ecoterrorism, and *monkeywrenching.

ecofarming An ecological approach to *farming that is designed to *conserve *biodiversity and *natural resources, for example by using organic rather than chemical *fertilizers and *pesticides. See also ALTERNATIVE AGRICULTURE.

ecofascism The insistence by some extreme *environmentalists that everyone should adopt *environment-friendly lifestyles, irrespective of their personal wishes, preferences, or needs.

ecofeminism A philosophy of respect for nature that is based on feminist philosophies of justice, egalitarianism, cooperation, and non-aggressive behaviour. It reflects a convergence of the *radical ecology movement and *feminism, and seeks to transform the traditional patriarchal socio-economic system which is based on male domination and emphasizes competition, dominance, and individualism.

ecoforestry A passive approach to *forest management that is *ecocentric and based on the values of *deep ecology.

ecogeographical rule A so-called 'rule' (in reality a general statement), proposed by 19th-century naturalists, that describes the association between climate and the morphology of warm-blooded animals (mammals and birds). See also ALLEN'S RULE, BERGMANN'S RULE, GOLGER'S RULE.

Ecoglasnost An *non-governmental environmental organization that was established in Bulgaria in 1989.

ecohouse An *environment-friendly house that is designed to be *sustainable, and built and operated to reduce environmental impacts to a minimum.

eco-label A label that is attached to products (such as paper products, textiles, detergents, paints, and electrical appliances such as refrigerators or dishwashers) that produce fewer *environmental impacts than other competing products, to help consumers make informed choices. See also ENVIRONMENTAL LABELLING.

ecological Relating to the interrelationships between *organisms and their *environment.

ecological amplitude The range of one or more environmental conditions in which an *organism or a process can function. See also TOLERANCE.

ecological approach An approach to the management of *natural resources that considers the relationships among all organisms (including humans) and their *environment.

ecological assessment An evaluation of how a *chemical of concern does or might affect plants and animals other than people and domestic species.

ecological balance A state of *dynamic equilibrium within a *community of organisms, in which *diversity (genetic, species, and ecosystem) remains relatively stable but can change gradually through natural *succession.

ecological community See COMMUNITY.

ecological development 1. A gradual process of environmental modification by organisms.

2. An approach to development that

takes into account ecological factors and the need to conserve *biodiversity and protect *ecological systems. *See also* SUSTAINABLE DEVELOPMENT.

ecological economics A branch of *economics that takes into account ecological principles and examines the economic values of *non-market ecological products and services. *See also* ENVIRONMENTAL ECONOMICS.

ecological efficiency The ratio between the *productivity at successive *trophic levels within a *food web, which is usually expressed as a percentage.

ecological energetics The study of how *energy is used within an *ecosystem, particularly by tracing the movement of energy through a *food web.

ecological entity Any particular part of an *ecosystem, including a *species, a group of species, an ecosystem function or characteristic, or a specific *habitat or *biome.

ecological equivalent Different species that occupy similar *niches in similar *ecosystems in different places.

ecological evaluation Determining the value of the functions of an *ecosystem that are provided by natural ecosystems, in monetary or other terms, to guide the planning and management of *nature conservation.

ecological exposure The exposure of a non-human organism to a chemical, radiological, or biological *agent.

ecological fallacy The mistake of assuming that a relationship that is found between variables based on aggregate or grouped data will also be found or will apply between individuals.

ecological footprint A measure of the amount of space or *environment that is necessary in order to produce the goods and services that are required to support a particular lifestyle. Also known as **footprint**.

ecological impact The effect that an activity (natural or otherwise) has on living *organisms and their *abiotic environment.

ecological indicator A characteristic of the *environment that can be measured and indicates the current state of ecological resources, or an ecological response to exposure to an *agent.

ecological integrity *See* ECOSYSTEM INTEGRITY.

ecological introduction *See* INTRODUCED SPECIES.

ecological land type A North American term for an area of land that has a distinct combination of natural, physical, chemical, and biological properties, which cause it to respond in a predictable way to particular *management practices.

ecological limit *See* CARRYING CAPACITY.

ecological management unit A North American term for an area of land that contains *soils with similar characteristics, enabling it to be managed as a single body.

ecological modernization The belief that *economic development and *environmental protection are compatible, that industrial economies can be reconciled with nature by means of markets, politics, and cultures, and that a sustainable balance can be achieved through *environmental management, environmental policy, cleaner technologies, and the 'greening' of institutions. *See also* SUSTAINABLE DEVELOPMENT.

ecological niche *See* NICHE.

ecological pyramid A diagram that shows the number of *organisms (*pyramid of numbers), total *biomass (*pyramid of biomass), and energy (*pyramid of energy), at each *trophic level within a *food chain.

ecological quality The quality of an *ecological system, in terms of biological, physical, and chemical conditions and of *environmental integrity.

ecological quality objective The desired level of the *ecological quality that is defined in a *management plan or policy.

ecological range See NATURAL RANGE.

ecological resilience See RESILIENCE.

ecological restoration The recovery of a damaged *ecosystem, either naturally or as a result of management that is designed to re-establish its structure and function. See also RESTORATION, REHABILITATION.

ecological risk See ENVIRONMENTAL RISK.

ecological risk assessment A formal process that evaluates the likelihood of adverse ecological effects as a result of *exposure to one or more *agents.

ecological risk characterization A process that evaluates the risk posed to a *population or *ecosystem by an *agent, based on results from studies of *exposure and *resilience.

ecological services See ECOSYSTEM SERVICES.

ecological species concept A definition of *species as a set of *organisms that is adapted to a particular set of resources (*niche) in the environment, which explains differences in form and behaviour between species as *adaptations to resource availability. Compare BIOLOGICAL SPECIES CONCEPT, CLADISTIC SPECIES CONCEPT, PHENETIC SPECIES CONCEPT, RECOGNITION SPECIES CONCEPT.

ecological sustainability The maintenance or restoration of the composition, structure, and processes of *ecosystems. See also SUSTAINABILITY, SUSTAINABLE DEVELOPMENT.

ecological system See ECOSYSTEM.

ecologically and scientifically significant natural area A North American term for land and water that retains most but not necessarily all of its natural character, which is significant for historical, scientific, palaeontological, or natural features.

ecologically sustainable use The use of a *species or *ecosystem within its natural ability for renewal or *regeneration, or its *carrying capacity.

ecology The study of the interrelationships between *organisms and their *environment, including all *biotic and *abiotic components. The term was first used in 1866 by the biologist Ernst Haeckel. More recently it is generally taken to mean the study of the structure and function of *nature. Core themes in modern ecology include: the relationship between *habitat stability, rate and direction of *evolution, and changes in *biodiversity; the relationship between environmental changes caused by human activities and by *extinction; the importance of diversity as an attribute of ecosystems and of the *biosphere, and the significance of losses of diversity; *carrying capacity and the extent to which useful productivity can be enhanced and sustained; and *resilience, the extent to which nature can restore itself.

Eco-Management and Audit Scheme (EMAS) A *European Union regulation that came into force in 1995 and is designed to promote continuous improvement in the environmental performance of industrial activities by allowing companies and services to certify that they have an appropriate *environmental management system (under *ISO 14001).

econometrics The application of mathematical and statistical models to economic theories and problems.

economic analysis Study of the economic effects of a particular activity or decision.

economic depletion The use of 80% or more of a *non-renewable resource.

economic development The process of raising the level of prosperity and material wealth in a society through increasing the productivity and efficiency of its *economy, particularly through an increase in industrial production. See also ECOLOGICAL MODERNIZATION, ECONOMIC GROWTH, SUSTAINABLE DEVELOPMENT.

economic evaluation Comparative analysis of alternative courses of action in terms of both their costs and consequences. See also CONTINGENT VALUATION METHOD, COST–BENEFIT ANALYSIS, HEDONIC PRICING, TRAVEL COST METHOD.

economic growth An increase in the amount of economic activity in a country, often expressed in terms of *gross national product.

economic impact The *impact or effect on an economy (of a place, region, or country) of a particular activity, project, or programme.

economic life The period of time over which an asset (such as a facility, structure, or source of resources) will have economic value and be usable.

economic scarcity The shortage of a marketable good or service, for which the price will rise so long as there is demand for it. See also RESOURCE SCARCITY.

economic system How a particular society distributes its resources in order to produce goods and services.

economics The study of the production, distribution, and consumption of resources, and the management of state income and expenditure. See also CLASSICAL ECONOMICS, ECOLOGICAL ECONOMICS, ENVIRONMENTAL ECONOMICS, MACROECONOMICS, MICROECONOMICS.

economies in transition (EIT) The Central European, Baltic, and former Soviet countries during their transition from centrally planned economies to market-based economies, which involves both political and economic reform.

economy The organized system for the production, distribution, and consumption or use of material goods and services.

ecopiracy See BIOPIRACY.

ecoregion Part of an *ecozone that has distinctive *climate, *physiography, *vegetation, *soil, water, and animals, and is defined by environmental, geological, and geographical factors. Contrast NATURAL REGION. See also BIOREGION.

ecosophy A philosophy of ecological harmony or equilibrium that was proposed by Arne *Næss, who stressed the need to consider ethics and values as well as the physical symptoms of environmental change.

ecosphere See BIOSPHERE.

ecosystem Short for ecological system, meaning the natural interacting *biotic and *abiotic system in a given area, which includes all of the *organisms (plants, animals, fungi, and microorganisms) that live in particular *habitat, along with their immediate physical environment. Examples include a *lake, *forest, or *drainage basin. The term was first used by British ecologist Arthur Tansley in 1935, who visualized ecosystems as being composed of two parts, the *biome and the habitat. In Tansley's view 'all parts of such an ecosystem—organic and inorganic, biome and habitat – may be regarded as interacting factors which, in a mature ecosystem, are in approximate equilibrium; it is through their interactions that the whole system is maintained'. Many ecologists regard ecosystems as the basic units of *ecology because they are complex, interdependent, and highly organized, and because they are the basic building blocks of the *biosphere. Also known as **biocoenosis or holocoen**.

ecosystem approach See ECOSYSTEM MANAGEMENT.

ecosystem diversity The variety of unique biological *communities or *ecosystems, in terms of species composition, physical structure, and processes. This is the highest level of *biodiversity. Compare SPECIES DIVERSITY, GENETIC DIVERSITY.

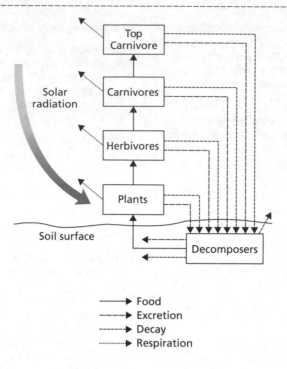

Food
Excretion
Decay
Respiration

Fig 6 Ecosystem

ecosystem dynamics The processes and adjustments that take place within an *ecosystem, including energy flow, *nutrient cycling, and vegetation *succession.

ecosystem integrity The ability of an *ecosystem to function healthily, continue to provide natural goods and services, and maintain *biodiversity. Also known as **ecological integrity**.

ecosystem management An integrated approach to the management of *ecosystems and *natural resources that seeks to balance ecological, economic, and social goals in a sustainable way, by respecting and protecting the natural integrity and processes of ecosystems, and through deliberate manipulation of ecosystem structure and/or function, and/or regulation of human uses of ecological systems. *Natural resource management up to the 1960s relied heavily on managing parts of ecosystems as more or less independent units. Thus, for example, *forest management, wildlife management, and water and soil resource management were widely practised, but as largely separate activities; integrated management of ecosystems was rare. One important development since the early 1970s has been the widespread adoption of an integrated ecosystems framework and perspective in natural resource management, for example for planning and managing protected areas such as *national parks and *wilderness areas.

ecosystem rehabilitation See ECO-SYSTEM RESTORATION.

ecosystem restoration The process of re-establishing, to as near its natural condition as possible, the structure, function, and composition of an *ecosystem.

ecosystem services Processes or materials such as clean water, energy, climate regulation, and *nutrient cycling that are naturally provided by *ecosystems. Also known as **ecological services**.

ecosystem structure The *biotic and *abiotic elements of an *ecosystem, and the relationships between them, particularly in terms of *trophic levels.

ecotage Ecological sabotage, in the form of direct action or sabotage in defence of nature. An aggressive form of *eco-activism. *See also* ECOTERRORISM, MONKEY-WRENCHING.

ecoterrorism An aggressive or violent form of *eco-activism that involves the use of *ecoterrorism. Also called **bioterrorism**. *See also* ECO-EXTREMISM, MONKEY-WRENCHING.

ecotone A transitional area between two adjacent *communities, *ecosystems, or *habitats; the boundary or border zone which contains *species from both ecological communities. Ecotones are far from fixed, and they can change location as the species in adjacent *biomes adjust to *environmental change. Ecologists believe that many of the ecotones we see today are still in the process of adjusting to *post-glacial climate change. It is possible, for example, that the northern edge of the *taiga is still advancing northwards as climate warms from the *Pleistocene ice age. Human factors can also cause shifts in the boundaries between plant communities. Over the last two centuries, for example, the boundary between short grass (in the east) and tall grass (in the west) in the Central Plains of North America has moved progressively westwards, partly in response to *grazing and other human pressures. *See also* EDGE.

ecotope An area that has uniform environmental conditions and characteristic plants and animals. Also known as a **biotope**.

ecotourism Nature-based tourism that is ecologically sustainable, environmentally sensitive, and often involves adventure travel, environmental education, and cultural exploration.

ecotoxicity The ability of a substance to have a long-term, *toxic effect on living *species and *ecosystems.

ecotoxicology The application of *toxicology to the natural environment.

ecotype A local *population of a widespread *species, that has adapted to a particular set of environmental factors.

ecowarrior Short for ecological warrior, someone engaged in extreme forms of *eco-activism.

ecozone A large area in which there are distinctive climate patterns, ocean conditions, types of landscapes, and species of plants and animals. *See also* ECOREGION.

ectocrine *See* ENVIRONMENTAL HORMONE.

ectoparasite A *parasite that lives on the outside of the host's body. *Contrast* ENDOPARASITE.

ectotherm *See* COLD-BLOODED.

ecumene All of the parts of the world that are inhabited or habitable by people.

edaphic Of or having to do with *soil.

edaphic factor Any property of *soil that influences the types of *plants that grow in an area, such as lack of *nutrients, presence of *toxic contaminants, *waterlogging, or moisture deficiency.

eddy A circular movement of water or air opposing the main direction of fluid flow. An eddy is formed, for example, where currents pass obstructions or between two currents that are flowing in opposite directions. *See also* MECHANICAL TURBULENCE.

eddy mixing Mixing of air or water that is caused by eddies.

edge A border or boundary between two different *ecosystems or *communities of plants or animals. *See also* ECOTONE.

edge effect The impact of a boundary between two different ecosystems or *communities of plants or animals on species composition, physical conditions, or other ecological factors.

edge species Species that prefer the greater diversity of *habitat factors that occurs at the boundary between adjacent *ecosystems.

EEA *See* EUROPEAN ENVIRONMENT AGENCY.

EEC *See* EUROPEAN ECONOMIC COMMUNITY.

Eemian An *interglacial optimum period between about 125 000 and 130 000 years ago. Part of the Pleistocene, when temperatures were warmer than today.

EEZ *See* EXCLUSIVE ECONOMIC ZONE.

effect *See* IMPACT.

efficacy A measure of effectiveness, or the ability to achieve a desired effect.

efficiency The ratio of the output to the input of any *system, usually expressed as a percentage.

efficient An activity that is performed successfully with a minimum of waste or unnecessary effort, or that produces a high ratio of results to resources.

effluent A waste product (such as water that contains *pollutants) that is released or discharged into the *environment. Means literally 'flowing out'. *Contrast* INFLUENT.

effluent based standard A *standard that sets the amount of *effluent that can be legally discharged into a *waterbody.

effluent fee A fee or tax that is paid by a polluter in exchange for the right to discharge an agreed amount of noxious *emissions into the air and water.

effluent limitation A restriction on the quantity, rate, and *concentration of a particular pollutant from a *point source that can be legally discharged into a waterbody. Also known as **effluent standard**.

effluent standard *See* EFFLUENT LIMITATION.

Ehrlich, Paul US entomologist (born 1932) who is best known as a researcher and author on the subject of human overpopulation. He co-wrote the book *The *Population Bomb* in 1968 and was a co-founder of the group *Zero Population Growth in 1968.

EIA *See* ENVIRONMENTAL IMPACT ASSESSMENT.

EIS *See* ENVIRONMENTAL IMPACT STATEMENT, ENVIRONMENTAL INFORMATION SYSTEM.

EIT *See* ECONOMIES IN TRANSITION.

Ekman effect The change in wind direction with increasing height above the ground, or change in *ocean currents with depth, that is determined by the *Coriolis force.

elastic limit The maximum *stress that a material can withstand without causing permanent *strain or deformation.

elasticity 1. The ability of a material to return to its original shape after it has been stretched or compressed.
2. In *economics, a measure of responsiveness of the quantity of a good or service that is demanded or supplied, to changes in price.

electric current A flow of electric charge through a wire or other material.

electric motor An engine that uses *electricity for doing work.

electric potential The difference in electrical charge between two points in a circuit, which can effectively push electric charges to different locations. Usually expressed in volts.

electric power The rate of energy transfer of *electricity.

electrical conduction The passage of a charge in an electric field.

electrical conductivity *See* CONDUCTIVITY.

electrical energy The energy that is produced by flow of electric charge in an electric field.

EL NIÑO–SOUTHERN OSCILLATION (ENSO)

A climatic phenomenon that occurs roughly every three to seven years, lasts between six and 18 months, and peaks around Christmas time (El Niño means 'boy child') in the surface oceans of the south-east Pacific. It is believed to be associated with a southward migration in the *Intertropical Convergence Zone (ITCZ). It involves weakening of easterly *trade winds and major warming of the equatorial waters in the Pacific Ocean, off the coasts of Peru and Ecuador, and brings nutrient-poor tropical water southward along the west coast of South America. It usually only affects the Pacific region, although major events can disrupt weather patterns over much of the Earth and El Niño has been linked to colder, wetter winters in parts of the USA, drier hotter summers in South America and Europe, and drought in Africa. Recent El Niño disturbances occurred in 1953, 1957–8, 1972–3, 1976, 1982–3, and 1997–8. In the years between ENSO events, the ITCZ remains in its more normal position much further north, the *trade winds are stronger, the South Equatorial Current is stronger, and the ocean surface waters off South America remain relatively cool. The southward migration of the ITCZ appears to be cyclical in nature, although the exact cause is not known. Sometimes it is a response to atmospheric disturbances associated with volcanic eruptions in the equatorial zone, as happened after the 1982 eruption of El Chichón volcano in Mexico. El Niño regularly brings heavy rain to the coast of Peru, because air movement across the warm ocean surface increases evaporation and the amount of water vapour in the atmosphere, and the moisture is condensed and precipitated when it reaches the coast (particularly by *orographic precipitation). Torrential rains across Peru are not uncommon during El Niño, and they can trigger landslides and flash floods in the Andes. Farmers benefit from El Niño because it waters their grazing land. But El Niño causes great hardship for the fishing people of Peru, because it disturbs and often eliminates the normal pattern of *upwelling which sustains the anchovy fishing industry by increasing nutrient supplies. Fish catches decline dramatically during ENSO events, with consequent impacts on food supplies and regional economies. When the El Niño flow pattern changes, the ENSO events can have widespread impacts on climatic and environmental systems. These impacts are not confined to the South Pacific, because they can be transferred around the world by adjustments to global climatic and environmental systems via so-called *teleconnections. Historical records of large-scale droughts and floods in India, Africa, Indonesia, and parts of China show close correlations with the chronology of El Niño events in South America at least as far back as 1750. *See also* LA NIÑA.

electricity A source of *energy that is made available by the flow of electric charge (*electrons) through a conductor.

electricity generation The process of producing *electricity or transforming other forms of energy into electricity. *See also* POWER PLANT.

electrodialysis A method for *desalination of water that involves passing an electric current through the *saline water in a container, in order to remove minerals from water.

electrolyte A non-metallic substance that conducts an *electric current in solution, (such as many *acids, *bases, and *salts) due to the presence of positive and negative ions.

electromagnetic radiation The energy that is transmitted through space in the form of *electromagnetic waves, which include light, radio waves, X-rays, and gamma rays. *See also* RADIATION.

electromagnetic spectrum The range of *electromagnetic radiation, characterized by frequency or *wavelength, that extends from very short wavelengths (including *ultraviolet, *X-rays, and *gamma rays), through visible light, to long wavelengths (*infrared). Different wavelengths have different properties and they affect people and *environmental systems in different ways. Long wavelengths have a low frequency and thus a relatively low energy, and as a result they are less damaging to people's health. Short wavelengths, on the other hand, have a high frequency, high energy, and are damaging to people's health.

electromagnetic wave A wave that is partly electric and partly magnetic, is emitted by vibrating electric charges, carries energy, and can travel through a vacuum (such as light waves, radio waves, and microwaves).

electron A negatively charged subatomic particle that orbits around the *nucleus of an *atom. *See also* PROTON, NEUTRON.

electro-osmosis The movement of liquid in a porous medium that is caused by differences in *electric potential.

electroplating The process of coating base metal with a thin layer of another metal by passing an *electric current through it.

electrostatic precipitator An electrical device used in *air cleaning, that removes *fly ash from *flue gas before it is released into the atmosphere from the stack/chimney in a *power plant. *See also* ADVANCED AIR EMISSION CONTROL DEVICE.

element One of more than a hundred known substances (92 of which occur naturally) that cannot be chemically changed or broken down, and that singly or in combination make up all matter. *See also* MACROELEMENT, MICROELEMENT.

elevation *See* ALTITUDE.

ELF *See* EARTH LIBERATION FRONT.

eligibility criteria The criteria that must be met by an *emissions reduction project, under the *Kyoto Protocol, in order to produce reductions which can be banked, traded, or offset against emissions.

eluvial horizon A *horizon in a *soil profile from which material has been removed, either in solution or suspension.

eluviation A *soil-forming process that involves the removal of fine particles of material in solution or suspension, usually from the *A-horizon. The material is usually deposited in an *illuvial horizon below.

EMAP *See* ENVIRONMENTAL MONITORING AND ASSESSMENT PROGRAM.

EMAS *See* ECO-MANAGEMENT AND AUDIT SCHEME.

embryo 1. An unborn or unhatched organism, the earliest stages of development until hatching or birth.
2. In cloud physics, a tiny ice crystal that grows in size and becomes an ice nucleus.

emergence The act of emerging or coming out, such as the appearance of a *plant above ground after *germination in the soil.

emergency A crisis that develops suddenly and unexpectedly, that usually involves danger, and that requires immediate action to prevent loss of life. Examples include *hurricanes, *tornados, *storms, *floods, *tidal waves, *tsunamis, *earthquakes, *volcanic eruptions, *landslides, *fires, and *nuclear accidents. *See also* CHEMICAL EMERGENCY, DISASTER.

emergency management *See* DISASTER MANAGEMENT.

Emergency Planning and Community Right-to-Know Act (EPCRA) A federal law (also known as SARA) that was enacted in the USA in 1986, requiring federal, state, and local governments and industry that are involved in either emergency planning and/or reporting of *hazardous chemicals to allow public access to information about the presence of hazardous chemicals in the community and releases of such substances into the environment.

emergency response Actions that are taken during and immediately following a *disaster, in order to minimize serious harm to people or the environment. Examples include search and rescue, relief services, and restoration of power, water, and telephone services. *See also* DISASTER MANAGEMENT.

emergent vegetation Aquatic plants that grow with their roots under water but their leaves and stems above the surface of the water.

emergent wetland A type of *wetland (such as a *marsh) that is dominated by grasses, sedges, rushes, forbs, and other rooted, water-loving *herbaceous plants that emerge from the surface of water or soil.

Emerson, Ralph Waldo US writer, leading exponent of *Transcendentalism (1803–82), and one of America's most influential thinkers and writers. Many of his core ideas are contained in the essay *Nature*. He inspired many other writers, including Henry David *Thoreau.

emigration The movement of individuals out of a country or area; the opposite of *immigration.

emission The release or discharge into the *environment of a substance, particularly an *air pollutant.

emission factor The relationship between the amount of *raw material that is processed or burned and the amount of *pollution it produces, which is used in preparing an *emissions inventory.

emission permit A non-transferable, non-tradable permission that is granted by a government to an individual firm, which allows it to emit a defined amount of a *pollutant in a given period of time. *See also* PERMIT.

emission rate The amount of a particular *pollutant that is discharged or emitted within a defined period of time.

emission source The place from which a particular *pollutant is released or emitted into the *environment. *See also* AREA SOURCE, NON-POINT SOURCE, POINT SOURCE.

emission standard The maximum amount of air pollutant (such as *carbon monoxide, *sulphur dioxide, or *particulates) that can be legally discharged from a single mobile or stationary *emission source.

emission tax A tax or surcharge (such as a *carbon tax) that is imposed on *emission sources, to provide incentives to firms and households to reduce their emissions, in order to control pollution.

emissions allocation An approach to the control of *greenhouse gas emissions, under the *Kyoto Protocol, that is based on allocating permits to an emissions source (such as a company with net emissions) by a regulatory body during a specific *Commitment Period. Permits are allocated mainly through *grandfathering or *auctioning. Also known as **emissions target**.

emissions allowance An approach to the control of *greenhouse gas emissions, under the *Kyoto Protocol, that divides the total permitted quantity of emissions of a particular gas into allowances that grant the holder the right to emit a specified amount of *pollution over a given period of time (for example 1 tonne a year). Allowances can be traded but the overall amount remains controlled. At the end of each *Commitment Period each holder must surrender sufficient allowances to cover their emissions during that period; if they produce less pollution than they are permitted to, they can trade or sell some of their emissions allowance, but if they produce more pollution they must trade or buy some allowance from another holder. Also known as **emissions quota**. *See also* EMISSIONS CAP.

emissions auctioning An approach to the control of *greenhouse gas emissions, under the *Kyoto Protocol, that is based on the trading of *emissions allowances by allowing *emission permits to be issued to polluters within a country, based on their willingness to pay for the permits. It may be combined with *grandfathering.

emissions banking Part of the *emissions trading procedures for *greenhouse gases that were agreed under the *Kyoto Protocol, in which *emissions allowances for future consumption or trading can be saved by holding reductions within a *Commitment Period, applying reductions earned from one period to another, or applying reductions earned before the first *Compliance Period via the *Clean Development Mechanism. *See also* FUTURES.

emissions borrowing An approach to the control of *greenhouse gas emissions, under the *Kyoto Protocol, that is based on borrowing from the future, by using *emissions reductions from future *Commitment Periods in order to meet current *emissions targets.

emissions bubble An approach to the control of *greenhouse gas emissions, under the *Kyoto Protocol, that treats two or more *emission sources as if they were one single source (the *bubble). This allows flexibility in applying pollution control technologies to whichever source within the bubble offers the most *cost-effective pollution control options, while ensuring that the total amount of emissions within the bubble does not exceed environmental requirements. *See also* EU BUBBLE.

emissions budget *See* ASSIGNED AMOUNT.

emissions buyer In *greenhouse gas *emissions trading, under the *Kyoto Protocol, a legally recognized entity (individual, corporation, not-for-profit organization, or government) that acquires credits, reductions, or allowances from another legally recognized entity, through a purchase, lease, trade, or other forms of transfer.

emissions cap A regulatory device that defines the maximum amount of emissions of *greenhouse gases that can be released into the *atmosphere within a given period of time, under the *Kyoto Protocol. A cap is a national *emissions allowance.

emissions inventory A list of the amounts of *air pollutants that are emitted by each particular source within a given area, which is used to establish *emission standards. The need for such inventories became obvious at the 1992 *Rio Earth Summit and during debate leading to the 1997 *Kyoto Protocol. An emissions inventory would usually include information on the chemical or physical identity of the pollutants included, the geographical area covered, the institutional entities covered, the time period over which the emissions are estimated, and the types of activities that cause emissions. Emission inventories are used in various ways. For example, they provide valuable inputs to air quality models, allow the success of air quality policies and standards to be evaluated, and enable regulatory agencies to establish compliance records with allowable emission rates.

emissions leakage The offsetting of emission reductions in one location (such as an *Annex I country) by an increase in emissions in another location (such as a non-Annex I country), under the *Kyoto Protocol.

emissions offset *See* OFFSET.

emissions quota *See* EMISSIONS ALLOWANCE.

emissions reduction Any decrease in the release of a *pollutant into the *environment. Usually used to refer to a decrease in emission of *greenhouse gases into the atmosphere in order to control *climate change and *global warming.

emissions reduction unit (ERU) A specified amount of *emissions reduction of *greenhouse gas that has been achieved and is used for *Joint Implementation, under the *Kyoto Protocol. *See also* CREDIT, CERTIFIED EMISSIONS REDUCTION.

emissions target *See* EMISSIONS ALLOCATION.

emissions trading A general term for the *flexibility mechanisms that were agreed under the *Kyoto Protocol and are designed to reduce *emissions of *greenhouse gases. It is a market-based system that allows polluters the flexibility to select cost-effective solutions to achieve the agreed emission goals, in which polluters that produce fewer emissions than they are allowed can sell or trade their excess capacity to others who do not, within the *bubble in which total approved emissions cannot be exceeded. *See* EMISSIONS AUCTIONING, GRANDFATHERING, INTERNATIONAL EMISSIONS TRADING.

emissivity The ability of a surface to emit heat by *radiation, compared with that which a *black body would emit under the same conditions.

emittance The amount of *radiant energy that a surface emits.

empirical A result based on experience, experiment, or observation, rather than on theory. *Contrast* THEORETICAL.

empowerment To increase the ability of an individual or community to do things for itself.

EMS *See* ENVIRONMENTAL MANAGEMENT SYSTEM.

emulsion A *colloid in which very small droplets of one liquid are suspended in another (such as oil in water), although the two liquids do not actually combine.

encapsulation The process in which *nutrients are wrapped in a *membrane to turn them into small pellets, for example to feed some larvae.

encephalitis Inflammation of the brain. *See also* SLEEPING SICKNESS.

enclosure An area of land that is walled or fenced.

end moraine A *moraine in the form of a ridge of *till that is deposited at or near the edge of an active *glacier. Also known as **recessional moraine** or **terminal moraine**.

endangered A *species that is in danger of *extinction if existing pressures on it (such as over-harvesting or habitat change) continue, and which is therefore likely to disappear if it is not offered adequate protection. A *threatened category defined by the *IUCN *Red Data Book. *Contrast* EXTINCT, EXTIRPATATION, THREATENED, VULNERABLE SPECIES.

Endangered Species Act (1973) US legislation that established a federal programme for the protection and *conservation of *threatened and *endangered species of fish, wildlife, and plants.

Endangered Species List A list of species that are considered to be in imminent danger of *extinction, and are protected under the US *Endangered Species Act.

endangerment assessment A study that is designed to determine the nature and extent of *contamination at a site on the *National Priorities List in the USA, under *CERCLA or *RCRA, and the risks that contamination poses for public health or the environment.

endemic *Native to or confined to a certain region.

endemic species A species whose distribution is restricted to a certain area. A *native species found only within a given area.

endocrine Any gland that introduces *hormones directly into the blood.

endocrine disruptor A chemical (such as *DDT and other *chlorinated hydrocarbon compounds) that disrupts the natural *hormones and thus disturbs the *endocrine system of an animal. Also known as **environmental hormone**.

end-of-the-pipe Technologies that reduce *emissions of pollutants after they have formed, such as *scrubbers on smokestacks and *catalytic converters on vehicle exhausts.

endogenous Internal, arising from within. *Contrast* EXOGENOUS.

endoparasite A *parasite that lives inside the body of its host. *Contrast* ECTOPARASITE.

endophyte An organism that lives at least part of its life cycle within a host plant, in a *parasitic or *symbiotic relationship.

endopterygota An insect whose larvae are physically different from the adults and lack wings. Wings then develop internally. *Contrast* EXOPTERYGOTA.

endoskeleton A skeleton or support structure that is inside the body of the organism. *Contrast* EXOSKELETON.

endotherm *See* WARM-BLOODED.

endothermic 1. A *chemical reaction that absorbs heat from its surroundings.
2. Animals that can independently generate and maintain their body temperature. *Contrast* EXOTHERMIC.

endrin A *pesticide that is *toxic to freshwater and marine aquatic life, and can affect human health via domestic water supplies.

energetics The study of how *energy is transferred and used in a *system, such as an *ecosystem.

energy Usable *power, or the capacity to do work, measured by the capability to do work (*potential energy) or the conversion of this capability into motion (*kinetic energy).

energy balance The balance between the total energy that enters, leaves, and accumulates within a system (such as an *ecosystem). Based on the *first law of thermodynamics, which says that energy may be transformed, but not created or destroyed.

energy budget A description of the overall *energy balance for a system (such as an *ecosystem), and of the components within it.

energy conservation Saving energy by eliminating wasteful use, making more efficient use of it, or reducing total use.

energy efficiency The amount of energy output (work done) relative to the amount of energy input (fuel consumed). Also known as **fuel efficiency**. *See also* ENERGY INTENSITY.

energy farming Using land to grow crops that provide fuel (such as fast-growing tree species). Also known as **energy plantation**.

energy flow The movement and loss of *energy through a *community or *ecosystem, via the *food web.

energy flux A measure of *energy flow, expressed as the amount of energy that moves through a given area in a unit of time.

energy intensity The amount of *energy used per unit of activity, such as litres of fuel per passenger-mile, or energy consumption per pound/dollar of *gross domestic product. Also known as **fuel intensity**.

energy plantation *See* ENERGY FARMING.

energy pyramid *See* PYRAMID OF ENERGY.

energy recovery The process of extracting useful *energy from *waste, such as the heat produced by *incineration or by harnessing *methane gas from *landfills.

energy resource *See* ENERGY SOURCE.

energy source Any material that is used to produce *energy, including *fossil fuels (*coal, *oil, *gas), *nuclear (*fission and *fusion), and *renewables (*solar, *wind, *geothermal, *biomass, *hydroelectric).

energy type The form in which energy is available, such as *chemical energy, *electrical energy, *heat energy, *kinetic energy, and *potential energy.

enforcement Legal methods that are used to obtain compliance with environmental laws, rules, regulations, or agreements and/or deal with violations.

engineering The application of science to the design, creation, and function of machines and structures.

englacial Within ice, usually inside a *glacier or *ice sheet. *Contrast* SUBGLACIAL, SUPRAGLACIAL.

englacial stream A *meltwater *river that flows inside a *glacier or *ice sheet.

enhanced greenhouse effect The increase in the strength of the natural *greenhouse effect that is caused by human activities, particularly the emission into the *atmosphere of large amounts of *greenhouse gases. *See also* GLOBAL WARMING.

enhancement An improvement, or increase in quality or value.

Enlightenment A broad intellectual movement in 18th-century Europe that advocated the use of reason in the re-evaluation of accepted ideas. Also known as the **Age of Reason**. *See also* TRANSCENDENTALISM.

enriched fallow A form of *agroforestry in which useful woody species (such as fruits, bamboos, and rattans) are sown or planted before or when cultivation ceases, so that products are available for household use or market. *See also* BUSH FALLOW, FALLOW.

enrichment The process of increasing concentration, such as the addition of *nutrients from sewage effluent or agricultural runoff to surface water (which causes *eutrophication) or the process of

increasing the proportion of uranium-235 in *nuclear fuel.

ENSO *See* EL NIÑO–SOUTHERN OSCILLATION.

enteric Relating to the intestines.

enteric bacteria *Bacteria that originate in the intestines of *warm-blooded animals.

enthalpy A measurement of the *energy content of a system.

entisol A recently developed *order of soil, that has no (or only poorly developed) soil *horizons. It often forms in recent *floodplains, under recent volcanic *ash and as wind-blown *sand.

entomology The study of insects.

entrain 1. To set in motion (for example, particles of *sediment).
2. To trap particles, gases, or liquid droplets in moving fluids, either mechanically (through turbulence) or chemically (through a reaction).

entropy A measure of the availability of a *system's energy to do work. A well-ordered and efficiently-functioning system has a low level of entropy.

environment 1. All of the external *abiotic and *biotic factors, conditions, and influences that affect the life, development, and survival of an *organism or a *community.
2. The natural world in which we live.

Environment Agency The public body (*statutory agency) that is responsible for protecting and improving the *environment in England and Wales.

environmental accounting An approach to *accounting that includes measures of the *environmental impacts and resource use of economic activities, as well as financial measures. Also known as **natural resource accounting**.

environmental activism Any form of activity by *environmentalists that is designed to raise public awareness of environmental issues. Extreme forms of activism include *monkey-wrenching and *EarthFirst, and more common forms include *lobbying and joining *pressure groups. Also known as **activism**. *See* GRASSROOTS ACTIVISM.

environmental agent *See* AGENT.

environmental analysis Assessment of the predictable long- and short-term environmental effects of a proposed action, and of reasonable alternatives to that action. *See also* ENVIRONMENTAL ASSESSMENT, ENVIRONMENTAL IMPACT ASSESSMENT.

environmental assessment An *environmental analysis that is required in the USA under the *National Environmental Policy Act, which analyses a proposed federal action for the possibility of significant *environmental impacts. If the environmental impacts are likely to be significant, the federal agency must then prepare a more detailed *environmental impact statement. *See also* FINDING OF NO SIGNIFICANT IMPACT.

environmental audit A formal independent evaluation (*audit) of compliance (by a company, plant, or agency) with environmental regulations, standards, and policies, or an assessment of the management of any impact on the environment. For example, pollution prevention initiatives require an audit to determine where wastes may be reduced or eliminated or where energy can be conserved.

environmental balance sheet A method of *environmental assessment that analyses the *life cycle of a product or process.

environmental capacity *See* CARRYING CAPACITY.

environmental capital *See* NATURAL CAPITAL.

environmental change Any alteration to the *environment or to an *environmental system.

environmental commitment *See* COMMITMENT.

environmental contamination *See* CONTAMINATION.

Environmental Defense A leading US-based non-profit organization that was founded in 1967 as the Environmental Defense Fund, and seeks innovative, equitable, and cost-effective solutions to serious environmental problems through a combination of science, economics, and law. It is committed to protecting the environmental rights of all people, including future generations, and includes the right to clean air, clean water, healthy food, and healthy *ecosystems.

environmental degradation Depletion or destruction of a potentially *renewable resource such as *air, *water, *soil, *forest, or *wildlife, by using it at a rate faster than it can be naturally renewed. *See also* SUSTAINABLE YIELD.

environmental determinism The view that was popular at the start of the 20th century, but is now regarded as too simplistic, that the most important control on human activities is the *environment.

environmental economics A branch of *economics that takes into account the current and future monetary costs and benefits of *environmental systems, human welfare, and *biodiversity. *See also* COST–BENEFIT ANALYSIS, ECOLOGICAL ECONOMICS, EXTERNALITY.

environmental education Formal and informal activities that are designed to promote people's understanding of, appreciation of, and care for the natural environment.

environmental effect *See* ENVIRONMENTAL IMPACT.

environmental equity The extent to which all groups of people in a region or country (regardless of race, ethnicity, economic status, or income) receive equal treatment and protection under environmental statues, regulations, and practices. Unlike *environmental racism, equity also considers the disproportionate burden of risk that any group of people (de-

ENVIRONMENTAL CRISIS

A term that is used to describe the sum of the environmental problems that we face today. Key contemporary environmental problems include the *greenhouse effect and *global warming, the hole in the *ozone layer, *acid rain, and *tropical forest clearance. New dimensions to the environmental crisis include emerging threats and the global nature, rapid build-up, and persistence of the problems. Whilst the problems appear to be largely physical (environmental), the causes and solutions lie much more in people's attitudes, values, and expectations. A number of factors have helped to create these problems, including developments in technology, which have given people a greater ability to use the *environment and its *natural resources for their own ends (particularly since the *Industrial Revolution); the rapid increase in human *population in recent centuries, which has significantly increased population densities in many countries and led to a significant rise in human use of natural resources; the emergence of *free market economies, in which economic factors play a central role in decision-making about production, consumption, use of resources, and treatment of *wastes; attitudes towards the environment, particularly amongst western cultures, which regard it as freely available for people to do whatever they like with; and the short-term time horizon over which many people, companies, and countries make decisions, which means that short-term maximization of *profit has generally been taken more seriously than long-term *sustainable use of the environment. There are many symptoms of the so-called 'crisis'. According to the *UNEP *Global Environment Outlook 2000* report: there will be a billion cars by 2025, up from 40 million since 1945; a quarter of the world's 4630 types of *mammals and 11% of the 9675 species of *bird are at serious risk of *extinction; more than half of the world's *coral is at risk from *dredging, diving, and global warming; 80% of *forests have been cleared; a billion city dwellers are exposed to levels of *air pollution that threaten human *health; the global population will reach 8.9 billion in 2050, up from 6 billion in the year 2000; global warming will raise temperatures by up to 3.6°C, triggering a 'devastating' rise in *sea level and more severe *natural disasters; and global use of *pesticides is causing up to five million acute poisoning incidents each year.

fined by gender, age, income, or race) is exposed to. Also known as **environmental justice**. *See* EQUITY.

environmental ethics A search for moral values and ethical principles in human relations with the natural world. *See also* ETHICS, LAND ETHICS, STEWARD-SHIP.

environmental exposure The *exposure of people to particular *pollutants.

environmental fate The destiny of a chemical or biological *pollutant after it has been released into the environment.

environmental gradient A continuum of environmental conditions, such as the progressive change in climate with increasing latitude.

environmental group Any group of people who voluntarily join together to pursue an environmental objective. Examples include the *Worldwide Fund for Nature and *EarthFirst.

environmental hazard *See* NATURAL HAZARD.

environmental heterogeneity The patchiness of the environment, from the scale of an individual plant to the pattern of vegetation across a region.

environmental history The history of natural and induced environmental change in a particular area.

environmental hormone *See* ENDOCRINE DISRUPTOR.

environmental impact Any *impact or effect (positive or negative) that an activity has on an *environmental system, *environmental quality, or *natural resources. Also known as an environmental effect.

environmental impact assessment (EIA) A formal assessment of the *environmental impacts that are likely to arise from major activities such as new legislation or a new policy, programme, or project. The results of the assessment are reported in the *environmental impact statement. EIA was first introduced in the USA in 1969 and has since been widely applied. It is being adopted in one form or another in an increasing number of countries as a basis for making informed and rational judgements about what sorts of developments are environmentally acceptable. *Strategic environmental assessment has developed out of EIA.

environmental impact statement (EIS) A detailed written statement that describes the results of an *environmental impact assessment of a proposed large-scale project, which usually includes a description of the project and of the environment that would be affected, an assessment of the important effects of the project on environment, a justification of the project, and a non-technical summary. An EIS is required for all major federal projects in the USA under the *National Environmental Policy Act, and for all major projects in the *European Union under a 1988 directive.

environmental indicator A measurement or statistic that provides evidence of environmental quality, such as which *threshold concentrations of different

*air pollutants are relevant to *human health and to *environmental stability, how best to measure the *biodiversity of species of *wildlife within *natural habitats, and how to express *land use change in the most useful ways. *See also* BIOINDICATOR, CORINE, EXPOSURE INDICATOR.

environmental informatics The use of computers and statistical methods in the classification, storage, retrieval, and analysis of environmental information. *See also* INFORMATICS.

environmental information *See* Appendix 1 for a list of useful online sources of information and data on particular environmental issues.

environmental information system (EIS) The storage and analysis of environmental data using a computer, which might include use of *geographical information systems (GIS), *remote sensing, and *simulation, which make it much easier to simulate how the environment works and how it might respond to different management *scenarios. Such analyses rely heavily on powerful computers that are used to store and manipulate databases and to run complex predictive *models. As computers become more and more powerful (between 1950 and 1990, for example, computers increased in speed by a factor of ten roughly every five years) it becomes possible to analyse bigger and bigger data sets and run more complex models in more sophisticated ways.

environmental justice *See* ENVIRONMENTAL EQUITY.

environmental labelling Attaching labels to products which advise consumers about environmental aspects of the production, use, or disposal of that product, such as the extent to which it is *biodegradable, can be *recycled, produces damaging *ozone, is *organic, is *toxic, and is based on *renewable materials. *See also* ECO-LABEL.

environmental lapse rate The average rate at which *temperature decreases with increasing *altitude in the *troposphere.

environmental law The body of official legal rules, decisions, and actions that relate to *environmental quality, *natural resources, and *ecological sustainability.

environmental literacy The level of understanding that an individual, group, or population has of ecological principles and the inter-relationships between people and *environment.

environmental management The deliberate *management and control of the *environment and of *natural resources systems, designed to ensure the long-term sustainability of development efforts.

environmental management system (EMS) A method or tool for systematically addressing environmental issues within an organization, which includes concrete objectives, plans of action, and responsibility for dealing with environmental matters. *See also* ECO-MANAGEMENT AND AUDIT SCHEME.

environmental monitoring Technologies, procedures, and protocols for collecting, analysing, interpreting, and reporting *environmental information, including the use of *continuous monitoring,*remote sensing and *satellites, *geographical information systems, and *environmental informatics. Great improvements have been made since the early 1990s in harmonizing practices and standards between countries, so that it is now becoming both possible and meaningful to collate national statistics into global summaries. A useful product of this new interest in producing and interpreting environmental information is the *state of the environment reports that many countries now produce regularly. Monitoring provides the data that are needed to facilitate the development, testing, and operation of complex *global models. It also provides *baselines against which to evaluate rates and pat-

terns of *environmental change, and which can be used to give early indications of natural environmental adjustments and possible risks. The *United Nations Environment Programme (UNEP) is responsible for coordinating the monitoring networks that provide invaluable information about the global environment. The data are collected mainly from surface stations around the world as part of the *Global Environmental Monitoring System (GEMS).

Environmental Monitoring and Assessment Program (EMAP) An *environmental monitoring system developed by the US *Environmental Protection Agency (EPA).

environmental movement A political movement that focuses on protecting the *environment, reducing environmental damage (such as *pollution), and reducing unsustainable use of *natural resources. *See also* ENVIRONMENTALISM.

environmental pathway Any route through the *environment along which *pollutants can move, such as air, water, and soil.

environmental philosophy *See* PHILOSOPHY.

environmental planning All planning activities that have the objective of preserving or enhancing environmental values or resources.

environmental policy 1. The official rules and regulations relating to the *environment that are adopted, implemented, and enforced by a government agency.
2. A statement by an organization of its intentions and principles with respect to environmental performance, which provides a framework for action and for the setting of environmental objectives and targets.

environmental politics *See* POLITICS.

environmental pollution *See* POLLUTION.

environmental protection Practices and procedures that are designed to avoid, minimize, eliminate, or reverse damage to the *environment and to *environmental systems. *See also* ECOLOGICAL MODERNIZATION, ENVIRONMENTAL IMPACT, SUSTAINABLE DEVELOPMENT.

Environmental Protection Agency (EPA) The federal agency in the USA that was created in 1970 and is responsible for efforts to protect human health and to control air and water *pollution, *radiation and *pesticide hazards, ecological research, and the disposal of *solid waste.

environmental quality The state of the *environment, generally in a particular place, defined either objectively (for example by the use of *environmental indicators) or subjectively (in terms of attributes such as beauty and attractiveness).

environmental racism Intentional or unintentional racial discrimination in environmental policy-making, enforcement of regulations and laws, and targeting of communities for the disposal of *toxic waste and siting of polluting industries. *See also* ENVIRONMENTAL EQUITY.

environmental refugee A migrant from an area that is threatened with or damaged by a major *natural hazard or *disaster. Under international law, people who flee their own country to escape from persecution, armed conflict, or violence can be granted refugee status, but those forced to leave for social or environmental reasons do not qualify.

environmental regime The *regime for managing the environment, either globally or at a national level, which includes environmental *regulations and *conventions, and the institutions through which they are managed.

environmental regulation The management of activities that affect the environment, through such means as *command and control and *market-based mechanisms.

environmental regulator An agency (such as national or local government) that oversees and applies the legislation and regulations governing the *environment.

environmental reporting The release of environmental statements by businesses that usually contain information on *environmental impacts, *environmental management initiatives and plans for future activities.

environmental resource *See* NATURAL RESOURCE.

environmental restoration The *cleanup and *restoration of sites that have been contaminated with *hazardous substances.

environmental review Study of a company's processes and procedures, to assess compliance with environmental laws and regulations and to review its environmental policy and best practice.

environmental risk Any source of harm or danger in the *environment, for example from *natural hazards, *pollution, or depletion of *natural resources.

environmental science The interdisciplinary study of *environmental systems, how they operate, how they interact with people, and how people interact with them.

environmental security The impact on security of the links between *environmental stress and violent conflict, particularly in *developing countries, caused, for example, by competition for scarce resources or exposure to serious natural hazards.

environmental site assessment The process of determining whether a particular plot of land has been contaminated.

environmental standard A limiting condition of *environmental quality, that is often expressed in numerical terms, has legal standing, and is designed to protect human health and well-being.

environmental stress Any form or level of *environmental change which particular organisms (including humans) find it difficult to adjust to.

environmental sustainability The long-term maintenance of *ecosystems and other *environmental systems for the benefit of future generations. *See also* SUSTAINABLE DEVELOPMENT, SUSTAINABILITY.

environmental system A *system that is based on the natural environment and includes *biotic and *abiotic components which interact. The major environmental systems are the *atmosphere (air), *biosphere (living organisms), *hydrosphere (water), *cryosphere (ice), *pedosphere (soil), and *lithosphere (rock). An *ecosystem is the term for a local environmental system.

environmental technology Any technology that is designed to control *pollution, treat or store waste materials, or cleanup contaminated sites. Examples include wet *scrubbers (air), soil washing (soil), granulated *activated carbon unit (water), and filtration (air, water). *Contrast* ALTERNATIVE TECHNOLOGY.

environmental treaty Any *treaty that deals with the *environment or with *natural resources. Appendix 2 contains a summary of international environmental treaties.

environmentalism 1. The beliefs behind an organized social movement of people who share a concern about solving problems of environmental *pollution and *natural resources. Environmental concern is reflected in a number of different ways, including membership of environmental *pressure groups and campaigning organizations (such as *Friends of the Earth and *Greenpeace), sympathy for and engagement with environmental *politics, *green consumerism, local *environmental activism, and adoption of *lifestyles that are *environment-friendly. Western environmentalism can be traced back to the emergence of concern about *nature and *natural landscape in the USA towards the close of the 19th century. The modern *environmental movement emerged during the 1960s, first in the USA and then in Britain. A number of books were particularly influential in orienting people's views and attitudes during this formative period, including *Silent Spring, *Blueprint for Survival, and *Limits to Growth. Environmentalism is founded on a number of concerns, including a reaction against *technocracy, a concern about the *welfare of deprived groups of people (particularly in *developing countries), a concern for wider issues of *equity and *justice, and a sense of personal responsibility to leave a worthwhile *environmental heritage for future generations.
2. The belief that *environment is more important than *heredity in determining intellectual growth in humans.

environmentalist A person who works or engages in activities that are designed to protect the *environment from destruction or *pollution.

Environmentally Sensitive Area (ESA) An area of land, designated by the UK government under *EU regulations, where landowners and tenants are eligible to claim for subsidies for certain types of management that will benefit *landscape or *wildlife features within the ESA. *See also* MANAGEMENT AGREEMENT.

environment-friendly Activities, products, or *lifestyles that are designed to have a small *environmental impact, such as restricting family size to a maximum of two children, using public rather than private transport, using renewable materials wherever possible, and engaging in *green consumerism. Also known as **green**.

enzyme A *protein that catalyses a biochemical reaction.

Eocene The second oldest of the five major *epochs of the *Tertiary period, from 56 to 35 million years ago, when the climate was significantly warmer and seas extended much further than today, and the first modern *mammals appeared.

eolian See AEOLIAN.

eolithic The earliest part of the *Stone Age, in which the earliest signs of human culture appear.

EOS See EARTH OBSERVATION SATELLITE.

EPA See ENVIRONMENTAL PROTECTION AGENCY.

EPCRA See COMMUNITY-RIGHT-TO-KNOW ACT.

ephemeral Transitory, *episodic, lasting for only a short period of time.

ephemeral flow Flow in a *river or *stream that only occurs during and immediately after *rain. Contrast INTERMITTENT FLOW, PERENNIAL FLOW.

ephemeral plant An *annual plant with a very short lifespan (a few weeks or a very few months), which is common in *desert climates.

ephemeral stream A stream or river that has no permanent flow, but flows only as the direct result of *rainfall or *snow melt.

ephemeral wetland A *wetland that fills with water in the *spring but is dry by the end of the *summer.

epibenthos All of the *organisms that live on the surface of the bottom of a *waterbody, such as the *sea-bed. See also EPIFAUNA, EPIFLORA.

epicentre The centre of an *earthquake, located on the ground surface directly above the *focus, where the most severe shock waves are usually experienced.

epidemic A widespread outbreak of an infectious *disease.

epidemiology The study of the *incidence, distribution, and control of *disease in a *population.

epidermis The outer layer of *cells on *plants and *animals which covers and protects the underlying dermis.

epifauna Animals that live on the floor of a *waterbody, such as the *sea-bed.

epiflora Plants that live on the floor of a waterbody, such as the sea-bed.

epilimnion The top layer of water in a *lake or *reservoir, which has the warmest water and a relatively uniform temperature, and which is mixed by the *wind.

epilithic Growing on stone or on rock surfaces.

epipelagic Relatively shallow water; the top 200 metres of the ocean, seas, and lakes.

epipelagic zone See PHOTIC ZONE.

epipelic At the *interface between sediment and water.

epiphyte A plant which does not root in *soil but uses the trunk or stem of another plant for support as it grows. It draws no nutrients from its *host plant, but gets its moisture and nutrients from the air, rainwater, and organic debris. This is a form of *commensalism and an example of *symbiosis.

epipsammic Attached to *sand particles.

episode See POLLUTION EPISODE.

episodic See INTERMITTENT.

epizoic Growing or living on the exterior of a living animal, as a non-parasitic organism.

epoch A subdivision of geological time that is shorter than a period. For example, the *Quaternary period is divided into the *holocene epoch and *pleistocene epoch.

equality The state of being the same in terms of quantity, value, or status. Contrast INEQUALITY.

equator An imaginary circle around the Earth that is equally distant from the North and South poles and defines the latitude 0°.

equatorial forest See TROPICAL RAIN-FOREST.

equatorial low See DOLDRUMS.

equatorial wet climate zone A hot, wet *tropical climate zone, with high temperatures (usually around 30°C) through the year, *precipitation (at least 60 mm of rainfall) every month, and *cloud cover on most days. *Rainforest is the typical natural vegetation.

equifinality A state or condition of a *system that can be created in more than one way, so it is often difficult to work out cause-and-effect associations between the many parts of a system.

equilibrium A state of balance in a *system that is produced and maintained by a variety of forces which may increase or decrease but they always cancel each other out, producing a *steady state. See also CHEMICAL EQUILIBRIUM.

equilibrium community A *community of *organisms that is subject to periodic disruptions (usually by fire), that prevent it from reaching a *climax stage. Also known as **disclimax community**.

equilibrium line The boundary in a *glacier between the *accumulation zone and the *ablation zone.

equilibrium response The anticipated change in *climate that would result from a change in *radiative forcing.

equilibrium species A *species whose *population exists in *equilibrium with available *resources, at a stable *density.

equilibrium state See STEADY STATE.

equilibrium theory A theory in *island biogeography that greater numbers of species are found on larger islands because the populations on smaller islands are more vulnerable to *extinction.

equilibrium warming commitment The increase in *temperature that would result at some point in the future if atmospheric concentrations of *greenhouse gases remained constant at the levels of a particular year. See also COMMITMENT.

equinox Either of the two days at which the Sun is directly overhead at the *equator, when daylight and darkness are of equal length. This is important in defining the *seasons. The *vernal equinox is usually 21 March; the *autumnal equinox is usually 23 September. See also WINTER, SPRING, SUMMER, AUTUMN/FALL.

equitability See EVENNESS.

equitable That which is fair, impartial, and just, and which provides equal opportunity for all. Contrast INEQUITABLE.

equity Fairness arising from the equal use and allocation of resources. See also ENVIRONMENTAL EQUITY. Contrast INEQUITY.

era A unit of geological time that is longer than a *period. Examples include the *Paleozoic era, the *Mesozoic era and the *Cenozoic era.

eradication The removal from the *environment of all units of an infecting *agent or *pathogen.

erg 1. A *desert surface that is covered by *sand dunes.
2. The work done by a force of one dyne (10^{-7} *joules) acting over a distance of one centimetre.

ergot A *fungus that affects *cereal plants.

Ericaceae Heathers.

erode To wear away by the process of *erosion. For example the effect of wind, water, and movement of *glaciers in moving material from the ground surface, which is then transported and deposited elsewhere. See also SOIL EROSION.

erodibility The ease with which a soil or rock can be *eroded.

erosion A group of natural *geological processes by which soil and rock material are loosened (*weathering) or dissolved (*solution) and then moved (*transportation) from their original location. The processes involve transporting agents such as running water, moving ice, or blowing wind, which are active within *rivers, *coasts and *oceans, *glaciers and *periglacial areas, and *deserts and semi-arid areas. Of the order of 10 million tonnes of *sediment is eroded from the world's continents each year, the vast majority (nearly 95%) of which is eroded by rivers. Much smaller quantities are eroded by wind and ice. Rates of erosion vary a great deal from place to place, reflecting variations in key controlling factors such as *climate (particularly temperature and rainfall), vegetation cover, changes in *land use, geology and soil type, and *topography (particularly the steepness and uniformity of a hillslope). Human activities can significantly alter the pace and pattern of erosion. The *United Nations Environment Programme estimates that, as a result of human activities world-wide, some 10 930 million square kilometres of land have been seriously damaged by *water erosion, 9.2 million square kilometres by *sheet and *slope erosion, and 1.73 million square kilometres by the development of *rills and *gullies. The main causes are clearance of natural vegetation and forest (43%), overgrazing (29%), poor farming practices such as cultivation of steep slopes (24%), and over-exploitation of natural vegetation (4%). *See also* ACCELERATED EROSION, NATURAL EROSION.

erosion surface Areas of land of similar altitude that might be the remnants of the long-term *erosion of the land surface and can help in the interpretation of the *denudation chronology of the area. *See also* CYCLE OF EROSION.

erosivity The ability of *rainfall to cause *erosion, which depends on rainfall intensity and drop size, as well as the material it lands on.

erratic A large particle of *rock that has been transported by a *glacier away from its source and deposited in a region of different rock. Also known as **exotic**.

ERS *See* EUROPEAN REMOTE SENSING SATELLITE.

ERTS *See* EARTH RESOURCES TECHNOLOGY SATELLITE.

ERU *See* EMISSIONS REDUCTION UNIT.

ESA *See* ENVIRONMENTALLY SENSITIVE AREA.

escape cover *See* COVER.

escape hypothesis *See* JANZEN–CONNELL ESCAPE HYPOTHESIS.

escapement 1. The number or proportion of elk or other *wildlife that survive the hunting season.
  2. The number or proportion of fish that avoid or escape from *fisheries and move offshore, where they eventually spawn.

escarpment A long cliff or relatively steep slope that faces in one direction, separating two generally level surfaces, often produced by *faulting or *erosion. Also known as **cuesta** or **scarp**.

esker A sinuous ridge of coarse gravel that is deposited by a *meltwater stream beneath a *glacier. *Contrast* KAME.

essential element A *chemical element that is required by all living organisms for normal growth. This includes the primary essential elements (*carbon, *hydrogen, *oxygen, *nitrogen, *phosphorus, and *potassium), secondary essential elements (*sulphur, *calcium, *sodium, *chlorine, and *magnesium), and the *trace elements (*iron, *boron, *manganese, *copper, *zinc, *iodine, *cobalt, *selenium, *chromium, *silicon, and *molybdenum).

establishment The successful growth and reproduction of plants and/or animals in a site that provides them with favourable conditions.

estate 1. An extensive tract of land, particularly in the country, that is owned and retained by the owner for his/her own use.

2. All of a person's assets and liabilities; everything they own.

estuarine Of or relating to an *estuary.

estuarine flow The seaward flow of freshwater over a deeper layer of seawater with higher *salinity, within an *estuary.

estuarine subtidal A *habitat of open water within an *estuary that is continuously covered by seawater.

estuarine wetland A *tidal *wetland within an *estuary.

estuary The wide mouth of a *river where it flows into the sea, in which fresh (river) water meets and mixes with salt (sea) water, and which is *tidal.

ethane (C_2H_6) A colourless, odourless *hydrocarbon gas that is extracted mainly from *natural gas and is used as a *fuel.

ethanol (C_2H_5OH) A colourless, flammable liquid produced by *fermentation of sugars (derived from agricultural products such as corn, grain, and sugar cane) which is used as an *alternative fuel and a fuel additive. It is the most widely used renewable *biofuel. Also known as **fuel ethanol**.

ethical values Statements of ethical principle that inform the private and social valuation of *environmental resources.

ethics The philosophical study of moral values and rules, that inform decisions about wrong and right. See also ENVIRONMENTAL ETHICS.

ethnobotany The study of what plants people use and how they use them.

ethnography The study of culture (including behaviour, beliefs, and attitudes), based on observation of and interaction with living people. See also ANTHROPOLOGY.

EU See EUROPEAN UNION.

EU Bubble A *greenhouse gas *emissions bubble in which members states of the *European Union have accepted an aggregated emissions reduction target and arrangements that allow the target to be shared among all of the countries within the *bubble.

EU Directive A legal *directive from the *European Union that is binding on member states but leaves the method of implementation for national governments to decide.

EU Habitats Directive See HABITATS DIRECTIVE.

EU Life Programme See LIFE PROGRAMME.

eugenic The creation of high-quality offspring. Contrast dysgenic.

eugenics The study of hereditary improvement amongst humans by controlled *selective breeding.

euhaline Fully saline; seawater with a *salinity of greater than 30 parts per thousand.

eukaryote An *organism that has a distinct *nucleus. This include all organisms other than *viruses, *bacteria, and *blue-green algae. See also PROKARYOTE.

eulittoral The intertidal zone.

euphotic zone The layer of water in a lake or sea (usually the top 80 metres) where sunlight is sufficient for *photosynthesis to occur. Also known as **photic zone** or **epipelagic zone**.

European Commission The administrative and executive institution of the *European Union, which is responsible for initiating proposals for legislation, ensuring the implementation of treaties to which the EU is party, executing EU policies, and representing the union in trade negotiations with non-member countries. See also COUNCIL OF MINISTERS.

European Community (EC) A regional organization created by the Treaty of Rome (1957), which provided for the gradual elimination of customs duties and other interregional trade barriers between member states in Europe. In 1993, with establishment of the *European

Union (EU), the EC became the customs union component of the EU.

European Economic Community (EEC) An economic union that was established in 1958 to promote trade in Western Europe by merging separate national markets into a single market that ensures the free movement of goods, people, capital, and services. Also known as the **Common Market**. Renamed *European Union in 1994.

European Environment Agency (EEA) The body of the *European Union that is responsible for supporting *sustainable development and helping to achieve significant and measurable improvement in Europe's *environment through the provision of timely, targeted, relevant, and reliable information to policy-making agents and the public.

European Remote Sensing (ERS) satellite A series of radar satellites launched by the European Space Agency during the 1990s.

European Union (EU) An economic association of European countries that was founded by the Treaty of Rome in 1957 and was known as the European Community before 1993. Its goals are a single market for goods and services without any economic barriers and a common currency with one monetary authority. In May 2004 the EU had 25 member states—Austria, Belgium, Cyprus (Greek part), Czech Republic, Denmark, Estonia, Finland, France, Germany, Greece, Hungary, Ireland, Italy, Latvia, Lithuania, Luxembourg, Malta, Netherlands, Poland, Portugal, Slovakia, Slovenia, Spain, Sweden, and the United Kingdom. *See also* COUNCIL OF MINISTERS, EUROPEAN COMMISSION.

eury- A prefix denoting the ability of an *organism to tolerate a wide range of changes in environmental conditions, such as *eurythermal. *Contrast* STENO-.

euryhaline The ability of an *organism to live in environments with a wide range of *salinity.

eurythermal The ability of an *organism to live in environments with a wide range of temperature.

eustasy The global position and changes in the position of *sea level.

eustatic A world-wide change in *sea level. *Contrast* ISOSTATIC ADJUSTMENT.

eutrophic Enriched in *nutrients. A body of water which has a high concentration of dissolved nutrients, is very biologically productive, and has large amounts of *algae, low water transparency, and low *dissolved oxygen. *Contrast* OLIGOTROPHIC. *See also* DYSTROPHIC, EUTROPHICATION.

eutrophication A common form of *water pollution which involves the *enrichment of a body of *freshwater with *nutrients such as *nitrate *fertilizers (washed from the soil by rain) and *phosphates (from fertilizers and *detergents in *municipal sewage). The pollution enriches the waterbody, and this encourages the rapid growth of aquatic plants and can cause excessive growth of *algae (*bloom) and vascular plants. This in turn reduces the availability of light and oxygen in the water, making it uninhabitable for some species. Some of the algae and *bacteria produce relatively large amounts of *toxins, which further disrupt the aquatic ecosystem. Since the early 20th century such blooms have been a regular occurrence in the most heavily polluted parts of the *Great Lakes in North America, but they have declined as *water quality has improved as a result of improved pollution control and water quality management strategies. Lakes can be classified according to their nutrient content as *oligotrophic (nutrient poor), *mesotrophic (moderately productive), or *eutrophic (very productive and fertile). Also known as **nutrient enrichment.**

evacuation A prolonged precautionary stay away from an area that is affected by a *hazardous material or a *natural hazard.

evaluation An examination or judgement about the worth, quality, significance, amount, degree, or condition of a project or object.

evaporation The process by which a liquid changes into a *vapour with the application of heat. *Contrast* CONDENSATION.

evaporation fog A type of *fog that is produced when sufficient *water vapour is added to the air by *evaporation, and the moist air mixes with relatively drier air. Also known as **frontal fog** or **mixing fog**.

evaporation pond An area where *sewage sludge is dumped and dried by *evaporation.

evaporite A type of *sedimentary rock that is formed by the *precipitation of minerals dissolved in water as a result of *evaporation. Examples include *rock salt and *gypsum.

evapotranspiration The process by which *water vapour is released to the atmosphere from surface water, soils, and plants through the combined effects of *evaporation and *transpiration.

even-aged stand A group of trees that are essentially the same age (usually 10–20 years). *Contrast* ALL-AGED STAND, UNEVEN-AGED STAND.

evenness The quality of uniformity and lack of variation. For example the degree to which all species in an area are equal in abundance and not dominated by one or a few species. Also known as **equitability**.

Everglades A large natural *freshwater *marsh which covers nearly 13 000 square kilometres of southern Florida in the USA. The wetland was created by regular overflowing from Lake Okeechobee during the wet season, which flooded the Everglades up to one metre deep in water. Most of the area is a wilderness of *swamp, *savanna, and *primary forest and it contains a rich diversity of wildlife, much of which is protected within the Everglades National Park (which covers 6100 square kilometres). The marsh is covered with dense saw grass which rises up to three metres above the water level, making access difficult in many places. Many small islands covered with thick bushes and trees (including cypress, mangrove, and palms) are scattered throughout the marsh, all of which are less than two metres above sea level.

EVOLUTION

A model which explains how all *organisms change over time, based on *survival of the fittest. The model is based on gradual change in the characteristics of a *population of animals or plants over successive generations, which explains how new species can be formed from more primitive forms through a series of very slow, gradual changes stretching over a long period of time. The scientific theory of evolution is usually traced to Charles Darwin's (1859) *On the Origin of Species by Means of Natural Selection.* *Darwin argued that varieties of plant and animal life evolve gradually over long periods of time, adapting themselves to their environment through the processes of *mutation and *natural selection. In natural selection, the organisms that are best fitted for survival (by being best adapted to their environment) breed and pass on their features to the next generation. A species slowly evolves over a long period, generation by generation, by *speciation. *Contrast* EXTINCTION. *See also* ADAPTATION, ADAPTIVE RADIATION, ARTIFICIAL SELECTION, SPECIATION, INDUSTRIAL MELANISM, PUNCTUATED EQUILIBRIUM.

evergreen Plants, shrubs, and trees with needles or leaves that remain alive and on the tree through the winter and into the next growing season, so it remains green throughout the year. *See also* CONIFEROUS. *Contrast* DECIDUOUS.

evolutionary change Changes (that can be *phenotypic or *genetic) that occur in an *organism from generation to generation through the exchange of *genes.

evolutionary classification A *classification of *organisms that is based on levels of evolutionary advancement or development.

ex situ Outside, off site, or away from the natural location. For example, biological material which is in a laboratory, collection, *botanical garden, *zoo, or *aquarium. Also known as **off-site**. *Contrast* IN SITU.

ex situ conservation An approach to the *conservation of *biodiversity that is based on keeping organisms and species alive by the deliberate removal of biological resources (seed, pollen, sperm, individual organisms) from their original habitat or natural environment, and protecting them elsewhere under controlled conditions. *Contrast* IN SITU CONSERVATION. *See also* GENE BANK.

exceedance A measured level of a particular *pollutant that is higher than the *ambient air quality standard.

exception In land *zoning, the granting of a special permit for a use which is outside of those normally approved for that type of zone.

exchange capacity A measure of the surface charge of a substance that is capable of absorbing ions. *See also* CATION EXCHANGE CAPACITY.

exclusion zone The area surrounding an operation which may threaten human health or well-being, from which non-essential personnel are kept out. Also known as a **hot zone**.

Exclusive Economic Zone (EEZ) A zone of the oceans over which a particular nation has claims or exclusive rights to explore, exploit, conserve, and manage *natural resources, as defined in the 1982 *United Nations Convention on the Law of the Sea.

exclusive use zoning A *zoning regulation that allows only one type of land use in a particular established zone or district.

exfoliation A *weathering process in which thin layers of rock peel off from the surface, which is often caused by the heating of the rock surface during the day and cooling at night leading to alternate expansion and contraction. It can also be caused by the release of pressure when previously covered rocks are exposed, for example by weathering and *erosion.

exhaust ventilation The mechanical removal of air from part of a building.

exhaustible resource Any *non-renewable resource (such as *minerals, non-mineral resources, and *fossil fuels) that are present in fixed amounts in the *environment.

exine The resistant outer shell of a spore or *pollen grain that can be preserved in *sediments and which can be identified in *pollen analysis.

existence value *See* NON-CONSUMPTIVE VALUE.

exobiology The study of life outside the *Earth.

exogenous External, arising from outside. *Contrast* ENDOGENOUS.

exopterygota Insects in which the newly born young physically resemble the adults, and wings develop externally. *Contrast* ENDOPTERYGOTA.

exoskeleton A skeleton or support structure that is outside the body of the organism. *Contrast* ENDOSKELETON.

exosphere The outer part of the *atmosphere that lies beyond the *thermosphere, above about 350 kilometres, where the air is extremely thin, but there are still traces of some gases (particularly *hydrogen) as far out as 8000 kilometres. The exosphere has no clearly defined outer limit, but instead fades off into the vacuum of *space.

exothermic A chemical reaction which gives off heat. *Contrast* ENDOTHERMIC.

exotic Not naturally occurring in a particular area. Also known as **alien**. *Contrast* INDIGENOUS, NATIVE. *See also* ERRATIC.

exotic species *See* INTRODUCED SPECIES.

exploitation 1. The killing, capture, or collection of *wild organisms for human use.
2. The use of a *natural resource for profit or benefit to humans.

exploitation competition Competition between two or more *organisms for the same limited *resource. *See also* INTERFERENCE COMPETITION.

exploratory well A *well that is drilled to find and produce oil or gas in an unproved area, to find a new reservoir of oil or gas in a field previously found to be productive, or to extend the limit of a known oil or gas reservoir.

exponential growth Growth at a constant rate of increase per unit of time, which can be expressed as a constant fraction or exponent and graphed as a *J-curve. *Contrast* ARITHMETIC GROWTH, GEOMETRIC GROWTH.

export To transport a commodity to another area, such as the transport of *municipal solid waste and *recyclables outside the locality where they originated.

exposed Visibly unprotected. In terms of wave *exposure, an open coast that is facing away from prevailing winds but has a long *fetch, where strong winds are common.

exposed group The members of a *population who have been exposed to a *pollutant or other *agent.

exposure 1. Contact between an *organism and a chemical or physical *agent, by swallowing, breathing, or direct contact (such as through the skin or eyes). Exposure may be either short term (*acute) or long term (*chronic).
2. The degree of wave action on an open shore, which is determined by the *fetch and the strength and duration of *winds. *Compare* EXPOSED, EXTREMELY EXPOSED, SHELTERED, ULTRASHELTERED, VERY EXPOSED, VERY SHELTERED.

exposure assessment An estimation of the magnitude, frequency, duration, route, and extent (number of people) of exposure to a chemical or physical *agent.

exposure event An incident of contact with a chemical or physical *agent.

exposure indicator An *environmental indicator that provides evidence of exposure to a chemical or biological *agent.

exposure level The amount or concentration of a chemical or biological *agent that is measured in or on an *organism.

exposure limit The regulated level of exposure a chemical or biological *agent that should not be exceeded in order to protect human health and well-being.

exposure pathway The route that a particular *pollutant takes from its source to people or other organisms, via soil, water, or food. *See also* EXPOSURE ROUTE, INDIRECT EXPOSURE PATHWAY.

exposure route The way that a chemical or pollutant enters an organism after contact (by ingestion, inhalation, or absorption through skin). *See also* EXPOSURE PATHWAY.

exposure–response relationship The relationship between *exposure level and incidence of *adverse effects.

extant Existing or surviving, still living at the present time. *Contrast* EXTINCT.

EXTINCTION

The permanent disappearance of a *species throughout its entire range, caused by the failure to reproduce and the death of all remaining members. Extinction has occurred throughout Earth history for entirely natural reasons, either because of major *catastrophes (such as meteorite impacts) or because species were unable to adapt quickly enough to natural *environmental change. Most recent extinctions are the result of *human impacts, amongst the most important of which are *habitat alterations, hunting and the introduction of alien species. Extinction is final, because once a species has become extinct its genetic composition is lost and cannot be replaced naturally, so it is lost to present and future generations. Since 1600 some 485 species of animals and 584 species of plants have become extinct, largely if not entirely as a result of human activities. More than a tenth of the world's plant species are heading towards extinction according to the first fully comprehensive study. The 1998 IUCN *Red List of threatened plants includes 33 798 species, of which 380 are extinct in the wild, 371 may be extinct, 6522 are endangered and the rest are vulnerable or rare. Around 90% of the species listed are *endemic to just one country, and those growing on isolated islands are especially vulnerable because they can be displaced by plants and animals introduced by humans. Ecologists fear that human activities could trigger another of the *mass extinctions that have occurred several times during the Earth's history. If this happened the consequences could be catastrophic, because environmental change is too rapid and time too short to allow natural rapid speciation. *Contrast* EVOLUTION, EXTIRPATION. *See also* BACKGROUND EXTINCTION, MASS EXTINCTION.

extended outlook A basic forecast of general weather conditions three to five days in the future.

extensive recreation *See* DISPERSED RECREATION.

extent In *conservation assessment, the size of site that is required to ensure that the unit to be managed is viable.

external cost *See* EXTERNALITY.

external forcing *See* RADIATIVE FORCING.

externality A consequence or impact of a decision about *resources that is not directly accounted for in the price paid for the resource, such as *pollution, loss of *wilderness, or *environmental change. Also known as **externalization of production costs**, or **social cost**. *See also* ENVIRONMENTAL ECONOMICS, INTERNALIZATION, MARKET FAILURE.

externalization of production costs *See* EXTERNALITY.

extinct No longer living anywhere. A *species for which there is no reasonable doubt that the last individual has died. A *threatened category defined by the *IUCN *Red Data Book. *Contrast* ENDANGERED, EXTIRPATION, THREATENED, VULNERABLE.

extirpated Locally extinct.

extirpation The disappearance of a *population or *species from an island, area, or region. Also known as **local extinction**. *Contrast* EXTINCTION.

extract To remove or take out.

extraction The act of removing a resource, such as felled timber from a forest or *groundwater from an *aquifer. Also known as **abstraction**.

extractive reserve A *conservation area in which certain kinds of *resource harvesting on a *sustainable basis are permitted.

extrapolation Estimating a value by projecting or extending known values, for example in order to predict outside a range of known conditions (such as predicting future values). *Contrast* INTERPOLATION.

extraterrestrial shortwave radiation The *radiation from the Sun that reaches the top of the Earth's *atmosphere. *See also* SOLAR CONSTANT.

extratropical A *mid-latitude *climate zone that lies poleward of the *tropics.

extreme environment Any *environment that has extremes in growth conditions for plants, including *temperature, *salinity, *pH, and *water availability.

extremely exposed In terms of wave *exposure, an open coastline which faces into the prevailing wind, is directly affected by both wind-driven *waves and ocean *swell, and has deep water close to the shore.

extremely hazardous substance Any of more than 400 *chemicals that are classified as *toxic by the *Environmental Protection Agency and listed under *SARA.

extremely sheltered In terms of wave *exposure, a fully enclosed *coast with a *fetch of no more than about three kilometres.

extremophile A *micro-organism whose optimum growth is under conditions of extreme *acidity or alkalinity, *salinity, *temperature, or *pressure.

extrinsic External, coming from outside. *Contrast* INTRINSIC.

extrinsic resource A *natural resource that has been adapted or modified by people in order to provide additional values, particularly for *recreation. Examples include historic sites and archaeological sites.

extrinsic value Value in terms of what something means to people. *Contrast* INTRINSIC VALUE.

extrusive *Igneous rock that has formed on the surface of the Earth, such as *basalt. *Contrast* INTRUSIVE.

exurbia The region surrounding a *city and its *suburbs where wealthier families tend to live. Also known as **suburbia**.

Exxon Valdez An *oil tanker that ran aground, spilling 300 000 barrels of oil in Prince William Sound in Alaska on 24 March 1989, creating an *oil slick that covered 12 400 square kilometres and polluting at least 1100 kilometres of coastline. Damage was extensive, partly because the area is ecologically sensitive and thus not able to withstand great stress. The problem was made worse by a lack of *disaster preparation and an inability to use containment methods, so large amounts of damaging *detergents were used to try to disperse the oil. The oil spill caused extensive damage to the ecology of the area.

eye The warm calm area at the centre of a *tropical cyclone, where the winds are light and skies are clear or partly cloudy.

eye wall The zone of powerful *updraft around the *eye of a *tropical cyclone, which has the heaviest *rain, strongest *winds, and worst *turbulence and causes a ring of heavy *thunderstorms.

eyrie The nest of a bird of prey, such as a hawk or eagle, which is usually built high in a tree.

F scale *See* FUJITA SCALE.

fabric The way in which particles and minerals are arranged in a rock, sediment, or soil. *See also* IMBRICATION.

facies The characteristics of a *sediment or *sedimentary rock unit, such as *mineralogy, colour, *texture, and *fossil content, which are indicative of the conditions when the rock formed.

facility A building or place that provides a particular service or is used for a particular activity.

factory farming Large-scale, industrialized *agriculture.

factory ship A ship that is equipped to process large quantities of fish or whale products at sea.

facultative Having the ability to live under different conditions, either with or without a particular environmental factor such as *oxygen.

facultative mutualist A beneficial *symbiont that associates with its *host but is also able to live apart from it. *Contrast* OBLIGATE MUTUALIST.

facultative organism An *organism (usually a *micro-organism) that is capable of adapting to either *aerobic or *anaerobic conditions.

faecal Relating to *faeces.

faeces Solid unabsorbed residue (waste matter) that is passed out of the digestive tract of an animal, though its anus. *See also* COLIFORM BACTERIA.

Fahrenheit (°F) A *temperature scale in which the freezing point of water is 32°F and the boiling point is 212°F at sea level. 32°F is equal to 0° *Celsius, and °C $= (°F-32)×5/9$. This scale is widely used in the USA.

fair A *weather condition in which there is less than four-tenths opaque cloud cover, no *precipitation, and no extremes in *temperature, *visibility, or *wind.

fall **1.** A *season between *summer and *winter, that is known as *autumn in most of the English-speaking world beyond North America.
2. A rapid process of *mass movement on a *hillslope.
3. Another name for *waterfall.

fall line The line of steepest descent on a slope.

falling tide *See* EBB TIDE.

fallout Solid material that falls to the ground from the sky, including *radioactive debris that falls after a nuclear explosion, and volcanic *ash that falls after a *volcanic eruption.

fallow The practice of leaving a patch of land idle and uncropped, either tilled or untilled, during much or all of a *growing season, in order to accumulate moisture, improve *soil structure, or restore nutrient content.

fallstreak *See* VIRGA.

family **1.** A social unit, including parents and children, which lives together and may include relatives.
2. The fifth highest (of seven) category in the scientific system of classification for organisms (*taxonomy), below *order and above *genus. Each order comprises more than one family, and each family comprises more than one genus.
3. A category in the *Comprehensive Soil Classification System that is intermediate between *great soil group and *soil series.

family planning A system of limiting family size and the frequency of child-bearing by the appropriate use of *birth control.

famine An acute shortage of food that leads to malnutrition and starvation.

fan A wedge-shaped body of sediment (usually sand and gravel), such as an *alluvial fan.

FAO See UNITED NATIONS FOOD AND AGRICULTURE ORGANIZATION.

farm Buildings and associated land that are used for *agriculture.

farm forestry See TREE FARMING.

farming See AGRICULTURE.

farmland *Arable land that is worked for *farming by ploughing, sowing, and raising crops.

farmstead Land that is occupied by the dwellings, barns, pens, corrals, gardens, and similar associated with an operating *farm.

farmyard manure The partly decomposed excreta of domestic animals mixed with straw or other litter. See also DUNG, SLURRY.

fast breeder See BREEDER REACTOR.

fast ice Ice in a lake, river, or the sea that is permanently attached to a *glacier or to the shore, the bottom, or in shallow water.

fatal accident An *accident that causes one or more deaths within a year.

fatality A death that is caused by an *accident or a *disaster.

fatality rate See DEATH RATE.

fault A large-scale linear fracture at the Earth's *crust, along which the rocks on either side have been displaced as a result of pressure in the adjacent bodies of rock. The displacement can be vertical, horizontal, or both at the same time. There are four common types of faults (*normal fault, *reverse fault, *strike-slip fault, and *thrust fault), which are produced in different ways and produce different landscapes. Distinctive large-scale *landforms (*graben and *horst) are created when two or more fault systems intersect one another. See also EARTHQUAKE.

fault line The contact zone between two sides of a *fault, along which the rocks on either side move.

fault plane/zone The part of the Earth's crust that is affected by *faulting. *Faults occur where such a zone intersects the ground surface.

fault-block mountain See HORST.

faulting A fracturing of the rocks on the Earth's *crust that produces a line (*fault line) along which significant movement takes place, which causes rocks on either side to be displaced vertically or horizontally relative to each other.

fauna 1. All forms of animal life that lives in a region, period, or special environment.

2. A record or book of animals.

Fauna & Flora International The world's longest-established international conservation body that was founded in 1903, and one of only a few whose mission is to protect the entire spectrum of *endangered species of animals and plants world-wide.

favourable condition The desirable state for a site to be in for *conservation purposes.

favourable status The desirable status of a *population for *conservation purposes.

FC See FLUOROCARBON.

feasibility study 1. An investigation to establish whether a project will work and achieve the desired results.

2. A study to evaluate alternative *remedial actions from a technical, environmental, and cost perspective, which normally recommends selection of a cost-effective alternative.

fecundity The state of being fertile; capable of producing offspring. See also FERTILITY.

Federal Emergency Management Agency (FEMA) The leading federal emergency planning agency in the USA that seeks to reduce the loss of life and protect property against all types of *hazards (including *disasters and urban riots) through a comprehensive, risk-based *emergency management programme.

federal land Land that is owned and managed by the US government, including *National Parks and *National Forests.

Federal Land Policy Management Act (FLPMA) (1976) US legislation that governs how the *US Bureau of Land Management manages, protects, develops, and enhances public lands. It specifically requires the Bureau to manage public lands for multiple use and sustained yield, for the benefit of both present and future generations. This act triggered the *Sagebrush Rebellion. It was partly replaced by the *Public Rangelands Improvement Act (1978).

Federal Water Pollution Control Act Amendments (1972) US legislation that established as a national objective the restoration and maintenance of the chemical, physical, and biological integrity of the nation's water, and required area-wide planning to prevent future *water pollution that could be associated with growth, development, and land use, including timber management.

feed Ground or processed grains that are fed to animals, including hay such as *alfalfa or *silage.

feedback An internal adjustment within a *system, where once a component reaches a certain level it inhibits (*negative feedback) or promotes (*positive feedback) further action. For example, a thermostat signals a boiler to turn on or off according to room temperature. See also HOMEOSTASIS.

feeding ground A place where animals feed naturally.

feedlot A confined area for the controlled feeding of large numbers of animals, which produces large amounts of animal waste that can flow into and pollute nearby *waterbodies.

feldspar A group of common *aluminium silicate minerals (the most common minerals in the world) that contain *potassium, *sodium, or *calcium, and that form rocks.

fell 1. To cut down standing vegetation, such as a tree.
2. Traditional British name for open *moorland.

felling Cutting down trees. See also CLEARING.

felsic A light-coloured *igneous rock rich in *silica minerals (such as *quartz and *feldspars), which is a major component of the Earth's *continental crust. The term is also used to refer to the minerals themselves. Contrast MAFIC.

FEMA See FEDERAL EMERGENCY MANAGEMENT AGENCY.

female In organisms with separate sexes, the one which is capable of giving birth or laying eggs. Contrast MALE.

feminism See ECOFEMINSM.

fen A type of *wetland dominated by *marsh-like vegetation, produced where slightly alkaline *groundwater emerges to the surface, that accumulates *peat deposits. *Bogs have similar types of vegetation but tend to be *acid.

fen peat *Peat that is *alkaline or non-acidic because of the presence of *calcium carbonate.

fencerow A row of *trees, *conifer, *shrubs, or groundcover plants that provides food and cover for *wildlife.

feral A *domesticated *animal that has adapted to living in the *wild.

fermentation A type of *anaerobic respiration by a living agent (such as yeast, *bacteria, or *mould), which breaks down complex *organic compounds into simple ones. For example, yeast converts sugar into alcohol and carbon dioxide. In *biotechnology, fermentation is the process of growing *microbes to produce chemical or pharmaceutical compounds.

ferric/ferrous Composed of and/or containing *iron.

fertilization 1. The union of male and female cells (sperm and egg; *gametes) to form a new individual.
   2. Adding *nutrients to soil or plants to stimulate growth.

fertilizer A substance (such as animal *manure or an artificial chemical, particularly one that contains *nitrogen, *phosphorus, and *potassium) that is added to *soil in order to increase its productivity for crops.

fertility The number of live births per female within a population, which is always lower than *fecundity. In 1998 the average fertility in the world was 2.9 children per woman; the average in the USA was 2.0, the averages in Italy and Spain were 1.2, and in the West African state of Niger it was 7.4. A population stops growing when it reaches the *replacement-level fertility, but this requires a *total fertility rate (TFR) higher than 2 mainly because some children die before they grow up to have their own children. The critical TFR is around 2.1 in a country with low mortality (such as the USA), but it can be higher than 3 in a country with high mortality (such as Sierra Leone).

fertility rate The number of children that are born in a given year, usually expressed per 1000 women in the reproductive age group.

fetch The distance of open sea over which the *wind may blow to generate *waves until it meets a coastline.

fetus *See* FOETUS.

field A cultivated area of land that is usually enclosed by a *hedge, wall, or fence, and is used for a particular crop or cropping sequence.

field capacity The amount of water that a particular *soil can hold after gravitational water has drained away.

This is mostly *capillary water. Also known as **field moisture capacity.**

field layer The vegetation at ground level in a *forest or *woodland.

field moisture capacity *See* FIELD CAPACITY.

field sport The pursuit, capture, and killing of *wildlife for entertainment. Also known as **country sports** and **game sport.** *See also* BIG GAME, FISHING, GAME, HUNTING, SHOOTING.

field tile A perforated pipe, usually made from plastic or clay, that is buried under the surface of the ground to improve soil *drainage. Also known as **drainage tile.**

fill Soil or sediment that is added to change the height of the land, for example to fill in a hole or to build an embankment.

filter A porous mesh that allows air or liquid to pass through but holds back solid particles.

filter feeder An *organism that filters out food particles (such as *plankton, *bacteria, or *detritus) from *freshwater or *seawater. Also known as **suspension feeder.**

filter strip *See* BUFFER STRIP.

filtration The process of removing *particulate matter from water by passing it through a porous medium such as sand or a *filter. Filtration does not remove *dissolved salts or *organic contaminants. *See also* DIRECT FILTRATION, FLOCCULENT.

final cut Cutting of the few original trees that remain in a *shelterwood cut, once new trees have been established.

final deal In *greenhouse gas *emissions trading, the fully matched and negotiated bids/offers that are under contract for delivery.

Finding of No Significant Impact (FNSI) A document prepared by a federal agency in the USA, based on the results of an *environmental assessment, that shows why a proposed action would not have a significant impact on the environment and would thus not require preparation of an *environmental impact statement.

fine filter approach An approach to the *conservation of *biodiversity that is directed toward particular habitats or species that may be *threatened or *endangered and might fall through the *coarse filter.

fine textured Having a smooth, *fine-grained structure.

fine-grained A *soil, *sediment, or *rock in which the grains or crystals are too small to be seen with the naked eye.

finished water *Water that has passed through all the processes in a *water treatment plant and is ready to be delivered to consumers. Also known as **product water**.

fipronil A type of *insecticide that works by disrupting normal nerve functions in the insect.

fire The rapid oxidation of a *fuel that results in the release of heat, light, and other *byproducts.

fire line See FIREBREAK.

fire management All activities that are associated with the management of fire-prone land, including the use of *fire to meet land management goals and objectives. See also FUEL BREAK, PRESCRIBED FIRE.

fire regime The role that *fire plays in an *ecosystem, which depends on the frequency and scale of fires, and may include proposals for the controlled use of fire in a given area. See also PRESCRIBED FIRE.

fire season The period(s) of the year during which fires are likely to occur, spread, and cause enough damage to require organized fire control.

firebreak A barrier, usually either a natural *clearing or one created by re-moving vegetation (*fuel break), that is designed to prevent or slow the spread of *fire. Also known as **fire line**.

fire-climax community A *climax community of vegetation that is maintained by periodic *fires. Examples include *grasslands, *chaparral shrubland, and some pine forests.

firestorm A raging *fire of great intensity that spreads rapidly.

firn *Snow on the surface of a *glacier that has remained from the previous year, which may be compact but not yet turned into ice. A transitional stage between snow and ice. Also known as **granular snow** or **névé**.

firn limit The dividing line between old ice and new snow on a *glacier, at the end of the melting season.

firn line The zone on a *glacier that separates the (upper) *accumulation area from the (lower) *ablation area.

first draw The *water that comes out when a tap is first opened, which is likely to contain the highest level of *lead contamination from plumbing materials.

First World The industrialized *capitalist or *market economy countries of Western Europe, North America, Japan, Australia, and New Zealand that were the first to industrialize. See also SECOND WORLD, THIRD WORLD.

fish A *cold-blooded *vertebrate animal that lives in water and breathes through gills; most fish have scales. See also GAME FISH, ROUGH FISH.

Fish and Wildlife Service See US FISH AND WILDLIFE SERVICE.

fish farm A place where aquatic plants and animals are grown commercially, in ponds, pens, tanks, or other containers. See also AQUACULTURE.

fish farming See AQUACULTURE.

fish kill The sudden death of most if not all of the *fish in a waterbody, caused by the introduction of *pollutants or a reduction in *dissolved oxygen concentration.

fish ranching A form of *aquaculture in which a *population of a *fish species (such as salmon) is held in captivity for the first stage of their lives, then released, and later harvested as adults when they return from the sea to their freshwater birthplace to *spawn.

fish trapping A method of small scale *fishing using traps and nets. Also known as **beach seining.**

fisheries habitat Any stream, lake, or reservoir that supports *fish, or has the potential to support fish.

fishery A place where *fish are caught and processed and sold.

Fishery Conservation and Management Act (1976) US legislation that provided for the conservation and management of fisheries, which established a 200-mile fishery conservation zone and Regional Fishery Management Councils comprising Federal and State officials, including the *US Fish and Wildlife Service. Also known as **Magnuson Fishery Conservation and Management Act.**

fishing Catching fish, for food or as a *field sport, in freshwater or at sea.

fissile 1. The ability of certain rocks to be split along *bedding planes.
2. *See* FISSIONABLE.

fission 1. The nuclear process in which the *nucleus of a particular *isotope splits into (usually) two nuclei of lighter elements, releasing great amounts of energy at the same time. This process is exploited in the *breeder reactor. *Contrast* FUSION.
2. Division of a cell into two cells by splitting.

fissionable The property of the nucleus of some atoms that allows them to split into smaller particles, and so be capable of sustaining a chain reaction of nuclear *fission. Also known as **fissile.**

fissionable isotope An *isotope that can undergo nuclear *fission when it is hit by a *neutron at the right speed. Examples include uranium-235 and plutonium-239.

fissure A long, deep, narrow opening in a *rock or *glacier (*crevasse), or an opening in a *volcano through which volcanic products can erupt.

fitness Adaptedness, or the ability of an *organism to survive and flourish in its current *environment, relative to the other organisms that are also there, which is measured ultimately by reproductive success.

fix To convert *carbon dioxide to carbohydrate (*carbon fixation), or *nitrogen dioxide to *ammonia (*nitrogen fixation).

fixed dune A sand *dune that has become stabilized by vegetation and is therefore largely protected from further *erosion and movement by the wind.

fjord A long, deep, steep-sided valley with a *U-shaped profile, that was formed by *glacial erosion and is now flooded by the sea.

flagellate A *micro-organism that has several hair-like projections on the cell surface, which it uses for locomotion or food gathering.

flagellate protozoa *Protozoa that have *flagella.

flagellum A thin whip-like appendage on a *motile cell, which is used for locomotion. The plural is flagella.

flagship species A small number of globally important *endangered species (including the giant panda, tiger, marine turtles, great apes, whales, elephants, and rhinos) which are well known and attract public interest in *conservation efforts both for themselves and other *species at risk.

flammable Any solid, liquid, vapour, or gas that will ignite easily and burn rapidly.

Flandrian *See* HOLOCENE.

flare 1. A sudden burst of flame.

2. A device that burns gases (such as *methane gas in a *landfill) to prevent them from being released into the *environment.

flaring The controlled burning of waste gases (for example, through a *chimney) before releasing them to the air.

flash flood A localized *flood that rises and falls quite rapidly with little or no advance warning, usually as the result of heavy rainfall over a relatively small area.

fledgling A young bird that has grown feathers.

flexibility mechanism The three co-operative mechanisms for reducing *greenhouse gas emissions that were agreed under the *Kyoto Protocol, which are *Joint Implementation, *emissions trading and the *Clean Development Mechanism. Also known as **Kyoto mechanism.** *See also* FUNGIBILITY.

flint A type of *chert that is found as *nodules or bands in *chalk and *limestone.

floc A collection of smaller particles that have come together into larger particles, such as the solids that are formed in *sewage as a result of biological or chemical action.

flocculation Chemical processes in which salts (*flocculent) are added to water to make *colloids aggregate into larger masses that are too heavy to remain suspended.

flocculent 1. A substance that is added to water to make particles clump together, in order to produce more effective *filtration.

2. Also used to describe the particles that clump together.

flock A group of *birds, sheep, or goats.

flood To fill quickly to beyond capacity, so that the fluid spills out. In a *river, it means a high flow that overflows from the *channel and inundates the *floodplain, which is normally dry. On a coast, it means seawater flowing over low-lying land that is normally dry. *See also* FLASH FLOOD, INUNDATION.

flood basin *See* FLOODPLAIN.

flood control An engineering scheme that is designed to protect areas of land from being flooded. Options include the construction of embankments (*levees and *dykes) and walls, *channel improvement, detaining part of the flood flow in a reservoir, and diverting some of the water into a bypass or *floodway.

flood conveyance The transport of floodwaters downstream, with little if any damage.

flood current The onshore movement of a *tidal stream, toward the *shore or up a *tidal river or *estuary.

flood damage The economic loss that is caused by a *flood, which includes direct damage due to *inundation, *erosion, and *sediment deposition, as well as emergency costs and business or financial losses.

flood forecast A prediction of the likely height, timing, and duration of a *flood, particularly the peak *discharge at a specified point on a stream, based on information about precipitation and/or snowmelt and the form of the drainage area. Also known as **flood prediction.** *See also* FORECAST.

flood frequency The probability (likelihood) that a *flood of a certain size will occur in a given year in a particular river or part of a river. *See also* RECURRENCE INTERVAL.

flood frequency analysis Analysis of *hydrograph records of *river flow to determine the *flood frequency at a particular location within a river system, based on either the *annual maximum series or the *partial duration series.

flood irrigation A method of *irrigation in which entire *fields are occasionally deliberately flooded.

flood peak The highest *stage (largest *discharge) reached during a particular *flood at a given point on a river. Also known as **peak discharge.**

flood prediction *See* FLOOD FORECAST.

flood protection *See* FLOOD CONTROL.

flood stage The water level (*stage) in a river or stream beyond which the flow starts to *flood adjacent land. *See also* BANKFULL.

flood tide The portion of the *tide cycle between low water and the following high water, during which the water is rising or levelling off. Also known as **rising tide.**

floodplain An area of flat, low-lying land adjacent to a *river, that is composed of *alluvium. It is normally dry but is covered by water during a *flood. Also known as **alluvial plain** or **flood basin.**

floodplain management A coordinated approach to the reduction of flood damage that usually includes emergency and contingency plans, *flood control works, and regulations to control current and future development in the flood plain.

floodplain zoning A type of *natural resource zoning that prohibits and restricts development within *floodplains, in order to prevent damage to property and risks to people. *See also* PROHIBITION ZONING.

floodproofing The process of protecting a building from *flood damage, for example by raising it above the ground, building a wall round it to keep out flood water, or using lower floors for activities that can tolerate *inundation.

floodway The *river channel and parts of the adjacent *floodplain that are required to carry flood water and must therefore not be built on or restricted.

flora 1. All forms of plant life that live in a region, period, or special environment. **2.** A record or book of plants.

floral region A large area that has a distinctive *vegetation.

floristic Referring to plants.

flow 1. The movement of a liquid or a gas. *See also* FLOW RATE. **2.** Rapid *mass movement of any *unconsolidated material. *See also* DEBRIS FLOW.

flow augmentation The addition of water to a stream from a storage reservoir in order to enhance flow, particularly to help fish migration.

flow duration The percentage of time that *discharge in a river exceeds a particular level, averaged over a number of years.

flow hydrograph *See* HYDROGRAPH.

flow meter A device that is used for measuring the flow or quantity of a moving fluid.

flow rate *See* VELOCITY.

flow regime The annual pattern of variations in *discharge in a river.

flow resource A *natural resource that is simultaneously used and replaced, which includes all *perpetual resources and *renewable resources.

flow velocity *See* VELOCITY.

flower 1. A *plant that is cultivated for its blooms or blossoms, its colours, and/or its scent. **2.** The reproductive part of a plant.

flowing well An *artesian well in which the pressure is large enough to make water flow out onto the land surface.

FLPMA *See* FEDERAL LAND POLICY MANAGEMENT ACT.

flue A vent or *chimney for a combustion device, through which smoke, gas, and fumes rise.

flue gas The air that is released by a *flue or *stack after combustion in the burner it is venting, which can contain *nitrogen oxides, *carbon oxides, *water vapour, *sulphur oxides, *particles, and many chemical *pollutants.

flue gas desulphurization A method of reducing the *sulphur content of *fossil fuels at combustion, in order to reduce *air pollution, that involves passing the *flue gas through a mixture of crushed *limestone and water. Also known as **flue gas scrubbing**. *See also* SCRUBBING.

flue gas scrubbing *See* FLUE GAS DESULPHURIZATION.

fluid Any substance, gas, or liquid that can flow freely.

fluidization Making particles float in a gas or fluid that is blowing upwards.

fluidized bed combustion A method of reducing the *sulphur content of *fossil fuels at combustion, in order to reduce *air pollution, that involves burning the fuel in a special furnace that lifts it in a stream of air and passes air through it while it burns. Also known as **fluidized bed technology**.

fluidized bed technology *See* FLUIDIZED BED COMBUSTION.

flume A natural or man-made structure that conveys *water.

fluorescence The property of absorbing light of a particular *wavelength and then emitting light of a different colour and wavelength.

fluoridation The addition of a *fluoride to the water supply, usually in order to prevent dental decay.

fluoride A compound that contains *fluorine.

fluorine (F) A non-metallic element, the lightest of the *halogens, usually in the form of a yellow, irritating, flammable gas which is one of the 100 most toxic substances known, and is persistent in the environment. Used commercially in the form of sodium fluoride.

fluorocarbon (FC) A compound that contains *fluorine and *carbon, including those which are chlorinated (*CFCs) and those which are brominated (*halons). They are inert and highly stable, and are used in refrigerators and aerosols.

flurry A sudden, brief, light *snowfall or gust of *wind.

flush 1. A sudden, rapid flow of water.
  2. A type of *fen that is irrigated by a *spring or soakway.

flushing rate *See* RETENTION TIME.

fluvent A *floodplain *soil, that usually has buried *horizons and amounts of *organic matter that decrease irregularly with depth.

fluvial Of or relating to *rivers.

fluvial process Geological work (*erosion, *transport, and *deposition of sediment) that is done by a *river.

fluvioglacial *See* GLACIOFLUVIAL.

flux Flow of energy, fluid, or particles per unit of area per unit of time.

fly ash Fine particles of *ash that are made from the burning of *coal, released in *flue gas, and carried in suspension in the air.

flyway A seasonal *migration route along which waterfowl travel between wintering grounds and nesting/breeding grounds.

FMD *See* FOOT AND MOUTH DISEASE.

FNSI *See* FINDING OF NO SIGNIFICANT IMPACT.

focal species A *species that is the focus of a *conservation effort.

focus The point within the Earth's *crust at which an *earthquake originates, which is located directly below the *epicentre. Also known as **hypocentre**.

fodder Coarse food that is composed of entire *plants or the leaves and stalks of a *cereal crop, and is fed to cattle and horses.

FOE *See* FRIENDS OF THE EARTH.

foehn (föhn) A warm, dry *adiabatic *wind which blows downslope on the *lee side of mountains. The term was originally applied only in the European Alps, but it is now used for all similar winds. *See also* CHINOOK, SANTA ANNA.

foetus The *embryo of a mammal (including humans) when the main adult features have developed.

fog A *cloud of tiny water droplets that is suspended in the lower atmosphere, produced by the cooling of moist air near the ground, which condenses to create a *visibility of 1 kilometre or less.

fog bank A dense mass of *fog surrounded by clearer air, often as viewed from a distance at *sea.

fogging A method of applying a *pesticide by rapidly heating the liquid chemical so that it forms very fine droplets like *smoke or *fog and is used to destroy mosquitoes, black flies, and similar pests.

fold A curve or bend in *stratified rocks that is usually caused by *compression pressure. The line along which a bed of rock folds is called its *axis, and the *limbs of the fold are the tilted beds that extend outwards from the axis. The fold can be upwards (*anticline) or downwards (*syncline).

fold mountain A *mountain or *mountain range (such as the Alps and the Himalayas) that is formed as a result of intense *folding of material within the Earth's *crust.

folding The process by which layered *sedimentary rocks in the Earth's *crust bend and buckle in response to *compressional forces, often caused by the convergence of two adjacent *tectonic plates.

foliage The leaves and needles that are growing on a *tree or *plant.

foliation The banding or *lamination seen in *metamorphic rocks, which is caused by the arrangement of *minerals in parallel, sheet-like layers as a result of *compression.

foliose Bearing leaves or resembling a leaf, particularly leaves or plants that are thin, flattened and lobe-like.

Food and Agriculture Organization *See* UNITED NATIONS FOOD AND AGRICULTURE ORGANIZATION.

food chain The transfer of food energy from plants through *herbivores to *carnivores. *Primary producers capture energy from the environment through *photosynthesis or *chemosynthesis, and they form the base of the food chain. Energy is then passed to *primary consumers (herbivores) and on to *secondary consumers (carnivores) and *tertiary consumers (top carnivores). For example, in a marine food chain *phytoplankton (primary producers) are eaten by *zooplankton (primary consumers), which are eaten by herring (secondary consumers), which are eaten by killer whales (tertiary consumers). Once they die, all of these organisms are in turn consumed and their energy transferred to *detrivores and *decomposers. *Compare* FOOD WEB.

Food Quality Protection Act (1996) US legislation that governs how the *Environmental Protection Agency regulates *pesticides, including a new requirement that a new safety standard (reasonable certainty of no harm) must be applied to all pesticides used on foods.

food waste Food that is discarded from kitchens, including restaurants, grocery stores, and homes.

food web A group of interconnecting *food chains within an ecosystem, which often has numerous organisms at each level.

foot and mouth disease (FMD) A severe disease that affects *ungulate *ruminant animals, particularly cattle, and is

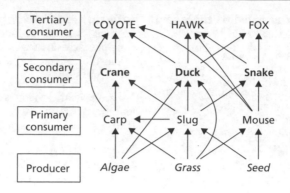

Fig 7 Food web

spread by a virus. Symptoms include fever, blisters, and damage to the animal's mouth and hooves, and it causes serious loss of production of meat and milk.

footprint *See* ECOLOGICAL FOOTPRINT.

forage 1. Grasses, herbs, and small shrubs that can be used as feed for livestock or wildlife.
2. The act of searching for food and provisions. *See also* FORAGING.

forage crop A crop of *cultivated plants that is grown to be grazed or harvested for use as *feed for animals.

forage production *See* FORAGE YIELD.

forage yield The weight of *forage that is produced within a particular area, per year. Also known as **forage production**.

foraging Searching or hunting for food, rather than growing food.

foraminifera Single-celled marine animals (*protozoa) that usually secrete a *carbonate shell, which is an important ingredient of *chalk.

foraminiferal ooze A *calcareous sediment composed of the shells of dead *foraminifera. *See also* PELAGIC SEDIMENT.

forb A *broadleaved *plant (such as a *wildflower) that has a soft rather than a woody stem, and is not a *grass or grass-like plant. *See also* HERB.

Forbes band *See* OGIVE.

forcing *See* RADIATIVE FORCING.

forecast To predict or estimate a future event or trend. *See also* FLOOD FORECAST, WEATHER FORECAST.

foreground The part of a scene or *landscape that is closest to the viewer. *See also* BACKGROUND, MIDDLEGROUND.

foreshore The lower zone of a *beach, that lies between *low water *spring tide level and *high water spring tide level.

forest An *ecosystem that is dominated by *trees and other wood plants. *See also* CONIFEROUS FOREST, DECIDUOUS FOREST, THORN FOREST, TROPICAL RAINFOREST.

Forest and Rangeland Renewable Resources Planning Act (1974) US legislation that provided for continuing assessment and long-range planning of the nation's *forest and *rangeland *renewable resources, under the jurisdiction of the Secretary of Agriculture.

Forest and Rangeland Renewable Resources Research Act (1978) US legislation that authorized more research on *renewable resources in *forest and *rangeland.

Forest and Rangeland Resources Extension Act (1978) US legislation that built upon the *McSweeny-McNary Act (1928) and authorized expanding the portion of the extension education programme that is dedicated to *renewable resources in *forest and *rangeland.

forest certification A process of labelling wood that has been harvested from a *forest that is responsibly managed.

forest clearance *See* DEFORESTATION.

forest conservation The deliberate *management of forests in order to conserve the *habitat and *natural resources. Strategies listed in *Agenda 21 include: planting new forests; increasing practical knowledge on the state of forests; increasing research into *forest products such as wood, fruits, nuts, dyes, medicines, gums, etc.; replanting damaged areas of woodland; breeding trees that are more resistant to environmental pressures; encouraging local business people to set up small forest enterprises; limiting and aiming to stop *slash and burn farming methods (the basis of *subsistence-level *shifting cultivation); keeping wood waste to a minimum; finding ways of using trees that have been burnt or thrown out; and increasing tree planting in urban areas.

forest cover A *forest stand that consists of a *plant community made up of trees and other woody vegetation which are growing together.

forest death *See* TREE DIEBACK.

forest ecology The relationships between *forest *organisms and their *environment.

forest fire An unplanned and uncontrolled *fire in an area of *forest.

forest floor The ground surface and *ground layer beneath a *forest *canopy, which often contains layers of fresh leaf and needle *litter, moderately decomposed organic matter, and well-decomposed organic residue (*humus).

forest   fragmentation Isolated patches of forest *habitat that survive when the intervening forest has been removed, either naturally (for example through *forest fires) or by clearance (*felling). *See also* FRAGMENTATION, HABITAT CONNECTIVITY.

forest grazing The combined use of a *forest or wood for both wood production and animal production by grazing of *forage shrubs.

forest health A measure of the resilience of a *forest to damage, reflected, for example, through its *biodiversity, its sustained provision of *natural resources for people to use, and its natural flows of *water and *nutrients.

forest litter The freshly fallen or only slightly decomposed plant material on the floor of a *forest, which includes foliage, bark fragments, twigs, flowers, and fruit. *See also* LITTER.

forest management The *management of *forest resources for specified purposes, which might include *nature conservation, *timber yield, or scenic variety.

Forest Pest Control Act (1947) US legislation that established a programme to protect forestlands under all ownerships from destructive forest insect and disease pests. This Act was repealed and replaced by provisions of the *Cooperative Forestry Assistance Act (1978).

forest plan A long-range plan for the management of particular *forestlands. *See also* NATIONAL FOREST LAND AND RESOURCE MANAGEMENT PLAN.

forest product All usable raw materials that a *forest makes available, including major products (such as *poles and *roundwood) and minor products (such as medicinals, gums, resins, oils, fungi, honey).

forest range *forestland that has a *herbaceous or *shrubby *understorey that provides *forage for grazing or browsing animals. Also known as **grazable woodland**.

forest resource The *natural resources and values associated with forestland, which includes *forest products and associated water, fish, game, scenic, historical, recreational, and geological resources.

forest sequestration The growth and use of *forests to remove *greenhouse gases from the atmosphere. *See also* SEQUESTRATION.

Forest Service *See* US FOREST SERVICE.

forest stand An area of forest.

Forest Stewardship Council (FSC) An independent, non-profit, *non-governmental organization, with international members from all areas of *forest management, whose aim is to ensure the sustainable management of global forestry reserves.

forest type A way of categorizing *forest that is usually defined by its *vegetation, particularly the *dominant tree *species.

forest use zoning A type of *natural resource zoning that restricts land uses to *forestry and related uses, such as timber production, *watershed protection, and *recreation.

forestation *See* AFFORESTATION.

forested wetland An area of tall woody vegetation in which the soil is at least periodically covered by water.

forestland A *land use category in the USA that includes *grassland, *shrubland, treeland, *wetland, and/or *barren land.

forestry The management of trees, forests, and associated resources, on a *sustainable basis, for particular purposes such as commercial timber production. *See also* SILVICULTURE.

formaldehyde (CH₂O) A colourless, strong smelling gas that is used in solution as a *disinfectant and *preservative.

formamidine A type of *insecticide which works by causing tremors, convulsions, and loss of appetite, suppressing reproduction, and disturbing insect flight.

formation 1. The basic unit for the naming of rocks in *stratigraphy, based on *lithology.
2. A group of *communities of plants and animals within a region that have a similar structure, *climate, and *environment. *See also* BIOME.

forward contract In *greenhouse gas *emissions trading, under the *Kyoto Protocol, an agreement to buy or sell an *emissions allowance at a certain time in the future for a certain price. Also known as **forward settlement, spot forward**.

forward settlement *See* FORWARD CONTRACT.

fossil The hardened (*lithified) remains, traces, or impressions of *plants, *animals, or other *organisms that existed in a past geological age and have been preserved embedded in *sedimentary rocks.

fossil air A sample of *air (for example in the bubbles trapped in an *ice core) that preserves the composition of the *environment at the time it was deposited.

fossil fuel A naturally occurring *fuel (such as *coal, *oil, and *natural gas) in the form of an *organic sedimentary deposit that contains *carbon or *hydrocarbon, is produced by the *decomposition of fossilized remains of plants and animals, and can be burned for heat or power. Fossil fuels emit *carbon dioxide into the air when they are burned, which contributes to the *greenhouse effect and *global warming. They are *non-renewable resources, because we use them much faster than they form naturally. Substitutes and alternative forms of energy are being actively sought and developed as a matter of urgency, before fossil fuel supplies run out completely or before they become so scarce that price rises make them non-viable as a major energy source. *Contrast* ALTERNATIVE ENERGY.

fossil fuel reserve The quantity of a particular *fossil fuel that is known to exist (based on geological and engineering evidence) and that can be recovered under current economic conditions and with available technology.

fossorial Adapted for burrowing or digging.

fouling community *Benthic organisms that are attached to submerged objects which have some economic importance, such as pilings or boats.

founder effect The decrease in *genetic diversity when a new *colony is formed by a small number of individuals from a larger *population elsewhere, because many rare and usually undesirable *alleles are excluded and a few that are carried by the founders become more frequent. See also GENETIC DRIFT.

Fourth World Very poor nations that have neither market economies nor central planning, and are developing very slowly if at all. See FIRST WORLD, SECOND WORLD, THIRD WORLD.

fowl A *domesticated bird (such as a chicken) that is kept for its eggs or meat.

fractional cloud cover The relative amount of *cloud cover at site, usually expressed in amounts from 0 to 1 in increments of 0.1.

fracture Any break in a rock including cracks, *joints, and *faults.

Fragile or Historic Lands Areas where uncontrolled or incompatible development could result in irreversible damage to important historic, cultural, scientific, or aesthetic values or natural systems which are of more than local significance. An *area of critical environmental concern in the USA.

fragility See VULNERABILITY.

fragipan A soil *horizon that is loamy, weakly cemented, has low permeability and high density, is compact and brittle, and is hard when dry.

fragment A small mass of *soil or piece of broken *rock.

fragmentation 1. The breakup of large *habitats into smaller, isolated fragments, which may or may not be connected by *corridors. See also CONNECTEDNESS, FOREST FRAGMENTATION, HABITAT CONNECTIVITY.

2. A type of asexual reproduction, in which a *fungus breaks into pieces, each of which is capable of forming a new organism.

Framework Convention on Biological Diversity See UNITED NATIONS FRAMEWORK CONVENTION ON BIOLOGICAL DIVERSITY.

Framework Convention on Climate Change See UNITED NATIONS FRAMEWORK CONVENTION ON CLIMATE CHANGE.

free air wind A *synoptic-scale *wind. Also known as **general wind.**

free atmosphere The part of the *atmosphere that lies above the *friction layer.

free convection *convection of a parcel of air in the *atmosphere that is triggered by intense heating of the Earth's surface by the Sun.

free enterprise system See MARKET ECONOMY.

free growth The ability of a tree or other plant to grow relatively free from *competition from adjacent vegetation.

free market A market in which buyers and sellers can exchange goods or services by competitive bidding which is open to all, without constraint. See also MARKET EQUILIBRIUM, MARKET ECONOMY.

free market environmentalism An ideology based on the belief that the *free market is the most effective means of preserving the health and sustainability of the environment, rather than relying on government intervention.

free product A *hazardous substance that exists as a separate liquid phase which is not dissolved in water, and which has been released into the *environment.

free trade Trade amongst countries that is free of such government interference as quotas, subsidies, and tariffs.

free-living An organism that lives independently, as opposed to a *parasite or an organism in a *symbiotic relationship.

freely drained A *soil or deposit that allows water to *percolate freely through it.

free-range Livestock and domestic poultry that are allowed to *graze or *forage, rather than being confined to a *feedlot.

freeze 1. To change *phase from a liquid to a solid.

2. Weather that is cold enough to cause *freezing over a wide area for a significant period of time, which often occurs as cold air moves into a region by *advection, causing freezing conditions to exist in a deep layer of surface air. Also called **advection frost**. *See also* HARD FREEZE.

freeze–thaw Process (particularly *weathering process) associated with daily and seasonal cycles of freezing and melting.

freezing The *phase change in a substance from a liquid to a solid state that occurs with cooling. *See also* FREEZING POINT.

freezing condensation A process that occurs in the clouds in which ice crystals trap *water vapour, grow larger and heavier, and begin to fall as *rain or *snow.

freezing drizzle *See* FREEZING RAIN.

freezing level The *altitude in the *atmosphere where the air temperature first drops below 0°C (*freezing point).

freezing nuclei Tiny particles that promote the freezing of *supercooled liquid droplets.

freezing point The temperature at which a substance changes from a liquid to a solid state. For water, the normal freezing point (unless it is *supercooled) is 32°F or 0°C.

freezing rain *Rain that falls as a liquid but freezes into *glaze on contact with the ground. *See also* ICE STORM.

frequency The number of occurrences within a given period of time or a given area. *See also* DENSITY.

freshet 1. The annual rise of streams in cold climates that is caused by *snow melt during the late winter or spring.

2. A flood that is caused by rain or melting snow.

freshwater All surface water, other than *seawater and *estuary water, which includes streams, rivers, lakes, ponds, and water in *wetlands. Freshwater covers about 2% of the surface of the Earth and generally contains less than 1000 milligrams per litre of dissolved solids. *See also* SALINITY, WATER RESOURCE.

friable *Soil that is easily crumbled into small fragments or powder under hand pressure.

friction The force that resists the relative motion between bodies that are in contact with each other.

friction layer The lower part of the *atmosphere, up to about 1 kilometre, in which air movement (*wind) is influenced by *friction of the Earth's surface, along with turbulence and surface heating. The *free atmosphere is above it.

Friends of the Earth (FOE) An international network of *grassroots environmental groups in 70 countries, which was founded in San Francisco in 1969 by David *Brower, and runs high-profile campaigns designed to raise public awareness of environmental issues. *See also* ENVIRONMENTALISM, PRESSURE GROUP.

fringing reef *Coral reef that is directly attached to the *shore, or separated from it by a shallow, narrow *lagoon. *Contrast* COASTAL REEF.

frog An *amphibian that has moist skin, a stout body, and long legs for jumping. *Contrast* TOAD.

frond The long, feathery leaf of a flowerless plant (such as a fern or palm), or the leaf-like blade of a *kelp plant or other sea plant.

front A sharp transition zone in the *atmosphere that separates adjacent *air masses which have different temperatures and origins. Also known as a **frontal system** or **frontal zone.**

front group A *pressure group that is structured to look like a voluntary association, but may in reality be controlled by a particular interest (such as a company, industry, or political party), and which may give the appearance of being set up to do one thing but actually be set up to do something else on behalf of its parent group. Examples of US environmental front groups include the *Climate Council, *Coalition for Vehicle Choice, *Global Climate Coalition, *National Wetlands Coalition, and the *Wise Use Movement.

frontal fog See EVAPORATION FOG.

frontal inversion A temperature *inversion that develops above a *frontal zone when cold air at the surface is overrun by warmer air above.

frontal lifting The lifting of warmer and more moist air over cooler, denser air. Frontal lifting occurs at a *front when cold air pushes under a warmer air mass at a *cold front, or when warm air flows over a *warm front with colder air below.

frontal rain See CYCLONIC RAIN.

frontal system See FRONT.

frontal thunderstorm A *thunderstorm that develops in response to forced *convection along a *front. Contrast AIR MASS THUNDERSTORM.

frontal zone See FRONT.

frontogenesis The process by which a *front is formed and strengthened. Contrast FRONTOLYSIS. See also CYCLOGENESIS.

frontolysis The process by which a *front weakens and subsides. Contrast FRONTOGENESIS.

frost A white deposit of ice crystals that forms on exposed surfaces when the temperature falls below *freezing point.

frost heave A process by which the soil surface is pushed up by the accumulation of ice in the soil below. A form of *mechanical weathering that occurs in cold climates and *periglacial environments. Also known as **frost weathering.**

frost point The temperature at which air reaches saturation with respect to *snow or *frost. See also DEW POINT.

frost weathering See FROST HEAVE.

frost wedging The process in which rocks are forced apart by the expansion of water as it freezes in fractures and pore spaces.

frostbite Human tissue damage, especially in the feet, face, and hands, that is caused by exposure to extreme cold.

frugivore An animal that eats fruit.

fruit The part of a plant that contains seeds, such as a berry or pod.

fruiting body The reproductive part of a *fungus (such as a mushroom) that produces, contains, and releases *spores. Also known as a **conk.**

fry Newly hatched fish.

fuel Any material (such as *wood, *coal, *oil, or *gas) that can be burned to produce heat or energy. See also FOSSIL FUEL.

fuel assembly A bundle of hollow, metal rods (fuel rods) that contain pellets of uranium oxide, used to fuel a *nuclear reactor.

fuel break A strip of land where the native vegetation has been permanently modified or replaced to assist *fire management. See also FIREBREAK.

fuel cell A mechanical device that uses *hydrogen or hydrogen-based *fuel (such as *methane) to produce an electric current. Fuel cells are clean, quiet, and very efficient sources of *electricity.

fuel cycle The stages in the total life of a fuel. For example, the fuel cycle of coal is extraction, transportation, combustion, air emissions, ash removal, transportation, and disposal. *See* also NUCLEAR FUEL CYCLE.

fuel desulphurization *See* DESULPHURIZATION.

fuel economy US term for the average mileage travelled by an automobile per gallon of gasoline (or equivalent amount of other fuel) consumed, measured in accordance with the testing and evaluation protocol set forth by the *Environmental Protection Agency. *See* also CORPORATE AVERAGE FUEL ECONOMY.

fuel efficiency *See* ENERGY EFFICIENCY.

fuel ethanol *See* ETHANOL.

fuel intensity *See* ENERGY INTENSITY.

fuel management Any planned activity (including cutting *firebreaks or starting *prescribed fires) that changes vegetation in order to reduce fire risk.

fuel rod *See* FUEL ASSEMBLY.

fuel switching A change from using one *fuel to another, usually in order to decrease emissions of *greenhouse gases or other *pollutants.

fuelwood Wood (including sawn branches, twigs, logs, and wood chips) that is harvested for use as *fuel.

fugitive dust Dust that comes from *non-point sources (such as open fields and roads) not from *point sources (such as industrial smokestacks).

fugitive emission The uncontrolled escape of a gas, liquid, solid, vapour, fume, mist, fog, or dust from process equipment, without going through a *smokestack.

fugitive species A species of plant or animal that is adapted to rapidly colonize newly disturbed habitats.

Fujita scale A system that is to classify *tornadoes based on wind damage. The scale is from F-0 for weak tornadoes (light damage caused by winds up to 32 metres per second) to F-5 for the strongest tornadoes (heavy damage from winds 116 metres per second). The scale is given in Appendix 5.

full cost pricing The pricing of commercial goods (such as electric power) in a way that includes not just the cost of production but also the cost of *externalities created by their production and use (which includes *pollution, *waste management, and environmental damage).

fumarole A vent on the side of a *volcano from which *volcanic steam and gases emerge.

fume A cloud of gas, smoke, or vapour that is suspended in a gas.

fumigant A chemical *pesticide that forms vapours (gases) that are used to destroy *weeds, plant *pathogens, insects, or other pests.

fumigation The destruction of any unwanted organisms by exposure to a poisonous gas or smoke.

fundamental niche The total range of environmental conditions that are suitable for the existence of a species, without effects of interspecific *competition and *predation. *See also* NICHE, POTENTIAL NICHE, REALIZED NICHE.

fungibility Interchangeability, such as the *flexibility mechanisms for reducing *greenhouse gas emissions that were agreed under the *Kyoto Protocol.

fungicide A chemical *pesticide that is used to control or kill *fungus. Also known as **fungistat**.

fungistat *See* FUNGICIDE.

fungus A simple type of plant (plural fungi) that contains no *chlorophyll and so does not carry out *photosynthesis, but instead derives *nutrients by secreting digestive *enzymes that release organic molecules from the tree, soil, or organism it is in contact with, which it then absorbs. Examples include mushrooms, moulds, mildews, and yeasts. In *taxonomy, the fungi comprise a separate *kingdom (the Fungi).

funnel cloud A rotating column of cloud beneath a *cumulus or *cumulonimbus cloud, which does not reach the ground.

furbearer Any mammal species or individual that is valued and sought for its pelt or fur.

furnace A *combustion chamber or an enclosed structure in which *fuel is burned.

furrow A long, narrow, shallow trench carved in the ground by a plough or other agricultural implement.

furrow irrigation A form of *irrigation in which water is allowed to flow along the *furrows between rows of crops.

fusion 1. The act of fusing or melting together.

2. The process in which *nuclei are joined together under intense heat to form a heavier nucleus, at the same time releasing great amounts of energy. *Contrast* FISSION.

futures A contract to buy or sell a specified resource at a fixed price at some agreed time in the future. *See also* EMISSIONS BANKING.

futurity In the future, yet to come.

FWS *See* US FISH AND WILDLIFE SERVICE.

G10 *See* GROUP OF 10.

G15 *See* GROUP OF 15.

G7 *See* GROUP OF 7.

G77 *See* GROUP OF 77.

G8 *See* GROUP OF 8.

gabbro A coarse-grained *intrusive volcanic rock that is formed at great depth within the *Earth from the slow crystallization of *magma.

gaging station A particular site on a river where variations in stage or *discharge are measured systematically, usually continuously.

gale A sustained strong wind, between a strong breeze and a storm, moving at speeds of between 34 and 47 knots (39 to 54 miles per hour or 63–87 kilometres per hour), which is force 7 to 10 on the *Beaufort scale.

gallery forest *See* RIPARIAN FOREST.

gallon A unit of liquid capacity. A British imperial gallon is 4.545 litres and a US gallon is 3.785 litres.

GAIA HYPOTHESIS

The idea that the *Earth should be viewed and treated as a living organism, which was first suggested by James *Lovelock in 1979 who named it after the Greek Earth goddess Gaea. This perspective assumes that the living (biotic) and non-living (abiotic) parts of the Earth *system are mutually interdependent, both having been influenced by the other, with the survival of each dependent on the survival of the other. According to the Gaia hypothesis, as life evolved on Earth it both created and was created by biological processes that have radically altered the chemistry of the planet and its *environment, in ways that promoted further biological *evolution of plants and animals on land and in the sea. This planetary *positive feedback gave rise to large-scale *dynamic equilibrium between life and the planet. There are a number of interesting implications of the Gaia hypothesis. For example: the whole system matters much more than its component parts; complexity in the environment is inevitable and fundamental, and much of it we simply do not understand; feedback links in environmental systems are critical, but whilst they are not always fully understood, they should form the basis of any human intervention (deliberate or accidental) in environmental systems; biotic and abiotic aspects of the environment affect each other, and the symbiosis is important but easy to disturb or destroy; the stability of global environmental systems has evolved over long periods of time; short-term changes must be seen against a background of long-term geological change and progression; it is unwise to radically modify global environmental systems and equilibrium, because we can't always properly predict the outcomes and consequences; the Earth has a great natural capacity to look after itself (self-repair), so long as change remains within critical limits (thresholds), many of which we simply don't know enough about; the Earth will probably survive, no matter what humans do to it, but its survival might not include humans.

galvanize To coat a metal (especially iron or steel) with zinc in order to prevent *corrosion.

game Animals (mammals, birds, or fish) that are hunted for food or sport. *See also* BIG GAME.

game bag The record of game taken from a particular place over a given period of time.

game fish Any species of *fish that is caught for sport. Examples include salmon, trout, black bass, and lake trout. Also known as **sport fish**. *Contrast* ROUGH FISH.

game reserve An area of wildland that is set aside for the protection of *wildlife, usually for tourism or hunting purposes.

game species Any species of wild animal that people hunt or fish for sport, recreation, and food, and for which hunting seasons, *bag limits, and other laws and regulations have been prescribed.

game sport *See* FIELD SPORT.

gamekeeper A person who is employed to take care of *game and *wildlife.

gamete A reproductive cell, either an egg (female gamete) or a sperm (male gamete). Also know as **germ cell**. *See also* FERTILIZATION, SEXUAL REPRODUCTION, ZYGOTE.

gamma diversity Regional-scale *species diversity, expressed as the total number of species in the different *ecosystems within a region. A measure of *biodiversity. Also known as **regional diversity**. *See also* ALPHA DIVERSITY, BETA DIVERSITY.

gamma ray Very short wavelength *electromagnetic radiation that can be produced by *radioactive decay. It is similar to X-rays, travels in straight paths at the speed of light, and penetrates matter (including lead) but does not make the material *radioactive. *See also* ELECTROMAGNETIC SPECTRUM.

gap analysis The identification of *species and *classes of organisms that are not currently protected, to find out where they need protecting, on the basis of comparing the actual distribution of species and vegetation classes with the areas that are preserved or managed for their protection.

gap formation The creation of a patch of *habitat that is different from the surrounding larger habitat, for example by *felling parts of a *forest.

gap wind A strong *wind that blows through low passes or breaks in *mountain barriers, for example where coastal mountain ranges meet the sea.

garbage A North American term for food waste, or *refuse in general.

garden A plot of ground on which plants (flowers, vegetables, fruits, or herbs) are cultivated. *See also* HORTICULTURE, MARKET GARDEN.

garden city A new town that has been designed with a special emphasis on natural landscaping and retaining a rural atmosphere.

gas 1. A substance that is not a solid or a liquid; matter that occupies all of its container regardless of its amount.
2. *See* NATURAL GAS.
3. North American term for *petrol; short for *gasoline.

gas chromatography An analytical separation technique in which the minor components in a mixture of gases are separated into their individual components. This involves entraining the mixture in a gas flow (usually helium) to move it over a solid adsorbent in a column; each gas is absorbed at a known rate.

gas chromatography-mass spectrometry An analytical technique for measuring the concentrations of various chemicals in samples of soil or water, by first separating them using *gas chromatography and then analysing them by *mass spectrometry.

gas solubility The ability of a gas to dissolve in another substance.

gas stripping The transfer of an undesirable gas from water to the *atmosphere.

gasification The conversion of solid or liquid *hydrocarbons (such as coal) into a *gas, for use as a *fuel.

gasohol A mixture of gasoline and ethanol, a *clean fuel produced by *fermentation of sugar cane.

gasoline A liquid petroleum *fuel that is composed of a mixture of small, light *hydrocarbons, produced by refining *crude oil, and used mainly by cars, trucks, and other motor vehicles. Known as petrol in the UK.

gastropod A member of the Gastropoda class of molluscs, which includes snails and slugs.

GATT *See* GENERAL AGREEMENT ON TARIFFS AND TRADE.

Gause's principle *See* PRINCIPLE OF COMPETITIVE EXCLUSION.

Gaussian curve A symmetrical bell-shaped curve representing the distribution of results from a normal sample population. Also known as the **normal distribution**.

GCM *See* GENERAL CIRCULATION MODEL.

GDP *See* GROSS DOMESTIC PRODUCT.

GEF *See* GLOBAL ENVIRONMENTAL FACILITY.

Geiger counter An electrical device that measures *radioactivity.

gene A distinct sequence of *DNA that forms part of a *chromosome, by which offspring inherit characteristics from a parent.

gene bank A facility established for storage in a viable form of individuals, tissues, or reproductive cells of plants or animals, so that the material can be used in the future when required, as part of an *ex situ conservation strategy.

gene fixation The condition in which a particular *allele becomes the only one that is present in a *population, because of either *natural selection or *genetic drift.

gene flow The movement of *genes from one population to another by movement of individuals, *gametes, or *spores.

gene frequency The relative occurrence of a particular *allele within a *population.

gene mapping Determination of the relative positions of *genes on a *chromosome and of the distance between them.

gene pool The stock of different *genes in an interbreeding *population at a given time.

gene splicing *See* GENETIC ENGINEERING.

gene therapy An approach to preventing and/or treating disease by replacing, removing, or introducing *genes or otherwise manipulating *genetic material.

genecology Study of the genetic variation among *populations of a *species, and its relationship with *environmental factors. *See also* AUTECOLOGY, SYNECOLOGY.

genera Plural of *genus.

General Agreement on Tariffs and Trade (GATT) A 1947 multilateral trade agreement that was designed to establish rules, reduce tariffs, and provide a setting for solutions to international trade problems. It was succeeded in 1994 by the *World Trade Organization.

general circulation The continuous circulation of *wind and *ocean currents that acts to control temperature differences between the *poles and the *equator.

general circulation model (GCM) A large, complex computer program of atmospheric behaviour that simulates the *global climate, based on mathematical equations derived from knowledge of the physics and chemistry that govern the Earth–atmosphere system. Such models are used to predict the likely speed and pattern of *climate change, given various assumptions and *scenarios.

general fertility rate The number of births per 1000 women of childbearing age (15–44) in a year, for an area, country, or the world.

General Mining Law (1872) US legislation that governs the mining of hard-rock *minerals (*gold, *silver, *copper, platinum, *uranium, and others) on federal lands, and is still in effect.

general plan See COMPREHENSIVE PLAN.

general systems theory A way of looking at the world or any part of it as an interacting set of parts, as *systems.

general wind See FREE AIR WIND.

generalist A species that can live in many different habitats and can feed on a number of different organisms. Compare SPECIALIST.

generating capacity The capacity of a *power plant to generate *electricity, which is usually expressed in megawatts (MW).

generation 1. All of the people who are alive at the same time or who are of a similar age.
  2. The production of heat or electricity.

generation time The average period of time between birth and reproduction in a given population.

generator A machine that produces electricity (*electrical energy) from mechanical energy.

genetic Relating to or carried by *genes; hereditary or inherited.

genetic analysis The study of how *traits, and the *genes that carry them, are passed from generation to generation, and how the interaction of genes and the *environment results in particular traits.

genetic assimilation The progressive disappearance of a *species as its genes are diluted through *crossbreeding with a closely related species. See also GENETIC DRIFT.

genetic base The total genetic variability that is available within a *cultivar or *species.

genetic code The *DNA sequence of a *gene, which determines the sequence of *amino acids in a *protein or *enzyme, and thus determines the functions of a living organism.

genetic damage Damage that is caused to the *chromosomes in the *germ cells by environmental factors (for example *radiation). Such changes can be passed on to offspring. Contrast SOMATIC DAMAGE.

genetic distance A measure of the genetic similarity between any pair of *populations, based on *phenotypic traits, *allele frequencies, or *DNA sequences.

genetic diversity The number of different types of *genes in a *species or *population, which is a measure of *biodiversity.

genetic drift Fluctuations in the frequencies of particular *genes within a *population over time that are caused by random events rather than by *natural selection, and can lead, over successive generations, to a progressive change in the genetic composition of the population. See also FOUNDER EFFECT, GENE FIXATION, GENETIC ASSIMILATION.

genetic effect A heritable effect, that is passed on to descendants via *genes.

genetic erosion The loss of genetic *diversity between and within populations of the same species over time, as a result of *environmental change or human impacts.

genetic manipulation See GENETIC ENGINEERING.

genetic material Any *organic material that contains functional units of *heredity.

GENETIC ENGINEERING

The selective, deliberate alteration of the genetic makeup (*DNA) of an organism by removing, modifying, or adding genes to a chromosome in order to change the information it contains, which enables cells or organisms to make new or different substances (*proteins) or perform new functions. Such manipulation is sometimes done for pure research purposes, to discover more about *genetics and the way it works. But, increasingly, the commercial benefits of genetic engineering are being exploited in the breeding of plants, animals, and bacteria for particular purposes which benefit humans (so-called 'designer breeding'). Such organisms with a foreign gene added are referred to as *transgenic. There are many potential applications of *biotechnology, including *hazardous waste disposal and the recovery and *recycling of natural resources. Some scientists fear that transgenic organisms could do more damage than good. There is also widespread public concern over the risk that new and harmful strains might be produced (accidentally or on purpose) when genes are transplanted between different types of bacteria. The release of such material into the environment could create a nightmare scenario. Genetically modified plants might breed with wild species, and thus spread their genes far and wide. Cultural problems surrounding biotechnology have to be addressed too, including *technology transfer between the developed and developing worlds. There is also a clear need for internationally accepted mechanisms to ensure that all humankind will benefit. Also known as **gene splicing, genetic manipulation, genetic modification, recombinant DNA technology.** *See also* BIOTECHNOLOGY, HUMAN GENOME PROJECT.

genetic modification *See* GENETIC EN-GINEERING.

genetic polymorphism The presence of several genetically controlled variants (for example, blue eyes versus brown eyes) within a population, which may be the result of chance processes, or may have been caused by external *agents (such as *viruses or *radiation).

genetic resource Genetic material of plants, animals, or micro-organisms that is of actual or potential value as a resource for humans. Examples include modern *cultivars and *breeds, primitive varieties and breeds, *landraces and wild/weedy relatives of *crop plants or *domesticated animals.

genetic use restriction technology (GURT) *See* TRAITOR TECHNOLOGY.

genetically engineered organism - *See* GENETICALLY MODIFIED ORGANISM.

genetically modified organism (GMO) An organism whose genetic makeup has been deliberately altered by inserting a modified *gene or a gene from another variety or species, in order to create or enhance desirable characteristics from the same or another species. Also known as a **genetically engineered organism.**

genetics The study of *genes, *inheritance, and variation in *organisms.

genome All of the *genetic information or hereditary material in the *chromosomes of a particular *organism.

genome project Scientific research that is designed to map and sequence the entire *genome of an organism, such as a human being.

genomics The study of *genes and their function.

genotype The genetic make-up or blueprint of an individual organism, the set of *genes that it possesses. *Contrast* PHENOTYPE.

genotypic adaptation A form of biological *adaptation that is genetically determined, occurs by *natural selection (because those individuals which have favourable genetically acquired traits tend to breed more successfully than those which lack these traits), and can be passed on to future generations. *See* ADAPTIVE STRATEGY.

genus The sixth highest (of seven) category in the scientific system of classification for organisms (*taxonomy), below *family and above *species. Thus each family comprises more than one genus, and each genus comprises more than one closely related species. Scientists refer to living things by a combined genus and species name, using Latin terms. The genus name is always capitalized, and the species name is usually not; both are normally put in italics. For example, the proper name for humans is *Homo sapiens* where *Homo* is the genus. The plural of genus is *genera.

geo- Relating to the *Earth.

geoarchaeology The study of sediments and deposits at *archaeological sites, in order to reconstruct local *environmental history.

geochemical Relating to the chemistry of rocks.

geochemistry The study of the chemical properties, abundance, and distribution of materials within the Earth's waters, *crust, and *atmosphere.

geochronology The science of *absolute and *relative dating of geological formations and events.

geocoordinates *See* GEOGRAPHICAL POSITION.

geodesic The shortest line connecting two points on the surface of a sphere, such as the *Earth.

geodesy The science of determining the size and shape of the *Earth, and the precise location of points on its surface.

geodetic Relating to or determined by *geodesy.

geodetic survey A precise survey of a large area of ground that takes into account the shape of the *Earth.

geographical information system (GIS) A computer system that is designed for the storage, manipulation, analysis, and display of large volumes of spatial data in a map format, with different characteristics (such as *soil type or *vegetation) stored as separate layers which can be combined to display interactions of characteristics. Often simply referred to as GIS.

geographical isolation The separation of a *population from the rest of its *species by a physical barrier such as a mountain range, ocean, or great distance.

geographical position The location of a point on the surface of the Earth, usually expressed in terms of *latitude and *longitude. Also known as **geocoordinates**. *See also* GLOBAL POSITIONING SYSTEM.

geographical range *See* NATURAL RANGE.

geographical variety A subdivision of a *species with distinct *morphology and a distinctive *geographical range, which is given a unique Latin name. A taxonomic variety is known by the first validly published name applied to it and that nomenclature tends to be stable. *See also* VARIETY.

geography The study of the Earth's surface and of the ways in which people, plants, and animals live on and use it.

geohydrology *See* HYDROGEOLOGY.

geoid The figure of the Earth assuming that its whole surface is at current sea level (that is, it ignores topography on land and below the sea), which provides the reference level for astronomic observations and for *geodetic surveying.

geoinformatics The use of computers and statistical methods in the classification, storage, retrieval, and analysis of information about the *Earth. *See also* INFORMATICS.

geological column The arrangement of *rock units in their proper chronological order, from youngest to oldest.

geological cycle *See* ROCK CYCLE.

geological erosion *See* NATURAL EROSION.

geological log A detailed description of all of the underground features (depth, thickness, and type of rock formations) that are discovered during the drilling of a *well.

geological process Any natural process that affects the surface or interior of the *Earth, including *erosion, *deposition, *folding, *faulting, *plate tectonics, and *sea-floor spreading. *See also* LANDFORM.

geology The study of the origin, history, and structure of the *Earth.

geomagnetic storm A worldwide disturbance of the Earth's *magnetic field which is created when particles sent from the Sun in *solar flares are drawn towards the Earth by its magnetic field. It often coexists with an *aurora.

geometric growth A pattern of growth that increases at a geometric rate over a specified time period, such as 2, 4, 8, 16 (in which each value is double the previous one). *Contrast* ARITHMETIC GROWTH, EXPONENTIAL GROWTH.

geomorphic cycle *See* CYCLE OF EROSION.

geomorphic process Processes that change the form or shape of the surface of the *Earth, including *desertification, *erosion, *fluvial processes, *glaciation, and *weathering.

geomorphology The study of *landforms and how they develop.

geophysics The study of the physical characteristics and properties of the *Earth, which includes *geodesy, *seismology, *meteorology, *oceanography, atmospheric electricity, terrestrial magnetism, and tidal phenomena.

geophyte A *perennial *herbaceous plant that reproduces by underground bulbs, tubers, or corms.

geopolitics The combination of geographical and political factors that affect a country or area.

geopotential height The height above sea level (in metres) at which the *atmosphere has a particular pressure.

georeferenced Data that are defined by *geographical position.

geoscience Study of the science of Earth systems.

geosphere The solid, *abiotic portion of the *Earth, excluding the *atmosphere, *hydrosphere, and *biosphere.

Geostationary Operational Environmental Satellite (GOES) A series of *geostationary satellites that were launched by the USA starting in 1968, and use a variety of *remote sensing devices for *weather forecasting and *environmental monitoring. GOES-E is positioned over the USA and South America, and GOES-W is positioned over the Pacific Ocean.

geostationary satellite A satellite that orbits the *Earth at the same rate that the Earth rotates and so remains over a fixed place above the *equator.

geostrophic wind A wind that flows parallel to *isobars, which is produced when the *pressure gradient is balanced by the *Coriolis force, usually at some height above the ground surface where *friction is reduced.

geosyncline A large *depositional basin on the ocean floor, in which *sediments accumulate to thicknesses of many kilometres.

geotactic Moving in response to the Earth's gravitational field.

geothermal Relating to or produced by the internal heat of the *Earth, which is generated by a number of processes, including *radioactive decay of rocks and *compression at great depth. *See also* HYDROTHERMAL.

geothermal belt An area where the *geothermal gradient is unusually steep, where *geothermal energy might be exploited.

geothermal energy Heat energy that is derived from the Earth's interior, either through *geysers, *fumaroles, *hot springs, or other natural *geothermal features, or the use of *geothermal heat to generate usable energy, for example through using heat from *hot rocks to drive turbines and produce electricity. To be useful the geothermal energy must be available in the form of super-heated water or steam. Hot water is pumped to the surface and converted to steam or run through a heat exchanger. Dry steam is pumped to the surface, and can then be directed through turbines to generate electricity. In volcanic regions such as New Zealand, Japan, and Iceland water heated beneath the ground may erupt at the surface as geysers, hot springs, or boiling mud. This heated material can sometimes be used directly, saving on the cost of pumping to the surface. By the early 1990s geothermal energy fields were being exploited in more than 20 countries, including New Zealand, Iceland, the USA, and the former Soviet Union. Geothermal energy is an important source of energy in some areas, particularly in Iceland and parts of New Zealand. Direct heat use of geothermal energy in the USA occurs mainly in the western States, particularly southern California. Geothermal power plants are usually relatively small, and serve local rather than regional or national needs. The fraction of total world energy needs that is provided by geothermal sources is currently very small, but rising rapidly.

geothermal gradient The natural increase in rock temperature with increasing depth within the Earth's *crust. Direct observation, in places such as deep mines and wells, has shown that this gradient varies from place to place, but the average is about 20–40°C for every kilometre of depth. In some places the geothermal gradient is much steeper than the norm, and this can be exploited as a source of *geothermal energy.

germ A *micro-organism that can cause *disease.

germ cell *See* GAMETE.

germicide A chemical *pesticide that is used to control or kill *germs or *micro-organisms.

germination The initial stages in the growth of a *seed to form a *seedling.

germplasm The hereditary *genetic material of which an organism is composed, which is transmitted from one generation to the next by the *gametes.

gestalt A pattern or structure whose qualities as a whole exceed the sum of its constituent parts.

geyser A violent ejection of super-heated water and steam from a hole in the ground. The underground reservoir consists of water-filled chambers that are connected by a central pipe. *See also* GEOTHERMAL GRADIENT, HYDROTHERMAL ACTIVITY, HOT SPRING.

Giardia lamblia A *protozoan *micro-organism that is frequently found in rivers and lakes, which (if not treated properly) can cause diarrhoea, fatigue, and cramps in humans who ingest it.

gigahertz (GHz) A unit of *frequency that is equal to 1000 million hertz (Hz).

gigatonne (Gt) A unit of mass that is used for large quantities (such as the amount of *carbon dioxide in the Earth's atmosphere). 1 Gt = 1000 million tonnes = 10^{15} grams.

gill A respiratory organ in aquatic organisms (such as fish) that is used for obtaining oxygen which is dissolved in the water.

gillnet A net that is set upright in water in order to catch fish by entangling their *gills in its mesh.

girdling A method of killing a *tree by removing a strip of bark from around its trunk or a branch, which interrupts the flow of *nutrients between the leaves and the rest of the tree.

GIS *See* GEOGRAPHICAL INFORMATION SYSTEM.

glacial 1. Relating to a *glacier.
2. A period when *ice sheets expand, and global climate is colder and drier. *See also* ICE AGE.

glacial deposit *Sediment that is transported by a glacier and left behind as a deposit when the ice melts. *See also* BOULDER CLAY.

glacial diamicton *See* BOULDER CLAY.

glacial drift *See* BOULDER CLAY.

glacial epoch *See* ICE AGE.

glacial erosion The processes of *erosion that are associated with *glaciers and *ice sheets, which include *quarrying and *abrasion. *See also* U-SHAPED VALLEY.

glacial flour Very fine particles of rock that are carried by a *glacier and contribute to *abrasion processes and *glacial polish.

glacial ice A very dense form of *ice that is formed from the accumulation and recrystallization of *snow, is relatively impermeable, and is much harder than snow, *névé, or *firn.

glacial lake A natural lake that is formed by the damming of *meltwater by *glacial deposits in front of a *glacier.

glacial maximum The position or time of the greatest advance of a *glacier, before its starts to recede and melt.

glacial meltwater *See* MELTWATER.

glacial milk Glacial *meltwater which is light coloured and cloudy because of the large quantities of *clay-sized *sediment that are held in *suspension within it.

glacial outwash Well-sorted sand, or sand and gravel, that is deposited by a *meltwater stream in front of a *glacier.

glacial polish The smoothing of a *bedrock surface as a result of *abrasion by *sediment that is transported on the bottom of a *glacier.

glacial rebound The *isostatic adjustment of an area after the retreat of a *continental glacier. Also known as **glacial uplift**.

glacial retreat The backward movement of the front of a *glacier, as the glacier melts and shrinks.

glacial surge A rapid forward movement of the front of a *glacier.

glacial trough *See* U-SHAPED VALLEY.

glacial uplift *See* GLACIAL REBOUND.

glacial valley A valley that is or was occupied and shaped by a *glacier. *See also* U-SHAPED VALLEY.

glaciation 1. The formation, movement, and recession of *glaciers or *ice sheets.
2. The geological processes of *glacial activity, including *erosion and *deposition, and the effects of such action on the Earth's surface.
3. The process by which cloud particles change from water drops to ice crystals.

glacier A large, slow-moving river of ice made of compressed snow, which moves downslope under its own weight and survives from year to year. The most common types are *alpine glacier, *continental glacier, *ice cap glacier, and *piedmont glacier. *See also* ICE CAP, ICE SHEET.

glacier budget The *mass balance of a glacier, which takes into account inputs (new snow and ice) and outputs (including *evaporation and *meltwater). The larger the positive balance the faster the glacier grows; when the mass balance is negative, outputs exceed inputs and the glacier melts and shrinks

glacier flow The slow but continuous downward or outward movement of *ice in a *glacier that is driven by *gravity. Also known as **ice flow**.

glacier wind A shallow downslope wind above the surface of a *glacier.

glaciofluvial Relating to meltwater coming from a glacier, which creates such features as *eskers and *kames, *outwash plains, *pro-glacial lakes, and *meltwater channels. Also known as **fluvioglacial.**

glaciology The study of the formation, movement, and properties of *glaciers and *ice in general.

glade An open clearing in a *woodland or *forest.

glass A solid that results from cooling of *magma that is too fast to allow crystals to grow.

glassiness The amount of glass in an *igneous rock.

glaze A layer or coating of *ice that is generally smooth and clear, and forms on exposed objects by the freezing of *raindrops, *drizzle, or *fog. *See also* ICING.

gleization *See* GLEYING.

glen Scottish name for a narrow, secluded *valley.

gley A dark grey to black *soil that forms slowly in a wet, poorly drained environment, has a thick *organic horizon over a horizon of chemically reduced *clay, and has low fertility.

gleying The *reduction of iron in an *anaerobic environment, which creates grey or blue colours mixed with rusty colours in soils. Also known as **gleization.**

global Involving or relating to the whole world.

Global 200 An initiative of the *Worldwide Fund for Nature (WWF) which seeks to ensure that the full range of *ecosystems will be represented in regional *conservation and development strategies. WWF have identified 867 terrestrial *ecoregions, as well as freshwater and marine ecoregions, and selected 233 of them (136 terrestrial, 36 freshwater, and 61 marine) to represent the Global 200 network of ecoregions which are most crucial to the conservation of global *biodiversity. *Contrast* BIODIVERSITY HOTSPOT.

Global 2000 A report on global environmental trends that was commissioned by US President Carter in 1977 and published in 1980, which concluded that by the year 2000 the world would be more crowded, more polluted, and more fragile if global trends continued.

global biodiversity The number of different *species in the world at a particular point in time. *See also* BIODIVERSITY.

Global Climate Coalition A US nonprofit *anti-environmental *front group comprising mainly US businesses that for reasons of self-interest oppose immediate action to reduce *greenhouse gas emissions. It disbanded in 2002 after losing most of its members.

global commons Natural systems and resources (such as the *atmosphere and the *oceans) that do not belong to any one country. *See also* COMMON RESOURCES, TRAGEDY OF THE COMMONS.

global dimming The gradual reduction in the amount of *sunlight reaching the Earth's surface, which is estimated at 2–3% per decade, believed to be caused by the increased presence of *aerosols and other *particulates in the *atmosphere.

Global Environment Facility (GEF) An international fund set up in 1991 under the umbrella of the *United Nations Conference on Environment and Development, by the *World Bank, the *United Nations Development Programme (UNDP), and the *United Nations Environment Programme (UNEP) to help meet the costs of *Agenda 21.

Global Environmental Monitoring System (GEMS) A global *environmental monitoring system that was established in 1973 under the *United Nations Environment Programme, which collects data relating to *atmosphere, *climate, *pollution, and *renewable resources.

global environmental research Research into environmental issues at the global scale, which reflects growing interest in two sets of processes. First, there are the natural *environmental processes that are global in scale. This includes *plate tectonics, *atmospheric circulation, *ocean currents, the *water cycle, and the *biogeochemical cycles. Second, there are induced environmental changes at local and regional scales which are becoming issues of global concern. Obvious examples include *air pollution, *climate change, *deforestation, *desertification, *ocean pollution, *ozone depletion, *soil erosion, and *tropical forest clearance.

Global Heritage Fund A private, non-profit, international funding organization, established in 2001, whose mission is the conservation of around 200 archaeological sites in the Americas, Asia, and the Pacific where the threats of neglect and destruction are most urgent, and which provide major opportunities for *sustainable tourism and development. *Contrast* WORLD HERITAGE FUND.

global positioning system (GPS) An instrument that utilizes *satellite signals to pinpoint exact *geographical position.

global rank (G-rank) The global *conservation status rank of a species, subspecies, or variety of plant or animal, based largely on the total number of known sites occupied by it world-wide, and the degree to which they are potentially or actively threatened with destruction. *See also* PROVINCIAL RANK.

global scale Large scale, at the scale of the whole world. Also known as **planetary scale**.

global warming potential (GWP) An index (created under the *Kyoto Protocol) that allows for direct comparison of the various *greenhouse gases, based on the *radiative forcing (amount of *global warming) that results from the addition of 1 kilogram of a particular gas to the atmosphere compared with 1 kilogram of *carbon dioxide. Carbon dioxide has a GWP of 1, and *methane has a GWP of 21 (it produces 21 times as much warming as carbon dioxide). Nitrous oxide has a GWP of 310, CFC-11 has a GWP of 5000 and CFC-12 has a GWP of 8500. Also known as **carbon dioxide equivalent**.

globalization Growth to a world-wide scale, the transition from national and regional economies to global economies.

globally rare A *native species that is both a *rare species and one that is found in only one or a few particular places.

globally threatened A *taxon that is threatened on a world-wide scale.

glucose A simple sugar that is a major energy source, and is made in plants by *photosynthesis and in animals from the *carbohydrates in food.

GMO *See* GENETICALLY MODIFIED ORGANISM.

GMT *See* GREENWICH MEAN TIME.

gneiss A hard, coarse-grained, banded *metamorphic rock that is composed of *quartz, *feldspar, amphibole, and *mica.

GNP *See* GROSS NATIONAL PRODUCT.

GOES *See* GEOSTATIONARY OPERATIONAL ENVIRONMENTAL SATELLITE.

gold (Au) A bright, yellow precious transition element that occurs naturally in rock, is widely mined, and has many uses, including currency, jewellery, printed circuits, and semiconductors.

Golger's rule A 19th century *ecogeographical rule that the drier the climate, the lighter the colour of animals relative to closely related taxa of more humid regions. Thus, for example, yellows and light browns are dominant in arid regions, and dark browns and blacks in humid regions. *See also* BERGMANN'S RULE.

Gondwanaland A large former continent that was located mainly in the

GLOBAL WARMING

An increase in the *temperature of the *troposphere, which has occurred in the past as a result of natural processes but is now believed to be accelerating as a result of increased *emissions of *greenhouse gases associated with the burning of *fossil fuels. Scientists expect global warming to change world *climate in a number of ways. For example, cold seasons are likely to become shorter and warm ones to become longer. Northern latitudes are likely to have wetter autumns and winters, and drier springs and summers. There would be more rainfall in the *tropics, and *subtropical areas could become drier. Global warming is likely to cause shifts in the main *climate zones around the world, and this will probably bring a rise in the frequency and intensity of *floods, *droughts, *typhoons, *tornadoes, and *hurricanes in many areas. This would further aggravate the losses and hardships caused by *sea level rise. *IPCC scientists have predicted that if emissions of greenhouse gases continue at present rates, the Earth's temperatures will increase on average by about 2.5°C by the year 2050. Different *scenarios—low, middle, and high—put the likely increase as high as 4.5°C or as low as 1.5°C. This is very much a global average, and regional variations may be much higher than this, particularly in polar and near-polar regions. Any rise of this level would far exceed natural temperature changes over the last 8000 years, which have been of the order of 1°C. Since the *Little Ice Age (between about AD 1450 and AD 1850) temperatures have risen by about 0.8°C. Records show that over the past 100 years global mean temperature has risen by between 0.3°C and 0.6°C, so the next 100 years is likely to have some climatic surprises in store for us! Warming by as much as 5.0°C would be greater than the difference between the depth of the last *ice age (about 18 000 years ago) and today. Most experts agree that a rise of between 1°C and 2°C above the average pre-industrial temperature is probably the most the Earth can cope with without massive and widespread environmental damage and disruption. If warming does occur as predicted, it will be relatively rapid. Most forecasts are of the order of about 0.3°C per decade, which is three times faster than *ecosystems are believed to be capable of adjusting to. Even reducing emissions of *carbon dioxide today might not be enough to prevent widespread environmental disruption and damage. Recent studies have suggested that the warming is likely to occur more rapidly over land than over the oceans. A further complication is that rises in temperature tend to *lag behind increases in greenhouse gases. It is likely, therefore, that initially the cooler oceans will absorb much of the additional heat, which would serve to decrease the warming of the atmosphere. Only when the oceans reach *equilibrium with the higher level of CO_2 will the full atmospheric warming occur. Global warming can probably best be slowed down by a combination of steps based on two broad *strategies—reducing emissions of greenhouse gases, and enhancing the terrestrial *sinks and stores for greenhouse gases (for example by planting *forests to absorb carbon from the atmosphere). The long-term and global character of the climate change problem requires an international long-term *strategy based on internationally agreed principles such as *sustainable development and the *precautionary principle. The *Kyoto Protocol is designed to reduce overall emissions of greenhouse gases by setting targets and allowing countries to buy, sell, and exchange permits to pollute the atmosphere. *See also* CLIMATE CHANGE, INTERGOVERNMENTAL PANEL ON CLIMATE CHANGE, UNITED NATIONS FRAMEWORK CONVENTION ON CLIMATE CHANGE.

*southern hemisphere and broke apart by *plate tectonics in the late *Palaeozoic, to form parts of what is today Africa, Australia, *Antarctica, South America, and India. *See also* LAURASIA.

gonochorism Describing a sexually reproducing species in which individuals are distinctly male or female. *Contrast* HERMAPHRODITE.

gorge 1. A deep *ravine with a river running through it.
2. A narrow pass between adjacent mountains. *See also* CANYON.

GPS *See* GLOBAL POSITIONING SYSTEM.

graben A *valley or *trough that is produced by *faulting and *uplift of adjacent blocks of the Earth's *crust. *Contrast* HORST, INTERMONTANE BASIN.

grade 1. The slope of the surface of the Earth.
2. Organisms that have the same morphological features but are not genetically related.
3. The size range of *sediment. For example sand grade.
4. Stream, or river grade is the balance between *erosion and *deposition.
5. The grade of a mineral ore is a description of its quality. For example high-grade ore.

graded A measure of the variation of sizes in *soil or an *unconsolidated *sediment; well-graded materials have many sizes, whereas poorly graded materials are more uniform in size.

graded bedding A layer of *sediment in which the coarsest particles are concentrated at the bottom and particle size decreases upward into fine *silt.

graded stream A stream that maintains an *equilibrium between the processes of *erosion and *deposition, and thus between *aggradation and *degradation.

gradient 1. The *slope of a surface, usually expressed as a percentage.
2. A variation in one quantity relative to another, such as a temperature gradient (how temperature changes with *elevation).

gradient wind A theoretical *wind that blows parallel to curved *isobars.

grading 1. Sorting and evaluation of a natural resource, such as grading fish by size or timber by quality.
2. Altering a land surface by cutting, filling, and/or smoothing in order to create a desired surface.

gradualism A model of *evolution that assumes slow, steady rates of change, such as *Darwin's concept of evolution by *natural selection. *Contrast* PUNCTUATED EQUILIBRIUM.

graft A transplant; a tissue or organ that is transferred from one individual plant or animal to another.

grain 1. The dry, starchy seed-like fruit that is produced by *cereal grasses.
2. The texture of wood, which is determined by the type of *xylem cells that it contains.

gram A metric unit of weight equal to one-thousandth of a kilogram.

Gramineae The *grass family of plants, which are mostly herbaceous but also includes some woody plants such as *cereals, bamboo, *reeds, and sugar cane.

graminoid Of *grasses (*Gramineae) and grass-like plants, including sedges and rushes (*marsh plants).

grandfathering In *greenhouse gas *emissions allocation, under the *Kyoto Protocol, an approach that allows existing emitters to continue their previous levels of emissions while new sources must meet new standards.

granite A hard, coarse-grained, light-coloured *intrusive *igneous rock that accounts for nearly 15% of the rocks exposed at the *Earth's surface.

granivore An animal that lives on a diet of grain or seeds.

G-rank *See* GLOBAL RANK.

granular activated carbon treatment A *wastewater filtering system that is used in small water systems and individual homes, which uses *activated carbon to remove organics.

granular snow See FIRN.

grass A member of the *Gramineae family of flowering plants, which have long, narrow leaves and stems that are hollow or pithy in cross section.

grass heath See ACID GRASSLAND.

grassland A plant *community dominated by various types of *grass, which is home to herds of *grazing animals. It is common in areas that are too dry during the summer months to support *forest or *shrubland. Temperate grassland vegetation in North America includes *bunchgrass, *California grassland, *desert grassland, *mixed grass, *Palouse prairie, *shortgrass, *sodgrass, and *tallgrass. Tropical grassland in Africa is called *savanna. See also RANGELAND.

grassroots Organizations or movements, people or society at a local level rather than at the centre of major political activity. Local, or person-to-person.

grassroots activism A form of *environmental activism that is local and based on *grassroots activities.

grassroots conservation *Conservation activities that are undertaken by individuals and citizen-based groups, usually on a voluntary basis.

graupel See SNOW PELLET.

gravel *Unconsolidated *sediment that is composed of particles of between 2 and 75 millimetres in diameter.

graveyard See BURIAL GROUND.

gravitational water *Soil water that is pulled down into *soil by gravity, moves between the soil particles, and fills the pore spaces between them.

gravity The force that pulls a body towards the centre of the Earth or another celestial body.

gravity transfer See MASS MOVEMENT.

gray (Gy) The SI unit for the *absorbed dose of *radiation. It replaced the *rad. 1 Gy = 100 rad.

gray water North American term for *sanitary water.

grazable woodland See FOREST RANGE.

graze To feed on vegetation, usually in a *field, *pasture, or *meadow.

grazer A *herbivore. See also PRIMARY CONSUMER.

grazing cycle The time elapsed between the beginning of one grazing period and the beginning of the next grazing period in the same paddock where the forage is regularly grazed and rested. One grazing cycle includes one grazing period plus one rest period.

grazing land Any land that is covered with vegetation and is suitable for *grazing by livestock.

grazing line See BROWSE LINE.

grazing season The period of time during which *grazing can normally be practiced each year.

Great Barrier Reef A chain of *coral reefs nearly 2010 kilometres long in the Coral Sea, off the north-eastern coast of Australia. This unique environment is the world's largest coral reef system, and it appears on the *Unesco World Heritage List. The reef is economically important to the state of Queensland and the country of Australia, because it supports significant tourism and fishing industries. Since the early 1960s the reef has suffered extensive damage, much of it caused by humans through intensive *recreational use, shipping accidents, *runoff from the mainland (which increases *turbidity, reduces *salinity, and introduces some *pollutants), and *climate change (via sea level rise, ocean warming, and increased air temperatures). The Great Barrier Reef Marine Park has been established to preserve the reef in perpetuity, and a comprehensive programme of *water quality management within the park is designed to protect the reef and the industries that depend on it.

Great Basin Desert A North American cold desert produced by the *rain shadow of the Sierra Nevada and Cascade Mountains, which has long, cold winters and much of its *precipitation falls as snow.

Great Lakes The five connected *lakes along the border of the USA and Canada which make up the world's largest body of *freshwater. They are Lake Ontario, Lake Erie, Lake Superior, Lake Michigan, and Lake Huron.

Great Plains A *prairie region in the western USA that extends from North Dakota south to Texas and from the Rocky Mountains east to western Minnesota and Missouri.

great soil group A broad group of *soils that share common characteristics which are usually associated with particular climates and vegetation types.

greater ecosystem An approach to *conservation in North America which is based on a *core protected area surrounded by a defined region of controlled development or some form of cooperative management. Examples include the Greater Yellowstone Ecosystem, the Greater Fundy Ecosystem, and the Greater Salt Lake Ecosystem.

green See ENVIRONMENT-FRIENDLY.

green consumerism Purchasing *environment-friendly products, such as vehicles that run on lead-free petrol, aerosols that don't contain *CFC propellants, wooden products that don't contain tropical hardwoods, and paper that is recycled.

green design Products and services that have been specifically designed to cause less harm to the environment.

green manure Any crop that is grown to be ploughed into the soil, in order to increase soil organic content and improve *soil quality.

Green Party An *environmentalist political party.

green plan An integrated environmental plan that is designed to reduce pollution and resource consumption, achieve *sustainable development, and *increase environmental restoration.

green politics An approach to politics in which environmental issues are taken seriously.

Green Revolution The development of *high yield cereal crops and their introduction to the less economically developed world from the 1960s, along with the use of machines, fertilizer, pesticides, irrigation, and the growth of hybrid varieties of rice, wheat, and corn. See also AGRICULTURAL REVOLUTION, INDUSTRIAL REVOLUTION.

green tax A tax or surcharge that is imposed on activities that pollute, deplete, or degrade the *environment.

greenbelt A *buffer zone that is designed to restrict development from particular areas.

Greenbelt Movement A tree-planting programme in Kenya that was initiated by Wangari *Maathai and the National Council of Women of Kenya (NCWK). It is based on turning small-scale farmers into *agroforesters, with a particular emphasis on training women to plant and cultivate seedlings in order to assist in *reforestation and generate a source of income for themselves.

greenfield Land in rural and urban areas that has not already been developed. See also BROWNFIELD.

greenhouse A building with glass walls and roof, in which plants are grown under controlled conditions. See also PASSIVE SOLAR HEATING.

greenhouse effect The mechanism in which radiation from the Earth's surface is trapped by the presence of *green-

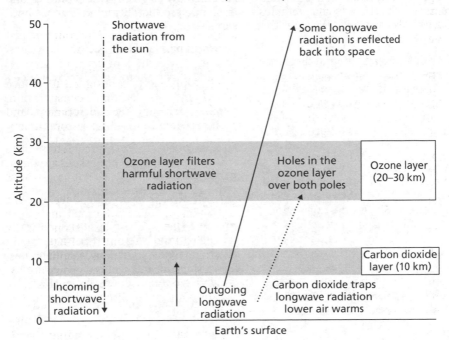

Fig 8 Greenhouse effect

house gases in the atmosphere, which warms the lower atmosphere and maintains temperatures suitable for the survival of *organisms. The Earth is a much cooler body than the Sun, and it radiates energy of much longer *wavelengths. Thus *insolation is of short wavelength and re-radiant energy from the Earth is of long wavelength. This long-wave energy from the Earth is stored for a time in the *atmosphere, trapped by gases that prevent it from escaping back into space. The trapped long-wave energy heats the atmosphere, producing a rise in the Earth's temperature—the greenhouse effect. Glass walls in a garden greenhouse trap heat in a similar way, hence the name, although the analogy is rather weak and much disputed. The widespread burning of *fossil fuels since the *Industrial Revolution has caused significant increases in atmospheric concentrations of the main *greenhouse gases (*carbon dioxide has risen by nearly 30%, *methane has more than doubled, and *nitrous oxide has risen by about 15%), which has enhanced the natural greenhouse effect and accelerated *global warming.

greenhouse gas A gas that traps heat in the *atmosphere, by absorbing radiation in wavelengths similar to those emitted by the Earth, which prevents excessive loss of terrestrial radiation and accompanying heat loss, and leads to *global warming. The most important greenhouse gases are *water vapour, *carbon dioxide, *methane, *nitrous oxide, and *CFCs. Others include *methyl chloroform and *carbon tetrachloride.

greenness The amount of living material (*biomass) in vegetation.

Greenpeace An international *non-government organization that works for environmental *conservation and the preservation of *endangered species.

greens 1. Leafy plants or their leaves and stems which are eaten as vegetables.
2. Individuals and political parties that have an environmental agenda.

greenspeak Environmental discourse. The term is most commonly used to describe the use of environmental terms and language by individuals or groups who are intent on portraying themselves and their causes as pro-environmental, even when they are not (*greenwash).

greenwash A term (combining green and whitewash) that *environmentalists use to describe the activity (for example by corporate lobby groups) of giving a positive public image to practices that are environmentally unsound. See also GREENSPEAK.

greenway See CORRIDOR.

Greenwich Mean Time (GMT) The mean solar time at 0° *longitude that passes through Greenwich, England, and is used as the basis for standard time throughout the world.

Greenwich Meridian The Prime Meridian, which passes through the original site of the Royal Observatory at Greenwich, England and was adopted in 1884 by a conference of nations as the initial or zero of *longitude. It is the reference for *Greenwich Mean Time, and for all longitude measurements. See also INTERNATIONAL DATE LINE.

gregarious Sociable, living in groups or communities. Contrast SOLITARY.

greywacke A coarse-grained *sandstone that is *cemented by more than 15% *clay minerals. Fragments are rounded to subrounded.

groin See GROYNE.

gross domestic product (GDP) The total monetary value of the goods and services that are produced by a country, within that country, over a given year.

gross national product (GNP) The total monetary value of all goods and services that are produced in a nation's economy over a given year. Unlike *gross domestic product, it includes goods and services produced abroad.

gross primary production The total amount or weight of organic matter that is created by *photosynthesis over a defined period of time.

gross primary productivity The total rate of *photosynthetic production of *biomass. Contrast NET PRIMARY PRODUCTIVITY.

ground cover Plants that are grown in order to protect the soil from *eroding. See also HERBACEOUS LAYER.

ground fire A fire that burns material on the ground such as *litter, grasses, and non-woody plants, and organic matter in the soil, but does not affect trees which have thick bark or high crowns.

ground flora Low plants (such as grasses and herbs) that grow at the lowest level of a plant community, such as on a forest floor. Also known as **ground layer.**

ground fog Any *fog that hides less than 60% of the sky. See also radiation fog.

ground game Rabbits and hares.

ground layer See GROUND FLORA.

ground moraine A thick layer of *till that has been deposited by a melting *glacier and often forms gently rolling ground.

ground truth The facts that are found when a location shown on a map, *air photograph, or *satellite image is checked on the ground, as *validation.

ground zero 1. The point on or above the ground at which a nuclear weapon explodes.
2. The common name for the site of the World Trade Center in New York after it was destroyed on 11 September 2001.

groundfish A species or group of fish that lives most of its life on or near the sea bed. Also known as **bottom fish** or **demersal fish.**

ground-lead logging A method of *cable logging in which timber is dragged from where it is cut to a collecting point by a powered cable at or near ground level. *Contrast* HIGH-LEAD LOGGING.

groundwater Water that occupies pores, cracks, and crevices in *rocks underground, below the *water table. World-wide between about 10% and 30% of river flow is accounted for by groundwater, and groundwater provides about 6% of the water and around 50% of the dissolved sediment that rivers input into seas and oceans. *See also* AQUIFER.

groundwater abstraction The *extraction of *groundwater, pumped from underground *aquifers, as a source of *freshwater. In many aquifers the groundwater has to be pumped out through *boreholes or *wells. As water is abstracted the *water table is lowered around the borehole. If rates of abstraction exceed rates of *groundwater recharge within an aquifer, the water table can fall across a wide area. Borehole yields decline when this happens, and wells begin to dry up. Sometimes it is necessary to sink boreholes deeper into the aquifer, which further lowers the water table. *Groundwater depletion follows, and water supplies are threatened if not reduced.

groundwater contamination *See* GROUNDWATER POLLUTION.

groundwater depletion A decline in the level of the *water table and availability of *groundwater within an *aquifer, often as a result of excessive *groundwater abstraction.

groundwater discharge North American term for *groundwater that enters coastal waters, having been contaminated by *landfill leachates, *hazardous wastes, and *septic tanks.

groundwater flow The movement of *groundwater through openings in *sediment and *rock within the zone of *saturation.

groundwater mining *See* AQUIFER DEPLETION.

groundwater pollution The ^pollution of *groundwater, for example by *intrusion of water into coastal *aquifers, or by leakage from *landfill sites or underground long-term storage of high-level *nuclear waste.

groundwater recharge The natural replenishment of *groundwater by the *water cycle, which raises the *water table.

groundwater remediation The treatment of *groundwater in order to remove *pollutants.

groundwater runoff *Groundwater that is discharged into a stream or river channel as spring water or *seepage water.

Group of 10 (G10) The ten most influential environmental *non-government organizations in the USA, which are the Environmental Defense Fund, Wilderness Society, Sierra Club, National Audubon Society, National Parks and Conservation Association, Natural Resources Defense Council, Defenders of Wildlife, Environmental Policy Institute, National Wildlife Federation, and Izaak Walton League of America.

Group of 15 (G15) A group of fifteen *developing countries that meet to discuss issues relating to the *Third World.

Group of 7 (G7) Seven of the largest industrialized countries (Canada, France, Germany, Italy, Japan, UK, and USA) whose leaders meet regularly in economic summit meetings.

Group of 77 (G77) A group of *developing countries that negotiate collectively on some international economic issues and political concerns about climate change and development, which was founded in 1964 under the auspices of the *United Nations Conference on Trade and Development (UNCTAD), with 77 initial members. The group now has 130 members.

Group of 8 (G8) A group that has replaced the *Group of 7, by also including the *European Union and Russia (which does not participate in all events). It holds an annual economic and political summit (the G8 Summit), comprising heads of government and international officials.

group selection 1. A method of tree harvesting in which patches of selected trees are cut in order to create a mosaic of openings in the forest *canopy and to encourage the growth of *uneven-aged stands.
  2. A theory which explains *altruistic behaviour in individuals in terms of benefits for the group.

grove A small *wood or group of trees that has no *undergrowth. *See also* COPPICE.

growing season The period of the year during which temperature remains high enough to allow plant growth, which is usually between the average date of the last *frost in *spring and the first frost in *autumn. Also known as **cropping season.** *See also* HARVEST.

grow-out The process of growing a plant in order to produce fresh viable *seed which can be used to evaluate the characteristics of its varieties.

growth band The annual growth pattern found in many organisms, such as snails. In *coral the band is caused by the secretion of *calcium carbonate. One yearly growth band contains two smaller bands, which represent winter growth and summer growth.

growth form The physical appearance and structure of an *organism.

growth rate The percentage rate at which something (such as an organism, population, or economy) is growing per unit time (normally a year). *See also* DOUBLING TIME.

growth ring The layer of wood growth that is formed by a tree during a single growing season, which includes both the *earlywood (which forms in spring) and the *latewood (which forms in autumn). Also known as **annual ring.**

groyne (groin) A narrow protective structure that is built out into the sea from the land, in order to restore an eroding beach by intercepting *longshore drift and trapping sand.

guild A group of species whose members exploit similar resources in a similar manner.

Gulf Stream A narrow, fast-flowing warm ocean *current which is part of the North Atlantic *gyre. It starts in the Gulf of Mexico around the Caribbean, and (driven by the westerly winds and deflected by the *Coriolis force) flows north-eastwards across the Atlantic and warms the climate of the east coast of the USA. The current splits around the latitude of Spain and part of it flows south, heating the coast of south-west Europe and eventually rejoining the Equatorial Current. Part of the current continues to flow towards the north-east, as the *North Atlantic Drift, and it brings relatively mild climatic conditions to Britain and north-west Europe.

gully A deep ditch caused by the *erosion of *soil on a *hillslope by running water.

gumbotill Glacial *till that is highly weathered, has a high *clay content, and becomes sticky and *plastic when wet.

Gunz The first period of *glaciation in Europe during the *Pleistocene *ice age, equivalent to the *Nebraskan in North America.

GURT *See* GENETIC USE RESTRICTION TECHNOLOGY.

gust A sudden, brief increase in wind speed, generally lasting less than 20 seconds.

guyot *See* SEAMOUNT.

GWP *see* GLOBAL WARMING POTENTIAL.

gymnosperm Any of the non-flowering seed plants. *Contrast* ANGIOSPERM.

gypsum An *evaporite mineral that is formed by the *evaporation of saltwater, and is used in making plasters and cements.

gypsy moth A European moth (*Lymantria dispar*) that was introduced into North America in 1869 and is a serious *defoliating *pest of *hardwood shade trees.

gyre A permanent large-scale circulation cell of water in the open *ocean, which is driven by prevailing *winds and the *Coriolis effect.

gyttja *Freshwater *anaerobic mud that contains abundant *organic matter.

g

habit The general growth pattern of a *plant, which includes creeping plants, trees, shrubs, and vines.

habitat The place or set of environmental conditions (*abiotic and *biotic) in which a *plant or *animal normally lives. For a *species to survive in a particular *ecosystem, its habitat must support a *population large enough to sustain itself by breeding. Habitats vary in size according to the species that occupy them. Carnivores usually occupy and require much larger habitats than *herbivores, because a *carnivore derives its food from a wider area. Each habitat has a particular environment, and habitats are usually defined on the basis of geology, vegetation, and location. Areas within a habitat that are occupied by a particular species or community are described as *microhabitats.

habitat action plan A *management plan that defines objectives and targets for the maintenance or enhancement of a particular habitat, and the actions that are regarded as necessary to achieve them. Also known as **habitat conservation plan**.

habitat connectivity The network of *corridors that links isolated *patches of *habitat, which allows organisms to move through an area. *See also* FRAGMENTATION.

habitat conservation plan *See* HABITAT ACTION PLAN.

habitat enhancement Any managed change to a *habitat that improves its value and increases its ability to meet the particular needs of one or more *species.

habitat fragmentation The breakup of a large *habitat into a number of separate, smaller remnants, for example because of *land use change. *See also* CORRIDOR.

habitat indicator Any physical attribute of the environment (such as soil *pH or the *salinity of a water body) that can be measured to characterize the conditions necessary to support an *organism, *population, or *community in the absence of *pollutants.

habitat loss The disappearance or conversion of natural *habitat, often as a result of human actions.

habitat patch An area of *habitat that differs from the surrounding area, and meets particular needs for particular *organisms.

habitat restoration Restoring a *habitat to its original *community structure, by removing *exotic species and/or reintroducing *native species. *See also* ACTIVE RESTORATION, NATURAL PROCESS RESTORATION.

habitat type An area of land that is capable of supporting a particular *climax plant *association.

Habitats Directive A major initiative of the *European Union that is designed to protect *biodiversity through the conservation of natural *habitats and *wild plants and animals, and the creation of a network of protected areas across the EU to be known as *Natura 2000 sites.

haboob A *dust storm or *sandstorm that is caused by cold *downdrafts from a *thunderstorm which turbulently lift dust and sand into the air. The term is mostly used in connection with the Sudan.

hadal zone The deepest part of the *ocean, at depths greater than the *abyssal zone or more than six kilometres, which includes the lower levels of *ocean trenches.

Hadean The oldest period of geological time, extending from the origin of the Earth about 4.5 billion years ago until the date of the oldest known rocks, about 3.9 billion years ago (the start of the *Archaean). There are no rocks on Earth this old.

Hadley cell An important cell in the *general circulation of the atmosphere. Warm air rises in the *tropics and it flows towards the poles, transporting heat energy, whereby the air cools and sinks at around 30° north and south. The air then moves back towards the *equator as wind (*trade winds).

haemoglobin A protein that is found in the red blood cells of mammals and other animals that carries *oxygen round the body.

hail Solid *precipitation that falls as ice particles from *cumulonimbus clouds. Individual particles are called hailstones, and the storm is called a hailstorm.

Haines index *See* LOWER ATMOSPHERE STABILITY INDEX.

half-life 1. The time required for half of the *atoms of a *radionuclide to decay, which is specific to each nuclide. *Radon-222 has a very short half-life of 3.82 days, so compared with radium-226 (half-life of 1622 years) or *thorium-232 (half-life of 14 million years), radon gas disappears very quickly.
2. The time required for a *pollutant to decrease in concentration by half. For example, the half-life of *DDT in the *environment is 15 years.
3. The time required for the elimination of half a total *dose of a substance from a body. Also know as **biological half-life.**

half-tide level *See* MEAN TIDE LEVEL.

haline The dominance of ocean salt in water. *See also* SALINE.

halite *See* ROCK SALT.

halo A circle of light around the Sun or Moon that is caused by the *refraction or *reflection of light by the ice crystals in *cirrus clouds.

halocarbon A compound that contains *carbon and at least one *halogen, and sometimes *hydrogen. *Chloro-fluorocarbons (CFCs) are halocarbons.

halocline A layer in a body of water at which *salinity and thus *density change sharply.

halogen Any of a group of five chemically related, non-metallic elements (*bromine, *fluorine, *chlorine, *iodine, and astatine) that can combine with metals to form salts or substitute for *hydrogen in many *organic compounds.

halogenated A compound in which a *halogen atom has been introduced, via a chemical reaction.

halomorphic soil A *soil that contains a significant proportion of soluble salts.

halon A synthetic *halogenated *hydrocarbon compound that consists of *bromine and other *halogens (which could include *fluorine and/or *chlorine), is used as a fire extinguishing agent, and persists in the atmosphere for a long time. Halons are much more effective in terms of *ozone destruction than *CFCs. *See also* OZONE DEPLETION POTENTIAL.

halophile A salt tolerant *organism, which is specially adapted to live in areas of high salt concentration. *See also* EXTREMOPHILE.

halophyte A plant that is adapted to grow in salty soil and air.

hamlet A *rural settlement that is smaller than a *village and contains a few houses.

hammada A surface of stones (pebbles, gravel, or boulders) on a *desert, that is formed by the washing or blowing away of the finer material.

hand-lining A traditional method of *fishing, using a hooked line which is held down in the water with a weight and pulled up by hand or with pulleys. Also known a **jigging.**

Hanford Nuclear Reservation The site of a nuclear facility in south-central Washington, USA which was established in 1943 (during World War II) as part of the Manhattan Project, to provide the *plutonium necessary for the development of nuclear weapons.

hanging valley A former *glacial *tributary valley that enters a larger glacial valley above the floor of the larger valley, high up on the valley wall. Waterfalls (such as Bridal Veil Falls in Yosemite) are common features of hanging valleys.

HAP See HAZARDOUS AIR POLLUTANT.

haploid Having one set of *chromosomes in the *nucleus of each cell as in mosses and many *protists and *fungi.

hard energy path An approach to *economic development that relies on conventional sources of *energy, including *fossil fuels and *nuclear energy, which create environmental problems including *pollution and *waste management. *Contrast* SOFT ENERGY PATH.

hard freeze A *freeze in which vegetation is killed and the ground surface is frozen solid.

hard mast The fruit or nuts of trees such as oak, beech, and walnut.

hard water *Alkaline water that contains dissolved mineral salts (particularly calcium and magnesium *bicarbonates), which reduces the cleansing power of soap and produces *limescale in hot water appliances. *See also* ION EXCHANGE TREATMENT.

hardboard A board-like building material that is made from wood fibre.

hardness 1. A measure of the amount of dissolved salts (particularly *calcium carbonate) in *alkaline water.
2. The degree of resistance of a *mineral to scratching, which is usually measured on the *Moh hardness scale.

hardpan A hard, relatively impervious, layer of soil in the lower *A-horizon or in the *B-horizon. It is difficult for plant roots to penetrate and is caused by cementation of soil particles with *organic matter or with materials such as *clay or *calcium carbonate. Also known as **clay pan**. *See also* IMPEDED DRAINAGE, INDURATED, IRON PAN.

hardwood *See* DECIDUOUS.

harmattan A dry *desert wind that blows from the *Sahara onto the coast of West Africa between November and March.

harpoon A spear with a barbed point and a strong line attached to it, which is used for catching large fish (such as giant tuna and swordfish) or whales.

harrow Farm equipment that is used to break up clumps of *soil on heavy ground.

harrowing A technique that breaks up the *soil in preparation for planting. Also known as **disking**.

harvest 1. The act of gathering a ripened *crop.
2. The season in which a crop is gathered. *See also* CROPPING SEASON.
3. The yield from plants in a single growing season. *See also* CROP YIELD.

harvest cut The removal of mature trees from a *forest. *Contrast* INTERMEDIATE CUT.

harvesting The gathering of a ripened *crop or the felling and removal of trees.

haustellate Relating to the mouthparts of an insect that are modified for sucking.

hay Grass that has been mowed and dried for use as *fodder.

hay meadow A *field of *grass which is cut two or three times in the summer to produce a crop of *hay for feeding to *livestock.

hazard A source of danger or disruption. *See also* NATURAL HAZARD.

hazard assessment The analysis and evaluation of the physical, chemical, and biological properties of a particular hazard. The term, as used in North America,

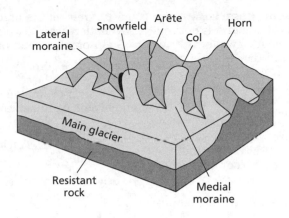

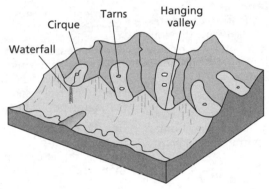

Fig 9 Hanging valley

is usually applied to human exposure to particular *agents such as *toxic substances. Also known as **hazards analysis, evaluation**.

hazard evaluation *See* HAZARD ASSESSMENT.

hazard identification The process of determining whether exposure to a particular *agent might affect human health.

hazard index The sum of all *hazard quotients for a specific pathway or *scenario, which takes into account the presence of multiple substances at one site, or exposures to the same chemicals through multiple media and pathways.

hazard mapping The process of establishing where and to what extent particular *hazards are likely to pose a threat to people, property, and the environment.

hazard quotient A measure of the health risk of exposure to a particular *toxic material, expressed as the ratio between average daily *dose and a reference dose defined from *toxicity tests.

Hazard Ranking System (HRS) A method used in the USA by the *Environmental Protection Agency for evaluating potential relative risks to public health and the environment from abandoned or uncontrolled *hazardous waste sites. The scale runs from 0 to 100; if a site scores 28.5 or more it is placed on the *National Priorities List. *See also* NATIONAL OIL AND HAZARDOUS SUBSTANCES CONTINGENCY PLAN, SITE INSPECTION.

hazardous air pollutant (HAP) Any *air pollutant that in the USA is not covered by *ambient air quality standards but which (as defined in the *Clean Air Act) may reasonably be expected to cause or contribute to irreversible illness or death. Examples include *asbestos, *mercury, *radionuclides, and vinyl chloride.

hazardous chemical Any chemical whose presence or use is a physical hazard or a health hazard. Examples include *flammables, explosives, and *acids.

hazardous material *See* HAZARDOUS SUBSTANCE.

hazardous substance Any substance that, because of its quantity, concentration, or physical, chemical, or infectious characteristics, has the potential to cause a physical or health hazard. Also known as **hazardous material**.

hazardous waste *Waste material that is reactive, *toxic, *corrosive, or otherwise poses a hazard to human health and the *environment. *Contrast* TOXIC WASTE.

hazardous waste landfill A *landfill site in which *hazardous waste is deposited and covered.

hazardous waste minimization Reducing the amount of *hazardous waste that is produced by a facility, via *source reduction or environmentally-sound *recycling.

hazards analysis *See* HAZARD ASSESSMENT.

haze A reduction in *visibility that is caused by fine dry particles of *dust and *pollutants.

HCFC *See* HYDROCHLOROFLUOROCARBON.

HDGCP *See* HUMAN DIMENSION OF GLOBAL CHANGE PROGRAMME.

head 1. The source of water from which a *stream arises. Also known as **headwater**.
2. The elevation of a water surface above a particular point, such as the height of water in a *reservoir above

the turbine that is used for generating *hydroelectricity.
3. Unsorted angular rock debris.

headland 1. A rocky *promontory that sticks out into the sea.
2. A strip of unploughed land at the end of a set of *furrows or along a fence.

headwall A steep slope at the head of a *valley, such as the rock cliff at the back of a *cirque.

headwater The source and upper part of a *stream. Also known as **head**.

health A state of complete physical, mental, and social well-being and not merely the absence of *disease or infirmity.

health assessment The process of collecting, analysing, and disseminating information on health status and health risks, which is used in prioritizing and managing public health problems.

health hazard A *chemical, a mixture of chemicals, or a *pathogen for which there is statistically significant evidence that *acute or *chronic effects may occur in any people who are exposed to it.

Healthy Forest Initiative (2003) An initiative from US President George Bush that is designed to improve regulatory processes to reduce the risk of catastrophic *wildfires and restore forest health.

Heartland Institute A US non-profit *anti-environmental *think-tank that was founded in Chicago, Illinois in 1984 and whose mission is to help build social movements in support of ideas that empower people. Such ideas include *market-based approaches to environmental protection, privatization of public services, and deregulation in areas where property rights and markets do a better job than government bureaucracies.

heartwood The wood in the centre of a tree that is composed of non-living cells and is usually darker than the *sapwood.

heat A form of *energy that is transferred from one body to another (by *conduction, *convection, *advection, and *radiation) because of a difference in *temperature, from regions of higher temperature to regions of lower temperature.

heat capacity The amount of heat energy that is required to to raise the average temperature of a mass, usually expressed in *Joules per *Kelvin (J K^{-1}). *See also* BRITISH THERMAL UNIT, CALORIE, SPECIFIC HEAT.

heat engine Any *engine that makes use of *heat to do mechanical work.

heat exhaustion A mild form of *heat stroke that causes faintness, dizziness, and heavy sweating in humans.

heat flux The amount of *heat that is transferred across a surface of given area in a given amount of time. Also known as **thermal flux**.

heat index (HI) The apparent *temperature that is created by the combined effect of high air temperatures and high *humidity levels.

heat island *See* URBAN HEAT ISLAND.

heat of fusion The amount of *heat energy that is required to change a substance from solid to liquid at constant temperature. Also known as **heat of melting**. *See also* LATENT HEAT.

heat of melting *See* HEAT OF FUSION.

heat of transformation *See* LATENT HEAT.

heat of vaporization The amount of *heat energy that is required to change a substance from liquid to gas at constant temperature.

heat stroke A condition that results from excessive exposure to intense *heat, causing high fever, collapse, and sometimes convulsions or coma in humans.

heat wave A continuous period of unusually and uncomfortably hot weather.

heath A plant *community that is dominated by low-growing shrubs with woody stems and narrow leaves (members of the *Ericaceae or *heather family), which often grow on acidic or upland soils.

heather A low shrub with small evergreen leaves and clusters of pink flowers, which was traditionally used in Britain to make brooms, thatch, and bedding, and for heating ovens. Also known as **ling**.

heathland Uncultivated land with *heather and *heath growing on *acidic, nutrient-poor, sandy soils.

heavy metal A *metallic element that has a high atomic weight, is often *toxic, and tends to accumulate in the *food chain. Examples include *arsenic, *cadmium, *chromium, *lead, and *mercury. *See also* BIOACCUMULATION.

heavy water Water that contains *deuterium, a heavy *isotope of *hydrogen, which is used in *nuclear reactors to slow down *neutrons.

heavy water reactor A nuclear *fission reactor that uses *deuterium-enriched water to *moderate the fission reaction.

hectare A unit of area that is equal to 10 000 square metres or 2.47 acres.

hedge A natural fence that is created by a row of closely planted shrubs or bushes.

hedgerow An untrimmed *hedge or barrier of bushes, shrubs, or small trees that are growing close together in a line, which creates a natural growing fence and provides *cover and a *corridor for *wildlife. *See also* LIVE FENCE.

hedonic pricing A method of *economic evaluation, based on the premise that the price of a good is partly determined by its characteristics or the services it provides. The approach seeks to value the individual characteristics of that good by studying how the price people are willing to pay for it changes when the characteristics change. It is widely used to explain variations in house prices in terms of variations in environmental quality (such as air pollution, water pollution, or noise) and environmental amenities (such as attractive views or access to recreational sites). *See also* TRAVEL COST METHOD.

helitanker A helicopter equipped with a specially designed tank that is used for transporting and dropping suppressants or retardants on *wildfires.

helium A natural, inert gas that is colourless, non-toxic, non-combustible, and lighter than air. It is the second lightest and second most abundant element. Roughly 25% of the Sun is helium, but it accounts for only 0.0005% of the Earth's *atmosphere. *See also* HIGH TEMPERATURE GAS-COOLED REACTOR.

hemimetabolism Incomplete or partial *metamorphosis in insects. *Contrast* HOLOMETABOLISM.

hemipelagic sediment A deep-sea *sediment that accumulates near the *continental margin, and is composed of material that comes from both land and sea.

hemisphere Half of the *Earth, defined either in terms of north and south of the *equator, or east and west of the *Greenwich Meridian.

HEP *See* HUMAN EXCEPTIONALISM PARADIGM, HYDROELECTRIC POWER.

herb Any small, non-woody *vascular plant, such as *grass or *forbs.

herbaceous Non-woody vegetation that dies back each season.

herbaceous layer The vegetation in the *understorey in a *forest that consists mainly of non-woody plants. Also known as **ground cover**.

herbage The edible *biomass of *herbaceous plant, on which grazing animals feed.

herbarium 1. A collection of dried, pressed, or preserved plant specimens (leaves, flowers, seeds, and stems).
2. The kitchen garden of an abbey or monastery.

herbicide A chemical *pesticide that is used to control or kill specific unwanted plants, particularly *weeds. *See also* CONTACT HERBICIDE, SYSTEMIC HERBICIDE.

herbivore An *organism that eats only *plants. *Granivores and *frugivores are special types of herbivores. *See also* PRIMARY CONSUMER.

herd 1. A group of *domesticated mammals (such as cattle or sheep) that are kept together and looked after by humans.
2. A group of wild animals of one species that remain together in a *colony, such as antelope, elephants, seals, or whales.

heredity The passing on of characteristics from one generation to another, which in organisms takes place via *genes. *See also* INHERIT.

heritability A measure of the extent to which a characteristic in an *organism is related to *genetic, inherited factors relative to the mean of the population.

heritable Something which can be *inherited.

heritage 1. Anything that is *inherited from an ancestor.
2. The overall natural and cultural inheritance of a country, including important natural or archaeological sites, historic buildings, customs, and traditions.

Heritage Foundation A US non-profit *anti-environmental *think-tank that was established in 1973 and whose mission is to promote conservative public policies based on the principles of free enterprise, limited government, individual freedom, traditional American values, and a strong national defence.

hermaphrodite An individual born with genitalia and/or secondary sexual characteristics which combine features of both sexes. For example, an earthworm. *See also* ANDROGENY.

hermatypic A *reef-building organism.

herpetofauna *Reptiles (such as snakes, turtles, and lizards) and *amphibians (such as frogs, toads, and salamanders).

herpetology The study of *reptiles and *amphibians.

hertz (Hz) An *SI measure of frequency (number of cycles per second).

heterogeneity Variety or diversity, lack of uniformity; the variety of qualities found in an environment (*habitat patches) or a population (*genotypic variation). *Contrast* HOMOGENEITY.

heterogeneous Varied, not uniform. *Contrast* HOMOGENEOUS.

heterotroph An *organism that is unable to synthesize organic compounds (and thus get its energy) from the *environment, so it must feed upon organic compounds that are produced by other *organisms. There are three types of heterotroph that derive energy from different sources, namely *saprophytes, *parasites, and *holozoic organisms. Also known as **consumer**. *Contrast* AUTOTROPH.

heterozygosity Having different *alleles (forms) of a particular *gene; genetic variability among *individuals within a *populations, and among populations.

HFC *See* HYDROFLUOROCARBON.

HI *See* HEAT INDEX.

hibernacula Secure places (such as *caves or *dens) where animals *hibernate during the winter in order to conserve energy.

hibernation A long-term, deep sleep during which some animals' heartbeat and respiration rates slow down to allow them to survive the winter cold period when there is little food available.

hide 1. The skin of a large, mature *animal.
2. A hut or tent that allows observers of *wildlife to conceal themselves.

hierarchy A series of ordered groupings of things within a system.

high *See* ANTICYCLONE.

high forest *Woodland that is managed to allow the majority of trees to reach maturity. *Contrast* COPPICE.

high latitude Towards the *poles, away from the *equator. *Contrast* LOW LATITUDE.

high occupancy vehicle (HOV) A passenger vehicle that carries more than a specified minimum number of passengers, such as a bus or a car used within a carpool. *Contrast* LOW OCCUPANCY VEHICLE.

high pressure An area of high *atmospheric pressure. Also known as **anticyclone** or **high**.

high risk community A community of people which is located close to numerous potential environmental and health hazards, which may result in high levels of *exposure to *contaminants or *pollutants.

high sea General term for international areas of the *oceans that lie beyond the legal control of any nation. Similar to *open sea.

high sulphur coal *See* NON-COMPLIANCE COAL.

high temperature gas-cooled reactor A nuclear *fission reactor that uses *helium gas to transfer heat from the *core to a steam generator.

high tide *High water. The highest level to which the *tide rises within the daily tidal cycle. *Contrast* LOW TIDE.

high water The maximum height or stage that is reached by rising water, such as a rising *tide at the *coast or a *flood in a *river.

high water mark The line along the coast to which the sea rises at *high tide. *Contrast* LOW WATER MARK.

high wind Sustained *winds of 40 miles per hour, mph (64 kilometres per hour, kmh) or greater which last for an hour or more, and/or *gusts of greater than 58 mph (93 kmh).

high yield variety A variety of *crop plant that produces a high *yield relative to other varieties. *See also* GREEN REVOLUTION.

high-density planting Closely planted *crops or *trees.

high-grading A method of *harvesting *natural resources that removes the biggest and best of a species (for example of trees in a forest, or fish in the sea) and ignores the rest.

highland See UPLAND.

high-lead logging A method of *cable logging in which timber is dragged from where it is cut to a collecting point by a powered cable above ground level. Contrast GROUND-LEAD LOGGING.

high-level waste The longest-lasting and most dangerous of all the *radioactive waste produced by the nuclear industry, which contains highly *radioactive, short-lived *fission products, *hazardous chemicals, and *toxic *heavy metals. Contrast LOW-LEVEL WASTE, INTERMEDIATE-LEVEL WASTE.

high-level waste facility A *facility that is designed to handle the safe disposal of spent *nuclear fuel, *high-level radioactive waste, and *plutonium waste.

high-level waste repository A place where highly *radioactive wastes can be buried and stored safely for tens of thousands of years, unexposed to *groundwater and *earthquakes. See also YUCCA MOUNTAIN.

hill A local, well-defined elevated area of land with a rounded top, smaller than a *mountain. See also HILLSLOPE.

hill farming *Farming in an *upland environment, where the land is sloping, *fields are often relatively small, *soil erosion is a common problem, and *yield is often relatively low because of harsh *climate and poor *soils.

hill prairie A type of dry *prairie that is found in well-drained sites at the tops of hills, bluffs, and ridges, and on steep slopes.

hillside See HILLSLOPE.

hillslope The side or slope of a *hill. Also known as **hillside**. See also COLLUVIUM.

hinterland The area that is *tributary to a place and linked to it by economic activities.

HIS See INTERNATIONAL HUMAN SUFFERING INDEX.

historic 1. Related to the known or recorded past, in times of written history. See also PREHISTORIC.
  2. Important or famous in history.

Historic District A land *zoning category in the USA that applies to an area in which the buildings have particular *historic, architectural, or cultural significance.

historic ecosystem The natural *ecosystem that currently exists or is known to have existed in the past, which guides *restoration activities.

historic land See FRAGILE or HISTORIC LANDS.

historic range The *natural range or geographical areas that a particular *species was known or believed to occupy in the past.

historic range of variability See RANGE OF VARIABILITY.

history The study of past events within the *historic period.

histosol A wet, organic *order of soil that is found in *bogs, *swamps, and *wetlands, and contains a large amount (usually more than 30%) of *organic material.

HNS Protocol See PROTOCOL ON PREPAREDNESS, RESPONSE AND CO-OPERATION TO POLLUTION INCIDENTS BY HAZARDOUS AND NOXIOUS SUBSTANCES (2000).

hoar A form of *frost that consists of a deposit of needle-like soft ice crystals formed on vegetation, the ground, or window-panes by direct *condensation from the air at temperatures below *freezing point. Also known as **white frost**.

hobby farming See AMENITY HORTICULTURE.

hogback A narrow ridge of *hills. See also ESCARPMENT.

holding pond A *pond or *reservoir that is usually made of soil or sediment, and is designed to store *polluted *runoff.

holding tank A container where *wastewater is stored before it is removed for treatment or disposal.

holding time The maximum amount of time that a sample (for example, of soil or polluted water) can be stored before analysis.

holism The idea that the whole is greater than the sum of the parts, so that studying a whole *system and its interrelationships is better than studying isolated parts of it. Contrast REDUCTIONISM.

holistic An explanation that attempts to explain complex phenomena in terms of the properties of the *system as a whole. See also HOLISM.

holistic management A *holistic approach to the management of natural resources, that takes into account interrelationships between people, the environment, and *ecosystems.

Holocene The *post-glacial epoch. The most recent phase of geological time within the *Quaternary, which began 10 000 years ago, after the end of the *Pleistocene ice age when glaciers retreated and sea level rose because of a warmer global climate. Climatic change in the northern hemisphere through the Holocene includes the *Climatic Optimum, the *Medieval Warm Period, and the *Little Ice Age.

holocoen See ECOSYSTEM.

holometabolism Complete *metamorphosis in insects. Contrast HEMIMETABOLISM.

holoplankton *Marine *organisms that spend all their life in water and not on or in the sea bed.

holozoic organism A *heterotrophic *organism that ingests complex organic matter, and can absorb large particles of undissolved food and soluble compounds. All higher animals (including humans) are holozoic.

home range An area in which an *animal normally lives, and from which it gets its food. Contrast TERRITORY.

homeland The country in which someone was born.

homeostasis The process of self-regulation which maintains *steady state within *environmental systems through *adjustment and *feedback.

homeotherm A *warm-blooded organism (such as a bird or mammal) that regulates its body temperature independently of changes in the temperature of its environment. Contrast ECTOTHERM, POIKILOTHERM. See also ENDOTHERM.

homestead The home and adjacent land that is occupied by a family.

hominid A primate of the family *Hominidae, which includes *Homo sapiens as well as extinct human ancestors.

Hominidae Modern humans and their extinct immediate ancestors. Erect, bipedal (walk on two feet) *primates with relatively large brains.

Homo erectus A species of *hominid that lived between 1.8 million and 300 000 years ago, stood upright, had a brain slightly smaller than that of modern humans, and was the first Homo species to migrate beyond Africa.

Homo habilis A species of *hominid that lived in East Africa between 1.9 and 1.8 million years ago. Homo habilis was the first hominid to make and use tools.

Homo neanderthalensis A species of *hominid that lived between 150 000 and 30 000 years ago (the Middle *Palaeolithic) in Europe and Western Asia. Once thought to be a geographical variant of Homo sapiens it is now regarded as a distinct species.

Homo sapiens Modern humans who appeared in the fossil record about 200 000 years ago.

homocentric A view of nature that only considers human, rather than plant or animal, needs. Contrast ECOCENTRIC. See also ANTHROPOCENTRIC.

homochromy Induced colour change in an organism based on perception of the background against which they are cultured.

homogeneity Uniformity.   *Contrast* HETEROGENEITY.

homogeneous Uniform. *Contrast* HETEROGENEOUS.

homosphere The lower part of the *Earth's *atmosphere, consisting of the *troposphere, *stratosphere, and *mesosphere.

Hooke's law The physical principle that the *stress within a solid is proportional to the *strain, so that the movement of an object is proportional to the pressure that is applied to it.

hoop-net A method of small-scale *fishing in which fish are caught in a net that is suspended from a hoop on a pole.

horizon 1. The line where the visible edge of the Earth's surface meets the sky.
2. A surface that separates two adjacent *beds in *sedimentary rocks.
3. *See* SOIL HORIZON.

hormone A chemical substance that is produced in the body by an *endocrine gland, is carried around the body in the bloodstream, and regulates the activity of other organs or cells in particular ways.

horn A sharp-edged mountain peak (such as the Matterhorn) that is formed by *glacial erosion when *cirques erode simultaneously on more than two sides of a mountain. *See also* ARÊTE.

horse latitudes The subtropical region at latitudes 30° to 35°, where winds are mainly light and the weather is hot, dry, and settled.

horst An upstanding block of the Earth's *crust that is bounded by *faults and has been uplifted by *tectonic forces. Also known as **fault-block mountain**.

horticulture The cultivation of plants, particularly fruit, flowers, ornamental plants and vegetables, usually for sale, either in *gardens and *smallholdings or on general farms.

host An animal or plant that supports a *parasite, although the host does not benefit and is often harmed by the association.

host country The country where the reduction, avoidance or *sequestration of *greenhouse gas *emissions takes place.

host preference The extent to which a *herbivore feeds on a host plant when it is offered a choice of host plants, either simultaneously or sequentially.

hot air In the context of *climate change negotiations, a broad term for reductions in *greenhouse gas emissions (for example, in the former Soviet Union) that are caused by economic crisis rather than intentional efforts to reduce emissions.

hot desert A *desert (such as the *Sahara in North Africa) that is located between about 20° and 30° north and south of the *equator, where *evaporation rates exceed *precipitation and temperatures can be very high. Desert surfaces heat up rapidly during the day in the scorching sun (daytime temperatures of up to 55°C in the shade are not uncommon) and lose heat rapidly at night by radiation back to the overlying air. Night-time temperatures can fall close to freezing. Hot *deserts make up about 20% of the Earth's land surface in two belts between about 20° and 30° north and south of the equator. These are surrounded by *semi-arid belts. Together they form a third of the Earth's land area and are home to nearly 1000 million people. *Contrast* COLD DESERT.

hot dry rock system An approach to the exploitation of *geothermal energy, in which cold water is pumped through fractured hot dry rock deep underground, is heated by contact with the rock, and is then extracted and used for heating and electricity generation.

hot rock An area of the Earth's *crust where there is molten *igneous rock close to the surface, which provides a source of *geothermal energy.

hot spring A natural *spring of hot mineral water, associated with *hot rock. *See also* GEYSER, HYDROTHERMAL ACTIVITY.

hot zone *See* EXCLUSION ZONE.

hotspot An area that has an unusually high level of some particular quality or activity, such as *air pollution, *biodiversity, or *volcanic activity. *See also* BIODIVERSITY HOTSPOT.

household A group of people who live together.

household waste Solid waste that is composed of *rubbish, may contain *toxic and *hazardous waste, and originates in a private home or apartment. Also known as **domestic waste, residential waste.**

HOV *See* HIGH OCCUPANCY VEHICLE.

HRS *See* HAZARD RANKING SYSTEM.

Human Dimension of Global Change Programme (HDGCP) A large-scale international social science research programme that ran through the 1990s, and sought to obtain a better understanding of the human causes of *global environmental change, and to formulate appropriate responses for reconciling *economic development and the maintenance of *environmental quality.

human ecology The study of the interactions between humans and the environment.

human equivalent dose A *dose which, when given to humans, produces an effect equal to that which is produced by a dose in *animals.

human exceptionalism paradigm (HEP) The view (*paradigm) that humans are different from all other organisms, all human behaviour is controlled by culture and free will, and all problems can be solved by human ingenuity and technology. *See also* ANTHROPOCENTRISM.

human exposure evaluation A description of the nature and size of the population that is exposed to a particular substance, and the magnitude and duration of *exposure.

human genome The complete *DNA sequence (*genome) for a human; all of the genetic materials that make up a human being.

Human Genome Initiative The original name for what became the *Human Genome Project.

Human Genome Project A major international research project that was established in 1990 (originally called the *Human Genome Initiative) to study differences in the genetic make-up of ethnic groups and to sample DNA from populations around the world. It seeks to identify the 60 000 to 80 000 *genes that are carried by humans, and help to find genetic causes and possible treatments of disease. *See also* GENETIC ENGINEERING, GENOME.

human health risk The likelihood or probability that a particular *exposure or series of exposures may have damaged, or will damage, the health of individual people.

human impact The result or *impact of human activities. *See also* ENVIRONMENTAL IMPACT.

humanitarian Devoted to the promotion of human welfare.

humic Relating to or derived from *humus.

humic acid A dark brown *organic substance which is the main constituent of *humus, and is soluble in water only at *pH values greater than 2.0.

humid Moist, damp.

humid continental climate A quite severe climate, with harsh winters, average temperatures below freezing (0°C) in several months, mild to warm summers, and adequate annual precipitation, which is found in mid-latitude continental areas of the northern hemisphere, between about 40° and 60°.

humid continental mixed forest A *temperate *forest *biome found in middle latitudes where *precipitation is greater than about 750 millimetres a year, and droughts and drying winds are rare. The natural *climax vegetation is *deciduous *broadleaved trees, including oak, maple, beech, hickory, and elm in the USA.

humid subtropical climate A mid-latitude *climate that is dominated by hot, humid summers and cool winters, and is often found on the eastern side of a *continent.

humid subtropical forest A *forest *biome of the *humid subtropical climate zone, which is dominated by *coniferous trees, has a dense *undergrowth and many climbing plants. Its *habitats support a wide variety of animals (particularly reptiles, mammals, and birds) and insects.

humid tropics Those parts of the *tropics that receive a lot of *rainfall during part of the year, and that have temperatures which are generally suitable for year-round *crop production.

humidification The addition of *water vapour to *air in order to increase its *humidity.

humidity The amount of *water vapour in a parcel of air, which can be expressed as *absolute humidity or *relative humidity.

humification The *decomposition of *organic material in the *soil into *humus, by *biochemical and *abiotic processes.

humus Partly decomposed *organic matter (from the bodies of dead plants and animals) within a *soil, that increases fertility and water retention in the soil, and improves soil texture. The decomposing humus is usually dark coloured (dark brown or black), so the uppermost *horizon of the soil (*A-horizon) where is accumulates it generally darker than the soil below. Also known as **soil organic matter**.

hunt To pursue, kill, or capture wild animals for food, *pelts, or as a sport.

hunter-gatherer A person in a society that obtains food by *hunting (fish and wild animals) and gathering (wild berries, fruits, fungi, nuts, leaves, and edible roots) rather than raising livestock or crops (*agriculture).

hunting The activity of finding and killing or capturing wild animals for food, *pelts, or as a *field sport.

hurricane 1. A tropical *cyclone that develops over the North Atlantic and Caribbean. A typical hurricane has a diameter of between 150 and 1500 kilometres, is circular in shape, and has a calm area (the *eye) of descending air at the centre which is surrounded by a zone of powerful up-draught (the *eye wall), having great turbulence, thick cloud cover, and torrential rainfall. Strong winds spiral inwards towards the centre of the storm (down the pressure gradient); this wind blows anticlockwise in the northern hemisphere and clockwise in the southern hemisphere. The strong winds cause extensive damage. The word hurricane is used in the North Atlantic Ocean, Caribbean Sea, Gulf of Mexico, and eastern North Pacific Ocean; a tropical cyclone is known as a typhoon in the western Pacific and a cyclone in the Indian Ocean.
2. Any storm with sustained winds of at least 119 kilometres per hour, kmh (74 miles per hour, mph), or force 12 on the *Beaufort scale. Most occur in tropical and subtropical areas, and the scale of a hurricane is usually defined using the *Saffir–Simpson scale. Hurricanes generally cause much more damage than *tornadoes because they are larger, last longer, and travel longer distances.

hurricane track The course followed by a *hurricane as it moves across the land and/or sea.

hurricane warning A warning (formal *advisory) that is given to inform the public and marine interests when it is likely that a *hurricane will strike an area within 24 hours.

hybrid An *organism that is the offspring of two different *varieties, *breeds, or *species. Hybrids are sterile and so unable to reproduce. For example a mule is the hybrid of a male donkey and a female horse. *See also* CROSSBREEDING.

hybridization The process of crossing individuals from *strains, *populations, or *species that are genetically different, which is usually done on purpose as part of a *selective breeding programme in order to produce *hybrids.

hydration The chemical combination of *water with a *solute.

hydraulic conductivity The rate at which water can move through a *permeable medium, such as soil. *See also* PERMEABILITY.

hydraulic geometry In a *river, the relationship between *discharge and channel width, depth, and velocity as discharge varies through time either at a particular site (the at-a-station hydraulic geometry) or in a downstream direction (the downstream hydraulic geometry), which can be shown in graphs and as equations.

hydraulic gradient The slope of the *water table, which determines the direction and rate of *groundwater flow.

hydraulic jump A rapid change in the flow depth of water in a *stream, which creates a *wave on the water surface. Also known as **standing wave**.

hydraulic mining A form of *placer mining in which water under pressure is used to break down *placer deposits (such as *china clay). *See also* PLACER MINING.

hydraulics 1. The study of the mechanical properties of liquids.
2. A movement or action that results from the flow of a liquid.

hydric Having or requiring an abundant supply of water. Also known as **hydrophilic**. *Contrast* HYDROPHOBIC.

hydric soil A *soil that is saturated with water long enough during the plant *growing season to become *anaerobic, and that supports *wetland vegetation.

hydro- Related to or produced by *water or the movement of water, as in *hydroelectricity.

hydrocarbon A naturally occurring compound of *hydrogen and *carbon. There are many different types of hydrocarbon, which can take the form of a gas, liquid, or solid; examples include *natural gas, *bitumen, and *petroleum. A *conventional or criteria pollutant. *See also* FOSSIL FUEL.

hydrochloric acid (HCl) An *aqueous solution of hydrogen chloride gas that is a strongly corrosive acid and is widely used in metal cleaning and electroplating.

hydrochlorofluorocarbon (HCFC) A partly *halogenated *chlorofluorocarbon compound that consists of *hydrogen, *chlorine, *fluorine, and *carbon. HCFCs are a replacement for *CFCs, but they are also *greenhouse gases that contain *chlorine and deplete stratospheric *ozone, but to a much lesser extent than CFCs. *See also* OZONE DEPLETION POTENTIAL.

hydrodynamic A body shape that is streamlined to allow easy movement through a liquid such as water. *Compare* AERODYNAMIC.

hydrodynamics The study of the motion of liquids.

hydroelectric power (HEP) Electricity that is generated by the passage of water through a *turbine, usually at a dam. Traditional HEP schemes make use of natural slopes and *topography to create a *head of water. In a typical scheme water is stored in a *reservoir, from where it drops under the influence of gravity down pipes into water turbines which are coupled to electricity generators. Steep mountain rivers with high *discharges provide ideal sites for hydropower developments. HEP provides more than 20% of the world's electricity, and it is now the largest renewable source of electricity around the world. Capacity has increased by 14-fold since 1950, but the spread between countries is very uneven. It is estimated that about a quarter of the world's HEP potential had been exploited by the mid-1990s. Large-scale growth of hydropower is unlikely to continue significantly in the future, partly because of the lack of suitable sites and the environmental and social impacts of such schemes. The high *capital cost per unit of electricity produced is a further constraint. Small-scale local hydropower schemes appear to have much greater potential, particularly in developing countries, because they are cheaper, quicker to install, easier to maintain, and cause less damage. This is a form of *hydropower that is *renewable, reliable, *sustainable, pollution-free, and environmentally sound. See also PUMPED STORAGE SCHEME.

hydroelectricity Electricity that is generated from flowing water via *turbines; a form of *renewable energy.

hydrofluorocarbon (HFC) A compound that consists of *hydrogen, *fluorine, and *carbon. A *fluorocarbon that is emitted as a *by-product of industrial manufacturing, is used as a *solvent and cleaner in the semiconductor industry, is an alternative to *CFCs, does not contain *chlorine or *bromine, and does not deplete the *ozone layer. HFCs are powerful *greenhouse gases, with *global warming potentials of between 140 and 12 100. See also OZONE DEPLETION POTENTIAL.

hydrogen (H) A colourless, odourless, tasteless, flammable gas that is the simplest, lightest, and most abundant of the elements, but is found only in trace quantities (about 0.00005% by volume) in the Earth's *atmosphere. It is a *macronutrient which is essential for plant growth; it combines with oxygen to form water and forms organic compounds (*hydrocarbons) with *carbon.

hydrogen sulphide (H_2S) A colourless, flammable, toxic natural gas that smells like rotten eggs and is emitted during organic *decomposition and produced as a *by-product of oil refining and burning. It is also found in North Sea gas and in volcanic emissions.

hydrogenation 1. A method of converting liquid fats to solid fats, by adding *hydrogen gas to a *hydrocarbon.
2. A method of converting coal to oil.

hydrogeological cycle See WATER CYCLE.

hydrogeology The study of the chemistry and movement of *groundwater. Also known as **geohydrology**.

hydrograph A graph that shows variations through time in river level or *discharge, in response to the rainfall input by an individual storm event. Also known as **storm hydrograph**.

hydrographic divide See DRAINAGE DIVIDE.

hydrography The study of large bodies of water on the *Earth. See also HYDROLOGY.

hydrological Relating to water flow.

hydrological budget An account of the inflow to, storage in, and outflow from a *hydrological unit over a given period of time, such as a year.

hydrological cycle See WATER CYCLE.

hydrological unit A *drainage basin or a subdivision of one, such as an *aquifer, soil zone, *lake, *reservoir, or *irrigation project.

hydrology The study of the properties, distribution, and circulation of *water on, over, and through the *Earth. See also HYDROGRAPHY.

hydrolysis The *decomposition of a chemical compound by reaction with *water, as in the *chemical weathering of *rocks. For example, in *granite the *feldspar minerals break down into clay minerals like *kaolinite.

hydrometeor Any condensed water particle in the atmosphere, including *rain, ice crystals, *hail, *fog, or *cloud.

hydrometer An instrument that is used for measuring the *specific gravity or density of a liquid.

hydromorphic soil *Soil that has developed under *waterlogged conditions.

hydrophilic *See* HYDRIC.

hydrophobic Water-avoiding, water-repelling, or non-soluble. *Contrast* HYDRIC.

hydrophyte A plant that is adapted to grow in or under the surface of water. Also known as **phreatophyte**. *Contrast* MESOPHYTE, XEROPHYTE.

hydropolitics The politics of allocating and managing *water resources, particularly between countries.

hydroponics A method of growing plants, especially vegetables, in water that contains essential mineral *nutrients, without the use of *soil.

hydropower *electricity or *power that is generated from the movement of *water; a *renewable source of *energy that emits no *greenhouse gases. Examples include *water mills, *hydroelectric power schemes, and *tidal power.

hydrosere A vegetation *succession from open water to *mire and *bog at the edge of a *lake.

hydrosphere The water on or around the surface of the Earth, which includes *oceans, *seas, *lakes, *rivers, *groundwater, and atmospheric moisture, and is cycled through the *water cycle. It is a major *environmental system that interacts with the *lithosphere, the *atmosphere, and the *biosphere.

hydrostatic Relating to the pressure or forces exerted by fluids in equilibrium.

hydrostatic equation An equation that represents the balance between *gravity and the vertical *pressure gradient force. If these forces are equal, there is *hydrostatic equilibrium and thus no vertical motion.

hydrostatic equilibrium A balanced state in which the outward pressure is balanced by the inward force of *gravity.

hydrostatic pressure The pressure that is exerted on water at rest. In *groundwater, it is the pressure at a specific elevation that is caused by the weight of water at higher levels in the same *zone of saturation. *See also* ARTESIAN WELL.

hydrothermal Relating to the hot water and steam that are generated by *igneous activity, trapped in fractured or porous rocks within the *Earth, and which contain dissolved *minerals. *See also* GEOTHERMAL.

hydrothermal activity A range of geological processes that involve the movement of hot *groundwater and steam, particularly the alteration and emplacement of minerals and the formation of *hot springs and *geysers. *See also* GEOTHERMAL.

hydrothermal mineral A *mineral that is precipitated from a *hydrothermal fluid.

hydrothermal vein A layer of *minerals that has been precipitated from a *hydrothermal fluid between layers of rock underground.

hydrothermal vent A fissure in a *mid-ocean ridge on the deep ocean floor that is created by *sea-floor spreading, where hot, *sulphur-rich water is released from rock that is heated by *geothermal activity.

hydroxide An ion (OH⁻) that is ionically bonded with a negative charge. It also forms compounds of an *oxide with water.

hydroxyl Alcoholic group (–OH) that consists of *hydrogen and *oxygen in a compound which is covalently bonded (so it has no charge).

hygrometer An instrument that is used to measure the amount of *water vapour (*humidity) in the *atmosphere.

hygroscopic Capable of absorbing and retaining *moisture, for example from *water vapour in the air.

hygroscopic nuclei See CONDENSATION NUCLEI.

hygroscopic water Water that is *adsorbed onto a surface from the *atmosphere. In soil, for example, this water is held as a very thin layer around each individual particle of soil; it is not very accessible to plants and at least some of it remains in the soil even after extreme *drought.

hypermetamorphosis A type of *metamorphosis that is typical of certain types of insects (such as some beetles), in which the larva changes into different forms during its development.

hyperparasite A *parasite that attacks another parasite inside a host, not the host itself.

hyperplasia An abnormal increase in tissue growth that is caused by an increase in the number of cells, by cell division. Contrast HYPERTROPHY.

hypersaline Containing excessive *salts, having high *salinity (greater than 35 parts per thousand).

hypertrophic Containing excessive *nutrients.

hypertrophy An increase in the size if an organ that is caused by an increase in cell size. Contrast HYPERPLASIA.

hypoallergenic Not likely to cause an *allergic reaction.

hypocentre See FOCUS.

hypolimnion The bottom layer of water in a thermally stratified lake. It is the densest layer, is usually the coldest layer in summer and warmest in winter, usually lacks oxygen, and is too dark to support *photosynthesis.

hypothecation The pledge of property and assets as collateral in order to secure a loan, which does not transfer title but does provide the right to sell the hypothecated property in the event of default.

hypothermia Lowering of core body temperature to below 35°C as a result of exposure to extreme cold, causing rapid, progressive mental and physical collapse. See also WINDCHILL.

hypothesis A statement of the expected relationship between things being studied, which is intended to explain certain facts or observations. An idea to be tested.

hypoxia Depletion of *dissolved oxygen in water and *sediments, relative to the needs of most *aerobic species.

hypoxic water Water which has a *dissolved oxygen concentrations of less than 2 parts per million (ppm), which makes it difficult for most aquatic life to survive and reproduce.

hypsithermal See ALTITHERMAL.

hypsometric curve A graph showing the proportion of a land surface that is higher or lower than a given level, *usually sea level.

IAEA *See* International Atomic Energy Agency.

IAM *See* integrated area management.

IBI *See* Index of Biological Integrity.

ICDP *See* integrated conservation and development programme.

ice Frozen water.

ice age A time of widespread *glaciation. *See also* pleistocene.

ice cap A large dome-shaped area of ice that covers a large area of land, such as a mountain peak or a polar region, and is not confined to valleys. Smaller than an *ice sheet but usually larger than a *glacier.

ice cliff A wall of ice where a *glacier meets the sea, for example at the edge of an *ice shelf. Also known as **ice front**.

ice contact deposit Sediment that is deposited when *meltwater flows over, through, or under the stationary front of a melting *glacier. *See also* ESKER, KAME, KAME TERRACE.

ice core A column of layered ice which has been extracted from a *glacier or *ice sheet, which can be used to reconstruct past changes in air chemistry and climate.

ice cover The extent (particularly the thickness) of ice on a land surface, or the proportion of a sea surface that is covered with sea ice.

ice fall A steeply sloping part of a *glacier, where the ice tumbles down and is broken by *crevasses. *See also* OGIVE.

ice floe A large piece of floating ice. *See also* ICEBERG.

ice flow *See* GLACIER FLOW.

ice fog A type of *fog that is composed of minute ice particles, occurs in very low temperatures under clear, calm conditions in polar latitudes, and can produce a *halo around the *Sun or *Moon.

ice front *See* ICE CLIFF.

ice jam A build-up of floating *ice that blocks a narrow river *channel, and can cause local *flooding during a thaw in late winter or early spring.

ice nuclei *See* DEPOSITION NUCLEI.

ice pellet *See* SLEET.

ice sheet A large, thick body of glacial ice (larger than an *ice cap) that is not confined by the underlying *topography. Examples include Greenland and *Antarctica today and much of North America and Northern Europe during the *Pleistocene *ice age. Also known as **continental glacier**.

ice shelf A sheet of very thick glacial ice which has a more or less level surface, with one side attached to the land but most of it is floating over an *ocean or large *lake.

ice storm *Freezing rain that results in a build-up of ice on trees, power lines, and roads.

ice stream The rapidly flowing body of ice within a valley *glacier, which does not mix with other ice streams it comes into contact with, and which often deposits *lateral moraine and/or *medial moraine at its sides.

iceberg A large piece of floating ice, most of which is below sea level, that has broken off a *glacier. Larger and deeper than an *ice floe.

ICES *See* International Council for the Exploration of the Sea.

ichthyofauna The fish population of a particular area or time.

icing The forming or deposition of ice (*glaze) on a solid object.

ICM See INTEGRATED CATCHMENT MANAGEMENT.

ICPD See INTERNATIONAL CONFERENCE ON POPULATION AND DEVELOPMENT.

ICSU See INTERNATIONAL COUNCIL OF SCIENTIFIC UNIONS.

identified resource Deposits of a specific mineral *ore or *fossil fuel whose location, quantity, and quality are known or have been estimated from direct geological evidence.

idiobiont A form of *parasitism. Idiobiont species parasitize later *host stages. *Contrast* KOINOBIONT.

IDNDR See INTERNATIONAL DECADE FOR NATURAL DISASTER REDUCTION.

IEA See INTERNATIONAL ENERGY AGENCY.

IET See INTERNATIONAL EMISSIONS TRADING.

IGBP See INTERNATIONAL GEOSPHERE-BIOSPHERE PROGRAMME.

igneous rock A type of *rock that is formed when molten *magma (that is created by *igneous activity) cools and solidifies, either underground or on the surface. Examples include *basalt, *rhyolite, *andesite, *lava, and *granite. Also known as **primary rock** or **volcanic rock**. *See also* ROCK CYCLE.

ignimbrite A *pyroclastic flow deposit that contains material varying in size from *ash to *pumice clasts; it may be *unconsolidated or *cemented.

ignitable Capable of burning or causing a fire.

IIASA See INTERNATIONAL INSTITUTE FOR APPLIED SYSTEMS ANALYSIS.

IISD See INTERNATIONAL INSTITUTE FOR SUSTAINABLE DEVELOPMENT.

IJC See INTERNATIONAL JOINT COMMISSION.

Illinoian The third period of *glaciation in North America during the *Pleistocene *ice age, equivalent to the *Riss glaciation in Europe.

illness The state of being sick, because of stress, *disease, *accident, or injury.

illuvial horizon A soil *horizon in which material that is transported downwards by *eluviation from a horizon above, in solution or in suspension, has been deposited and accumulates. *See also* B-HORIZON.

illuviation A *soil-forming process that involves the movement of material (*humus, chemical substances, and fine *mineral particles), in solution or in suspension, down through the *soil profile from the *A-horizon and its deposition in the *B-horizon below.

imago The adult stage of an* insect.

imbrication The layering of inclined particles of *sediment or rock fragments against each other, rather like roof tiles overlap and lie parallel to one another. *See also* FABRIC.

immature soil A soil that lacks a well-developed *profile, usually because it has not had enough time for one to develop by normal *soil-forming processes.

immigration The movement of people or *organisms into a country or area; the opposite of *emigration.

imminent threat In *risk assessment, a high probability that exposure to an *agent is actually occurring or is likely to occur in the very near future.

immiscible Incapable of being mixed or blended together to form a *homogeneous mixture, such as oil and water. The opposite is *miscible.

immobilization 1. preventing movement to allow natural healing to take place.

2. the conversion of an element from the *inorganic to the *organic form, in the tissues of *plants or *micro-organisms, which makes it unavailable to other organisms or plants.

immunity A natural or acquired resistance (for example to a particular *disease) that is provided by a person's immune system. *See also* NATURAL IMMUNITY.

immunization The process by which a susceptible individual is protected against the adverse effects of infection by a disease-causing *micro-organism.

immunology The scientific study of *immunity and the immune system.

immutability The ability to withstand change, or the quality of being incapable of *mutation.

IMO *See* INTERNATIONAL MARITIME ORGANISATION.

impact 1. The direct or indirect changes, whether beneficial or adverse, that result from a specific act or series of acts, or a project or programme. Also known as **effect**. *See also* ECONOMIC IMPACT, ENVIRONMENTAL IMPACT, SOCIAL IMPACT.
2. The force of impression of one thing on another, such as a *meteorite hitting the surface of the *Earth.

impact assessment *See* ENVIRONMENTAL IMPACT ASSESSMENT.

impair To damage or make worse or less effective.

impairment Any adverse or damaging *impact on an *environment or *ecosystem which makes it less suitable for an intended use.

impeded drainage A restriction of the downward movement of water in *soil, which is usually caused either by *waterlogging or by the existence of a *hardpan.

imperilled species A general term used in North America which includes *endangered, *threatened, and *species at risk and *species of concern.

impermeable Not easily penetrated. An impermeable substance does not permit fluids to pass through it. *Contrast* PERMEABLE. *See also* IMPERVIOUS.

impervious A surface through which little or no water will move due to lack of pore space. *Contrast* PERVIOUS. *See also* IMPERMEABLE.

imports 1. commodities (goods or services) that are bought from a foreign country.
2. *Solid waste materials and *recyclables that have been transported from where they originated to another place for processing or final disposal.

important species *Species or *biotopes which are rare, have a restricted distribution, and/or are in decline; species which are listed for protection on *statutes, *directives, and *conventions.

impoundment A body of water (such as a *reservoir) that is confined by a barrier such as a dam or floodgate.

improved grassland Grassland in which yield is deliberately increased by *drainage, application of *fertilizers and/or *herbicides, *ploughing, and reseeding with fast-growing *varieties.

in situ On site or in its natural location. *Contrast* EX SITU.

in situ conservation The *conservation of *species in their natural *habitat. *Contrast* EX SITU conservation.

in situ treatment The treatment of a particular *waste material on site, where it is generated, rather than transporting it elsewhere for treatment.

in vitro In an artificial environment outside a living *organism. *Contrast* IN VIVO. *See also* CRYOPRESERVATION.

in vivo Within a living *organism. *Contrast* IN VITRO.

inbred Produced by *inbreeding.

inbreeding The *mating of close relatives, i.e. individuals who are likely to share some of their *genes due to common ancestry, which reduces genetic diversity. *See also* BREEDING, OUTBREEDING.

inbreeding depression The accumulation of harmful genetic traits (through random *mutations and *natural selection) that results from *inbreeding in a small *population, and that decreases the viability and reproductive success of enough individuals to adversely affect the whole population.

in-bye Enclosed *grassland that is located close to a *farm.

incentive-based regulation A government *regulation (such as a tradable *emissions allowance) that is designed to induce changes in the behaviour of individuals or firms, in order to produce environmental, social, or economic benefits that would otherwise be prescribed by *legislation.

inceptisol A young *order of soil in the early stages of *pedogenesis, in which the *horizons are starting to develop.

incidence The frequency of new *occurrences of a condition (such as the number of new cases of a *disease) in a *population over a defined period of time. *Contrast* PREVALENCE.

incidence rate The ratio of new cases (for example, of a *disease) in a *population to the total population at risk over a defined period of time.

incidental catch/take *See* BYCATCH.

incineration The process of burning *solid waste and other material, under controlled conditions, to produce *ash. *See also* MASS BURN, WASTE DISPOSAL.

incineration at sea The disposal of *toxic waste by burning at sea using specially designed *incinerator ships.

incinerator A furnace or chamber in which waste material is destroyed by burning. *See also* AFTERBURNER.

inclination The degree of dip or tilt from the vertical. *See also* DIP.

inclusion *See* XENOLITH.

income The flow of money, goods, or services that is created by the productive use of assets.

incommensurable Values or benefits that are not measurable by a common

standard (for example in dollars or cubic metres), and thus cannot be objectively compared. *Contrast* COMMENSURABLE.

incompatible Not *compatible, unable to exist together harmoniously.

incompatible use Different uses of land or other resources which cannot exist together in the same area because one inhibits or adversely affects another. *Contrast* COMPATIBLE USE.

incompatible waste A *waste material that is unsuitable for mixing with another material because it may react to create a *hazard.

increment 1. An increase in quantity, by a factor of one unit.
2. The increase in diameter, height, volume, weight, or value of individual trees or crops over a defined period of time.

incrementalism An approach to decision-making which is based on making decisions one at a time, each one designed to deal with short-term imperfections in an existing policy, rather than establishing long-term future goals.

incubation period The period of time between initial exposure to an infectious *agent and the appearance of the first sign or symptom of *disease. Also known as **latent period**.

Independent Commission on International Development A major international *commission, chaired by German Chancellor Willy Brandt, that examined the interrelationships between *environment and *development. It reported in 1980, coined the terms North (*developed countries) and South (*less developed countries), and called for the cancellation of old debts among countries of the *Third World.

indeterminacy Unpredictability in outcome, because a very large number of interrelated factors are involved and/or because understanding of the particular *system is still quite limited.

indeterminate species A *threatened category of species defined by the

*IUCN *Red Data Book as 'taxa known to be *extinct, *endangered, *vulnerable, or *rare but there is not enough information to say which of the four categories is appropriate'.

index fossil A *fossil that is specific to one geographical area or geological time and so can be used to identify and date rocks. Also known as **zone fossil**.

Index of Biological Integrity (IBI) A measure of *biological integrity for a particular site that is based on integrating a number (usually at least seven) of *indices describing site conditions.

index of sustainable economic welfare A measure for assessing the strength of an *economy and human well-being within a country, which (unlike *gross national product and similar *indicators) includes measures of environmental damage and reductions in *environmental quality.

index species *See* INDICATOR SPECIES.

index/indices A numerical scale (such as the *Celsius temperature scale) that is used to compare variables with one another or with a reference number.

Indian summer A period of unusually warm weather in mid to late autumn, with clear skies and cool nights, which usually follows a period of cool weather.

indicator 1. An *organism, *species, *community, or aspect of the environment whose characteristics show the presence of specific environmental conditions or *pollutants. *See also* BIOINDICATOR, ENVIRONMENTAL INDICATOR, POLLUTION INDICATOR, PROXY INDICATOR. **2.** A substance that shows a visible change, usually of colour, at a known point in a chemical reaction.

indicator species A *species whose presence or absence is an *indicator of environmental conditions in a *habitat or *community. Also known as **index species, management indicator species**.

indigenous Naturally occurring in a particular area. *Contrast* EXOTIC. *See also* NATIVE.

indigenous knowledge *See* TRADITIONAL KNOWLEDGE.

indigenous peoples The original or natural inhabitants of a country. For example, Native Americans are the indigenous peoples of the USA. Also known as **aborigines, native peoples, tribal peoples**.

indigenous species *See* NATIVE SPECIES.

indirect competition The use of a *resource by one individual in such a way that it reduces the availability of that resource to others.

indirect discharge The introduction of *pollutants into a *municipal waste treatment system from any non-domestic source, such as an industrial or commercial *facility.

indirect effect A secondary effect or *impact which occurs elsewhere and/or after the initial action. Examples include *acid rain, *bioaccumulation, and *global warming.

indirect exposure pathway An *exposure pathway that contains at least one intermediate release to any media between the source and the point of exposure. For example, *chemicals of concern can be transported by water from the *soil through *groundwater to the point of exposure.

individual risk The probability that a particular *individual within a *population will experience an adverse effect.

individual variation The diversity of *phenotypes within a *population.

Indomalayan realm A *biogeographical realm that is almost entirely *tropical forest, throughout much of South East Asia.

indoor air The breathable air within a building or structure.

indoor air pollution *pollution of *indoor air by chemical, biological, or physical *contaminants, such as *radon gas, *carbon monoxide, or tobacco smoke.

indoor climate The temperature, humidity, lighting, air flow, and noise levels inside a building or structure.

induced recharge Replenishing a waterbody or *aquifer using water from somewhere else.

induction *See* INDUCTIVE REASONING.

inductive reasoning Inferring general principles from specific examples. Also known as **induction**. *Compare* DEDUCTIVE REASONING.

indurated A *soil or *sedimentary rock which has become hardened or *cemented, and will not soften when wetted. *See also* HARDPAN, IRON PAN.

industrial biocatalyst A *biocatalyst (particularly an *enzyme) that is used in industrial processes.

industrial biotechnology The application of *biotechnology to create new and alternative products for consumers, such as chemicals, textiles, food and animal feed, pulp and paper, energy, metals, and minerals.

industrial chemical Any *chemical that is used in industrial processes, including *pharmaceuticals, *plastics, and *enzymes.

industrial ecology A framework by which industry and organizations can reduce their *environmental impacts by treating their operations as *ecosystems and monitoring and reducing the flow of materials and energy between different parts of their operation. This is designed to use materials more efficiently, reduce *waste, and prevent *pollution.

industrial forestry Large-scale, commercial tree planting in order to produce *timber and other wood products (such as wood chips).

industrial melanism The progressive *adaptation of a species, over a number of generations, to changing environmental factors. The classic example is the adaptation of the peppered moth (*Biston betularia*) to increased sooty *air pollution in Manchester between 1848 and 1895; white moths on dark surfaces were eaten by *predators, so darker moths became more common. *See also* EVOLUTION, MELANISM, NATURAL SELECTION.

industrial process waste Waste materials that are produced during manufacturing operations.

Industrial Revolution The rapid development of industry that began in Britain in the late 18th and early 19th centuries, assisted by the introduction of machinery and the development of a market economy, and which spread to other countries.

industrial sector That part of the *economy that is based on manufacturing industries, mining, construction, agriculture, fisheries, and forestry. *See also* COMMERCIAL SECTOR, RESIDENTIAL SECTOR, TRANSPORTATION SECTOR.

industrial sludge A semi-liquid residue or *slurry that remains after the treatment of *industrial water and *wastewater.

industrial source reduction Measures and equipment that reduce the amount of a *hazardous substance, *pollutant, or *contaminant that is produced by industry and enters any waste stream or is released into the *environment, in order to protect the environment and *public health.

industrial timber Trees grown for wood that is used for *lumber, plywood, veneer, particleboard, chipboard, and paper. Also known as **roundwood**.

industrial waste Any unwanted materials from an industrial operation or activity, which includes liquid, *sludge, and solid and *hazardous waste.

industrialization The development of *industry on a large scale. A stage in the development of the *economy and *society in a particular country during which resources are shifted from *agriculture to *manufacturing. *See also* DEMOGRAPHIC TRANSITION.

industrialized countries Nations (such as the USA, Japan, and the countries of Europe) which have economies based on industrial production and which use large amounts of energy (particularly *fossil fuels and *nuclear power).

industry The organized process of *manufacturing goods for sale.

inequality The state of being different in terms of quantity, value, or status. Contrast EQUALITY.

inequitable That which is unfair, partial, and unjust, and which does not provide equal opportunity for all. Contrast EQUITABLE.

inequity Unfairness arising from the unequal use and allocation of resources. Contrast EQUITY.

inert A material that is very stable and does not readily take part in chemical reactions with other substances.

inert gas Any one of six gases (*helium, *neon, *argon, *krypton, *xenon, and *radon) that are almost completely chemically inactive. Also known as **rare gas** or **noble gas**.

inertia The tendency of an object to remain at the same velocity, or at rest, unless a force acts on it.

inerts A collective term for *non-biodegradable products (such as *glass or *plastics) that are contained in wastes.

infant mortality The number of infants who die within the first year of life, per 1000 live births.

infanticide The killing of young by the parent or sibling.

infauna *benthic organisms that live in the *sediment on the floor or a water-body such as a lake or sea.

infection An incident in which an *infectious agent is transmitted and invades the body of people or animals.

infectious agent Any micro-organism (such as a *pathogenic *virus, *parasite, or *bacterium) that has the ability to invade body tissues, multiply, and cause disease.

infectious disease A *disease that is caused by *pathogenic organisms.

infectious waste *Hazardous waste that is capable of causing *infections in humans. This includes contaminated animal waste, human blood and blood products, isolation waste, and pathological waste and discarded syringes, needles, and blades. Also known as **red bag waste**.

inference A logical conclusion that is supported by evidence.

infest 1. To occupy in large numbers. 2. To live on a host. See also PARASITE.

infestation A troublesome invasion by *parasites or *pests.

infill The process of developing open zones within an established area (such as *brownfield sites within a town) before developing outside the established area (such as a *greenfield sites).

infiltration The movement of water from the ground surface into a *soil or into a *porous *rock or *sediment. Contrast PERCOLATION.

infiltration capacity The maximum rate at which water can enter the soil at a particular point, under a given set of conditions.

infiltration rate The amount of water that enters the soil at a particular point in a given period of time.

inflation The average rate at which the general price of goods and services increases through time.

inflow 1. The natural flow of water into a *lake or *reservoir from upstream *tributaries. 2. The flow of rainwater into a *sewer system from drains and sewers.

influent 1. A stream that enters a large or larger river. 2. A waste product (such as water that contains *pollutants) that enters a particular part of the environment, often accidentally. Means literally 'flowing in'. *Contrast* EFFLUENT.

informal outdoor recreation *Recreation activities that take place outside and are not directly managed. *See also* DISPERSED RECREATION, DEVELOPED RECREATION.

informal settlement Houses (for temporary or permanent use) which have been built on land without formal planning approval.

informatics The use of computers and statistical methods in the classification, storage, retrieval, and analysis of data, information, and knowledge. *See also* BIOINFORMATICS, DATA MINING, GEOINFORMATICS.

information *Data that have been transformed through analysis and interpretation into a form which is useful for drawing conclusions and making decisions. *See also* MANAGEMENT INFORMATION.

information management The administration, use, and transmission of information and the use of theories and methods of *information science to create, modify, or improve systems for handling information. *See also* KNOWLEDGE MANAGEMENT.

information science The study of the creation, use, and management of information.

infrared (IR) *Radiant energy that has a *wavelength of 1000 to 0.77 *micrometres within the *electromagnetic spectrum, between *visible light and *microwaves. Most of the energy that is emitted by the Earth and its atmosphere is infrared. Gases such as *water vapour, *carbon dioxide, and *ozone absorb infrared radiation, which creates the *greenhouse effect. Also known as **thermal radiation** and **longwave radiation**.

infrastructure The basic services and facilities (such as roads and sewers) that are needed for the development and growth of an area or activity.

ingestion Taken into the body via the mouth, after which it is digested by swallowing. *Contrast* ABSORPTION.

inhalation Breathing in a substance in the form of a gas, vapour, fume, mist, or dust.

inherent *See* INTRINSIC.

inherent value *See* INTRINSIC VALUE.

inherit 1. To receive *genetic material from parents through biological processes. *See also* HEREDITY.
2. To receive an item from a previous owner (often a relative) upon their death. *See also* LEGACY.

inholding General term for private land within public parks, forests, or wildlife refuges.

injection well A *well used that is for injecting water or other fluid into a groundwater *aquifer, by pressure.

injection zone A geological formation that receives fluids through a *well.

inland waters Waterbodies (such as *lakes, *streams, *rivers, *canals, waterways, inlets, and bays) that have no direct access to the ocean.

inlet An entrance to a *waterbody, such as an *estuary or *bay.

inner core *See* CORE.

innoculation *See* VACCINATION.

innovative treatment technology New methods to effectively treat *hazardous waste, which can offer cost-effective, long-term solutions to *cleanup problems, may provide an alternative to land disposal or incineration, and are often more acceptable to local communities than some established treatment technologies.

inorganic A substance (such as *metals, *minerals, and *rocks) that does not contain *carbon and is not derived from biological material. *Contrast* ORGANIC.

inorganic chemistry The chemistry of substances that do not contain *carbon. *Contrast* ORGANIC CHEMISTRY.

inorganic compound Any compound that does not contain *carbon. *Contrast* ORGANIC COMPOUND.

inorganic nitrogen *Nitrogen which is in the *mineral state and can be readily used by plants. *Contrast* ORGANIC NITROGEN.

inorganic waste Waste (such as *sand, *glass, or any other *synthetic material) that consists of materials other than plant or animal matter.

input Energy, matter, or information that is put into a *system in order to achieve a result. *Contrast* OUTPUT.

INQUA *See* INTERNATIONAL UNION FOR QUATERNARY RESEARCH.

inquiline An *organism that lives on or in the body of another, or in its nest or abode, without benefit or damage to the other organism from which is gets shelter and sometimes food.

insect A small, air-breathing, *invertebrate animal (*arthropod) that has three body parts and three pairs of jointed legs; some species of insect have wings. Examples include bees, ants, and moths.

insecticidal oil A light-weight oil that smothers plants, insect eggs, and larvae on contact, which offers an environmentally sound means of controlling *insects.

insecticidal soap A *soap that is derived from fats or oils that have been treated with a specific kind of alkaline substance (such as *potash), which paralyses and eventually kills insect pests (such as aphids) when sprayed on them. A common form of *insecticide.

insecticide A chemical *pesticide that is used to control or kill *insects.

insecticide resistance *See* PESTICIDE RESISTANCE.

insectivorous Insect-eating, feeding largely or exclusively on *insects.

inselberg A steep-sided *hill composed mainly of hard rock, which rises sharply above a *plain in tropical and subtropical areas.

inshore Towards the *shore, or close to a shore, where waves are transformed by interaction with the sea bed. *Contrast* OFFSHORE.

insolation Incoming *solar radiation, the *radiation from the *Sun that enters the Earth's *atmosphere. Around a third is returned to *space (21% is reflected back into space from *clouds, 6% is reflected back from the ground, and 5% is scattered by tiny particles in the air as *diffuse radiation); half reaches the ground (30% by *direct radiation and 20% by diffuse radiation); and 18% is absorbed in the atmosphere (3% by clouds and 15% by *dust and the *ozone layer).

insolation weathering A type of *physical weathering which involves repeated heating and cooling of rock over daily cycles, progressively breaking apart the grains of rock.

insoluble Not capable of being dissolved. *Contrast* SOLUBLE.

instability A condition that exists when a rising parcel of air in the *atmosphere continues to rise because it is warmer, and thus lighter, than the air around it (the opposite of *stability). *Precipitation can only occur in such unstable air conditions, which allow a parcel of air to cool to the *dew point, *condensation to occur, and *water vapour to condense around the *condensation nuclei to produce the initial tiny water droplets that eventually coalesce (through mixing of the air) into *raindrops.

institutional waste *Waste material that is generated at institutions such as schools, libraries, hospitals, and prisons.

instream flow The quantity of water that is required in a stream in order to meet the needs of fish, wildlife, and recreational users.

instream use Water that is used within the *stream channel, for uses such as *hydroelectric power gener-

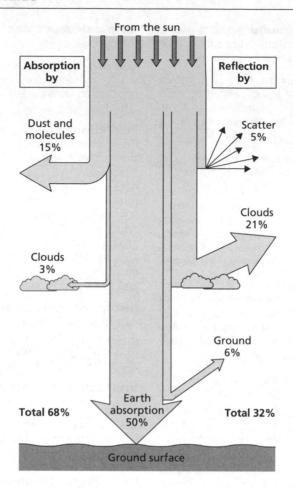

Fig 10 Insolation

ation, navigation, *fish farming, and *recreation. *Contrast* OFFSTREAM USE.

instrumental value The *value or worth of objects that provide a means to some desirable end, that satisfy some human needs and wants. *See also* INTRINSIC VALUE.

insufficiently known species A category of species defined by the *IUCN *Red Data Book as taxa that are suspected but not definitely known to belong to any of the at-risk categories (*endangered, *extinct, *extirpated, *threatened, *vulnerable), because of lack of information. *See* UNKNOWN (THREAT) CATEGORY OF SPECIES.

intake 1. A structure through which water enters a *reservoir.
2. The process (*ingestion or *inhalation) by which a substance enters a body without being absorbed through its exterior. *See also* ADMINISTERED DOSE.

intangible Incapable of being touched or seen, having no physical or material presence.

intangible resource Assets such as open space, attractiveness, diversity, and satisfaction that are not material or physical. Also known as **intangibles**.

intangibles *See* INTANGIBLE RESOURCE.

integrated area management (IAM) An approach to the management of *natural resources in which a specific area is zoned and regulated for a variety of uses (such as research, *conservation, tourism, harvesting, *hunting, or *fishing) that is compatible with the management goals for the area.

integrated assessment An assessment that brings together data about the past, current, and future condition of *ecosystems, the impact of humans on the ecosystems, and the effects of *environmental change on humans. This form of analysis is regularly carried out to predict the likely impact of environmental change (such as *global warming) and to evaluate policy responses to it. See also ENVIRONMENTAL ASSESSMENT.

integrated catchment management (ICM) A *systems approach to the management of *natural resources within a *drainage basin, which includes consideration of all environmental, economic, and social issues within an overall management plan. See also INTEGRATED WATER RESOURCE MANAGEMENT.

integrated conservation and development programme (ICDP) An approach to *environmental management that is based on linking nature conservation in *protected areas (such as *national parks) with local social and economic development, with a view to making *biodiversity conservation more effective, increasing local community participation in conservation and development, and increasing economic development for the rural poor.

integrated exposure assessment An estimate of the total *exposure of a person or population to a particular *toxic chemical, through all media, over a given period of time.

integrated pest management (IPM) An *ecological approach to the management of *pests that relies on natural mortality factors (such as natural enemies, weather, and cultural control methods) and carefully applied doses of *pesticides.

integrated pollution control (IPC) An approach to the control of *pollution by industry in the UK, in which any person who carries out a prescribed process must obtain authorization from the *environmental regulator, and must comply with conditions set by the environmental regulator. This has been replaced since 2000 by *Pollution Prevention and Control.

Integrated Pollution Prevention and Control (IPPC) Directive An *EU *directive that was introduced in 2000, which promotes an integrated approach to controlling the *environmental impacts of certain industrial activities, and involves determining the appropriate controls for industry to protect the environment through a single permit process. See also POLLUTION PREVENTION AND CONTROL.

integrated resource management The comprehensive management of two or more *natural resources (including water, soil, timber, range, fish, wildlife, and recreation) in the same area, in order to optimize the overall benefits on a sustainable basis. Also known as **integrated resource planning, multiple-use management**.

integrated resource planning See INTEGRATED RESOURCE MANAGEMENT.

integrated waste management (IWM) An approach to the *sustainable management of *waste, based on environmentally sound processes such as *source reduction, *recycling, *resource recovery, *incineration, and *landfill.

integrated water resource management (IWRM) An approach to the *sustainable management of *water resources in a *drainage basin which recognizes the need to preserve the integrity of the basin *water cycle, and respects the interrelationships between different parts of that cycle. See also INTEGRATED CATCHMENT MANAGEMENT, INTEGRATED RESOURCE MANAGEMENT.

integrity Completeness or totality, unmodified and not reduced. *See also* BIOLOGICAL INTEGRITY.

integumentary system In animals, the covering of skin (which includes hair, nails, scales, and feathers) which supports and protects the body.

intellectual capital The overall body of *knowledge that exists within an organization in terms of data, information, and wisdom, which is held in the minds and experience of people, creates intelligence, and helps to drive creativity and innovation.

intellectual property *Knowledge that stems from the exercise of the mind, is recognized and protected under law (for example by patent, trade mark, design, copyright, and plant breeder's rights), and can be given a monetary value and made subject to contractual agreements such as licensing.

intellectual property right (IPR) The major legal mechanisms by which *intellectual property is protected (particularly copyright, patent, and trademark), which enable owners to select who may access and use their property, and thus protect it from unauthorized use.

intensive agriculture *Agriculture which involves intensive management of land, designed to maximize output through the use of chemicals (*fertilizers) and machinery, the reseeding of *grassland, clearance of unproductive areas (such as *hedges and small *copses), and drainage of wetland. *See also* AGRICULTURAL INTENSIFICATION.

intensive recreation High-density *concentrated recreation activities, such as developed campgrounds and picnic sites, swimming beaches, hiking trails that are major access routes into wilderness areas, and areas used by all-terrain vehicles. *Contrast* DISPERSED RECREATION.

interbasin transfer The diversion and conveyance of water from one *drainage basin to another, usually via pipes and *aqueducts. Also known as **trans-basin diversion**.

interception The *precipitation that lands on the leaves and branches of plants, and is then evaporated back to the atmosphere as *water vapour.

intercropping Growing two or more crops in the same field at the same time, either mixed together or in alternating rows, in order to protect the soil and use land efficiently. Also known as **interplanting**.

interdisciplinary Applying the knowledge and skills from different academic disciplines or subjects that are normally regarded as distinct, to the same task or project. *See also* MULTIDISCIPLINARY.

interest group *See* PRESSURE GROUP.

interface A shared *boundary, for example between liquids, solids, and gasses, or between two systems (for example where a river meets the sea).

interference competition *Competition in which one *species prevents another from having access to a limiting *resource. *See also* EXPLOITATION COMPETITION.

interflow Water that moves downslope within a *soil, through the *pore spaces within the *soil structure. *Contrast* THROUGHFLOW.

interfluve *See* DRAINAGE DIVIDE.

intergenerational equity The principle that the present generation should pass on to future generations enough *natural resources and sufficient *environmental quality that they can enjoy at least a comparable quality of life, and inherit a healthy and sustainable *environmental heritage. It seeks a fair distribution of the costs and benefits of a long-term *environmental policy, when costs and benefits are borne by different generations. For example, the costs of policies designed to address the problem of *global warming by reducing *emissions of *greenhouse gases will be largely carried by the present and next few generations, whereas generations in the more distant future will enjoy the long-term benefits.

interglacial A period of increased temperatures between *ice ages, or between major phases of *glaciation or ice advance. The present *Holocene period is an interglacial. *See also* INTERSTADIAL.

Intergovernmental Panel on Climate Change (IPCC) A major international scientific collaboration between hundreds of specialists from around the world, that focuses on the likelihood and probable nature of induced *climate change, based largely on forecasts from *general circulation models. It was established in 1988 by the *World Meteorological Organization and the *United Nations Environment Programme to assess the scientific, technical, and socioeconomic information needed to understand the risk of human-induced climate change. The IPCC has three working groups dedicated to particular aspects of climate change. *Working Group I: The science of climate change.* This group works to assess the available information on the science of climate change, and is concerned with developments in the scientific understanding of past and present climate, of climate variability, of climate predictability and of climate change modelling (including feedback from climate impacts), progress in the modelling and projection of global and regional climate and sea level change, observations of climate, including past climates, and assessment of trends and anomalies and gaps and uncertainties in current knowledge. *Working Group II: Impacts, adaptation vulnerability.* This group's work focuses on assessing the scientific, technical, environmental, economic, and social aspects of the vulnerability (sensitivity and adaptability) to climate change of, and the negative and positive consequences for, *ecological systems, socioeconomic sectors, and human health, with an emphasis on regional sectoral and cross-sectoral issues. *Work-*

ing Group III: Mitigation of climate change. This group is concerned particularly with the scientific, technical, environmental, economic, and social aspects of mitigation of climate change. The IPCC has concluded that 'the balance of evidence suggests that there is a discernible human influence on global climate'.

Interior Department *See* US DEPARTMENT OF THE INTERIOR.

interior drainage A *drainage pattern in which streams converge in a closed *basin and evaporate without reaching the sea.

intermediate A *crown class of trees in a *forest.

intermediate cut The removal of immature trees from a *forest, often by thinning, before the major harvest, in order to improve the quality of the remaining forest stand. *Contrast* HARVEST CUT.

intermediate disturbance hypothesis The belief that, within a given *ecosystem, the highest species *diversity exists under conditions of intermediate disturbance because too much disturbance causes local extinction (*extirpation) and at low levels of disturbance competitive species exclude others.

intermediate rock *Igneous rocks that have a *silica content between the extremes of the *mafic and *felsic rocks, and weather down to produce quite fertile soils. Intermediate between *basic and *acidic rocks.

intermediate technology *Technology that can be made at an affordable price, by ordinary people, using local materials, in ways that minimize damage to people and the environment. Also known as **appropriate technology**.

intermediate-level waste *Radioactive waste that contains higher levels of *radioactivity than *low-level waste, and less than *high-level waste.

intermittent Occurring from time to time. Also known as **episodic**.

intermittent flow River flow that is irregular and only occurs after heavy rain or *snow melt. *Contrast* EPHEMERAL FLOW, PERENNIAL FLOW.

intermittent stream A *stream that flows only at certain times of the year when it receives water from upstream, from heavy rain or *snow melt.

Intermodal Surface Transportation Efficiency Act (ISTEA) (1991) US legislation that authorized funding for highways, highway safety, and mass transportation between 1991 and 1997, when it expired, but much of the programme was carried forward by the *Transportation Equity Act for the 21st Century.

intermontane basin A low-lying area (*basin) between mountain ranges, such as a *graben.

internal combustion engine An engine in which the fuel burns inside the engine itself, as opposed to a steam engine in which fuel is burned in a separate furnace. Most vehicles use *gasoline as fuel, and emit *air pollutants such as *carbon monoxide, *nitrogen oxides, and reactive *hydrocarbons.

internal cost The direct cost (monetary or otherwise) that is met by those who use a resource.

internal dose *See* ABSORBED DOSE.

internal energy The *energy which a substance possesses because of the motion and configuration of its *atoms, *molecules, and *subatomic particles.

internal migration The movement (*migration) of people or animals within a country or defined region. *Contrast* INTERNATIONAL MIGRATION.

internal trading In *greenhouse gas *emissions trading, internal trading is trading within a company which allows the company to trade *emission permits between its business units in order to maximize cost-effective emission control.

internal waters Any body of surface water that is under the exclusive control of a coastal nation, including *bays, *estuaries, and *rivers. *Contrast* INTERNATIONAL WATERS, INTERSTATE WATERS.

internalization In *accounting, the process of including in financial balance sheets the cost factors which were previously either not calculated (for example the cost of *environmental damage) or met by society at large (for example the health costs associated with pollution). *See also* EXTERNALITY.

International Atomic Energy Agency (IAEA) A United Nations agency that was formed in 1957, is based in Vienna, Austria, and is responsible for supervising the use of nuclear material, developing safety standards, preventing the proliferation of *nuclear weapons, and promoting the peaceful use of *nuclear energy.

International Bank for Reconstruction and Development *See* WORLD BANK.

International Conference on Population and Development (ICPD), Cairo (1994) A major international conference organized by the *United Nations that met in Cairo, Egypt in September 1994, and discussed the global dimensions of *population growth and change. The conference agreed goals for 2015 that are designed to improve individual and family well-being and enhance the status of women. The goals include universal access to *family planning and primary school education, increased access by girls and women to secondary and higher education, and reductions in infant, child, and maternal mortality. The conference is widely credited as a major turning-point in establishing international consensus on effective ways of slowing the pace of population growth and improving *quality of life, by addressing root causes of unwanted *fertility.

International Conference on Population, Mexico City (1984) An international conference organized by the *United Nations which built on the success of the *World Population Conference in Bucharest 1974, was attended by representatives from 150 countries, provided a further impetus for the *World Population Plan of Action, and was a key milestone in the development of the 1994 *International Conference on Population and Development (ICPD) held in Cairo.

international convention See CONVENTION.

International Convention for the Prevention of Pollution from Ships (1973) An international *convention that covers the prevention of pollution of the marine environment by ships, from operational or accidental causes. The convention was modified by a protocol in 1978.

International Convention for the Regulation of Whales (1946) An international *convention that was signed in Washington, DC in December 1946, and is designed to safeguard whale stocks for future generations. It set up the *International Whaling Commission and introduced regulations concerning: the conservation and utilization of whale resources, relating to protected and unprotected species; open and closed seasons; open and closed waters; the designation of sanctuary areas; size limits for each species; time, methods, and intensity of whaling (including the maximum catch of whales to be taken in any one season); types and specifications of gear and apparatus and appliances that may be used; methods of measurement; and catch returns and other statistical and biological records.

International Convention on Oil Pollution Preparedness, Response and Co-operation (1990) An international *convention that provides a global framework for international co-operation in combating major incidents or threats of marine *pollution.

International Convention on the Control of Harmful Anti-fouling Systems on Ships (2001) An international *convention that prohibits the use of harmful *organotins in antifouling paints used on ships, and establishes

a mechanism to prevent the potential future use of other harmful substances in antifouling systems.

International Convention Relating to Intervention on the High Seas in Cases of Oil Pollution Casualties (1969) An international *convention that affirms the right of a coastal state to take such measures on the *high seas as may be necessary to prevent, mitigate, or eliminate danger to its coastline or related interests from pollution by oil or the threat of such pollution, after a maritime accident.

International Council for the Exploration of the Sea (ICES) An intergovernmental organization concerned with marine and fisheries science. It was established in Copenhagen in 1902 to promote the exchange of information on the sea and its living resources and to promote and coordinate marine research between member countries.

International Council of Scientific Unions (ICSU) A non-governmental organization that was founded in 1932 to bring together scientists in international scientific endeavours in order to help members address major international, interdisciplinary issues.

international credit In *greenhouse gas *emissions trading, under the *Kyoto Protocol, a method by which emissions credits or permits can be transferred between countries through *Joint Implementation, *Clean Development Mechanisms, and *international emissions trading. Also known as **international permit**.

International Date Line A line of *longitude that lies generally 180° east and west of the prime meridian (*Greenwich Meridian); the date is one day earlier to the east of the line.

International Decade for Natural Disaster Reduction (IDNDR) An international programme of the *United Nations, between 1990 and 2000, that sought to harness the political resolve, experience, and expertise of each country in order to reduce the loss of life, human suffering, and economic losses caused by *natural hazards. Some scientists point out that many of the IDNDR targets complement the objectives of *Agenda 21, although few planners and decision-makers yet build upon the links between successful *disaster management and *sustainable development.

international emissions trading (IET) The international trading of *greenhouse gas permits (*assigned amount units), under the *Kyoto Protocol, which allows developed countries to meet their assigned amounts. See also EMISSIONS TRADING.

International Energy Agency (IEA) An intergovernmental organization that was formed in 1973 by major oil-consuming nations in order to manage the security of *energy supply, *economic growth, and environmental *sustainability through energy.

International Geosphere-Biosphere Programme (IGBP) A major international programme of research that began in the early 1980s and was designed to describe and understand the interacting physical, chemical, and biological processes that regulate the Earth's *environmental systems. It was established by the *International Council of Scientific Unions (ICSU) in 1986, and its activities are focused on a series of core projects, which are the International Global Atmospheric Chemistry Project, Stratosphere–Troposphere Interactions and the Biosphere, the Joint Global Ocean Flux Study, the Global Ocean Euphotic Zone Study, Land–Ocean Interactions in the Coastal Zone, Biospheric Aspects of the Hydrological Cycle, Global Change and Terrestrial Ecosystems, Past Global Changes, and Global Analysis, Interpretation and Modelling. *Mission to Planet Earth is an important component of the IGBP.

International Human Suffering Index (HIS) A composite index of economic and social well-being that can be used to show changes in human well-being, based on a number of key variables including life expectancy, daily calorie intake, clean drinking water, infant immunization, secondary school enrolment, *GDP per capita, rate of inflation, communications technology, political freedom, and civil rights.

International Institute for Applied Systems Analysis (IIASA) A non-governmental research organization located near Vienna in Austria, that conducts *interdisciplinary scientific studies on environmental, economic, technological, and social issues in the context of the human dimensions of global change.

International Institute for Sustainable Development (IISD) A non-governmental organization based in Winnipeg, Canada, that advances policy recommendations on international trade, economic instruments, *climate change, and *natural resource management in order to make development *sustainable.

International Joint Commission (IJC) An independent binational organization established by the *Boundary Water Treaty of 1909, whose purpose is to help prevent and resolve disputes relating to the use and quality of *boundary water between Canada and the USA.

International Maritime Organisation (IMO) An intergovernmental body that is concerned with setting international *standards in the marine environment, particularly to improve maritime safety and to prevent marine *pollution. The IMO is responsible for a number of UN conventions on marine pollution—the *Convention on the Prevention of Marine Pollution by Dumping of Wastes and Other Matter (1972), the *International Convention for the Prevention of Pollution from Ships (1973), the *International Convention on the Control of Harmful Anti-fouling Systems on Ships (2001), the *International Convention on Oil Pollution Preparedness, Response and Co-operation (1990), the *International Convention Relating to Intervention on the High Seas in Cases of Oil Pollution Casualties (1969), and the *Protocol on Preparedness, Response and Co-operation to Pollution Incidents by Hazardous and Noxious Substances (2000).

international migration The movement (*migration) of people or animals across a national border, from one country to another. *Contrast* INTERNAL MIGRATION.

International Organization for Standardization (ISO) A network of national standards institutes from 145 countries. *See also* ISO 9000, ISO 14001.

international permit *See* INTERNATIONAL CREDIT.

International Planned Parenthood Federation (IPPF) An international organization made up of autonomous family planning associations (FPAs) in over 180 countries around the world, which are concerned with *family planning and sexual and reproductive health.

international pollution *See* TRANS-FRONTIER POLLUTION.

international regime An international agreement between countries.

International Union for Conservation of Nature and Natural Resources (IUCN) *See* WORLD CONSERVATION UNION.

International Union for Quaternary Research (INQUA) An international scientific organization that was founded in 1928 for the study of *Quaternary geology, climate, biology, and anthropology.

international waters The open seas outside the territorial waters of any individual nation. *Contrast* INTERNAL WATERS, INTERSTATE WATERS.

International Whaling Commission (IWC) An international organization that was set up under the *International Convention for the Regulation of Whales in 1946 in order to regulate the whaling industry and conserve *whales.

interplanting *See* INTERCROPPING.

interpluvial A period during which the *climate is relatively dry, lasting for decades or longer, between *pluvials.

interpolation Estimating values between known numerical values, for example in order to draw an *isopleth. *Contrast* EXTRAPOLATION.

interpretation 1. An explanation of the meaning of data.
2. The provision of scientific and academic information about a site or area in formats (such as fixed signs) that are meaningful to non-specialists and help people to understand and appreciate the natural and cultural heritage.

interseeding Sowing seeds into existing vegetation.

interspecific competition *Competition for resources between individuals of different species in a given *community. *Contrast* INTRASPECIFIC COMPETITION.

interstade A brief period of glacial retreat (*deglaciation) caused by warming within a glacial stage. *Contrast* STADE.

interstadial A relatively short phase (50–200 years) of warmer *climate within a *glacial period; cooler and shorter than an *interglacial.

interstate pollution Pollution (such as *acid rain) that moves across the borders between states within a country. *Contrast* TRANS-FRONTIER POLLUTION.

interstate waters Bodies of *surface water that flow across or form part of state or international boundaries, such as the *Great Lakes and the Mississippi River. *Contrast* INTERNAL WATERS, INTERNATIONAL WATERS.

interstices The space between two or more objects, such as the *pore spaces between the individual grains in a *soil or *rock.

interstitial Relating to *interstices.

intertidal The zone of the *shore that lies between the *high and *low water marks, which is submerged at *high tide and exposed at *low tide.

Intertropical Convergence Zone (ITCZ) The zone of converging *trade winds along the *equator, which causes rising air currents, low *atmospheric pressure, and often a continuous band of *clouds or *thunderstorms. It provides the upward motion in the *Hadley cell.

intolerant In ecology, unable to withstand adverse environmental conditions such as shade or drought. *Contrast* TOLERANT.

intraspecific competition *Competition for *resources between members of the same *species within a *community. *Contrast* INTERSPECIFIC COMPETITION.

intrazonal soil A *soil that is untypical of its *climate zone because it has been strongly affected by more local factors such as *topography and parent material. *Contrast* ZONAL SOIL, AZONAL SOIL. *See also* SOIL ORDER.

intrinsic Inside, or belonging to something by its very nature. Also known as **inherent**. *Contrast* EXTRINSIC.

intrinsic resource *See* NATURAL RESOURCE.

intrinsic value Ethical *values or rights that exist as an *intrinsic characteristic of a particular thing or class of things simply because of their existence. Also known as **inherent value**. *Contrast* ACQUIRED VALUE, EXTRINSIC VALUE, INSTRUMENTAL VALUE.

introduced species A *species that has been brought into an area where it does not naturally occur, outside its known *historic range, usually by humans, and which is able to survive and reproduce. Also known as **alien, exotic species,** or **non-native species**. *See also* INVASIVE SPECIES, NATIVE SPECIES.

intrusion A body of *igneous rock that is formed by the injection of molten *magma into pre-existing rocks beneath

the *Earth's surface. Also known as a **pluton**.

intrusive Molten *magma that is forced into cracks between layers of other rock. Also known as **plutonic**. *Contrast* EXTRUSIVE.

inundation The process of being covered with standing or slow-moving water. *See also* FLOOD.

invader *See* INVASIVE SPECIES.

invasive species An aggressive *introduced species which spreads and dominates its new location, competing with and often replacing *native species and proving difficult to remove. Also known as **invader**. *See also* INTRODUCED SPECIES.

inventory A detailed list of items, such as the register of *sources and *sinks of *greenhouse gases in a particular country.

inversion An increase in air temperature with height. This can cause a layer of warm air in the *atmosphere that prevents the rise of relatively cool air, which traps *pollutants beneath it and can cause an air *pollution episode. *See also* ADVECTION INVERSION.

invertebrate An animal that has no backbone or spinal column. Examples include *arthropods, *insects, and *molluscs. *See also* MACROINVERTEBRATE, MICROINVERTEBRATE. *Contrast* VERTEBRATE.

iodine (I) A non-metallic *halogen element that is required by humans in small amounts for healthy growth and development, and is often used in medicine as an antiseptic. It kills bacteria and prevents algal growth, is found naturally in *seawater, and is necessary in small quantities for reef invertebrates such as *corals and clams.

ion An *atom or group of atoms that has acquired an electric charge by the loss or gain of one or more *electrons.

ion exchange A reversible *chemical reaction between a solid and a solution by means of which *ions of the same charge may be exchanged between the two.

ion exchange treatment A common method for making *hard water softer, based on adding calcium oxide or calcium hydroxide which increases *pH to a level at which *metals *precipitate out.

ionic bond A force that holds together two electrically charged atoms (*ions).

ionic strength A measure of the concentration and charge of *ions in a solution, which affects the *solubility of *compounds.

ionization The production of *ions.

ionizing radiation High-energy radiation (with wavelengths shorter than those of visible light, e.g. *gamma rays, *X-rays, or *ultraviolet) that is capable of causing *ionization in the matter through which it passes and can damage living tissue. *See also* IRRADIATION.

ionosphere The outer parts of the *atmosphere, between about 60 and 1000 kilometres above the Earth, above the *mesosphere.

IPAT An equation (I = P × A × T) that was developed by Paul *Ehrlich and John Holdren in 1972 to describe how impact (I) or *environmental change is a function of population size (P), affluence (A), and *technology (T). Although the I = PAT formula is a useful way of studying the relationship between factors that govern environmental change, critics have pointed to two important weaknesses in this sort of approach. First, the factors contributing to any particular impact can vary a great deal, depending on the *environmental impact in question (for example, different factors contribute to depletion of the *ozone layer and to loss of *biodiversity). Secondly, the equation suggests that the three factors (P, A, and T) operate independently, whilst in reality they may well interact with each other.

IPC *See* INTEGRATED POLLUTION CONTROL.

IPCC *See* INTERGOVERNMENTAL PANEL ON CLIMATE CHANGE.

IPM *See* INTEGRATED PEST MANAGEMENT.

IPPF *See* INTERNATIONAL PLANNED PARENTHOOD FEDERATION.

IPR *See* INTELLECTUAL PROPERTY RIGHTS.

IR *See* INFRARED.

IRA Inventoried roadless area in US *national forest lands, defined under the *Roadless Area Review and Evaluation survey.

iridescence Brilliant spots or borders of colours (most often red and green) that can be seen in high- or medium-level *clouds (particularly in thin *cirrus clouds), which are caused by *diffraction of light by small cloud particles.

iron (Fe) A *malleable metal that occurs naturally and is the fourth most abundant *element in the Earth's *crust. It is a *trace element (*micronutrient) for plants and an essential *nutrient for humans.

Iron Age The *prehistoric period of human culture which began in Europe around 1000 BC (after the *Bronze Age), during which *iron was the main material used for making tools and weapons.

iron formation A *sedimentary rock that has a high *iron content, usually more than 15%, as iron sulphide, oxide, carbonate, or silicate.

iron oxide (Fe$_2$O$_3$) A compound, whose proper name is iron (III) oxide, that consists of *iron combined with *oxygen, which forms the rust that is visible on iron or steel left exposed to oxygen and moisture in the air. *See also* LATERIZATION.

iron pan Soil in which *iron compounds have been washed from the upper *horizon and deposited as a *hardpan in a lower horizon. *See also* INDURATED.

iron triangle A US term for a close relationship between an agency, a congressional committee, and an interest group that can grow into a mutually advantageous alliance.

irradiated food Food that has been briefly exposed to *radioactivity (usually *gamma rays) on purpose, in order to kill *insects, *bacteria, and *mould, and to allow storage without refrigeration.

irradiation Exposure to *ionizing radiation for medical purposes, to sterilize milk or other foodstuffs, or to induce chemical changes (such as the vulcanization of rubber).

irreplaceability In *conservation assessment, *habitat features, *biotopes, or *species that cannot be replaced if they are destroyed in some way.

irretrievable Not retrievable, impossible to recover; a *natural resource (such as a *stream, *fishery, *wetland, or *wildlife habitat) which would be lost forever as a consequence of a particular activity. One of the categories of *impact that must be included in *environmental impact statements in the US under the *National Environmental Policy Act. The resource is irretrievable, but the action need not be *irreversible.

irreversible Not reversible; a category of *impact relating to non-renewable resources, or actions that can be renewed only after a long period of time (such as loss of soil productivity), that must be included in *environmental impact statements in the US under the *National Environmental Policy Act.

irrigation The application of *water to *land (by a sprinkler, ditch, or *canal), particularly in an *arid area, in order to supply the water and *nutrients that plants need to grow properly. By 1990 about 15% of all the farmland in the world was irrigated, although the proportions varied from 6% in Africa to 31% in Asia. The total area under irrigation increased by more than a third between 1970 and 1990, with most of the increase concentrated in developing countries. Output from irrigated land is more than double that from the same land unirrigated, and one-fifth of the world's food is grown on irrigated land. Two common problems created by irrigation are *waterlogging and *salinization of soils,

often as a result of excessive inputs of water into soil systems with inadequate drainage, which cause a decline in *crop productivity and *yield.

irrigation efficiency The amount of *water that is stored in the crop *root zone of a soil, compared with the amount of *irrigation water that has been applied.

irrigation return flow Surface and subsurface water that leaves a *field after *irrigation water has been applied, and returns to the normal *water cycle.

irritant A *substance that produces an irritating effect when it comes into contact with the skin, eyes, nose, or respiratory system.

irruptive growth See MALTHUSIAN GROWTH.

island An area of land, smaller than a *continent, which is surrounded by water.

island arc A curved chain of *volcanic islands that develops between an *ocean trench and a continental *landmass, and forms above the *subduction zone at a *convergent plate boundary.

island biogeography The study of the relationship between the area of an island and the number of *species that live on it, based on the effect of factors such as rates of *colonization, *immigration, and *extinction, and size, shape, and distance from other inhabited regions. See also EQUILIBRIUM THEORY.

ISO See INTERNATIONAL ORGANIZATION FOR STANDARDIZATION.

ISO 14001 The international standard for *environmental management systems and good environmental practices. See also ECO-MANAGEMENT AND AUDIT SCHEME.

ISO 9000 A set of certification standards for quality management systems.

isobar A line on a *weather map which connects points of equal *atmospheric pressure, measured in *millibars.

isobaric saturation point *See* DEW POINT.

isochron 1. A line on a map which connects points which have the same time or time interval.
2. In geology, a line of equal age on a graph.

isohyet A line on a *weather map which connects points of equal *rainfall over a given period of time.

isolated showers *Showers that cover less than 15% of an area.

isolated system A *system that has no exchange of *energy or *matter across its boundaries.

isoline *See* ISOPLETH.

isomer A chemical *compound that has the same number and kinds of *atoms as another compound, but a different structural arrangement of the atoms.

isopleth A line on a map which connects points of a constant value, such as *isobar, *isochron, *isohyet, *isotach, or *isotherm. Also known as **isoline**.

isostasy The mechanism by which the *lithosphere floats on the *asthenosphere, with *continents and *mountains supported by low-density 'roots' within the Earth's *crust; changes such as loading and unloading by ice or water bodies, cause depression or rebound of the lithosphere. *See also* PLATE TECTONICS.

isostatic adjustment The *warping of part of the Earth's *crust as a response to the redistribution of weight, for example associated with large-scale *glaciation and *deglaciation. Also known as **isostatic rebound**. *Contrast* EUSTATIC.

isostatic rebound *See* ISOSTATIC ADJUSTMENT.

isotach A line on a *weather map that connects points of equal *wind speed.

isotherm A line on a *weather map that connects points of equal *temperature.

isothermal At a constant *temperature, with height or through time.

isotonic Two *solutions, or a *cell and a surrounding solution, which contain the same *concentration of a *dissolved substance.

isotope One of two or more *atoms of the same *element that have the same number of *protons (and hence the same chemical properties) but a different number of *neutrons and thus different *atomic weights.

isotope analysis A technique that is used to reconstruct past *climate change, based on detailed analysis of the *oxygen isotope composition of *ice cores, the remains of marine organisms, and *calcite deposits in *caves.

isotropy Identical in all directions. *Contrast* ANISOTROPY.

ISTEA *See* INTERMODAL SURFACE TRANSPORTATION EFFICIENCY ACT.

Itai-Itai disease Health problems in humans occurring as a result of *cadmium poisoning, which causes softening of the bones and kidney failure. First observed in Japan in 1950.

ITCZ *See* INTERTROPICAL CONVERGENCE ZONE.

IUCN *See* WORLD CONSERVATION UNION.

IUCN management category Categories for the management of protected areas which were developed by the IUCN (*World Conservation Union) and are widely used around the world. The ten categories are: Category I, *Scientific Reserve/Strict Nature Reserve; Category II, *National Park; Category III, *Natural Monument/Natural Landmark; Category IV, *Managed Nature Reserve/Wildlife Sanctuary; Category V, *Protected Landscape or Seascape; Category VI, Resource Reserve; Category VII, *Natural Biotic Area/Anthropological Reserve; Category VIII, *Multiple-Use Management Area/Managed Resource Area; Category IX, *Biosphere Reserve; Category X, *World Heritage Site.

IWC *See* INTERNATIONAL WHALING COMMISSION.

IWM *See* INTEGRATED WASTE MANAGEMENT.

IWRM *See* INTEGRATED WATER RESOURCE MANAGEMENT.

Janzen–Connell escape hypothesis The hypothesis that isolated individuals of a given plant *species will escape their specialist *herbivores and thus survive better than individuals growing together in clumps. Also known as the **escape hypothesis.**

J-curve A growth curve that describes *exponential growth in a *population, so-called because of its shape. *Contrast* S-CURVE.

jet streak The region of fastest wind flow within the *jet stream.

jet stream A narrow, meandering band of very fast *wind high in the *troposphere, where temperature gradients are particularly strong, and which generally flows from west to east over the mid-latitudes.

JI *See* JOINT IMPLEMENTATION.

jigging *See* HAND-LINING.

joint A small-scale surface fracture or break within rocks which are subjected to *compression pressure or *tensional pressure beyond their *plastic limit, without movement (unlike a *fault).

joint and several liability A legal standard for the *cleanup of *Superfund sites in the USA, which makes all owners or users of a site potentially liable for cleaning up a site that has become contaminated over a number of years.

Joint Implementation (JI) A mechanism, established under the *Kyoto Protocol, which allows a *developed country to acquire emission credits to assist in meeting their assigned amounts (*emissions reduction unit) when it helps to finance projects that reduce net emissions in another developed country, including *countries with economies in transition. This allows industrialized countries to meet their obligations for reducing their *greenhouse gas emissions by receiving credits for investing in emissions reductions in *developing countries. *See also* ACTIVITIES IMPLEMENTED JOINTLY.

jökulhlaup A destructive *flood caused by the sudden release of glacial

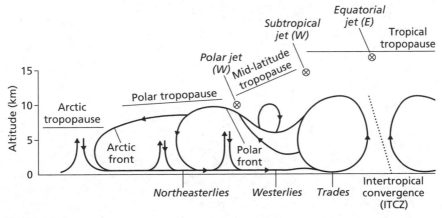

Fig 11 Jet stream

*meltwater, often associated with a *subglacial *volcanic eruption.

joule (J) A unit of energy; 1 joule is the work done when a force of 1 newton moves a point to which it was applied 1 metre in the direction in which the force is acting.

judicial precedent A previous decision set by a court, which can be reversed only by a higher court.

jungle Non-scientific name for impenetrable *rainforest.

Jurassic The middle *period of *geological time in the *Mesozoic era, between about 208 and 145 million years ago, named after the Jura Mountains between France and Switzerland. It is characterized by the existence of dinosaurs and the appearance of the earliest *mammals and *birds.

jurisdiction The territory within which legal power can be exercised.

JUSSCANNZ A group of developed countries outside the *European Union, which share information and discuss matters of common interest (including the *Kyoto Protocol). It stands for Japan, the US, Switzerland, Canada, Australia, Norway, and New Zealand. Iceland, Mexico, the Republic of Korea and other invited countries may also attend meetings.

justice Fairness, the quality of being fair or just.

juvenile Youthful, before reaching sexual maturity at the adult stage.

juvenile gas Gas that comes to the surface of the Earth for the first time from deep inside.

juvenile hormone A natural *hormone that enables an immature insect to develop normally into an adult. An immature insect will not develop properly if it is exposed to synthetic juvenile hormones applied as a *pesticide. Also known as **neotenin**.

juvenile hormone analogue A synthetic *juvenile hormone that is used as a *pesticide, which reduces the reproductive capacity of the insect population and results eventually in eradication.

kairomone A chemical that is emitted by an *organism (such as a plant) and triggers a response in an individual of another species (such as an insect) which benefits the emitting organism. For example, a plant scent that makes it easier for an insect *pest to identify and thus avoid the plant.

kame An undulating mound of *sand and *gravel that has been deposited in an irregular pattern by *meltwater adjacent to a *glacier or *ice sheet. *Contrast* ESKER.

kame delta A ridge of *sediment in the form of a steep, flat topped hill, that is deposited by *meltwater in front of a retreating continental *glacier. *See also* KAME TERRACE.

kame terrace A continuous line of *kame that stretches along much of a valley side.

Kansan The first and oldest period of ice advance (*glaciation) in North America during the *Pleistocene *ice age; equivalent to the *Gunz in Europe.

kaolin *See* CHINA CLAY.

kaolinite A mineral that is produced by *chemical weathering of *feldspar and consists of aluminium silicate; the main source of kaolin/*china clay.

karst A type of *landscape that results from the *chemical weathering and collapse of *carbonate rocks (such as *limestone and *dolomite), which creates such features as *sinkholes, *caves, and underground drainage. *See also* DOLINE.

karyotype The characteristic *chromosomes of a *species.

katabatic wind Cool dense air driven downslope by gravity beneath warmer

lighter air. Examples include a *land breeze, *sea breeze, *mountain breeze, or *valley breeze. Also known as an **air drainage wind** or **mountain breeze**. They often have local names such as *mistral, *bora, or *taku. *Contrast* ANABATIC WIND.

kata-front A weather *front in the *atmosphere in which warm air is given little opportunity to rise over cold air because of prevailing downward movement of air from a higher level. *Contrast* ANA-FRONT.

kelp Any of a variety of brown seaweeds which contain *trace minerals, *vitamins, and *micronutrients. They are used in Japanese cookery, and can be ground up and used to enrich poor soil.

kelp forest A *marine *ecosystem that is dominated by large *kelps. These forests are restricted to cold and temperate waters, and are most commonly found along the western coasts of *continents.

Kelvin (K) The *SI unit of temperature which is equal in magnitude to a degree *Celsius; temperature expressed in Kelvin is the temperature in °C + 273.15. The temperature of absolute zero is 0 K. Note that degree symbols (°) are not used in the Kelvin scale. The Kelvin scale is used at extremely low temperatures, mostly by physicists but rarely by climatologists or meteorologists.

kerbside collection The collection of *compostable or *recyclable material and/or trash at individual homes and businesses, which is then taken to a waste processing facility.

kerbside recycling scheme A *recycling scheme that involves collecting, sorting, and processing *solid waste material (such as glass, paper, *aluminium, and some plastics) which households have left in special containers by the side of the road in front of their properties.

kerogen The solid, *bituminous, waxy substance in *oil shales which can be extracted to produce *shale oil.

kettle hole An enclosed depression that is created by the melting of a mass of ice trapped in *glacial deposits, and which is often occupied by a *lake.

key facility A term used in the USA to describe public facilities which tend to induce *development and *urban development. Examples include major airports, road interchanges and highways, recreational lands, and energy production facilities such as major powerplants, transmission lines, oil and gas pipelines, and refineries.

key species In land or conservation management, the major plant or animal *species that should be taken into account.

keystone species A *species that interacts with many other species in a *community, and whose loss would cause a greater than average change in the *populations of other species or in *ecosystem processes.

Kilauea One of the world's most active *volcanoes, located in Hawaii, whose quiet non-explosive eruptions are increasing the size of the island as the *lava descends to the sea.

kilobar (kb) A unit of *pressure equal to 1000 *bars. See also MILLIBAR.

kilometre (km) A metric unit of distance equal to 1000 metres or 0.62 miles.

kilowatt (kW) A unit of electrical power equal to 1000 watts (roughly 1.341 horsepower).

kilowatt hour (kWh) The work performed by 1 *kilowatt of power applied for 1 hour.

kinetic energy *Energy which is possessed by a body because of its motion, such as a rock rolling down a hill, the wind blowing through the trees, or water flowing over a dam.

kingdom The highest and broadest category in the scientific system of classification for organisms (*taxonomy), a major division of living organisms. All organisms are classified into one of five kingdoms: *Monera (*bacteria or *prokaryotes), *Protoctista, *Fungi, *Plantae, and *Animalia. See also BIOLOGICAL CLASSIFICATION.

kitchen garden A small *garden in which vegetables are grown.

knick-point A discontinuity in the *long profile of a *river, which usually develops where downcutting is inhibited by an *outcrop of resistant *rock, causing a step in the long profile which through time will move upstream as it erodes.

knoll 1. A small natural *hill.
2. A small *reef within a *lagoon.

knot A measure of speed, usually applied to wind or a boat, which is equal

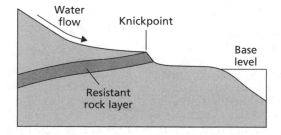

Fig 12 Knick-point

to one *nautical mile per hour, mph (0.51 metres per second, m s^{-1}).

knowledge Information, gathered from experience and reasoning, that has been interpreted and can be used.

knowledge management The systematic management and use of the *knowledge (collective data, information, and body of experience) within an organization, to enhance the performance of the organization.

knowledge transfer The effective sharing of ideas, *knowledge, or experience between people, companies, or organizations.

known resource A *natural resource (such as a *mineral deposit) that has been located but not yet fully mapped, but which is likely to become economically viable in the foreseeable future.

koinobiont A form of *parasitism. Konobiont species parasitize earlier host stages, host activity continues following attack, and association of the host and parasite is prolonged. *Contrast* IDIOBIONT.

Köppen classification system A system for classifying *climates, and defining *climate zones, that was developed by W. Köppen and is based mainly on annual and monthly averages of *temperature and *precipitation.

Krakatoa A small, steep, forested, uninhabited *volcanic island in Indonesia, between Java and Sumatra, which erupted suddenly and violently in 1883, throwing an estimated 17 cubic kilometres of rock, *ash, and *pumice high into the *atmosphere (which shaded out the Sun in many places for months afterwards), and triggering a catastrophic *tsunami that flooded coastal towns and villages on the nearby coasts and drowned 36 000 people.

krill Small shrimp-like *planktonic *crustaceans that form an important part of the *food chain in *Antarctic waters.

krypton (Kr) An *inert, colourless, non-toxic *gas that occurs naturally in the Earth's *atmosphere in trace quantities. It is one of the six inert gases, and is used for insulating windows.

K-selection *Natural selection that favours species that are adapted to the maximum survival of individuals, at the expense of species with a high reproductive capacity. *Contrast* R-SELECTION.

K-strategy A life strategy in which *species are best adapted to live in a stable *habitat with a high level of *competition between species, so they are larger, live longer, and produce small litters of large offspring compared with other species. *Contrast* R-STRATEGY.

K–T boundary In geology, the boundary between the rocks of the *Cretaceous and the *Tertiary periods 65 million years ago, which marks the time of a *meteorite impact on Earth, which caused *mass extinction of species including the *dinosaurs.

Kuznets curve A bell-shaped curve that shows the relationship between *development and *environmental degradation, which has *per capita income as the horizontal axis (x), and per capita pollution as the vertical axis (y). As per capita income rises per capita pollution

KYOTO PROTOCOL

An international agreement that sets limits on the *emission of *greenhouse gases into the *atmosphere, in order to reduce the threat of *global warming by *air pollution. It was drawn up under the *United Nations Framework Convention on Climate Change (UNFCCC), drafted during the *Berlin Mandate process and agreed in Kyoto, Japan, on 11 December 1997 by the *signatory countries to the UNFCCC. The protocol specifies targets for the level of emission reductions, the methodologies, and the deadlines that signatory countries have to achieve. The main components of the agreement are that *developed countries should commit to reducing their emissions of six key greenhouse gases by an average of at least 5%, national *emissions targets must be achieved by 2008–12, and countries must have some flexibility in how to reduce their emissions through a *Clean Development Mechanism, *international emissions trading, and *Joint Implementation. Some flexibility was agreed that would allow countries to emit more *greenhouse gases if they planted more trees (that act as *carbon sinks), or could demonstrate that they had reduced pollution in other countries (joint implementation, for example through *emissions trading). Under the Kyoto Protocol, rich industrial *developed countries agreed to reduce their emissions of greenhouse gases by an average of 5.2% by 2010. *Developing countries were not set formal emission limits, partly because to do so would unfairly inhibit their pursuit of *economic growth.

also increases, but as people gain the disposable income that is necessary to value such things as clean air and water, per capita pollution peaks and the reduces as per capita income continues to rise.

Kyoto mechanism *See* FLEXIBILITY MECHANISM.

LA21 *See* LOCAL AGENDA 21.

La Niña The opposite of *El Niño, when a major cooling occurs in the equatorial waters in the Pacific Ocean and causes shifts in normal *weather patterns. La Niña means 'the little girl'.

laccolith A large lens-shaped mass of *igneous *intrusive rock that has pushed up the overlying rocks; similar in form to but much smaller than a *batholith.

lacustrine Of or relating to a *lake.

lacustrine deposit *Sediment that has been deposited in a *lake.

ladder fuel *Vegetation (including shrubs and small trees) that is located below the *crown level of *forest trees, which can spread *fire from the forest floor to tree crowns.

lag The period of time that elapses between the occurrence of an event and its resulting impact. Also known as **lag time**.

lag time *See* LAG.

lagoon A body of water that is cut off from a larger body (such as the *ocean) by a *coral reef or *sandbar.

lahar A *mudflow or lubricated *landslide of volcanic debris, caused when melting snow, rain, or water from a volcanic steam or a crater lake mix with ash or volcanic fragments on the sides of a *volcano.

LAI *See* LEAF AREA INDEX.

lair A place where wild animals live.

lake A body of (usually fresh) *water that is entirely or nearly entirely surrounded by land.

lake breeze A *wind that blows onshore from the surface of a large *lake, which is caused by the difference in the rates of heating of water and land. A type of *local convective wind.

Lake Nyos A lake in the volcanic crater Iwi at Nyos in Cameroon, where a sudden release of *toxic *volcanic gas occurred on the night of 21 August 1986. More than 1700 people and all animal life within 14 kilometres of the lake were killed, but plant life was mostly unaffected. Survivors suffered symptoms compatible with an asphyxiant gas like *carbon dioxide. Two different explanations have been offered for this extremely rare type of event – that it was the result of a *phreatic eruption of *groundwater that was exceptionally rich in CO_2 or it was a result of *overturn of water in the 220 metre deep lake.

lake-effect storm A heavy localized snowstorm that can form on the downwind side of a large lake (such as the *Great Lakes) in late *autumn/fall and early *winter, as cold, dry air picks up moisture and warmth from the unfrozen body of water.

Lamarck, Jean-Baptiste A French naturalist (1744–1829) who proposed that the *evolution of *species is caused by the *inheritance of *acquired traits.

laminar flow A type of *flow that involves smooth movement without any turbulence. *Contrast* TURBULENT FLOW.

lamination Layered or in layers. *See also* STRATIFICATION.

land The solid, dry surface of the Earth, or any part of it. *Contrast* OCEAN.

land application The discharge of *wastewater onto the ground, for treatment or reuse. *See also* IRRIGATION.

land ban The phasing out of disposal on land of most untreated *hazardous wastes, as required by *legislation.

land breeze A coastal *breeze that blows from land to sea, usually at night. A type of *local convective wind. Also known as **offshore breeze**. *Contrast* SEA BREEZE.

land bridge A natural connection between two land masses (for example, between Alaska and Siberia across the Bering Strait, or between France and England across the English Channel) that has been removed by *sea level rise, crustal *warping, or the movement of *crustal plates.

land capability The inherent ability of *land to be used for agriculture without permanent damage (such as *soil erosion).

land capability classification A way of grouping *soils according to their suitability for intensive use (growing *crops and *pasture plants) on a *sustainable basis. *See also* LAND EVALUATION.

land capability map A map, produced from a *land capability classification, which shows the distribution of *land capability units.

land capability unit A group of *soils that are very similar in their inherent suitability for intensive use and in their responses to soil management.

land class A way of grouping areas of land on the basis of slope. For example, the US Forest Service uses three classes: 0–35%, 36–55%, and greater than 55%.

land classification A way of grouping areas of land on the basis of physical characteristics (including vegetation) without an explicit assessment of *land capability.

land cover *See* LAND USE.

land cover change Converting land from one use to another (such as from *grassland to *cropland), or changing the condition within a particular land use (such as thinning the trees in a *forest).

land development The process of improving an area of land and making it more useful, by installing streets, sewers, utilities, and other infrastructure.

land disposal restriction The requirement that *hazardous wastes must be treated before disposal on land, in order to prevent the *contamination of *soil and *groundwater by hazardous materials.

land ethic The principles and values that guide the *sustainable use and treatment of *land and other *natural resources. *See also* ENVIRONMENTAL ETHICS, STEWARDSHIP.

land evaluation The formal assessment of the suitability of land for different types of use (for example for *agriculture, *forestry, or *recreation). Also known as **terrain evaluation**.

land farming A process of *waste disposal and *bioremediation in which *hazardous wastes are deposited on or in *soil and allowed to *decompose naturally by the action of *microbes.

land management The intentional process of planning and managing the activities that affect *land use, in order to achieve stated objectives.

land reclamation The process of creating new dry land (for example from the *sea-bed), or of restoring *productivity to land that has been damaged (for example by *erosion or *mining activities). *See also* RECLAMATION.

land reform The redistribution of land by the government, from rich land owners to the people who actually work the land, in order to improve the distribution of income and promote *sustainable *rural development.

land tenure The right to own land, or to occupy and use exclusively a particular piece of land.

land use The purpose for which an area of land is being used, such as residential, agricultural, commercial, retail, or industrial. Also known as **land cover, land utilization**.

land use allocation Committing a particular area of land to one or more specific uses, for example for *residential development or as a *nature reserve.

land use capability An assessment of the ability of a particular area of land to sustain a range of *land uses.

land use change Converting an area of land from one *land use to another, usually on purpose.

land use plan A plan which outlines proposed future *land uses and their distribution in a particular area.

land use planning Local development of a plan, usually by a local government, for the use and development of land within its boundaries, taking into account existing *land use, *land capability, and local policies and strategic objectives.

land use survey A systematic survey of land use in a particular area, usually shown both in map form and statistically.

land utilization See LAND USE.

land value An assessment of the worth of a piece of land, expressed either in terms of money or more broadly in terms of factors such as *environmental quality, or *amenity or *heritage interest.

landfill A place or a process for the disposal of non-hazardous *waste, based on burying it in depressions in the ground then compacting it to reduce the volume and finally covering it with soil and landscaping it to look like part of the surrounding land. See also SANITARY LANDFILL, SECURE LANDFILL.

landform A natural feature of the surface of the land that has been created by *geological processes, which can vary in scale from the continental (for example the Andes *mountain chain which runs through South America) to the local (for example a *hillslope, river *meander, or *beach). See also GEOMORPHOLOGY.

landing In *forest management, a place where cut *timber is collected before being transported away from the forest.

landings The number of *fish that are landed (caught and brought back to the docks and marketed), either in a particular place or for an entire country.

landline A property boundary in the USA, for example for *National Forest land.

landlocked An area that is totally enclosed by land, with no direct access to the sea.

landmark Any readily identified structure or *landform (such as a particular building or mountain) on land that can be used in determining a location or direction, or that commemorates a historic event or achievement (such as the Statue of Liberty in New York).

landmass A large continuous area of land, such as a *continent or large *island.

landnám A local phase of *forest clearance by humans during *prehistory, detected in *pollen diagrams, and indicating brief agricultural use which was often followed by abandonment and then natural regeneration of *secondary woodland.

landrace A primitive variety of crop (*cultivar) or animal (*breed) that through time has been genetically improved by traditional agriculturalists to make it better adapted to local environmental conditions, but has not been influenced by modern breeding practices.

Landsat Land Remote-Sensing Satellite, a US *satellite used for *remote sensing of *natural resources on Earth, originally called *Earth Resources Technology Satellite (ERTS). It uses two sensors: the *multispectral scanner (MSS) and the *thematic mapper (TM).

landscape *Scenery, either natural (*natural landscape) or modified by human activities (*cultural landscape); often used to refer to the expanse of scenery that can be seen from a single *viewpoint. *See also* SEASCAPE, SOUNDSCAPE, TOWNSCAPE, WILDSCAPE, WOODSCAPE.

landscape architecture Arranging and designing land and buildings, to optimize human benefit.

landscape ecology Principles and theories for understanding the structure, functioning, and change of *landscapes and *ecosystems over time.

landscape indicator An index or measurement of some aspect of the *landscape of a particular place (such as extent of *forest cover), based on mapped or remotely sensed data, that can be used to describe *land use and monitor *environmental change.

landscape planning Part of the *land use planning process which seeks to balance decisions relating to large-scale physical factors (such as *ecosystems and *drainage basins) and broader aesthetic and cultural values (such as historical importance and *amenity value).

landscaped Deliberately improved by gardening or by *landscape architecture.

landscaping Improving the appearance of an area of land by planting (including lawns, trees, and shrubs), adding decorative features (such as structures of buildings), and/or altering the level or shape of the ground.

landslide A sudden, rapid form of *mass movement of material down a *hillslope, often after heavy *rain or as a result of an *earthquake; slower than an *avalanche, but often equally destructive. Also known as **slide** or **landslip**.

landslip *See* LANDSLIDE.

landspout A violently rotating column of air (like a small weak *tornado but not formed by a storm-scale rotation) which is sometimes seen beneath *cumulonimbus or towering *cumulus clouds; the land equivalent of a *waterspout.

landward Towards *land. *Contrast* SEAWARD.

lapilli Cinders or small lumps of *lava that are thrown up by a *volcanic eruption.

lapse rate The rate at which air temperature decreases with altitude in a stationary parcel of air, because of *adiabatic expansion. The normal lapse rate is about 6.5 degrees Centigrade per kilometre, although it varies from place to place and through time. *See also* DRY ADIABATIC LAPSE RATE, SATURATED ADIABATIC LAPSE RATE.

large woody debris Part of a *tree (either dead or alive), including branches and roots, usually longer than one metre, which rests in a *stream or on a *forest floor.

large-lot zoning A type of *open space zoning which prohibits the subdivision of land into individual plots smaller than a defined size (usually an acre or more).

larva The young, immature stage in the life cycle of some animals, such as fish and *amphibians. Larvae develop into the adult form by *metamorphosis; for example caterpillar larvae turn into butterflies.

larval paedogenesis Reproduction by larvae.

Last Glacial Maximum The last prolonged period of *cold climate and *glacier development before the present day, around 18 000 years ago.

late forest succession The stage of *succession in a forest in which most of the trees are mature, for example in *old growth forest. *Contrast* EARLY FOREST SUCCESSION.

late successional reserve (LSR) North American name for an *old growth forest or old *clearcut that has been set aside as a wildlife habitat, in which *logging is only permitted if it will help the reserve to reach old growth characteristics more quickly.

late-glacial The final part of the last *glacial stage, between about 16 000 and 11 500 years ago.

latency The period of time between first exposure to an *agent (such as something that causes a *disease) and the appearance of a health effect.

latent Potentially existing, but not currently active.

latent heat The heat energy that is transferred when *water changes state between *gas (*water vapour), *liquid (water droplets), and solid (*ice). Also known as **heat of transformation**.

latent heat flux The transfer of *latent heat energy via the global circulation of air and water, and through the processes of *evaporation and *condensation.

latent period *See* INCUBATION PERIOD.

lateral Sideways, as opposed to up and down (*vertical).

lateral accretion *Deposition by a *river on the inside corner of *meander bends, which creates *point bars and helps to build the *floodplain.

lateral moraine A type of *moraine that is deposited at the side of a valley *glacier, often composed of rock particles that have fallen off the sidewalls of the valley as a result of *frost wedging.

laterization A *soil-forming process that involves the deposition of a *hardpan made of metallic oxides (*laterite) in the *A-horizon, which most commonly occurs in humid tropical and subtropical areas, where precipitation is high.

laterite A hard subsurface deposit (*hardpan) of oxides of *aluminium and *iron that is found in tropical soils where the *water table fluctuates with seasonal changes in precipitation.

latewood The portion of the annual *growth ring of a tree that is formed when growth slows down (during summer and autumn, or during the dry season), after the *earlywood formation has ceased. It is usually denser, darker, and stronger than earlywood.

latex A white, milky fluid that is derived from the rubber tree and is used in a wide variety of consumer products, including rubber gloves, tubing, and rubber bands. Also known as **rubber**.

Latin name The name for a *species of plant or animal that is used around the world. The naming system, developed in the mid-1700s by *Linnaeus, uses Latin and has two parts; the first part, which is capitalized, is the *genus or family name, and the second part is the name of the species or variety. Both parts are always written in italics. For example, the beech tree is the *Fagus sylvatica*. Also known as **scientific name, species name**. *Compare* COMMON NAME.

latitude Distance on the Earth, measured in degrees, north or south of the *equator (0°); the *poles are 90°N and 90°S.

latosol A zonal *order of soil that develops under forested, tropical, humid conditions. It is usually red in colour, contains a *hardpan of *laterite, and has a deeply weathered *profile and relatively low mineral content, and thus low fertility.

Laurasia A large former continent in the northern hemisphere, believed to be part of the fragmentation of the *Pangaea supercontinent which broke apart by *plate tectonics. It is thought to have comprised modern-day North America, Greenland, Europe, and Asia. *See also* GONDWANALAND.

lava The term given to *magma when it reaches the Earth's surface through *volcanic vents and fissures, which then cools and solidifies into hard *igneous rock.

lava cone *See* VOLCANO.

lava flow A stream of molten *lava that flows from a volcanic *vent.

law 1. A collection of rules, imposed by government, which govern a particular type of activity and apply over a certain territory (such as a country). *See also* COMMON LAW.
2. A general principle or description of

how a natural phenomenon will occur under certain circumstances.

law of conservation of energy See THERMODYNAMICS, FIRST LAW.

lawn A patch of grass that has been cultivated and is regularly mowed.

laws of ecology Four so-called laws (which are really general principles) proposed by Barry *Commoner in the book *The Closing Circle: Nature, Man, and Technology* (1971). They are: everything is connected to everything else; everything must go somewhere; nature knows best; there is no such thing as a free lunch.

LBAP See LOCAL BIODIVERSITY ACTION PLAN.

LC50 The concentration of a given substance in air that will kill 50% of a group of test animals with a single *exposure (usually 1–4 hours). Also known as **lethal concentration fifty**. See also LD50.

LCL See LIFTING CONDENSATION LEVEL.

LD50 The size of the single *dose of a given substance that causes the death of 50% of an animal population from exposure by any route other than *inhalation, with a given period of time; the lower the LD50, the more *toxic the substance. Also known as **lethal dose fifty, median lethal dose**.

LDC See LESS-DEVELOPED COUNTRY.

Le Chatelier's principle When an external force is applied to a *system in *equilibrium, the system will undergo *adjustment in order to minimize the effect of the force.

leach To remove *soluble components from a solid by *infiltration of a *solvent. For example, to leach *salts from *soil by dissolving them in water, which *percolates through soil.

leachate A solution formed by *leaching.

leachate collection system A system that gathers *leachate (usually from *landfill) and pumps it to the sur-face for treatment, in order to prevent the spread of *contamination.

leaching The process by which *soluble *minerals (*humus and *inorganic *nutrients) in soil are dissolved in percolating water, and can be washed out from the *A-horizon and transferred within the soil. See also HARDPAN.

lead (Pb) A naturally occurring, grey-white *metal that is soft, *malleable, *ductile, and resistant to *corrosion. It was used in *gasoline, is still used in paints and plumbing compounds, and is both a *criteria air pollutant and a *toxic air contaminant. Human exposure to it can damage the brain and nervous system, especially in children. See also LEAD POISONING.

lead agency The agency that is identified as having primary responsibility for responding to a particular *disaster.

lead poisoning Damage to the body (specifically the brain) that is caused by breathing or swallowing substances that contain *lead, or by absorbing lead through the skin.

lead time The period of time between the start and end of an activity, such as the interval between introducing pollution control legislation and a reduction in pollution being observed.

leaded fuel *Petrol to which lead has been added in order to raise the octane level. Contrast UNLEADED FUEL.

leaf The flat, green structure that grows from the stem or stalk of a *plant, in which *photosynthesis and *transpiration take place.

leaf area index (LAI) The area of leaf that is exposed over a unit area of land surface.

leaf litter The accumulation of dead leaves and vegetation on a *soil surface or *forest floor.

League of Conservation Voters (LCV) A US-based non-profit lobby group that was founded in 1970 by leaders of the *environmental movement following the first *Earth Day, and which is devoted to making the Congress and the White House more pro-environmental. It holds Congress and the Administration accountable for their actions on the environment through the *National Environmental Scorecard.

leaking underground storage tank (LUST) A US term for any underground container that is used to store chemicals (such as *gasoline, diesel fuel, or home heating oil) and which is damaged and leaking its contents into the ground; these may *contaminate *groundwater.

lectin A complex *molecule that is made by both plants and animals, which has both *protein and sugars and is able to bind to the outside of a cell and cause *biochemical changes within it.

lee *See* LEEWARD.

lee wave An air wave that occurs in the *lee of a *mountain range when rapidly flowing air is lifted up the steep front of the range. Also known as **mountain wave**.

leeward In a downstream or downwind direction, or the side of something that is sheltered from the wind or water. Also known as **lee**. *Contrast* WINDWARD.

legacy Something that is passed down in a will, or left behind from the past. *See also* INHERIT.

Legionella A type of *bacterium that thrives in central heating and air conditioning systems and can cause legionnaires' disease.

legislate To make *laws, or bring into effect by *legislation.

legislation The process by which government makes *laws.

legume A plant (such as peas, beans, alfalfa, soybeans, and clover) that has pods containing one or more seeds as fruit. It also has *nitrogen-fixing *bacteria in its roots. These increase the nitrogen content of *soil and thus increase

soil *fertility, which is why legumes are a common component of crop *rotation. Also known as a **pulse**.

leisure Spare time that is available for relaxation.

lentic water A standing or non-flowing body of water, such as a *lake, *reservoir, *pond, or *swamp.

lenticular In the shape of a lens.

Leopold, Aldo US ecologist, forester, and environmentalist (1887–1948) who many people regard as the father of wildlife ecology. He was a gifted writer, and is best known as author of the book *A *Sand County Almanac* (1948), in which he outlined his philosophy (*land ethic) of living in harmony with the land and with nature.

less developed country (LDC) *See* DEVELOPING COUNTRY.

lestobiosis *See* CLEPTOBIOSIS.

lethal concentration fifty *See* LC50.

lethal dose fifty *See* LD50.

leukaemia A *cancer of the blood-forming cells in bone marrow, which causes excessive production of white blood cells and weakens the body's immune system.

levee **1.** A ridge of fine *alluvium that is deposited on a *floodplain close to the *channel by a river under *flood conditions.
2. An artificial embankment constructed alongside a river (such as much of the lower Mississippi) designed to contain flood flows and reduce *flooding.

level of concern (LOC) The *threshold concentration in air of an extremely *hazardous substance, above which even a single short exposure can kill or seriously injure humans. *See also* VULNERABLE ZONE.

ley An area of temporary *grassland that is sown with *grass and/or *herbs to produce grazing and *hay or *silage, usually for a period of up to ten years, before being ploughed and cropped as part of a *crop rotation.

liana A woody-stemmed vine-like climbing *plant which roots in the ground and grows by supporting itself on *trees or *shrubs.

lichen A composite *plant that is made up of an *alga or a *cyanobacterium and a *fungus growing together in a *symbiotic relationship. Lichen grows in patches on tree trunks, bare ground, rocks, and walls, and can be a sensitive *indicator of *pollution.

life cycle 1. The stages through which an *organism passes during its existence, from birth to death. *See also* LIFE HISTORY.
2. Stages in the total life of a product, from raw materials, design, and manufacture to use, disposal, or *recycling.

life cycle analysis A process for evaluating and reducing the inputs and outputs of material and energy at each stage in the *life cycle of a product, from raw material to final disposal.

life cycle cost The total cost of a good or service over its entire *life cycle, including the cost of disposal.

LIFE Environment Part of the *LIFE Programme that is sponsored by the *European Union, which includes actions designed to implement EU *environmental policy and legislation, and to demonstrate and develop new methods for the *protection and the enhancement of the *environment.

life expectancy The length of time that an individual *organism (such as a person) is expected to live. *See also* LIFESPAN.

life form The characteristic structure of a *plant or *animal in a particular *environment. *Contrast* GROWTH FORM.

LIFE Fund A *European Union (EU) fund that has been established to assist the development and implementation of the EU's *environmental policies.

life history A descriptive account of the complete *life cycle of an *organism.

LIFE Nature Part of the *LIFE Programme sponsored by the *European Union. It includes actions which are aimed at the *conservation of natural *habitats and the wild *fauna and *flora of EU interest, and supports implementation of the EU *nature conservation policy and the *Natura 2000 network.

LIFE Programme A major initiative (L'instrument Financier pour l'Environnement) which is part of *European Union *environmental policy and finances project in three areas: *LIFE Nature, *LIFE Environment, and *LIFE Third Countries.

life table A statistical table that summarizes the *mortality and *survivorship characteristics of a given population, broken down by *age class.

LIFE Third Countries Part of the *LIFE Programme sponsored by the *European Union, which includes actions that provide technical assistance for promoting *sustainable development in third countries.

life zone A region or belt that is defined by climate, altitude, or latitude, in which plants and animals are well adapted to the *environment and across which the *species composition and *biodiversity are relatively uniform.

lifespan The theoretical maximum number of years that the most healthy individuals within a *species can be expected to live. *Life expectancy in humans is lower than this because it reflects the real-life conditions within a population, which includes stresses such as *disease, *malnutrition, *environmental risk, and social tension. *See also* LONGEVITY.

lifestyle The mode or pattern of living that is adopted by an individual, which is reflected in their activities, interests, and opinions.

lifetime The period of time during which something is functional, such as between birth and death in a person, plant, or animal. *See also* RESIDENCE TIME.

lifetime exposure The total amount of *exposure to a particular substance that a human is likely to receive in a lifetime (usually taken as 70 years).

lifting condensation level (LCL) The level to which a parcel of dry air must be lifted before it cools *adiabatically and becomes *saturated because of *condensation. Also known as **condensation level**.

light A type of *radiant energy with a *wavelength of 0.39 to 0.77 *micrometres within the *electromagnetic spectrum, which is visible to humans.

light non-aqueous phase liquid (LNAPL) Undissolved chemicals, such as petroleum products (*hydrocarbon fuels and lubricating oils), which float on the surface of *groundwater and do not mix with it.

light pollution High levels of artificial light that prevent viewing of the night sky.

light wave *Radiation in the middle part of the *electromagnetic spectrum, between shortwave (X-rays) and longwave (radio waves). This includes *ultraviolet (4 to 400 nanometres), visible light (400 to 750 nanometres), and *infrared (750 to 1 million nanometres).

lightning An electrical discharge from a *thunderstorm that causes a flash of light to run between *clouds or from a cloud to the ground, and is often accompanied by *thunder. The lightning is seen before the thunder is heard because light travels faster than sound.

light-water reactor A type of *nuclear power plant that uses ordinary water as the cooling medium and slightly enriched *uranium as a fuel. The *boiling water reactor and the *pressurized water reactor, the most common types of commercial reactor used in the USA, are light-water reactors.

lignified Made hard like *wood as *lignin is deposited in the cell walls.

lignin A naturally occurring *organic chemical in the *cell walls of *plants, that give strength and rigidity to the plant.

lignite A soft, brownish, low-grade *coal that is formed during the middle stages of the *coal cycle.

limb 1. A main branch that grows from the trunk or a bough of a tree.
2. A jointed appendage of an animal (such as an arm, leg, or flipper), that is used for locomotion or grasping.
3. The tilted beds of a *fold in rock that extend outwards from the *axis.

lime 1. A small, green, tart citrus fruit, and the tree on which it grows.
2. A common name for calcium oxide (*quicklime); it is made from heating calcium carbonate ($CaCO_3$) and is used to make calcium hydroxide (*slaked lime) for neutralizing *acidic soil.

limescale The hard, white, chalky deposit that is often found in kettle, hot water boiler, and old pipes where *hard water has evaporated.

limestone A *carbonate *sedimentary rock, composed mainly of *calcium carbonate and smaller amounts of magnesium carbonate, usually in the form of the mineral *calcite. Commonly used as a building stone.

limestone pavement A flat, bare area of *limestone that has been smoothed by *glacial action, and has deep fissures crossing the surface in both directions.

limestone scrubbing An *air pollution control process in which *sulphur gases are passed through a solution of *limestone and water, before being released into the *atmosphere through a *smokestack, in order to reduce the sulphur content. Also known as **scrubbing**.

limit of detection The level below which a test (or *assay) cannot measure the presence of a *substance.

limiting factor An environmental factor (such as the availability of the *nutrient *nitrogen in a particular *soil) that limits the growth and survival of a *species. *See also* PRINCIPLE OF LIMITING FACTORS, TOLERANCE LIMIT.

Limits to Growth A report written on behalf of the *Club of Rome by Dennis and Donella *Meadows in 1971. Using dynamic computer models of various possible future *scenarios for world growth it predicted that serious *natural resource scarcity would follow if world *population and resource use continued to grow. It had a significant impact on attitudes towards the *environment, and on the development of modern *environmentalism. *See also* NEO-MALTHUSIAN, SUSTAINABLE DEVELOPMENT.

limnetic Of or relating to deep open waters, such as large *lakes.

limnetic zone The open water zone to the depth of effective light penetration.

limnology The study of the physical, chemical, hydrological, and biological properties of freshwater.

lindane An *organochlorine *pesticide that is slowly *biodegradable, can be distributed through domestic water supplies, damages human *health, and is *toxic to *freshwater fish and *aquatic life.

line *See* LINEAGE.

line thinning The act of removing (*thinning) particular rows of *trees in a *forest or *plantation, for example every fifth row. *See also* MECHANICAL THINNING, SELECTIVE THINNING.

lineage The *offspring or *descendants of one *individual. Also known as line.

liner A protective layer, often made of plastic or dense *clay, that is installed along the bottom and sides of a *landfill in order to prevent or reduce the flow of *leachate into the environment.

ling *See* HEATHER.

Linnaean classification A hierarchical method (developed by *Linnaeus) for naming *organisms in which each individual is assigned to a *species, *genus, *family, *order, *class, *phylum, and *kingdom, and some intermediate classificatory levels. *See also* LATIN NAME, TAXONOMY.

Linnaeus, Carolus A Swedish botanist (1707–78) who developed the hierarchical system of *biological classification that groups similar-looking *species into *genera, genera into *families, families into *orders, and so on. He also established the *Latin name system for giving *species names which is used around the world.

lipid solubility The maximum concentration of a *chemical that will dissolve in fatty substances, and that can disperse through the *environment via uptake in living tissue.

lipids A group of *organic substances (such as fats and steroids) that are insoluble in water but soluble in solvents such as alcohol and ether.

liquefaction The process by which *soil or *sediment is transformed temporarily from a solid into a fluid state, when the *cohesion of *particles is lost, for example during and immediately after an *earthquake.

liquefied natural gas (LNG) *Natural gas (usually *methane) that has been converted to a liquid state, either by refrigeration or by pressure, in order to reduce its volume and thus make it easier to store or transport.

liquefied petroleum gas (LPG) A mixture of light *hydrocarbons (such as *propane and *butane) that has been converted to a liquid state, either by refrigeration or by pressure, in order to reduce its volume and thus make it easier to store or transport.

liquid A substance in the fluid state whose shape alters to that of the container it is in, but whose volume does not change.

listed species Any *species of fish, *wildlife, or *plant that has been officially designated by an agency (such as a government department) as being *endangered or *threatened, and is therefore given enhanced protection.

listed waste In the USA, *wastes that are listed by the *Environmental Protection Agency as *hazardous by definition, even if they do not always exhibit the defined characteristics of hazardous waste.

listing The formal process by which an agency adds *species to the list of *endangered and *threatened wildlife and plants.

lithic Made of stone.

lithification The *cementation of *sediment after deposition. *See also* DIAGENESIS.

lithology The description of the *mineral composition and *texture of *rocks and *sediments.

lithosphere The outer, solid part of the *Earth, including the *continental crust, *oceanic crust, and uppermost *mantle, which is generally around 100 kilometres thick and is where *faulting and *earthquakes occur.

lithostratigraphy The organization of *rock or *sediment *strata into units based on their *lithology and *stratigraphic position relative to other units. *Contrast* BIOSTRATIGRAPHY.

lithotroph An *organism that synthesizes all of its *organic molecules from inorganic sources, such as *carbon dioxide or *bicarbonate.

litre A unit of volume that is equal to one cubic decimetre, roughly 1.76 pints.

litter *Organic matter in the form of plant debris such as leaves and twigs that falls off living and dead plants, accumulates on the *soil surface, and is broken down by soil organisms to become *humus within the soil. *See also* DECOMPOSITION.

litterfall Leaves, twigs, and other plant material that falls to the ground and contributes to the *litter layer.

Little Climatic Optimum A period of relatively mild, warm, dry climate between AD 900 and 1300 and the warmest period since the *Climatic Optimum. Vineyards flourished in southern England and the seas around Iceland and Greenland were relatively ice-free. Also known as the **Medieval Warm Period**.

Little Ice Age A period of very cold climate that lasted from about AD 1550 to about 1850 in Europe, North America, and Asia, as climate became much cooler after the *Medieval Warm Period. During this period, global temperatures were at their coldest since the beginning of the *Holocene, *valley glaciers grew rapidly, and there was renewed *periglacial activity, especially in the Alps, Scandinavia, and Greenland. The coldest conditions of the last 560 years were between 1570 and 1730 and in the 1800s. The Little Ice Age seems to have been associated with longer, more severe winters. In the late 1600s the River Thames completely froze over in London during some particularly cold winters, allowing Frost Fairs to be held on the ice. Tree rings in the south-eastern USA preserve evidence of the Little Ice Age there, too. The cold conditions did not only exist in the northern hemisphere. In southern South America tree-ring evidence indicates a long, cold, moist period from AD 1270 to 1660, which was most pronounced between 1340 and 1640, and glaciers in many places advanced between the late 1600s and early 1800s. The climate has was much warmer during the 20th century.

littoral Of or relating to the the side of a waterbody, such as a *lake or the *intertidal zone. *See also* SHORE.

littoral drift The movement of *sand and other material along the *shoreline, in the *littoral zone, under the influence of waves and currents.

littoral zone 1. The margin of a freshwater body, extending out from the shore to the limit of attached or rooted plants, where light can penetrate to the bottom.

2. The area between the high- and low-water marks on the seashore. Also known as **intertidal zone**.

live fence A *field *boundary formed by planting a line of closely spaced trees or shrubs. Also known as **living fence**. *See also* HEDGE, HEDGEROW.

livestock *Domesticated animals (such as beef cattle, dairy cows, goats, sheep, pigs, chickens, and turkeys) that are kept for meat or dairy production, usually on a *farm or *smallholding.

livestock unit A method for describing different types and age groups of *livestock, based on energy requirements, using standard ratios; one livestock unit is equal to one Friesian dairy cow.

living collection An approach to the *conservation of *biodiversity that is based on using off-site methods (such as *zoos, *botanical gardens, *arboretums, and *captive breeding programmes) to protect and maintain biodiversity.

living fence *See* LIVE FENCE.

living landscape A *landscape that changes with the changing needs of people and *wildlife for space and resources.

living modified organism (LMO) Any *genetically modified organism that has been produced through the use of *recombinant DNA technology, whose genetic material does not occur naturally by mating or natural recombination.

LLRW *See* LOW-LEVEL RADIOACTIVE WASTE.

LMO *See* LIVING MODIFIED ORGANISM.

LNAPL *See* LIGHT NON-AQUEOUS PHASE LIQUID.

LNG *See* LIQUEFIED NATURAL GAS

LNR *See* LOCAL NATURE RESERVE.

loading rate The rate at which materials (such as *suspended sediment, *nutrients, or *contaminants) are transported into a waterbody.

LOAEL *See* LOWEST OBSERVED ADVERSE EFFECT LEVEL.

loam A loose, *permeable *soil composed of *clay, *sand, and *silt, that is often very *fertile and good for growing most crops.

lobbying Activities engaged in by individuals or organizations which are designed to influence the thinking and voting of legislators or other public officials for or against a specific cause. *See also* ENVIRONMENTAL ACTIVISM.

LOC *See* LEVEL OF CONCERN.

Local Agenda 21 (LA21) A programme of action by *local government in the United Kingdom that is designed to promote *sustainable development at a local community level, under *Agenda 21.

local authority An administrative unit of *local government in the United Kingdom that governs local services such as education, housing, and social services.

local biodiversity The number of different *species in a particular area at a particular point in time. *See also* BIODIVERSITY.

Local Biodiversity Action Plan (LBAP) A *biodiversity action plan (BAP) that is produced for an area, within an overall national BAP.

local convective wind A local, *diurnal, thermally driven *wind that is generated over a relatively small area and is influenced by local terrain. Examples include *sea breezes and *land breezes, lake breezes, diurnal mountain wind systems, and *convection currents.

local diversity *See* ALPHA DIVERSITY.

local extinction *See* EXTIRPATION.

local government The administrative functions of regional areas which are carried out by locally elected political bodies, under the overall authority of central (national) government.

Local Nature Reserve (LNR) A *nature reserve in the United Kingdom which has been designated by a *local authority under the National Parks and Access to the Countryside Act (1949).

Local Plan A detailed *land use plan that is prepared and adopted by a local planning authority in the United Kingdom, is accordance with the policies of a *Structure Plan.

local wind A *wind that blows over a relatively small area, often caused by factors such as mountain barriers or large *waterbodies.

loch Scottish word for a *freshwater *lake, or a long narrow *inlet of the *sea.

locus The position of a *gene on a *chromosome.

lode A clearly defined vein of rich *ore within a *rock formation, which extends through a continuous zone or belt. Also known as **mining lode**. *See also* MOTHER LODE.

lode mining The *mining of a valuable mineral *lode which is found between other distinct mineral or rock units, and consists of several *veins that are spaced closely enough together so that all of them, together with the intervening rock, can be mined as a single unit.

loess A *soil or *deposit composed of small *silt-sized particles that were transported by the *wind to their present location, for example after a *glacial period. *See also* AEOLIAN.

log 1. A cut segment of the trunk of a *tree.

2. A record, for example a daily record of a ship's speed and progress or a record of *geological data collected in a *borehole.

logged-over forest A North American term for a *forest in which most or all of the commercially valuable *timber has been removed.

logging 1. Cutting down *trees for *timber. *See also* AERIAL LOGGING, CABLE LOGGING, GROUND-LEAD LOGGING, HIGH-LEAD LOGGING, SKYLINE LOGGING, TRACTOR LOGGING.

2. Storing information about events, or storing the record of a *log.

logging concession A *concession of land for *logging operations.

logging debris *See* LOGGING RESIDUE.

logging residue The residue that is left on the ground after timber cutting, which includes unused logs, uprooted stumps, broken branches, bark, and leaves. Also known as **logging debris** or **slash**.

logistic growth Growth rates that are regulated by internal and external factors which establish an *equilibrium with *environmental resources. *See also* S-CURVE. *Contrast* J-CURVE.

London Convention *See* CONVENTION ON THE PREVENTION OF MARINE POLLUTION BY DUMPING OF WASTES AND OTHER MATTER (1972).

long profile A section along a river *channel which shows how the *slope of the river changes downstream. Many rivers have concave long profiles because headwaters are steep and slope decreases downstream. Also known as **longitudinal profile**. *See also* BASE LEVEL, WATERFALL.

long term An extended period of time; in environmental terms often defined as of the order of 10–15 years. *Contrast* MEDIUM TERM, SHORT TERM. *See also* LONG-RANGE PLANNING.

Long Term Ecological Research Network (LTER) A collaborative scientific programme in the USA, funded by the National Science Foundation, which brings together scientists and students who are studying *ecological processes over long time and broad geographical scales.

longevity The *lifespan of an *organism, or age at death. *See also* SURVIVORSHIP.

longitude Distance on the Earth's surface east or west from a reference line (usually the *Greenwich Meridian), measured in degrees. The *International Date Line is longitude 180° east or west. *Contrast* LATITUDE.

longitudinal dune A long sand *dune that is aligned parallel to the direction of the prevailing *wind.

longitudinal profile *See* LONG PROFILE.

longleaf pine savanna A park-like type of vegetation, shaded by tall pines, kept open by frequent *wildfires, with a diverse layer of grasses and flowering plants including rare orchids, which was once common along the gulf coastal plain of Louisiana in the USA but has been reduced to isolated fragments as a result of *land use changes.

longline A long fishing line at sea, that stretches for tens of kilometres and is baited with hundreds of hooks and held by floats. This type of line often catches many *non-target species (such as tuna, sharks, and dolphins) so has a large unplanned *bycatch.

long-range navigation (Loran) A system of navigation over long distances, in which *latitude and *longitude are determined from the time displacement of radio signals from two or more fixed transmitters.

long-range planning Strategic planning over an extended time period, usually at least 50 years. *Contrast* SHORT-RANGE PLANNING. *See also* LONG TERM.

Long-Range Transport of Air Pollutants (LRTAP) A major international research programme that was launched in 1972 by the *Organisation for Economic Co-operation and Development and is designed to measure air quality at 70 sites in order to develop a better understanding of the dispersion (particularly the *trans-frontier movement) of *air pollutants such as *acid rain.

longshore Along and parallel to the *coastline.

longshore current A current that moves parallel to the *shore, and is formed from the momentum of breaking *waves that approach the shore at an angle.

longshore drift The processes by which *sediment is transported along the *coast by *wave action and *longshore currents. *See also* SPIT, TOMBOLO.

long-term effect An effect that generally occurs after a *lag of about 20 years.

long-term monitoring An *environmental monitoring system that is designed to continue over a period of decades.

longwave radiation *See* INFRARED.

loop highway A scenic drive which begins and ends at the same place, in the form of a circuit (which is not necessarily circular).

lop To cut *branches off a standing *tree.

LORAN *See* LONG-RANGE NAVIGATION.

Los Alamos A town in New Mexico, USA, which in 1942 was chosen as a *nuclear research site where the first atomic bombs were produced.

Los Angelization *See* SUBURBAN SPRAWL.

lot A North American term for a plot of land.

lotic Relating to flowing water such as rivers and streams. *See also* LENTIC WATER.

Love Canal An area near Niagara Falls, New York State, USA, which between 1920 and 1953 was used as a site to dump *hazardous wastes. In 1953, the canal area was covered with soil and presumed safe, but officials began to investigate the dumping of the hazardous wastes and local residents started to report serious health problems. The site was declared to be a health hazard, and the area was evacuated. Since then, the site has been cleaned up using the *Superfund, and people began moving back to live in the area in the early 1990s.

Lovelock, James The British engineer and science writer (born 1919), who was the first to measure *ozone-depleting *CFCs in the air at very low concentrations, and proposed the *Gaia hypothesis.

low The centre of an area of low pressure in the *atmosphere. Also known as a **depression**.

low flow Below-average depth or volume of water in a river, as occurs for example during a *drought or prolonged dry period.

low latitude Close to the *equator, away from the *poles. Contrast HIGH LATITUDE.

low pressure See DEPRESSION.

low risk A *species that is regarded as having a relatively low risk of *extinction. Also known as **near threatened**. Contrast ENDANGERED, VULNERABLE SPECIES, RARE.

low sulphur coal See COMPLIANCE COAL.

low tide Low water; the lowest level to which the *tide falls within the daily tidal cycle. Contrast HIGH TIDE.

low water mark The line along the coast to which the sea drops at *low tide. Contrast HIGH WATER MARK.

low-emission vehicle Vehicles which emit relatively little *air pollution compared with conventional *internal combustion engines.

lower atmosphere stability index A measure of the stability and dryness of the air, which is used to indicate the potential for wildfire to grow and spread through an area.

lowest achievable emission rate The lowest possible rate of *emission for a particular type of *facility and a specific *air pollutant.

lowest observed adverse effect level (LOAEL) The lowest *concentration of a *hazardous substance at which it is possible to detect a statistically or biologically significant increase in the frequency or severity of an *adverse effect between an *exposed population and a *control group.

lowland Low-level country. Contrast HIGHLAND, UPLAND.

low-level radioactive waste (LLRW) *Radioactive waste that is generated by hospitals, research labs, and some industries, and is less hazardous than *high-level radioactive wastes that are generated by *nuclear reactors. Also known as **low-level waste**.

low-level waste See LOW-LEVEL RADIOACTIVE WASTE.

LPG See LIQUEFIED PETROLEUM GAS.

LRTAP See LONG-RANGE TRANSPORT OF AIR POLLUTANTS.

LTER See LONG TERM ECOLOGICAL RESEARCH NETWORK.

LULU Locally unwanted land uses such as toxic waste dumps, incinerators, smelters, airports, freeways, and other sources of environmental, economic, or social degradation. See also NIMBY.

lumber See TIMBER.

LUST See LEAKING UNDERGROUND STORAGE TANK.

lustre The way that *light is reflected from the surface of a *mineral. Some minerals have a glassy (vitreous) lustre, some have a metallic lustre (like polished metal), and some have a dull (earthy) lustre.

lysimeter A device for measuring potential and actual *evapotranspiration in a column of *soil.

m.y. An abbreviation for million years.

maar A flat-bottomed *volcanic crater that was formed by an explosion of trapped gases and is often filled with water.

Maathai, Wangari Muta A Kenyan environmental and political activist (born 1940) who helped to establish the *Greenbelt Movement, and in 2004 became the first African woman to receive a Nobel Peace Prize for 'her contribution to sustainable development, democracy and peace'.

MAB *See* MAN AND THE BIOSPHERE PROGRAMME.

MAC *See* MARGINAL ABATEMENT COST, MAXIMUM ALLOWABLE CONCENTRATION.

Mach A measurement of speed relative to the speed of sound. 1 Mach is the speed of sound, which varies from about 1225 kilometres per hour (km h^{-1}) in warm air at sea level to about 1060 kilometres per hour in very cold air at altitude.

mackerel sky An appearance of the sky, caused by *altocumulus clouds, which is greenish-blue in colour and wavy in form, and looks like the markings on a mackerel fish.

macro- Large scale. *Contrast* MESO-, MICRO-.

macrobenthos The larger organisms of the *benthos, generally longer than 0.5 millimetres. *Contrast* MEIOBENTHOS, MICROBENTHOS.

macroburst A large scale *downburst of wind in the *atmosphere, which usually lasts for 5–20 minutes and has a radius of more than four kilometres. Strong macrobursts can cause damage similar to that of an F2 tornado on the *Fujita Scale. *Contrast* MICROBURST.

macroclimate The *climate of a large area (up to hundreds of square kilometres), as distinguished from the smaller scale *microclimates and *mesoclimates within it. Also known as **regional climate**.

macroeconomics The study and analysis of the whole *economy of a country or region, that considers issues such as such as *economic growth, inflation, unemployment, and economic fluctuations. *Contrast* MICROECONOMICS.

macroelement *See* MACRONUTRIENT.

macroevolution Large scale biological *evolution, that extends over geological eras and results in the formation of new taxonomic groups through *adaptive radiation.

macrofauna Small animals (such as earthworms and large *arthropods) that are longer than about 1 millimetre and can thus be seen with the naked eye. *Contrast* MESOFAUNA, MICROFAUNA.

macroinvertebrate Any *invertebrate that is large enough to be seen with the naked eye (with a body longer than 2 millimetres). This includes *crustaceans (such as crayfish), *insects, and worms. *Contrast* MICROINVERTEBRATE.

macroinvertebrate community index (MCI) A method for assessing the quality of a stream or river, based on the presence or absence of particular types of *invertebrate (*indicator species) on the stream bed, which have different abilities to tolerate *pollution.

macronutrient A *nutrient or chemical element that is required in relatively large amounts by plants because it is vital for healthy plant growth. After *carbon, *hydrogen, and *oxygen the three most important ones are *nitrogen, *potassium, and *phosphorus, which are the usual ingredients of *fertilizers; others include *calcium, *magnesium, and *sulphur. Also known as **macroelement** or **major element**. Contrast MICRONUTRIENT.

macroparasite A relatively large *parasite. Contrast MICROPARASITE.

macrophyte Any *plant that is large enough to be seen with the naked eye. The term is often applied to *aquatic plants such as aquatic mosses, *ferns, and rooted plants.

macroscopic Large enough to be visible to the naked eye, usually longer than about 1 millimetre. Contrast MICROSCOPIC.

MACT See MAXIMUM ACHIEVABLE CONTROL TECHNOLOGY.

Madrid Protocol on Environmental Protection to the Antarctic Treaty See ANTARCTIC TREATY.

MAFF In the UK, the Ministry of Agriculture, Fisheries and Food, now known as *DEFRA.

mafic A dark-coloured *igneous rock (such as *basalt) that generally contains little *silica but a great deal of *magnesium and iron-rich minerals; found mainly in *oceanic crust. Contrast FELSIC.

magma Molten rock found beneath the Earth's crust from which *igneous rocks are formed, and which becomes *lava when it reaches the surface via a volcanic vent. It may contain some solid particles and gases.

magma chamber A chamber or large space within the *lithosphere in which *magma is stored.

magnesium (Mg) A silver-coloured, naturally occurring metallic element that is abundant in the environment (it accounts for 2.09% of the Earth's *crust, particularly the *oceanic crust), contributes to *hard water in high concentrations, is not generally considered toxic, and is a *macronutrient for animals and humans.

magnet Any material that has a permanent *magnetic field.

magnetic Capable of being magnetized, or having the properties of a *magnet.

magnetic anomaly Differences from place to place in the strength and direction of the *magnetic field, compared to the average.

magnetic field A field of force that is generated by electric currents. The Earth's magnetic field is believed to be generated by the rotation of the molten iron–nickel *outer core around the solid *inner core, like a dynamo. See also PALAEOMAGNETISM.

magnetic north The direction to which the needle of a magnetic compass points, which is towards the northern *pole of the Earth's *magnetic field rather than towards *true north.

magnetic polarity The direction, north (normal) or south (reversed), in which a magnetic compass needle points.

magnetic pole The end of a *magnet (or either end of the *Earth's axis) where the strength of the *magnetic field is greatest.

magnetic reversal The process by which the *Earth's magnetic north pole and its magnetic south pole reverse their positions over time. See also PALAEOMAGNETISM.

magnetic separation The process of separating *ferrous materials from mixed *municipal waste streams using *magnets.

magnetic storm A sudden, worldwide disturbance of the Earth's *magnetic field that is caused by emission of particles from the Sun (*sunspots).

magnetic susceptibility The degree of magnetization of a material in response to a *magnetic field, which depends mostly on the concentration of ferromagnetic minerals within it.

magnetics The study of the properties of *magnets and *magnetic fields.

magnetometer An instrument for measuring the magnitude and direction of the Earth's *magnetic field.

magnetopause The boundary layer in space (about 63 000 kilometres from Earth in the direction of the Sun), between *magnetosphere and the *solar wind, where the Earth's *magnetic field balances the pressure of the solar wind.

magnetosphere The region of space that exists above the *atmosphere and within the *magnetopause, that is under the direct influence of the Earth's *magnetic field.

magnetostratigraphy See PALAEO-MAGNETISM.

magnox An *alloy of *magnesium oxide and *aluminium.

magnox reactor A redundant type of commercial *nuclear reactor that used *magnox as fuel cladding, was designed in the UK, and was capable of producing *plutonium for nuclear weapons. A total of 26 were built in the UK. By the end of 2003 eight magnox reactors were still in operation, and all are planned to be closed by 2010.

Magnuson Fishery Conservation and Management Act See FISHERY CONSERVATION AND MANAGEMENT ACT.

mainstream flow The portion of the flow in a fluid that is away from the surrounding surface, and thus not under the influence of the *boundary layer.

major element See MACRONUTRIENT.

major recreation facility A North American term for developed *recreation facilities that are costly to build and run, and are heavily used, such as campgrounds, picnic areas, swimming areas, boating sites, and *interpretation sites. Contrast MINOR RECREATION FACILITY.

malaria An infectious *disease that is caused by a parasite (*Plasmodium*) transmitted by the bite of infected mosquitoes. It is common in hot countries, its symptoms include recurring chills, fever, and sweating, and it can be fatal. Also known as **marsh fever**.

male In organisms with separate sexes, the one which produces sperm. Contrast FEMALE.

malformation Something abnormal; in humans, a structural defect that is present at birth, occurs infrequently, and is due to abnormal development.

malignant Cancerous; a growth in which cells reproduce faster than normal cells. Contrast BENIGN.

malleable Capable of being shaped or bent or drawn out. See also DUCTILE.

mallee An Australian name for grassy, open *woodland *habitat that is found in many *semi-arid areas.

malnourishment See MALNUTRITION.

malnutrition Undernourishment caused by insufficient food and/or a *diet that does not contain enough *protein, essential fats, vitamins, *minerals, and other *nutrients that are required for good health. Also known as **malnourishment**.

Malthus, Thomas An English economist (1766–1834) who argued, in *An Essay on the Principle of Population* (1798), that increases in the human *population would outgrow increases in the resources available to support it, because populations increase *geometrically while food supplies increase *arithmetically, and that population is kept in check by a combination of war, *famine, and *disease. He assumed, based on his observations of 18th century English society, that if population growth continued unchecked, then population would outstrip the food available and this would cause widespread famine and death. He also described a *feedback mechanism: when the population became too large for the available food supply, increased mortality would reduce the population to the level that could be sustained by the amount of food produced.

Malthusian growth A pattern of growth in a *population (human or animal) in which a *population explosion is followed by a *population crash. Also known as **irruptive growth.**

mammal A class of *warm-blooded *vertebrate animals that breathe air, have hair, feed their young with the mother's milk, have four types of well-developed teeth, and usually have four well-developed limbs with toes that have nails, claws, or hoofs.

Man and Nature A book by George Perkins *Marsh (1864) which described and illustrated many ways in which human activities change and damage environmental systems, and was a catalyst for the development of reforestation, watershed management, soil conservation, and nature conservation.

Man and the Biosphere Programme (MAB) An interdisciplinary programme of research and training, established by *UNESCO in 1971, designed to develop the basis for the *sustainable use and *conservation of biological diversity, and for the improvement of the global relationship between people and their environment.

managed burn *See* PRESCRIBED BURN.

Managed Nature Reserve/Wildlife Sanctuary An *IUCN Management Category (IV) for protected areas, designed to ensure the natural conditions necessary to protect nationally significant species, groups of species, biotic communities, or physical features of the environment.

Managed Resource Area *See* MULTIPLE-USE MANAGEMENT AREA/MANAGED RESOURCE AREA.

management The act of directing and controlling the affairs of a business, organization, or other body to ensure that they operate efficiently and effectively,

in order to accomplish agreed objectives. *See also* RESOURCE MANAGEMENT.

management agreement An agreement between the owner of a property and the company or individual who is responsible for managing it. In the UK, it is a written contract agreement between an official body (such as *DEFRA) and landowners and tenants regarding their management of land, which is a prerequisite of certain incentive payments. *See also* ENVIRONMENTALLY SENSITIVE AREA.

management area An area of land, or a group of areas which are not necessarily contiguous, which have common *management objectives.

management indicator species (MIS) *See* INDICATOR SPECIES.

management information *Information that is required or useful for *management decision-making.

management information system (MIS) A system for collecting and analysing *information, usually computer-based, that is designed to help *management decision-making.

management objective A high-level statement of what is desired, or a description of an intended outcome, which is usually measurable, time limited, specific, and practical.

management plan A written plan that describes the overall guidelines within which an activity or project is organized, administered, and managed to ensure that agreed *management objectives are achieved in a timely manner.

management practice A specific action, measure, or treatment which is part of a *management plan and is designed to meet one or more stated *management objectives.

management prescription Definition of the most appropriate *management practices for use in a specific situation, or in a specific area, in order to achieve agreed *management objectives.

m

management programme The *management practices that are required to achieve agreed *management objectives within a management *plan.

management science *See* OPERATIONAL RESEARCH.

management unit A defined area of land or water to which specific *management decisions apply. Also known as **management zone**.

management zone *See* MANAGEMENT UNIT.

management-by-objective An approach to *management (within an organization or agency) which involves the definition of goals and responsibilities, in terms of measurable results, and the use of these measures to guide the operation of the organization/agency and the assessment of the contribution of different parts of the organization/agency to meeting the specified goals.

managerialism Looking at the behaviour of an organization or activity from the perspective of a manager who is responsible for delivering agreed *management objectives, with an emphasis on measuring performance, efficiency, and effectiveness against targets.

mandible The lower jaw, that is used for biting or chewing.

manganese (Mn) A brittle, silvery *metal that is associated with iron *ores, is used in the manufacture of *steel (to improve hardness and resistance to corrosion), and is an essential *element (*micronutrient) for plant and animal life.

manganese nodule A small, rounded *concretion that is found on the deep *ocean floor and may contain up to 20% *manganese and smaller amounts of *iron, *copper, and *nickel *oxides and *hydroxides.

mangel *See* MANGROVE FOREST.

mangrove A tropical *evergreen tree or shrub which grows in coastal areas, particularly on tidal flats and *estuaries. It has stilt-like roots which knit together to form dense thickets, which can trap coastal sediment and encourage coastal deposition, allowing the mangrove to extend further out from the shore. Mangrove swamps provide important habitats for aquatic and amphibious species.

mangrove forest A *shoreline *ecosystem that is dominated by *mangrove trees, with associated *mud flats. Also known as **mangel** or **mangrove swamp**.

mangrove swamp *See* MANGROVE FOREST.

manifest system *See* CRADLE-TO-GRAVE SYSTEM.

man-made disaster *See* TECHNOLOGICAL DISASTER. *Contrast* NATURAL DISASTER.

mantle The hot, pliable zone within the *Earth's interior that surrounds the partially molten *core and underlies the thin, cool, outer *crust. The mantle is about 2900 kilometres thick, and is made from rock that is relatively dense. It is often described as solid, but in reality it behaves like a *plastic and can flow. It is composed mainly of minerals that are *silicate compounds (combining silica and oxygen). *Iron is the main metal, but there is also a great deal of *magnesium. The composition and behaviour of material within the mantle varies with distance from the core (below) and crust (above), and two main zones can be distinguished within it—the *asthenospere and the *lithosphere. The sharp and well-defined junction between the mantle and the crust is called the *Mohorovicic discontinuity.

manual separation The sorting by hand of material in *waste that can be *recycled or turned into *compost.

manufacturing The conversion of raw materials into finished goods for sale.

manure The waste products of *animals that are sometimes used as a form of organic *fertilizer on *farms, *allotments, or *gardens. *See also* FARMYARD MANURE, GREEN MANURE.

maquis The name given to *scrub vegetation, consisting of shrubs and isolated trees, in the *arid *biome in France.

marble A *metamorphic rock that is formed by applying great heat and pressure to *limestone.

mare's-tail cirrus (Ci) Long, white, hair-like wisps of *cirrus clouds, which are composed mainly of ice crystals. They are thicker at one end then tapering to the other and look like the tails of female horses.

marginal abatement cost (MAC) The additional cost that is associated with reducing *emissions of a particular *pollutant by a set amount.

marginal cost The additional cost that is associated with producing one more unit of a particular good or service.

marginal land Land which barely pays the cost of working it because of physical limitations such as infertility, *drought, or shallow soils.

marginal value The additional value of a resource that is associated with an incremental change in the quantity of it that is available.

mariculture *See* AQUACULTURE.

marine Of or relating to the *sea or *ocean. Also known as **maritime**.

marine climate A *climate that is dominated by the proximity of the land to the *ocean, which keeps winters relatively mild and summers cool.

marine ecosystem A *saltwater *aquatic *ecosystem that includes *estuaries and *coastal areas, along with the *open sea and *oceans.

marine intertidal A *coastal *saltwater *wetland *habitat that is flooded by *tidewaters.

Marine Mammal Protection Act (1972) US legislation that provides for the protection and *conservation of *marine *mammal species by prohibiting the taking (harassing, hunting, capturing, or killing) of marine mammals, and prohibiting the import of any marine mammal product or any fish that has been associated with the taking of marine mammals.

Marine Nature Reserve (MNR) A statutory marine protected area in Great Britain, declared under the Wildlife and Countryside Act 1981, which is designed to protect marine flora, fauna, or geological or *physiographic features in the area, and to provide opportunities for study and research.

marine pollution *Pollution of the seas and oceans.

Marine Protected Area (MPA) An area of land or sea that is dedicated to the protection and maintenance of *biodiversity, and of natural and cultural resources, and managed through legal or other effective means.

Marine Protection, Research, and Sanctuaries Act (1972) US legislation that regulates the dumping of any material in the ocean which might adversely affect human health, marine environments, or the economic potential of the ocean. Also known as the **Ocean Dumping Act**.

marine resource Any *natural resource that is found in, on, or under the *sea or *ocean, including fish and *minerals.

marine sanctuary A designated area of the sea or ocean that serves as a *nature reserve, which people can access. Non-extractive activities like boating, swimming, and snorkelling are allowed, but all types of fishing, including trawling, and all collecting and extractive activities are prohibited.

marine subarctic climate zone A cold *subpolar *climate zone, in which temperatures remain cold throughout the year. Natural vegetation in this zone is either *tundra (where it is coldest) or *boreal forest (where it is milder).

marine system The open *ocean over the *continental shelf, where conditions are affected by *waves, *currents, and oceanic *tides.

marine west coast climate zone A *mid-latitude climate zone with a warm and wet climate, which is found at higher *latitudes than the *Mediterranean climate zone. Forest cover often develops because of the heavy rainfall and moderate temperatures.

marine west coast forest A coniferous *temperate forest *biome that is found along the marine west coast of some continents, including the Pacific Northwest of North America, northwest Europe, southern Chile, and New Zealand. The biome often includes trees that are taller than 30 metres, have diameters greater than two metres, and are well over 1000 years old.

maritime *See* MARINE.

maritime air Air with a high moisture content, whose humid characteristics were developed over a large *sea or *ocean.

maritime polar air mass A mass of cool, humid air that forms over the cold ocean waters of the North Pacific and North Atlantic.

maritime tropical air mass A mass of warm, humid air that forms over *tropical and *subtropical oceans.

market barrier In *economics, any factor or condition that prevents or inhibits *market equilibrium. As applied to *air pollution control, for example, this includes policy and legal frameworks that prevent the adoption of cost-effective technologies or practices that could reduce *emissions of *greenhouse gases.

market cost The cost that a consumer pays to purchase a particular good or service, which usually fails to take into account *environmental degradation and other *externalities. *See also* TRUE COST.

market economy An *economy in which most goods and services are produced by the private sector rather than the public sector, and in which the prices of goods and services are determined by *supply and *demand, usually under *market equilibrium. *See also* MARKETPLACE.

market equilibrium A situation in which the quantity of a particular good or service that people are willing to buy (*demand) equals the quantity that people are willing to sell (*supply). This is the basis of the *free market system, in which there are neither monopolies nor government interventions. *See also* MARKET ECONOMY.

market failure Any situation where a *free market fails to produce the best use of scarce *resources, which arises when *market prices are not equal to the *social cost of resources, because of *externalities. *See also* NON-MARKET DAMAGE.

market forces The interaction of *supply and *demand which determines *market equilibrium and shapes a *market economy.

market garden A plot of land on which fruit and vegetables are cultivated for sale. *See also* HORTICULTURE, SMALLHOLDING.

market price The price at which both buyers and sellers of a good or service in a *market economy are willing to do business with each other, when *supply and *demand are equal. Also known as **market value**.

market value *See* MARKET PRICE.

market values Values that can be expressed in financial or monetary terms.

market-based incentive *See* MARKET-BASED MECHANISM.

market-based mechanism A mechanism (such as *command and control, and *environmental regulation) that is designed to influence *market forces in order to manipulate *market equilibrium, with a view to improving *environmental protection or reducing *environmental damage. Also known as **market-based incentive**.

marketplace A situation (an actual place or any means of contact) in which two or more people agree to buy and sell a product or service. *See also* MARKET ECONOMY.

mark–recapture *See* CAPTURE–RECAPTURE METHOD.

marl An *alluvial *clay or *deposit that contains a relatively large amount of *calcium carbonate, and can be used as a *fertilizer on *soils that are deficient in *lime.

marl lake A *freshwater *lake in which the concentration of dissolved *calcium carbonate is greater than 100 milligrams per litre.

marram grass A type of *grass that has an extensive, tightly knit underground root system and can survive in the relatively salty, nutrient-poor free-draining sand that is typical of *beach dunes.

marsh A low-lying *wetland habitat with grassy vegetation, often close to a river or lake, that is frequently flooded to a shallow depth. Marshes can be *freshwater or *saltwater, tidal or non-tidal. Also known as **marshland**.

marsh fever *See* MALARIA.

Marsh, George Perkins A US diplomat and classical scholar (1801–82) who published the influential book *Man and Nature* in 1864 and is regarded by some as the father of the *environmental movement.

Marshall, Robert A US conservationist (1901–39) who founded the Wilderness Society. He was among the first to suggest that large tracts of Alaska be preserved, helped to shape the *US Forest Service's policy on *wilderness designation and management, and wrote with conviction on all aspects of *conservation and *preservation.

marshland *See* MARSH.

marsupial A *mammal (of the order Marsupialia) which carries its young in a pouch. Examples include opossums, kangaroos, and wombats.

Marxism The economic and political theories of Karl Marx (and Friedrich Engels), based on the belief that human actions and institutions are driven mainly by economic factors (particularly by a profit motive), that social change is driven mainly by class struggles, and that *socialism will ultimately replace *capitalism.

mass balance The relative balance between the *input and *output of material and processes within an *open system, such as the snow and ice within a *glacier, or the water or sediment moving through a *lake or *river.

mass burn The deliberate *incineration of unsorted *solid waste material.

mass extinction A catastrophic, widespread event in which a large proportion (up to 90%) of species become *extinct in a relatively short time compared with normal background extinction. There have been at least five such mass extinctions, including the wholesale extinction of the *dinosaurs, other big reptiles, and many marine invertebrates at the *K–T boundary. In the short term mass extinctions have a catastrophic impact on *biodiversity. But the long-term impact on biodiversity is generally limited, because mass extinctions have usually been followed by phases of rapid *speciation. Surviving species generally evolve rapidly, by *adaptive radiation, to occupy the *niches vacated by the species that became extinct.

mass movement The process by which unconsolidated material moves down *hillslopes in the form of a *flow, *slide, *fall, or *creep, as a result of *gravity. Also known as **mass wasting**. *See also* AVALANCHE, LANDSLIDE.

mass selection A method of plant *breeding in which seed from a number of individuals is selected to form the next generation, with selection criteria relaxed until later generations and crosses being created at random.

mass spectrometry An instrument that is used to identify and measure the chemicals that are present in a substance by their mass and charge, based on very small samples. *See also* GAS CHROMATOGRAPHY-MASS SPECTROMETRY.

mass transit A form of public transportation in which buses, trains, trams, and other forms of transportation are used to efficiently move large numbers of people, with reduced overall *environmental impacts.

mass wasting *See* MASS MOVEMENT.

massif A block of the Earth's *crust that is bounded by *faults and has been moved to create one or more peaks of a *mountain range. It is usually more rigid than surrounding rocks.

massive rock A solid mass of rock that is homogeneous, with no joints, cracks, *foliation, or bedding.

mast The *fruit and *seed of shrubs, woody vines, trees, cacti, and other non-herbaceous vegetation (including acorns, beechnuts, and chestnuts) that is available as food for *wildlife.

master plan *See* COMPREHENSIVE PLAN.

mating In animals, the act of pairing a male and female for sexual reproduction.

matrix In geology, the small-grained material that is found in the spaces between larger-grained materials in a rock.

matter Anything that takes up space and has mass.

mature soil A well-developed *soil that normally has a *soil profile with clearly defined *horizons.

mature timber A stand of *trees that is fully grown, especially in height, and is in full seed production. Also known as **old growth**, or **late successional**.

mature tree A *tree that has reached the desired size or age for its intended use, which varies from species to species.

maturity 1. Part of the *life cycle of an organism, when it has become adult, between *youth and *old age.
2. The middle stage in the *cycle of erosion, in which the landscape has high relief and well-developed drainage. *Contrast* OLD AGE, YOUTH.

Mauna Loa A volcano in Hawaii which is intermittently active and last erupted in 1984.

Mauna Loa record The record of measurements of atmospheric *carbon dioxide concentrations which has been collected by scientists at the Mauna Loa Observatory on *Mauna Loa in Hawaii, since March 1958. This is the longest reliable daily record of atmospheric carbon dioxide measurements in the world.

Maunder minimum The period from 1645 to 1715 when solar activity was very low and few *sunspots were observed.

maximum achievable control technology (MACT) Emissions limitations based on the best demonstrated *control technology or practices, taking into account cost and technical feasibility, which produces a very high level of *pollution control.

maximum allowable concentration (MAC) The highest concentration of a particular *pollutant that workers can safely be in contact with in their work environment.

maximum contaminant level (MCL) The highest allowable concentration of a given *contaminant in *potable water.

maximum exposure limit (MEL) The highest level of exposure to a particular *pollutant to which a person may safely be exposed.

maximum permissible dose (MPD) The highest amount of *radiation to which a person may safely be exposed.

maximum sustainable yield (MSY)
The maximum crop or *yield that can be taken on a sustained basis from a particular *population in a defined area (such as a portion of the *ocean) without driving it towards *extinction. *Contrast* OPTIMUM YIELD.

maximum tolerated dose (MTD)
The highest dose of a drug, drug combination, or other treatment that an individual of a particular species of animal (including humans) can safely withstand over a major portion of its lifetime.

MCI *See* MACROINVERTEBRATE COMMUNITY INDEX.

McIntyre-Stennis Act (1962) US legislation that established a cooperative research programme in *forestry for state land-grant colleges and universities.

MCL *See* MAXIMUM CONTAMINANT LEVEL.

McMansion North American pejorative term for a large house built on a small plot of land, often constructed cheaply but including ostentatious traditional features. Also known as **starter castle**.

McSweeney-McNary Act (1928) US legislation that authorized a comprehensive research programme for the *US Forest Service. This act was repealed and replaced by the Forest and Rangeland Renewable Resources Research Act (1978).

MDC *See* MORE DEVELOPED COUNTRY.

meadow A *field in which *grass or *alfalfa are grown to be used as *pasture or made into *hay. *See also* HAY MEADOW.

Meadows, Dennis US economist, co-author (with Donella Meadows) of the book *Limits to Growth*, and member of the *Club of Rome.

mean annual runoff The average amount of water that flows down a particular river, per year, expressed either as a depth (in millimetres) of water spread evenly across the entire *drainage basin, or as a volume (in cubic metres) of water flowing past a given point.

mean annual temperature The average *temperature of the air over a particular location, over the entire year.

mean daily temperature The average *temperature of the air over a particular place, over a 24-hour period.

mean low water The long-term average height of all low waters at a particular place.

mean sea level The long-term average height of the water level of the sea surface at a particular place.

mean tide level The average of mean high water and mean low water. Also known as half-tide level.

meander A curve or bend in a *river channel, which forms naturally by *fluvial *erosion. *See also* FLOODPLAIN, OXBOW LAKE.

measure 1. To determine the size of something.
2. A reference point to which other things can be compared.
3. An action that can be taken, for a defined purpose.

mechanical aeration The use of mechanical energy to inject air into water, to cause *waste material in the water to absorb *oxygen and *decompose.

mechanical separation The use of mechanical means to sort waste material into separate components.

mechanical site preparation Any activity that involves the use of mechanical machinery to prepare a site for other uses (including building).

mechanical thinning An approach to *thinning trees in a *forest which is based on removing an agreed proportion of the trees irrespective of tree size or

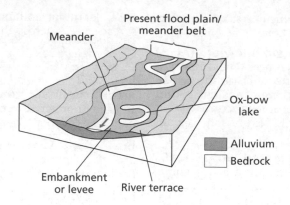

Fig 13 Meander

quality, for example by removing every second tree. *See also* LINE THINNING, SELECTIVE THINNING.

mechanical turbulence Turbulent *eddy motions in the air which are caused by physical obstructions, such as trees, buildings, and mountains.

mechanical weathering The process by which frost action, growth of salt crystals, absorption of water, and other physical processes break down rock into smaller fragments, without involving any chemical change. Also known as **physical weathering**. *Contrast* CHEMICAL WEATHERING. *See also* WEATHERING.

media Plural of *medium.

medial moraine A type of *moraine that forms in the centre of a *glacier or *ice stream downstream from the confluence of neighbouring valleys, where adjacent *lateral moraines join together.

median lethal dose *See* LD50.

median tolerance limit The concentration of a particular *toxic substance at which 50% of the animals in a test are able to survive for a specified period of *exposure.

mediation An informal, voluntary process for resolving disputes, in which parties are encouraged to discuss issues openly, all decisions are reached by con-

sensus, and any participant can withdraw at any time.

medical waste Any *solid waste that results from medical activity, including hospital waste. This includes human pathological wastes, human blood and blood products, used or unused sharps (syringes, needles, and blades), certain animal waste, and other waste that is contaminated with *infectious agents or *pathogens.

medicinal Something used to prevent or cure illnesses and diseases.

medicinal and aromatic plant material Whole plants and plant parts (including seeds and fruits) that are used primarily in pharmacy and in making perfumes.

medieval The period of history between the Fall of the Roman Empire (AD 300) and the Renaissance (1450) in Europe. Also known as the **Middle Ages**.

Medieval Warm Period *See* LITTLE CLIMATIC OPTIMUM.

mediterranean climate A *climate that has mild, wet winters and warm to hot, dry summers, and is usually located in middle *latitudes on the western side of *continents. Also known as **dry-summer subtropical climate**.

medium A substance through which something is transported, such as air or water through which *contaminants are distributed. The plural of medium is media.

medium term A period of time, in environmental terms often defined as between about three and five years. *Contrast* LONG TERM, SHORT TERM.

megacity *See* MEGALOPOLIS.

megalopolis The largest type of *urban settlement, with a population of more than 10 million inhabitants, which may contain some open land, and is formed when large *conurbations or *metropolises grow together and link up. In 1950 there were only two cities in the developing world with populations of more than five million each—Shanghai (China) and Buenos Aires (Argentina). The *United Nations estimates that by the year 2000 there were around 35 such cities, including 16 with more than 10 million people. The number of megacities continues to rise. In 1960 only New York and Tokyo housed more than 10 million people each, yet by 1997 there were 17 megacities (13 of them in developing regions). If recent trends continue, there are likely to be 26 megacities by 2015, 22 of them in developing regions (and 18 of these 22 will be in Asia). Between them these 26 megacities are expected to house more than a tenth of the world's population. Also known as **megacity** or **supercity**.

megaplankton Large *planktonic organisms, at least two millimetres in size.

megatonne (Mt) A unit of weight equal to one million *tonnes.

megawatt (MW) A unit of electrical power equal to 1000 kilowatts (one million *watts).

meiobenthos Organisms of the *benthos which are between 0.1 and 0.5 millimetres in size. *Contrast* MACROBENTHOS, MICROBENTHOS.

meiofauna Animals which are between 0.1 and 0.5 millimetres in size.

meioflora Plants which are between 0.1 and 0.5 millimetres in size.

MEL *See* MAXIMUM EXPOSURE LIMIT.

Melaleuca tree An Australian shrub (*Melaleuca quinquenervia*), also known as the tea tree, whose extracted oil has antiseptic and healing properties. It has invaded *wetlands of south Florida, displacing *native species, reducing the quality of the wildlife *habitat, and changing local *hydrology, soils, and fire regimes.

melanin The substance that gives colour to the skin, hair, and parts of the eye, and protects against the damaging effects of *ultraviolet radiation.

melanism The process by which a *species adapts to changing environmental conditions by producing more *melanin in the skin for increased coloration in the body. *See also* INDUSTRIAL MELANISM.

melanoma The most serious form of *cancer of the skin, which is highly malignant.

melittophily *Pollination by bees.

meltdown In a *nuclear reactor, a situation in which uncontrolled nuclear chain reactions cause the *core to melt.

melting A *phase change from a solid to liquid.

melting point The *temperature at which a substance changes from a solid to a liquid state, by *melting.

meltwater Water that is produced by the melting ice of a *glacier or a *snowbank.

meltwater channel A channel that is eroded by glacial *meltwater, either under the glacier or along its side. Also known as **overflow channel**.

membrane A thin film, usually *semipermeable.

Mendes, Chico A Brazilian rubber tapper, trade unionist, and environmental activist (1944–88) whose full name was Francisco Alves Mendes Filho. He fought to stop the logging of the Amazon *rainforest to create land for cattle ranching, founded a national union of rubber tappers to try to preserve their profession and the rainforest that it relied upon, and was murdered in 1988 by ranchers opposed to his activism.

Mercalli scale A scale that is widely used to describe the *intensity of *earthquakes. *See also* MODIFIED MERCALLI INTENSITY SCALE. *Compare* RICHTER SCALE. The scale is given in Appendix 7.

Mercator projection A projection of the *Earth onto a cylinder, in order to produce a flat map, in which areas appear greater the farther they are from the *equator. It is the most commonly used map projection in atlases.

merchantable timber Commercially valuable *timber, which can be sold at profit.

mercury (Hg) A naturally occurring element which is present in various *ores. It is a toxic *heavy metal which is liquid at room temperature. Mercury is used as an industrial chemical in a range of processes (such as the production of *chlorine), and in instruments such as *thermometers and *barometers. It does not *biodegrade naturally in the environment, and once released remains indefinitely but can be *biotransformed into various different states. It can *bioaccumulate in the environment and is highly toxic if inhaled or swallowed.

mercury barometer A type of *barometer used for measuring atmospheric pressure in which mercury rises or falls in a graduated glass tube as the pressure of the atmosphere increases or decreases.

mercury poisoning A *toxic condition in humans that is caused by swallowing or inhaling *mercury, and can cause vomiting and diarrhoea and kidney problems that may lead to death. *See also* MINAMATA DISEASE.

mere A shallow *freshwater *lake.

meridian A north–south line between the *North Pole and the *South Pole, at right angles to the *equator, which connects all places with a given *longitude.

meridional Along a meridian or line of longitude. *Contrast* ZONAL.

meromictic A permanently stratified lake in which complete mixing of the water does not naturally occur.

meroplankton *Organisms that spend part of their time as *plankton but also spend some time as *benthos, such as planktonic larvae of benthic invertebrates.

mesa A broad, flat-topped hill surrounded by cliffs, capped with a protective layer of resistant rock, common in *semi-arid areas.

mesic A moderately moist *habitat.

mesic prairie A type of *prairie that is found on relatively well drained sites which have high moisture available through most of the growing season.

meso- Medium-scale. *Contrast* MACRO-, MICRO-.

mesoclimate The *climate of an area up to several square kilometres in area; larger than a *microclimate but smaller than a *macroclimate.

mesocycle The deep rotating column of air within a *supercell thunderstorm.

mesocyclone *See* SUPERCELL.

mesofauna Small animals, such as worms and insects, which are bigger than *microfauna but smaller than *macrofauna.

mesoflora Small plants, which are bigger than *microflora but smaller than *macroflora.

mesohaline *Brackish water with a *salinity of between 5 and 18 parts per thousand (‰).

Mesolithic The middle *Stone Age, following the *Palaeolithic, which began in Europe about 8500 BC, during which cultures were reliant on hunting, fishing, and gathering.

mesopause The upper boundary of the *mesosphere with the *thermosphere, at a height of about 80 kilometres above the ground.

mesopelagic The zone of *oceans and *seas between about 200 and 1000 metres below the surface, between the upper sunlit (*photic) zone above and the dark ocean depths below, where some light can still penetrate to allow limited *photosynthesis.

mesophyte A plant that is adapted to grow in fairly moist conditions. *Contrast* HYDROPHYTE, XEROPHYTE.

mesoscale meteorology The study of weather systems that range in size from a few kilometres to about 100 kilometres in scale. This includes local winds, *squall lines, *thunderstorms, and *tornadoes. *Contrast* SYNOPTIC METEOROLOGY. *See also* MESO-.

mesosphere 1. The layer within the upper *atmosphere that lies above the *stratosphere, stretching from about 50 kilometres to the *mesopause (about 80 kilometres above the ground), in which temperature decreases quite sharply with height down to about −90°C.
2. The lower *mantle within the Earth's *crust.

mesotrophic A moderately productive *habitat (such as a *lake or *reservoir) that has a moderate supply of *nutrients. *Contrast* OLIGOTROPHIC, EUTROPHIC. *See also* EUTROPHICATION.

Mesozoic The *era of geological time after the *Palaeozoic era and before the *Cenozoic era, about 245 to 65 million years ago, when there is evidence of the first *mammals, birds, and flowering plants.

mesquite A small, thorny hardwood tree that is found in arid parts of North America.

metabolic activity The chemical changes that occur in a living animal or plant as a result of *metabolism. In animals, this includes synthetic reactions (such as the manufacture of *proteins and *fats) and destructive reactions (such as the breakdown of sugars into *carbon dioxide and *water). In plants, it includes *photosynthesis.

metabolic engineering A process that is used to deliberately modify *organisms so they will produce small molecules and chemicals, or *biochemicals.

metabolic rate The rate of *biochemical reactions within an *organism, which can be estimated by measuring food consumption, energy released as heat, or oxygen used in *metabolic activity.

metabolism The biochemical processes through which a living cell or organism converts food to energy and then gets rid of waste products, thus allowing it to continue to function and grow. This involves two sets of processes: *catabolism (the breakdown of complex substances into simple ones, which releases energy) and *anabolism (the building up of complex substances from simpler ones, with involves absorption or storage of energy).

metabolite A product of *metabolism, including intermediate and waste products, or something which takes part in the reactions.

metal A dense, opaque element that is usually a lustrous solid and is a good conductor of heat or electricity. Eighty per cent of the known elements are metals. Examples include *iron, *lead, *copper, *aluminium, *silver, and *gold.

metalimnion The middle layer of a thermally *stratified *lake or *reservoir, between the *epilimnion (above) and the *hypolimnion (below), in which temperature decreases rapidly with depth. Also known as **thermocline**.

metamorphic rock Any *rock that has been altered by *metamorphism, involving heating and pressure. Also known as **tertiary rock** because metamorphic rock is derived from primary *igneous rocks and secondary *sedimentary rocks.

metamorphism Changes in the composition and texture of *rock which are caused by pressure, heat, and chemical reactions beneath the surface of the Earth, and result in a more compact and more highly *crystalline rock. *See also* CONTACT METAMORPHISM, REGIONAL METAMORPHISM.

metamorphosis 1. In general, a major change in appearance or character.
   2. In biology, a stage in the *life cycle of certain animals during which the *larva rapidly transforms into an adult, such as from a tadpole to a frog or from a caterpillar to a butterfly. *See also* HYPERMETAMORPHOSIS.

metapopulation A set of partially isolated *populations that belong to the same species, between which individuals can freely migrate.

meteor An extraterrestrial mass which has reached the Earth's *atmosphere from outer space and is heated to glowing hot by friction as it passes through the atmosphere. A meteor that reaches the Earth's surface is a *meteorite.

Meteor 3M-1 A *satellite remote sensing system launched in 1998 which monitors the distribution and concentration of *aerosols and chemical compounds in the *atmosphere.

meteorite An extraterrestrial mass which has reached the Earth from outer space, without burning up in the atmosphere. *See also* METEOR.

meteorology The study of the Earth's *atmosphere and of the *weather of the lower atmosphere (below about 100 kilometres).

Meteosat A *geostationary weather *satellite launched by the European Space Agency (ESA) and now operated by Eumetsat, that is positioned over Africa and Europe and provides weather imaging of the Earth at both visible light and *infrared wavelengths.

methane (CH_4) A colourless, non-poisonous, flammable *hydrocarbon gas created by *anaerobic *decomposition of organic compounds. It is the main component of *natural gas and can be used as a *fuel. The main sources of methane are *landfills, coal *mines, rice paddy fields, natural gas systems, and livestock (such as cows and sheep). It is a *greenhouse gas with a *global warming potential of 23. Concentrations of methane in the atmosphere have risen steadily since the 1950s, and they are increasing more than twice as fast as CO_2. Analysis of *ice cores reveals natural atmospheric concentrations of methane of about 0.3 ppmv (parts per million by volume) at the peak of glaciation. Current levels are around 1.7 ppmv, compared with around 350 ppmv for *carbon dioxide. Also known as **swamp gas**.

methane recovery The capture and reuse of *methane *emissions (particularly from animal wastes, *landfill sites, and coal *mines) in order to reduce the *greenhouse effect and decrease *global warming.

methanol A colourless, nearly odourless, *volatile liquid *alcohol *fuel that is produced mainly from *natural gas and is used as an *alternative fuel or *biofuel, and as a fuel additive.

methoxyfenozide A type of *insecticide which works by accelerating the *moulting process.

methyl bromide (CH_3Br) A poisonous gas or liquid that contains carbon, hydrogen, and bromine, and is used as an effective pesticide. The bromine is an *ozone depleter which is harmful to the *stratospheric ozone layer.

methyl chloroform (CH_3CCl_3) An industrial chemical that contains *carbon, *hydrogen, and *chlorine, and is used as a solvent, aerosol propellant, *pesticide, and degreasing agent. It is a *greenhouse gas and *ozone depleter and production of it is now banned under the *Montreal Protocol.

methyl isocyanide (C_2H_3N) A persistent and particularly *toxic *gas which attacks many different organs in the human body. It is the gas that killed and injured many people in the *Bhopal accident.

metric ton A tonne, equal to 1.1 tons.

metropolis The largest *town or *city in a region or country, which dominates its economic and cultural life.

mica A group of *silicate *minerals that contain varying amounts of *aluminium, *potassium, *magnesium, *iron, and *water, and that split into thin, tough, smooth plates. They are common constituents of *igneous and *metamorphic rocks. See also VERMICULITE.

micro- Very small scale. Contrast MACRO, MESO-.

microbe See MICRO-ORGANISM.

microbenthos The smaller organisms of the *benthos, generally shorter than 0.1 millimetres. Contrast MACROBENTHOS, MEIOBENTHOS.

microbial Relating to microbes or *micro-organisms.

microbial contamination *Contamination of a *medium (such as a waterbody) by *pathogenic *micro-organisms. See also BIOLOGICAL CONTAMINANT.

microbial digestion The natural processes by which *micro-organisms break down and use a *substance.

microbial growth The activity and growth of *micro-organisms such as *bacteria, *algae, *diatoms, *plankton, and *fungi.

microbial pesticide A *micro-organism that is used to kill a particular *pest, to which it is *toxic, but which poses little health risk to humans.

microbiology The study of *micro-organisms.

microburst A small scale *downburst of wind in the *atmosphere, which usu-

ally lasts for 2–5 minutes and has a radius of less than 4 kilometres. Contrast MACROBURST.

microclimate The *climate of a small area (such as a building, city, or valley) that may be different from that in the surrounding area (*mesoclimate) or region (*macroclimate).

microcosm A miniature model (such as a laboratory test) that is representative of a larger object (such as the natural environment).

microeconomics The study and analysis of the individual parts of the *market economy (such as companies or households), particularly in terms of the market process and how it works. Contrast MACROECONOMICS.

microelement See MICRONUTRIENT.

microenvironment See MICROHABITAT.

microevolution *Evolutionary changes on the small scale, such as changes in *gene frequencies within a population in response to changing local *environmental conditions. Contrast MACROEVOLUTION. See also ADAPTATION, EVOLUTION, NATURAL SELECTION.

microfauna Animals (less than 0.1 millimetres long) that are invisible to naked eye and are visible only through a microscope; for example protozoa and nematodes. Contrast MACROFAUNA, MESOFAUNA.

microflora Small plants that are invisible to the naked eye and can be seen only through a microscope. Contrast MACROFLORA, MESOFLORA.

microhabitat The immediate *environment in which an *organism lives, where factors such as moisture and light may be different from those in the surrounding area. Also known as **microenvironment** or **microsite**.

microinvertebrate Any *invertebrate organism that is too small to be seen with the naked eye (with a body length of less than 1 millimetre). *Contrast* MACROINVERTEBRATE.

micrometeorology The detailed study of weather at a particular location, over an area up to 2 kilometres wide.

micrometre *See* MICRON.

micron A unit of length equal to one millionth (10^{-6}) of a metre. Also known as **micrometre**.

micronutrient A *nutrient or chemical *element that is required for plant growth, but only in relatively small quantities. Examples include *iron, *zinc, and *copper. Also known as **microelement** or **trace element**. *Contrast* MACRONUTRIENT.

micro-organism Any living organism (including *bacteria, *viruses, *protozoa, and some types of *fungus and *algae) that is too small to be seen by the naked eye, and is only visible through a microscope. Also known as **microbe**.

microparasite A small *parasite. *Contrast* MACROPARASITE.

micropropagation The use of *biotechnology to grow large numbers of plants from very small pieces of plants, including single cells, tissues, and seeds by *in vitro methods.

microrelief Small-scale variations in the ground surface within a particular area. Also known as **microtopography**.

microscopic Too small (usually smaller than about 1 millimetre) to be seen by the naked eye. *Contrast* MACROSCOPIC.

microseism A weak more or less continuous vibration of the ground that is strong enough to be detected by *seismographs, and which is caused by *waves, *wind, or human activity, not by an *earthquake.

microsite *See* MICROHABITAT.

microsporidia Important primary *pathogens of *arthropods, the smallest of the *eukaryotes, closely related to *fungi.

microtopography *See* MICRORELIEF.

microwave *Electromagnetic radiation which has a longer *wavelength (1 to 35 GHz within the *electromagnetic spectrum) than visible light, and is used in radar (longer waves) and in cooking (shorter waves). A type of *radiant energy.

Mid-Atlantic Ridge An important *sea-floor spreading zone in the Atlantic Ocean that drives the lateral movement of crustal plates. *See also* MID-OCEAN RIDGE, PLATE TECTONICS.

Middle Ages *See* MEDIEVAL.

middle latitudes *See* MID-LATITUDE.

middle term planning *See* MID-RANGE PLANNING.

middleground The part of the *scenery or *landscape that is mid-distance away from the viewer, further than the *foreground but closer than the *background.

mid-latitude The region of the world that lies between 30° and 50° *latitude. Also known as **middle latitudes**.

mid-latitude climate The main *climate zones within the *mid-latitudes are the *mediterranean, *marine west coast, *humid subtropical, *humid continental, and the *mid-latitude steppe and desert.

mid-latitude steppe and desert climate zone A *mid-latitude *climate zone that is found in continental interiors, further from the coast than the *humid continental and *humid subtropical climates, and often cut off from maritime air masses by mountain ranges. The *climate is drier (particularly in winter) than in the tropics, and temperatures vary more throughout the year (temperatures below freezing are not uncommon in winter).

midnight dumping General term for the illegal disposal of *hazardous wastes in remote locations, often at night.

mid-ocean ridge An extensive *mountain chain that can rise thousands of metres above the surrounding *ocean floor, created by the rise of *magma along submarine *constructive plate boundaries, resulting in *sea-floor spreading. The ridges are part of a continuous system which runs for 60 000 kilometres through all the oceans, and includes the *Mid-Atlantic Ridge. *See also* PLATE TECTONICS.

mid-range planning Planning over a period of between about 10 and 25 years into the future, which is intermediate between *short-range planning and *long-range planning. Also known as **middle term planning** or **mid-term planning**.

mid-term planning *See* MID-RANGE PLANNING.

midwater The water that lies below the surface but above the bottom of a *waterbody.

migmatite A rock that contains both *igneous and *metamorphic materials, that was formed under immense heat and pressure, and often contains large *mineral crystals.

migrate To move from one place to another and settle there, permanently, periodically or seasonally.

migration 1. The movement of people or animals, including *immigration, *emigration, *net migration, *internal migration, and *international migration. 2. The stage in vegetation *succession at which the first plant seeds arrive on bare ground.

migration corridor *See* CORRIDOR.

migration pathway Any of the routes along which a *contaminant may move through the *environment, for example through *soil, in *groundwater, in surface water, and through the *air.

migratory The seasonal movement (*migration) of people, animals, birds or fish from one region to another.

migratory bird sanctuary In North America, a *nature reserve designed and set aside for the protection of *migratory birds.

Milankovitch cycle The cyclical changes in the orbit of the *Earth around the *Sun which are believed to cause long-term changes in the Earth's climate, according to the *Milankovitch theory. These are determined by variations in the elliptical shape of the Earth's orbit around the Sun (over a 95 000-year cycle), variations in the tilt of the Earth's axis of rotation (over a 42 000-year cycle), and variations in the time of year when the Earth is closest to the Sun (*perihelion) (over a 21 000-year cycle).

Milankovitch theory The theory, put forward by Yugoslav astronomer Milutin Milankovitch, that long-term changes in the Earth's *climate are caused by variations in the amount of *solar radiation that is received at the Earth's surface, which are due to cyclical changes in the relationship between the *Earth and the *Sun. *See also* MILANKOVITCH CYCLE.

mill 1. Machinery (such as rotating *millstones) that grinds grain into flour or meal.
　2. A building that contains such machinery.

millennium A period of 1000 years.

millibar (mb) A unit of atmospheric pressure that is equal to a thousandth of a *bar. Normal pressure at the Earth's surface is about 1013 mb.

millstone One of a pair of heavy, flat, disc-shaped stones that are rotated against one another to grind grain in a *mill, or used separately for sharpening steel blades

millstone grit Coarse-grained *sedimentary rock, often comprising *sandstones, grits, and *conglomerates with alternate *shales, that can be used for making *millstones.

milpa *See* SLASH AND BURN.

mimicry The similarity (for example in appearance) of two *species, for evolutionary advantage. *See also* BATESIAN MIMICRY, MÜLLERIAN MIMICRY.

Minamata disease A crippling form of *mercury poisoning which was first detected among people who ate fish from the mercury-contaminated waters of Minamata Bay off Japan in the 1950s.

Mindel The second phase of *glaciation in Europe during the *Pleistocene *ice age, equivalent to the *Kansan in North America.

mine An underground excavation from which mineral or rock is extracted. *See also* MINING, QUARRY.

mine drainage Any water that is released from a mining operation. *See also* ACID MINE DRAINAGE.

mine spoils Mining waste from open-pit mining, rather than from hard rock mining. *See also* MINE TAILINGS.

mine tailings The solid waste that is separated and left after raw minerals or *ore from hard rock mining have been processed. Also known as **mining refuse**. *Contrast* MINE SPOILS.

mined-land reclamation The return of *land to another use, often to the original pre-mining use, after *mining activities have ceased.

mineral A solid, naturally occurring, inorganic substance (such as *copper) with a definite chemical composition, a specific internal crystal structure, and characteristic physical properties. *See also* COMMON VARIETY MINERAL.

mineral cycle *See* BIOGEOCHEMICAL CYCLE.

mineral oil In general, any oil that is made from minerals. Particularly, a thick, greenish-brown, flammable liquid that is found underground in permeable *organic *sedimentary rock, which can be refined to produce a number of valuable products including *oil and *petrol (gasoline). Also known as **crude oil** or **petroleum**.

mineral resource Any of the various naturally occurring substances such as coal, *mineral oil, *metals, *natural gas, *salt, *sand, *stone, and *water, which are extracted from the Earth's *crust and converted into products useful for humans. Mineral resources are normally classified as being either *metallic (such as *iron and *aluminium ores) or *non-metallic (such as *fossil fuels, sand, and salt).

mineral rights The ownership of all rights to *gas, *oil, or other *minerals as they naturally occur at or below the surface of a particular area of land.

mineral soil Any *soil that is composed mainly of *inorganic material (such as weathered rock, or *sand, *silt or, *clay materials) rather than *organic matter.

mineralization 1. The breakdown of *organic compounds to their *inorganic forms (for example, proteins to nitrates and phosphates), as a result of *microbial *decomposition.
  2. In *fossil formation, the replacement of *organic parts by *inorganic materials.

mineralogy The study of the structure and properties of *minerals.

minerogenic Any *clastic sediment which is composed of individual *mineral particles. They are usually classified according to grain size, such as *sand or *clay.

minimum acceptable flow In the UK, the minimum amount of water that should be allowed to flow down a particular *river in order to meet the requirements of existing lawful uses of the water (for *agriculture, industry, water supply, or other purposes) and the requirements of *navigation, fisheries, or land drainage. Called **minimum flow** in the USA.

minimum flow The US name for *minimum acceptable flow.

minimum tillage An approach to the *conservation of *soil and *water resources in which the crop *residue or *stubble is left on the surface rather than ploughed under, in order to minimize the number of times that a field is tilled.

minimum viable population The smallest number of individuals of a *species in a particular locality that could reasonably be expected to survive in the long term.

mining The removal of *minerals (such as *coal, *gold, or *silver) from the ground. *See also* AREA MINING, AUGER MINING, CONTOUR MINING, LODE MINING, OPEN CAST MINING, PIT MINING, PLACER MINING, STRIP MINING.

mining debris Solid waste from mining activities, which includes the *overburden, *mining refuse, and *mining spoil. Also known as **mining waste**.

mining lode *See* LODE.

mining refuse *See* MINE TAILINGS.

mining spoil The overlying material that is removed during *mining in order to gain access to the *ore within the *mineral material below.

mining waste *See* MINING DEBRIS.

minor element *See* MICRONUTRIENT.

minor recreation facility North American term for developed *recreation facilities that are relatively cheap to build and run, and are not heavily used, such as observation sites, playgrounds, fishing sites, trailheads, and minor *interpretation sites. *Contrast* MAJOR RECREATION FACILITY.

minority group Any group of people (often defined by race or ethnicity) whose members have significantly less control or power over their own lives than the members of a dominant or majority group have over theirs.

Miocene An *epoch of geological time during the early *Tertiary period, from 23 million to 5 million years ago, during which many modern mammals first appeared.

mirage An optical illusion in which reflections of distant objects are distorted by atmospheric *refraction caused by a layer of hot air.

mire A soft, wet *peatland plant community that develops on waterlogged land.

MIS *See* MANAGEMENT INFORMATION SYSTEM, MANAGEMENT INDICATOR SPECIES.

miscible Capable of being mixed or blended together to form a homogeneous mixture, such as water and *methanol. The opposite of miscible is *immiscible.

misfit stream A small meandering *river within a larger meandering *valley which appears too big to have been formed by river erosion and may be *glacial in origin. Also known as an **underfit stream**.

Mission to Planet Earth A coordinated international plan, as part of the *International Geosphere-Biosphere Programme (IGBP), to provide the *satellite platforms and instruments, data and information systems, and related scientific research which are necessary to support the programme. It examines critical interactions between the Earth's physical, chemical, biological, and social systems.

mist A suspension of fine water droplets in the atmosphere. *Contrast* FOG.

mistral A strong, cold, dry, northerly, *katabatic wind which blows offshore along the north coast of the Mediterranean in winter.

mite A tiny *arthropod that may be a *parasite on plants and animals, and feeds on plants, other mites, or small insects.

mitigate To make less severe, or to lessen the seriousness or extent of.

mitigation Any actions that are taken to avoid or minimize negative *environmental impacts. This can take various forms, including avoiding the impact by not taking a certain action; minimizing impacts by limiting the scale of the action; rectifying the impact by repairing or restoring the affected environment; reducing the impact by taking protective steps; and compensating for the impact by replacing or providing substitute resources.

mitigation bank A site where *habitats (such as *wetlands) are restored, created, or preserved to serve as compensation for habitats that are going to be lost to *development elsewhere in a region.

mitochondrion A part of a *cell (*organelle) that is responsible for carrying out *aerobic respiration in cells. The plural is mitochondria.

mixed cloud A *cloud that contains both water drops and *ice crystals.

mixed cropping A *cropping system in which two or more *crops are grown on the same area of land at the same time, or with only a short interval between. See also INTERCROPPING, MULTIPLE CROPPING, RELAY CROPPING.

mixed farming Growing *crops and feed and *livestock on the same *farm at the same time.

mixed grass A North American native *prairie *grassland that is dominated by short and tall grass (including needle grasses and wheat grasses), with some tall shrubs and trees in moist areas and river valleys.

mixed layer 1. The upper part of a *waterbody, that is mixed by *wind and *wave action.
2. The layer of the *atmosphere that lies immediately above the ground, in which *air pollutants are well mixed by *turbulence.

mixed stand A *stand that contains two or more *species of *tree.

mixed waste A mixture of *radioactive waste and *hazardous waste.

mixing cycle The pattern of seasonal variations in temperature conditions and mixing status in a large body of *freshwater (such as a *lake or *reservoir), from *spring (mixed), through *summer (*stratified) and *autumn (*mixed), to *winter (stratified) pattern. See also MONOMICTIC, TURNOVER.

mixing depth The depth in a *waterbody, or height in the atmosphere, of the *mixed layer, in which the water or air is evenly mixed by *wind action.

mixing fog See EVAPORATION FOG.

mixing ratio The mass of a particular gas (such as *water vapour) in a parcel of air, divided by the mass of the dry air.

mixture An aggregate of two or more substances that are not chemically united.

MMI See MODIFIED MERCALLI INTENSITY SCALE.

MNC See MULTINATIONAL CORPORATION.

MNR See MARINE NATURE RESERVE.

mobile Capable of spontaneous movement, able to move freely.

mobile dune A coastal *sand dune that is becoming stabilized by the growth of *vegetation, but can still be affected by *wind erosion. Contrast FIXED DUNE.

mobile incinerator system A *hazardous waste *incinerator that can be transported from one site to another.

mobile source A pollution source (usually of *air pollutants) that moves. Mobile sources are divided into two groups: road vehicles, which includes cars, trucks, and buses, and non-road vehicles, which includes trains, planes, and lawn mowers.

mobility The ability of a *chemical element or *pollutant to move through the *environment, for example through a *food chain, or from a *waterbody into the *sediment beneath it.

mobility transition Change through time in the relative sizes of the rural-to-urban and urban-to-urban migration flows during the course of *urbanization within a country. *Contrast* DEMOGRAPHIC TRANSITION.

model A representation or description of a complex process or object, usually on a smaller scale and/or in a simplified form.

modelling The act of representing something, usually on a smaller scale, for example by use of statistical analysis, computer analysis, or a physical model.

moderate 1. Not extreme.
2. To lessen the intensity of, or slow down.

moderately exposed In terms of *wave exposure along a *coastline, coasts which face away from prevailing winds and do not have a long *fetch, but where strong *winds can be frequent.

moderately stratified estuary An *estuary in which the seaward flow of low-*salinity water and moderate vertical mixing result in a modest vertical gradient of salinity.

moderator 1. Material used to control the *fission reaction in *nuclear/thermal reactors.
2. A person who acts as a controlling influence during debates.

modernization theory The theory that *less-developed countries will follow the course of industrial development experienced by the *developed countries.

modified Mercalli intensity scale (MMI) The most widely used form of the *Mercalli scale, devised by American seismologists in 1932, for describing the *intensity of *earthquakes in a qualitative way, based on the relative amount of damage that structures undergo during an earthquake, using a scale from I to XII. The scale is given in Appendix 7.

Moh hardness scale A ten-point scale for measuring the *hardness of *minerals, based on the ability of a specimen to scratch another specimen on the scale.

The full scale is: talc 1 (softest), gypsum 2, calcite 3, fluorite 4, apatite 5, orthoclase 6, quartz 7, topaz 8, corundum 9, and diamond 10 (hardest).

Moho *See* MOHOROVICIC DISCONTINUITY.

Mohorovicic discontinuity The boundary between the Earth's *mantle and the *crust, named after the Yugoslav geophysicist who first suspected its presence in 1909. It follows variations in the thickness of the crust, and is found roughly 32 kilometres below the continents and about 10 kilometres below the oceans. *Seismic waves speed up noticeably when they reach this junction. Also known as the **Moho**.

moisture Wetness, for example as *water vapour in the *atmosphere or as a *condensed liquid on the surface of an object.

moisture content 1. The amount of *water that is present in the air, or in a material such as *wood or *soil, where it is usually expressed as a percentage of the oven dry weight of that material.
2. The water equivalent of snow on the ground.

Mojave Desert A North American hot desert that occupies a significant portion of southern California and parts of Utah, Nevada, and Arizona, and has scattered creosote bushes, Joshua trees, and sagebrush.

molecular biology The study of the structure, function, and composition of biological *molecules within *cells.

molecule A collection of two or more *atoms held together by chemical bonds.

mollisol A *soil order in the *Comprehensive Soil Classification System. These soils are dark coloured, have upper *horizons that are rich in *organic matter, and form mainly under *grassland.

mollusc A soft-bodied, aquatic *invertebrate animal, that is often protected by a shell. Examples include the snail, clam, mussel, squid, and octopus.

molten In a liquid state, having been melted by heat.

molybdenum (Mo) A naturally occurring silvery-grey metallic *trace element, one of eight *micronutrients that are essential to plant health, and mined for use in hardening steel and cast iron.

momentum The product of the mass and velocity of a moving body.

monadnock An isolated *hill or *mountain that rises above a flat landscape or *peneplain.

Monera In *taxonomy, the *kingdom of *prokaryotes (*organisms without a distinct nucleus).

monitor-and-modify approach See ADAPTIVE MANAGEMENT.

monitoring The collection and analysis of information on a regular basis in order to check performance against objectives, or compliance with a predetermined standard. Contrast SURVEILLANCE.

monitoring well A *well that is used to measure *groundwater levels or to collect samples of *water quality, for example at a *hazardous waste management *facility or a *Superfund site.

monkey-wrenching Popular name for environmental sabotage, involving such activities as driving large spikes in trees to protect them from loggers, vandalizing construction equipment, pulling up survey stakes for unwanted developments, and destroying billboards. See also ECO-ACTIVISM, ECO-EXTREMISM, ECOTAGE, ECOTERRORISM.

monocline A one-sided *fold in *stratified rocks, caused by relatively weak *compression pressure, usually found on the outer edges of tightly folded areas like *fold mountains.

monocotyledon A flowering *plant which has a seed with a single seed-leaf (cotyledon), flower parts arranged in threes, and leaves with parallel veins. Examples include grasses, lilies, and palms.

monoculture *Crop cultivation in which a single *variety of plant is grown in a *field. Contrast POLYCULTURE.

monolectic A *pollinator which collects *pollen from only one species of plant. Contrast OLIGOLECTIC, POLYLECTIC.

monomictic A relatively deep *lake or *reservoir which does not freeze over during winter, and undergoes a single *stratification and *mixing cycle during the year.

monophagous An animal that feeds on or uses a single species of *host plant or animal. Contrast POLYPHAGOUS.

monophyletic A group of *organisms originating from a single ancestor. A monophyletic group is called a *clade.

monoxenous A *parasite that lives within a single host during its whole *life cycle. Contrast OLIGOXENOUS.

monsoon A large-scale seasonal reversal of *winds and air pressure systems that occurs in some *tropical areas (particularly southern Asia, but also in Africa) and is caused by the different rates at which the *oceans and *continents heat and cool. The wind blows from land to sea in the winter and from sea to land in the summer. Monsoon circulations are associated with the movement of pressure cells and the intertropical front. When the *Intertropical Convergence Zone moves much further north of the equator during the northern hemisphere summer, it pulls the southeast *trade winds with it. These are deflected to the right by the *Coriolis force as they cross the equator, turning them into southwesterly winds which blow across the Indian subcontinent, drawn in by an intense thermally induced *low pressure area across northwestern India and Pakistan (caused by rapid summer heating). Heavy rains during the monsoon often cause extensive and destructive *flooding.

monsoon climate zone A *tropical climate zone, dominated by a winter–summer reversal of air flow (*monsoon) in the tropics, and best developed in

South Asia. *Precipitation is less than 60 millimetres during at least one month in the year, *temperatures remain fairly high throughout the year, and there is pronounced seasonality (stormy, cloudy, wet summers, and dry winters).

monsoon forest A type of *forest that grows in the *tropics, adjacent to the *tropical rainforests. Most are in South East Asia, including northeastern India, Burma, Thailand, Cambodia, Laos, and Vietnam. Many species are found in both the monsoon forest and the tropical rainforest, but in the monsoon forest they have adapted by *natural selection to cope with *drought during the dry season. *Adaptations include a period of dormancy, shedding of leaves, and wider spacing between trees (which compete for water). Monsoon forests also have a much denser *ground layer of shrubs, because the wider spacing of trees allows more sunlight to reach the forest floor. There are fewer tree species in the monsoon forest than in the rainforest, and this makes commercial timber extraction easier and more cost-effective. Some of the rich teak forests in South East Asia are extensively worked, and in danger of disappearing altogether. Other monsoon forests (particularly in India) are being cleared to create land for agriculture and settlements, leaving small patches in remote areas.

monsoon season The rainy season in southern Asia and India, from about April to October, when the southwest *monsoon blows moisture-laden air from over the Indian Ocean, which brings heavy rains.

montane Relating to *mountains or mountainous country.

montane forest A forest type of vegetation in which the principal trees are *conifers (such as Ponderosa pine), which grows below the *alpine zone on mountains.

month A measure of time based on how long it takes the Moon to circle around the Earth, which is roughly 30 days. A calendar month varies between 28 and 31 days.

montmorillonite A very plastic *clay; an *aluminium *silicate which is the principal constituent of *bentonite.

Montreal Protocol *See* MONTREAL PROTOCOL ON SUBSTANCES THAT DEPLETE THE OZONE LAYER.

Montreal Protocol on Substances that Deplete the Ozone Layer (1987) An international *protocol that was drawn up in 1987 and came into force in 1989 to protect the *ozone layer in the *stratosphere from depletion. It built upon the *Vienna Convention for the Protection of the Ozone Layer (1985) and governs the phasing out of production and use of ozone-depleting substances such as *CFCS. The 39 *signatory countries agreed to freeze production of *CFCs at 1986 levels, and to decrease production by 20% by 1993 and by half by 1999. The Montreal Protocol was widely welcomed, because it provided the international community with a mechanism for protecting the ozone layer that is effective (the measures are likely to work, particularly as substitutes for CFCs are developed and make it easier to reach the *emissions reduction targets), equitable (the measures share responsibility for solving the problem amongst all countries concerned), and dynamic (the measures get tougher through time, and can be further adapted if new scientific evidence comes to light). Also known as **Montreal Protocol**.

moon A large body (natural *satellite) that orbits around a *planet. The Moon is a natural satellite of the Earth. Unlike the Earth, the Moon has no atmosphere or water. The Moon is much smaller than the Earth, barely a quarter of the size, with a diameter of 3476 kilometres. Its surface gravity is only about one-sixth that of the Earth. The Moon orbits in a west–east direction, about 385 000 kilometres from the Earth, and each orbit takes 27.32 days (a sidereal month) to complete. It also spins on its axis, with one face permanently turned towards the Earth. On Earth the *tides (cyclic rise and fall of sea level) are caused by the gravitational pull of the Moon and Sun, so the movements of the Moon affect the oceans which cover two-thirds of the Earth's surface. The Moon might also be implicated in global climate change, because recent research has shown that the Moon affects temperatures on Earth through the influence of the alignment of Earth, Moon, and Sun on the tides.

moonlight The light of the Earth's *moon.

moor See MOORLAND.

moorland Unenclosed land in an *upland area that supports upland *heath, *blanket bog, and upland *grassland, with a moist peaty soil covered with *heather, *bracken, and *moss.

mor A layer of *organic matter that develops beneath *conifer forest and is associated with *acidic soils. Contrast MULL.

moraine The rocks, boulders, and debris that are carried and deposited by a *glacier or *ice sheet. The main types of moraines are *lateral moraine, *ground moraine, *medial moraine, and *terminal moraine.

moral Arising from a sense of right and wrong.

morality A recognized code of conduct or set of ethical principles that guide actions and relationships, based on the difference between right and wrong.

morals The ethics, values, principles, and customs of a person or society.

moratorium A temporary prohibition or suspension of a particular activity.

morbidity The rate of incidence of a particular *illness or *disease within a given *population.

more developed country (MDC) A highly industrialized country that has high *per capita incomes, low *birth rates and *death rates, low *population growth rates, and high levels of *industrialization and *urbanization. Examples include the USA, Canada, Japan, and many countries in Europe. Contrast LESS DEVELOPED COUNTRY.

morph A distinct, readily observable type of a particular *species.

morphogenesis The evolution of form, such as the development of *landforms in a *landscape or the change in shape or structure of an *organism as it grows.

morphogenetic region A *region that has distinct *landscape characteristics which have been produced by dominant *geological processes which are determined by climatic conditions.

morphological map A map of variations in the structure and form of the land surface, and of the *landforms on it.

morphology The study of the structure and form of objects (such as *organisms and *landforms), without regard to function.

morphometry The quantitative expression of *morphology, for example by measuring the depth, surface area, and volume of a lake.

morphostratigraphy The organization of *rock or sediment *strata into units based on their surface morphology (*landforms).

mortality The *death rate of a population, normally expressed as the number of deaths per thousand per year. Contrast NATALITY.

mosaic 1. An interlocking group of *habitats that cover a particular area.
2. A disease in plants, where patches of yellow develop on the leaves.

moss 1. Any of about 9500 species of small, simple, flowerless green plants (*bryophytes) which grow in moist places such as *moorland.
2. Another term for *mossland.

mossland A term used in northern Britain to describe a lowland *raised bog habitat. Also known as **moss**.

Mother Earth A general term that is widely used to refer to the Earth as our home planet, particularly by those who view it in a spiritual way.

mother lode The principal *lode of an *ore in a region.

mother-of-pearl cloud See NACREOUS CLOUD.

motile Able to move at will, self-propelled. *Contrast* SESSILE.

mottling The marble-like pattern that occurs in *soils as a result of periodic fluctuations in the *water table.

mould 1. The common name for a *fungus that grows in a filamentous fashion and reproduces by means of *spores. All moulds are fungi, but not all fungi are moulds. There are over 20 000 species of mould.
2. The impression of a *fossil in a deposit.

moulin A wide, vertical shaft in a *glacier, created by *meltwater, by which the meltwater enters the ice.

moult The seasonal or periodic shedding of old skin, feathers, or hair to allow new growth.

Mount St Helens A *volcano within the Cascade volcanic group in southern Washington State, USA, which had been *dormant since 1875 but erupted violently on 18 May 1980. Before the eruption *magma was pushed upwards beneath Mount St Helens, causing the mountain to bulge and the ground surface to tilt. The explosive eruption released pressure on the magma, causing a massive landslide and a powerful lateral blast which deposited ash and blew down trees over a wide area.

mountain A high area of land that rises steeply above its surroundings, usually has a sharply pointed top, and is larger than a *hill.

mountain breeze See KATABATIC WIND.

mountain chain See MOUNTAIN RANGE.

mountain glacier A *glacier which forms in an area of *mountains. Also known as alpine glacier or valley glacier. *Contrast* ICE CAP, ICE SHEET.

mountain meteorology The study of *weather phenomena associated with *mountains or *topographically complex areas.

mountain range A long chain of *mountains connected together by continuous high ground. Also known as mountain chain.

Mountain States Legal Foundation A US non-profit *anti-environmental group public interest legal centre that is dedicated to individual liberty, the right to own and use property, and ethical government and the free enterprise system.

mountain top removal A method of *mining which involves removing all of the *overburden covering a *coal seam, to allow all of the *mineral to be recovered. The *spoil is transported to a nearby hollow to create a valley fill. A form of *open cast mining.

mountain wave See LEE WAVE.

mountain wind system The system of *diurnal winds that forms in an area of complex terrain, consisting of mountain–plain, along-valley, cross-valley and slope wind systems.

Mountains of California A book by John *Muir which was published in 1894 and describes the geology, landscape, and natural history of the Sierra Nevada mountains.

MPA See MARINE PROTECTED AREA.

MPD See MAXIMUM PERMISSIBLE DOSE.

MSS See MULTISPECTRAL SCANNER.

MSW See MUNICIPAL SOLID WASTE.

MSY See MAXIMUM SUSTAINABLE YIELD.

MTD *See* MAXIMUM TOLERATED DOSE.

muck 1. The general term (particularly in North America) for a highly decomposed layer of *organic material in a wet organic *soil.
2. *Manure.

mud Soft wet *soil.

muddling through *See* INCREMENTALISM.

mudflat A flat area along the coast or in an *estuary, that is covered with a thick layer of *mud or *sand, and is usually under water at *high tide.

mudflow A rapid *mass movement process in which *fine-grained *sediment (*clays) is saturated with water and moves downhill by *gravity. Common after prolonged or unusually heavy rain in areas with little or no protective vegetation cover.

mudslide A *landslide of *mud. Faster, more sudden, and more damaging than a *mudflow.

mudstone A fine-grained *clastic *sedimentary rock composed of compacted and hardened *clay, similar to *shale but with less developed *lamination.

Muir, John US naturalist, traveller, and writer (1838–1914) who was born in England and became America's most famous and most influential conservationist. He lived in the Yosemite Valley in California and studied it in great detail and was the author of the book The *Mountains of California. He played a key role in the formation of Yellowstone and Yosemite National Parks and was one of the founders of the *Sierra Club, becoming a hero of the early American conservation movement.

mulch A protective covering for *soil; any natural or artificial substance that is spread or allowed to remain on the soil surface in order to conserve soil moisture, shield soil particles from *erosion, prevent freezing of plant roots, and control the growth of *weeds.

mulching The process of spreading a *mulch over a soil surface.

mull A type of *humus or layer of dark *organic matter that forms in freely drained, neutral to *alkaline *soils with good *aeration, mainly because of earthworm activity. *See also* MOR

Müllerian mimicry A form of *mimicry in which two *species, both of which are unpalatable and have some form of defence mechanism (such as poisonous stingers), evolve to resemble each other. *Contrast* BATESIAN MIMICRY.

multicellular Composed of many *cells.

multidisciplinary The coordinated application of several academic disciplines or subjects, in order to achieve a common goal. *See also* INTERDISCIPLINARY.

multimedia approach An approach to *environmental management that is based on monitoring and managing several environmental *media (such as air, water, and land) at the same time.

multinational corporation (MNC) A company that operates, produces, and sells products in many countries, and is not wholly subject to the laws of any one nation. *Contrast* TRANSNATIONAL CORPORATION.

multiparasitism The process whereby a *host is subjected to parasitism by more than one type of *parasite.

multiple cropping Growing more than one crop on the same piece of land. *See also* INTERCROPPING, MIXED CROPPING, SEQUENTIAL CROPPING.

multiple use The management and use of a *natural resource for more than one purpose. For example, using an area of land for the grazing of *livestock, protection of *watersheds, *conservation of *wildlife, *recreation, and timber production, or using a body of water for recreation, *fishing, and *water supply.

Multiple Use-Sustained Yield Act (1960) US legislation that declares that the purposes of the *National Forest System include outdoor recreation, range, timber, watershed, and fish and wildlife. The act directs the Secretary of Agriculture to administer national forest *renewable surface resources for *multiple use and *sustained yield.

multiple-use forestry Any forestry activity that is designed to meet two or more management objectives, including providing products, services, or other benefits. Also known as **multipurpose forestry**.

multiple-use management See INTEGRATED RESOURCE MANAGEMENT.

Multiple-Use Management Area/ Managed Resource Area An *IUCN Management Category (VIII) for protected areas, designed 'to provide for the sustained production of water, timber, wildlife, pasture, and outdoor recreation, with the conservation of nature primarily oriented to the support of economic activities (although specific zones may also be designed within these areas to achieve specific conservation objectives)'.

multipurpose forestry See MULTIPLE-USE FORESTRY.

multispectral scanner (MSS) A *radiation sensor system that is often carried in aeroplanes or *satellites (such as *Landsat), and records information about the same location in several *wavelength bands at the same time. This information shows the amount of energy reflected in particular wavelengths from specific ground locations, which can be decoded to indicate such things as land use, vegetation cover, crop growth, and health.

multistorey cropping The growing together of a number of different plant species (*annuals and *perennials) which are of different heights, so that several *canopy layers are formed. See also MULTIPLE CROPPING.

multivoltine Having many broods and generations in a year or season. See also VOLTINISM. Contrast BIVOLTINE, UNIVOLTINE.

municipal Belonging to or characteristic of a *municipality.

municipal discharge The discharge of *effluent from *wastewater treatment plants which receive wastewater from households, commercial establishments, and industries.

municipal sewage *Sewage that originates from households, commercial establishments, and industries.

municipal solid waste (MSW) Urban *refuse, including residential, industrial, and commercial wastes, that is collected for *landfill. It does not include agricultural and wood wastes or residues.

municipal waste combustor A facility in which recovered *municipal solid waste is converted into a usable form of *energy, usually via *combustion. Also known as a **waste-to-energy facility**.

municipality A *city, *town, or *village that enjoys self-government in local matters.

mutagen A substance that can induce *mutation. See also CARCINOGEN, TERATOGEN.

mutagenic Capable of causing *mutations in an organism. Mutagenic substances may also be *carcinogenic.

mutant A variant *organism that differs from its parent because of a change in its genetic material, caused by *mutation.

mutate To undergo *mutation and produce a *mutant.

mutation Any change in the *genotype of an *organism that is permanent and can be inherited by future generations, which occurs either by chance or a result of an external influence. Contrast IMMUTABILITY.

mutualism A mutually beneficial relationship (*symbiosis) between two different species, such as the *pollination of flowers by honey bees.

mycorrhizae *Mutualism between a *fungus and a *plant, in which the fungus facilitates the uptake of *nutrients and water by the plant.

myiasis The *infestation of the living tissue of humans or animals by fly larvae.

myophily *Pollination by flies.

myrmecophily *Pollination by ants.

m

NAAQS *See* NATIONAL AMBIENT AIR QUALITY STANDARD.

nacreous cloud A type of *cloud that has a soft, pearly lustre and forms at altitudes of about 25 to 30 kilometres above the Earth's surface. Also known as **mother-of-pearl cloud**.

Næss, Arne A Norwegian philosopher, alpine climber, and influential environmental thinker (born 1912), who coined the term *deep ecology to express a vision of the world in which we protect the environment as a part of ourselves, never in opposition to humanity, based on a personal philosophy that he called *ecosophy.

nanometre A millionth of a millimetre (10^{-9} metres). *See also* ANGSTROM.

nanoscale On the scale of large molecules, measured in *nanometres.

nanotechnology Any *technology that is applied at the *nanoscale, particularly the development of new materials and processes (such as thin films, fine particles, or miniature machines) by manipulating molecular and atomic particles.

NAPL *See* NON-AQUEOUS PHASE LIQUID.

nappe A complex type of *fold in rock, in which an *overturned fold is detached and broken.

NASA The US National Aeronautics and Space Administration, which was created in 1958, and whose mission is to advance and communicate scientific knowledge and understanding of the Earth, the solar system, and the universe and use the environment of space for research.

natal dispersal The movement of *organisms away from the places where they were born. *See also* DISPERSAL.

natality The *birth rate of a population, normally expressed as the number of births per thousand per year. *Contrast* MORTALITY.

natatorial Specialized for swimming.

National Ambient Air Quality Standards (NAAQS) US *air quality standards, established by the *Environmental Protection Agency under the *Clean Air Act, that apply for outdoor air throughout the country.

National Biodiversity Network (NBN) An initiative to set up a network of local record centres across Britain to collect and collate biological and geological observations and information from both amateur and professional naturalists, biologists, and ecologists.

National Center for Atmospheric Research (NCAR) A federally funded research and development centre in the US, based in Boulder, Colorado, whose mission is to improve our understanding of *atmospheric systems.

national conservation strategy A plan that highlights country-level environmental priorities and opportunities for the sustainable management of *natural resources, following the example of the *World Conservation Strategy published by the *World Conservation Union (IUCN) in 1980.

National Contingency Plan (NCP) *See* NATIONAL OIL AND HAZARDOUS SUBSTANCES CONTINGENCY PLAN.

national delegation One or more officials who are empowered to represent and negotiate on behalf of their government in international discussions and at the drafting of international *conventions.

National Emissions Standards for Hazardous Air Pollutants (NESHAPS) The US *emissions standard, defined by the *Environmental Protection Agency, for an air pollutant that is not covered by *NAAQS but may cause an increase in fatalities or in serious, irreversible, or incapacitating illness. Primary standards are designed to protect human *health, secondary standards to protect public *welfare (including buildings, visibility, crops, and domestic animals).

National Environmental Policy Act (NEPA) (1969) US legislation that requires that environmental considerations be incorporated into all federal policies and activities, and requires all federal agencies to prepare *environmental impact statements for any actions significantly affecting the environment. It led to the setting up of the *Council on Environmental Quality.

National Environmental Scorecard A method of measuring the environmental actions of members of the US Congress that was developed in 1970 by the *League of Conservation Voters, and is based on the environmental voting records, based on a consensus of experts from 19 respected environmental and conservation organizations who selected the key votes on which Members of Congress should be graded.

National Estuary Program (NEP) A national programme that was established under the US *Clean Water Act Amendments of 1987 in order to develop and implement *conservation and *management plans to protect *estuaries, restore and maintain their chemical, physical, and biological integrity, and control pollution from *point sources and *non-point sources.

National Fire Danger Rating System A system for rating *fire danger in the USA, based on the environmental factors that affect the *moisture content of *fuels. Fire danger is rated daily over large administrative areas, such as *National Forests.

National Fish and Wildlife Foundation (NFWF) A non-profit charitable organization in the USA which is dedicated to the *conservation and management of *fish, *wildlife, and *plant resources, and the *habitats on which they depend.

National Forest Public land (mostly *forest, *range, or *wildland) in the USA which is administered by the *Forest Service of the US Department of Agriculture, under a programme of multiple use and sustained yield for *timber harvesting, *grazing, *conservation of *wildlife, *watershed protection, and *outdoor recreation purposes. *Mining is also allowed within National Forests.

National Forest Land and Resource Management Plan (NFLRMP) A plan that guides the management of a particular *National Forest and establishes management standards and guidelines for all lands of that National Forest. Also known as a **forest plan**.

National Forest Management Act (1976) US legislation that reorganized, expanded, and amended the *Forest and Rangeland Renewable Resources Planning Act (1974) and is the primary statute relating to the management of renewable resources on National Forest lands. It requires the Secretary of Agriculture to assess forest lands, develop a management programme based on *multiple-use, *sustained-yield principles, and implement a resource management plan for each unit of the *National Forest System.

National Forest System The national system of *forest land management in the USA, which includes all National Forest lands reserved or withdrawn from the public domain or acquired through purchase, exchange, donation, or other means, as well other land that is administered by the Forest Service. The system includes state and private forestry programmes.

National Grassland An area or region of *grassland in the USA that has been designated for *conservation by the government.

National Historic Preservation Act (1966) US legislation that provides for managing *cultural resources on federal lands and that established procedures for determining relative significance among cultural resources.

National Monument An area in the USA that is owned by the federal government and administered by the *National Park Service, for the purpose of preserving and making available to the public objects of scientific and historical interest that are located on federal lands.

National Nature Reserve (NNR) A site in the United Kingdom that has been declared as a *nature reserve by English Nature (or its predecessors or national equivalents) and is either owned or controlled by English Nature or held and managed by approved bodies such as wildlife trusts. The US equivalent is the national wildlife refuge.

National Oceanic and Atmospheric Administration (NOAA) A branch of the US Department of Commerce, established in 1970. NOAA is the parent organization of the *National Weather Service.

National Oil and Hazardous Substances Contingency Plan (NOHSCP/NCP) The policy directive in the USA that guides the selection of sites to be cleaned up under the *Superfund Programme and the programme to prevent or control spills into surface waters or elsewhere. Also known as the **National Contingency Plan**. *See also* HAZARD RANKING SYSTEM.

National Park An *IUCN Management Category (II) for protected areas, designed 'to protect natural and scenic areas of national or international significance for scientific, educational and recreational use'. According to the IUCN, a national park is 'a relatively large area—(a) where one or several *ecosystems are not materially altered by human exploitation and occupation, where plant and animal species, *geomorphological sites and *habitats are of special scientific,

educative, and recreational interest or which contain a natural landscape of great beauty, and (b) where the highest competent authority of the country has taken steps to prevent or eliminate as soon as possible exploitation or occupation in the whole area and to enforce effectively the respect of ecological, geomorphological, or aesthetic features which have led to its establishment, and (c) where visitors are allowed to enter, under special conditions, for inspirational, cultural, and recreative purposes'.

National Park Service (NPS) The agency of the US Department of the Interior which is responsible for the administration of *National Parks, *National Monuments, and historic sites.

National Parks and Access to the Countryside Act (1949) The national legislation under which Britain's first *National Parks were established.

National Priorities List (NPL) A formal list of the most serious uncontrolled or abandoned *hazardous waste sites in the USA that pose a serious threat to human health and/or the environment, which are candidates for long-term *cleanup using money from the *Superfund trust fund. The list is maintained by the *Environmental Protection Agency, is updated at least once a year, and contains about 1200 hazardous waste sites.

national rank (N-rank) A term used by *Conservation Data Centers in the USA, and by *NatureServe, to refer to the national *conservation status rank of a *species or *ecological community.

National Recreation Area An area of federal land in the USA that has been set aside by Congress for *recreational use by members of the public.

National Register of Historic Places A national list of districts, sites, buildings, structures, and objects that are significant in history, architecture, archaeology, engineering, and culture in the USA.

National Registry of Natural Land-marks A register of areas in the USA that possess such exceptional values or qualities for illustrating or interpreting the natural *heritage of the nation that they are considered to be of national significance.

National Response Center (NRC) The federal operations centre in the USA that receives notifications of all releases into the environment of *oil and *hazardous substances. It is operated by the US Coast Guard, which evaluates all reports and notifies the appropriate agency.

national sovereignty The right of individual nations to look after their own interests and to manage resources within their territorial borders to suit their own purposes.

national water quality standard The maximum allowable levels of *contamination in the USA for a range of *chemicals, *metals, and *bacteria, as defined by the US Safe Drinking Water Act.

National Weather Service (NWS) The branch of the *National Oceanic and Atmospheric Administration that is responsible for providing weather, hydrological, and climate forecasts and warnings for the USA and its territories.

National Wetlands Coalition A US non-profit *anti-environmental *front group that advocates a balanced federal policy for conserving and regulating the nation's *wetlands. Members of the coalition own or manage wetlands; they include local governments, ports, water agencies, developers, agriculture groups, electric utilities, oil and gas developers, producers, the mining industry, banks, environmental and engineering consulting firms, and Native American groups.

National Wild and Scenic Rivers System A series of *wild rivers and *scenic rivers in the USA that 'possess outstanding remarkable scenic, recreational, geologic, fish and wildlife, his-toric, cultural, or other similar values' and are protected 'for the benefit and enjoyment of present and future generations' under the *Wild and Scenic Rivers Act.

National Wilderness Preservation System A system of wild federal land and *National Forest land in the USA that was established by the *Wilderness Act (1964), which was to be managed 'for the use and enjoyment of the American people in such manner as will leave them unimpaired for future use and enjoyment as wilderness…'.

National Wildlife Federation (NWF) The largest and oldest protector of *wildlife in the USA, founded in 1936 as an environmental advocacy and educational group.

national wildlife refuge The US equivalent to a *National Nature Reserve in the United Kingdom.

National Wildlife Refuge System A national system of *nature reserves in the USA, which is managed by the *US Fish and Wildlife Service, and has the mission 'to administer a national network of lands and waters for the conservation, management, and where appropriate, restoration of the fish, wildlife, and plant resources and their habitats within the USA for the benefit of present and future generations of Americans'.

native Naturally occurring in a particular place; belonging there by birth or origin. Also known as **indigenous**.

Native American A tribe, people, or culture that is *native to the USA.

native element An *element (such as *gold) that occurs by itself in *rocks, uncombined.

native metal A natural *deposit of a *metallic element in pure metallic form, which is neither *oxidized nor combined with *sulphur or other elements.

native pasture Land that is used for *grazing, on which the *climax plant community is *forest, but which is used and managed primarily to produce native or naturalized plants for *forage.

native peoples *See* INDIGENOUS PEOPLES.

native species A species that is within its known *natural range, and occurs naturally in a given area or habitat, as opposed to an *introduced species or *invasive species. Also known as **endemic species, indigenous species**. *Contrast* NON-NATIVE SPECIES.

Natura 2000 A network of areas within the *European Community that is designed to conserve *natural habitats and *species of plants and animals which are *rare, *endangered, or *vulnerable. *See also* HABITATS DIRECTIVE.

natural Occurring in *nature; caused by *natural processes. *Contrast* SYNTHETIC.

natural air pollutant Any of a variety of substances that are released from *natural sources or processes, and which can locally overload the atmospheric system. These include *dust, *gases, and *aerosols from *volcanic eruptions, *forest fires, and *sea spray.

Natural Area An area identified as having significant or unique natural *heritage features, with boundaries based upon the distribution of *wildlife and of natural features rather than administrative borders. The term is used in the UK, the USA, and Canada.

Natural Area Preserve An area of land in the USA that is designated by a public or private agency specifically to preserve a representative sample of an *ecological community, primarily for scientific and educational purposes, in which commercial exploitation is generally not allowed and general public use is discouraged. Also known as **Natural Reserve**. *See also* RESEARCH NATURAL AREA.

natural barrier A natural feature, such as a dense *stand of trees, that will restrict the free movement of *animals.

Natural Biotic Area/Anthropological Reserve An *IUCN Management Category (VII) for protected areas, designed 'to allow the way of life of societies living in harmony with the environment to continue undisturbed by modern technology'.

natural capital The stock of *natural resources and environmental assets within an area, country, or the world, which includes *water, *soil, *air, *plants, *animals, and *minerals. Also known as **environmental capital**.

natural community *See* ECOLOGICAL COMMUNITY.

natural disaster A disaster that is caused by a *natural hazard. *Contrast* TECHNOLOGICAL DISASTER.

natural disturbance regime The historic patterns (frequency and extent) of natural processes such as fire, insects, wind, and *mass movement that affect the *ecosystems and *landscapes in a particular area. *See also* DISTURBANCE.

natural environment *See* ENVIRONMENT.

natural erosion The gradual *erosion of the land surface under natural conditions, without any human activity or impact. Also known as **geological erosion**.

natural flow The flow of a *river as it would be under natural conditions, unaltered by upstream diversion, storage, import, export, or *consumptive use.

natural forest *See* PRIMARY FOREST.

natural gas A mixture of gaseous *hydrocarbons (*fossil fuel), chiefly *methane (CH_4), *ethane (C_2H_6), *propane (C_3H_8), and *butane (C_4H_{10}), which is trapped in porous rocks beneath the ground and is often found in association with reserves of *oil. It is a *clean fuel which burns without smoke or soot and has a high heat value, and is used as a fuel. It is now one of the world's three main fossil fuels, along with *coal and oil. Most is extracted from offshore wells (in the North Sea, for example) or from land-based wells in the USA and the Middle East. North America and the Middle East hold about 40% of the world's known recoverable gas resources, and the former Soviet Union holds a similar amount. It provides around a third of the energy used in the USA.

natural greenhouse effect *See* GREENHOUSE EFFECT.

natural habitat As defined by the EU *Habitats Directive 'natural habitats means terrestrial or aquatic areas distinguished by geographic, abiotic, and biotic features, whether entirely natural or semi-natural'.

natural hazard A process or event in the physical environment that is not caused by humans, is usually not entirely predictable, but can injure or kill people and damage property. Examples include natural processes such as *volcanoes, *earthquakes, violent *storms (including *hurricanes and *tornadoes), river *flooding, *storm surges, *droughts, *avalanches and *landslides, and *sea level rise. Three trends have increased the problems of coping with natural hazards in recent decades—*population increase, *human impacts on *environmental systems, and *technological hazards. The net effect of these changes, particularly in recent decades, has been an increase in exposure to many hazards, and increased potential for catastrophic losses. Also known as **environmental hazard**. *See also* INTERNATIONAL DECADE FOR NATURAL DISASTER REDUCTION.

natural heritage All of the living *organisms, ecological *communities, *ecosystems, and natural areas that we inherit and leave to future generations.

Natural Heritage Network The network of *Conservation Data Centers and *Natural Heritage Programs throughout the Americas, which use the same methodology and database to monitor changes in *biodiversity in their jurisdictions.

Natural Heritage Program *See* CONSERVATION DATA CENTER.

natural history The systematic study of *nature, including animals, plants, minerals, and other natural objects.

natural immunity Inherited *immunity of an individual to a particular *disease.

natural increase The surplus of births over deaths in a *population over a given period of time (usually a year), normally calculated as crude *birth rate minus crude *death rate.

Natural Landmark *See* NATURAL MONUMENT/NATURAL LANDMARK.

natural landscape *Landscape that has not been modified by human activities. *Contrast* CULTURAL LANDSCAPE.

Natural Monument/Natural Landmark An *IUCN Management Category (III) for protected areas, designed 'to protect and preserve nationally significant natural features because of their special interest or unique characteristics'.

natural nidus The *niche of a disease.

natural process A process that results from natural forces rather than human activities.

natural process restoration The restoration of *habitat by natural processes (such as flooding or fire), which allows native vegetation to region and inhibits the growth of *exotic plants.

natural range The geographical area over which a *species has naturally lived in recent times (since about 5000 years before the present), excluding any changes to that range that result from human activities. Also known as **ecological range** or **geographical range**. *See also* HISTORIC RANGE, INTRODUCED SPECIES, NATIVE SPECIES.

natural range of variation *See* RANGE OF VARIABILITY.

natural regeneration The natural regrowth of vegetation (for example the renewal of a forest by natural seeding) without any human interference.

natural region An area of land that is defined by physical features (such as a surrounding mountain range) and contains similar plant and animal species. Natural regions and *ecoregions often overlap, but they are not the same.

Natural Reserve *See* NATURAL AREA PRESERVE.

natural resource Any feature of the *natural environment that is of value in meeting human *needs, including *renewable resources (such as vegetation, water, soil, and wildlife) and *non-renewable resources (such as oil, natural gas, coal, and iron ore). Also known as **environmental resource, intrinsic resource**.

natural resource accounting *See* ENVIRONMENTAL ACCOUNTING.

natural resource accounts (NRAs) The publication of information about the level of *natural resource use, existing stocks of natural resources, and environmental degradation, within a region or country.

natural resource management The *management of *natural resources on a *sustainable basis, usually in ways that meet multiple objectives including the *conservation of *wildlife and *ecosystems, and the minimizing of *environmental impacts and *environmental change.

natural resource zoning All types of land use *zoning (such as *agricultural zoning, *floodplain zoning, and *forest use zoning) which restrict the uses that are made of *natural resource land, to protect the resource base.

Natural Resources Defense Council A US organization founded in 1970 to protect *natural resources.

natural sciences Those sciences that study the natural environment, including *astronomy, *biology, *chemistry, *geology, *physics, *oceanography, and *meteorology.

natural selection The process in which, for a given *species, the individual *organisms that have the most appropriate *adaptations to environmental (including social) change are more successful than other individuals that don't have them. The adapted individuals tend to survive better and produce more offspring; those that are not able to adapt disappear. Over time, natural selection helps species become better adapted to their environment. It is the driving force behind the *evolution of species, was first suggested and illustrated by Charles *Darwin, and is also known as survival of the fittest.

natural vegetation The natural state and *species composition of *vegetation in an area, as it is or would be if unmodified by human activities.

naturalized A previously *exotic or *foreign species which has become established in a particular *ecosystem and is now commonly found within it.

naturally occurring background level *Ambient concentrations of *chemicals that are present in the *environment and have not been influenced by humans.

naturalness In *conservation assessment, the extent to which a location and its associated *biotopes are unaffected by human activities.

natural-technological disaster A *natural disaster that creates a technological emergency, such as an urban *fire that results from *seismic motion, or a chemical *spill that result from *flooding. *See also* DISASTER, TECHNOLOGICAL DISASTER.

Nature An essay by Ralph Waldo *Emerson in 1836, in which he outlined his views of the links between people and nature.

nature The natural physical world including *plants, *animals, and *landscapes, particularly those parts that remain in a primitive state, unchanged by humans.

Nature Conservancy, The (TNC) A private, international *conservation group based in the USA which was established in 1951, whose mission is 'to preserve plants, animals, and natural communities that represent the diversity of life on earth by protecting the lands and waters they need to survive'.

nature conservation The *conservation of *wildlife, *biodiversity, and *natural ecosystems, by regulating the human use of ecosystems and *natural resources and promoting *sustainable development.

nature reserve An area of land that is set aside and managed in ways that benefit *nature conservation, usually by limiting human access and use. Nature reserves are usually much smaller than *National Parks or *Biosphere Reserves, and many *endangered species of plants and animals are protected inside them.

nature study The study of *plants and *animals in their natural *habitats.

NatureServe A non-profit *conservation organization. It provides the scientific information and tools needed to help guide effective conservation action, partly through an international network of natural heritage programmes which provide information about *rare and *endangered species and threatened *ecosystems. Also known as the **Association for Biodiversity Information**.

nautical mile (nm) The distance between each minute of *latitude, which is about 1.85 kilometres. There are 60 nautical miles to one degree of latitude.

navigable waters *Waterways that are deep and wide enough to allow *navigation by all or specified vessels.

NBN *See* National Biodiversity Network.

NCAR *See* National Center for Atmospheric Research.

NCP *See* National Contingency Plan.

Neanderthal A *hominid, similar to but distinct from modern humans, that lived in Europe and western Asia between about 150 000 and 30 000 years ago.

neap tide A *tide that occurs every 14 to 15 days, coincides with the first and last quarter of the Moon, and has a small tidal range. *Contrast* spring tide.

near infrared *Electromagnetic *radiation with a wavelength from 750 to 2500 *nanometres, which is usually created by molecular vibrations.

near threatened *See* low risk.

Nearctic realm A *biogeographical realm that covers most of North America and contains a variety of *biomes including *tundra, *grassland, *deciduous forest, *coniferous forest, *chaparral, and *desert.

Nebraskan The earliest period of *glaciation in North America during the *Pleistocene *ice age, equivalent to the *Gunz in Europe.

necrophage A *scavenger that feeds on animal carcasses (dead meat) rather than live *prey.

necrosis The death of plant or animal cells or tissues.

necrotic Dead and discoloured.

nectar The sweet, sugary liquid that many *flowers produce in order to attract the *insects that assist *pollination.

nectarivore An animal that eats *nectar.

needs The fundamental requirements of people, including basics such as food, clothing, shelter, drinkable water, and breathable air. Wants and desires go beyond the basics, and reflect taste and fashion as well as the need to survive.

negative charge An electrical charge that is created by having more *electrons than *protons.

negative feedback An interaction (*feedback) that reduces or dampens the response of a system to change. *Contrast* positive feedback.

nekton *Aquatic *organisms that swim and can move long distances to feed or breed. *Contrast* PLANKTON.

Nelson, Gaylord A US politician (1916–2005) who came up with the idea for *Earth Day in 1969 when he was a Senator from Wisconsin.

nematode A microscopic free-living worm that has an unsegmented body. Nematodes are commonly found in *marine and *freshwater *habitats, in *soil, and as *parasites of plants and animals.

neo-classical economics A branch of *economics that studies the allocation of scarce *resources between competing uses and users, based on principles of *market equilibrium and profit maximization.

neo-Darwinism A modern *Darwinian theory that explains new *species in terms of genetic *mutations and *natural selection.

neoendemic The formation of a new *species that has evolved relatively recently and is locally distributed, as a result of divergent *adaptation of existing species to differing environmental conditions.

Neogene A period of geological time within the *Tertiary subera, between about 24 and two million years ago, which comprises the *Miocene and *Pliocene epochs.

neoglaciation A series of relatively small *glacier advances in the *northern hemisphere during the last few thousand years of the *Holocene.

Neolithic The New *Stone Age, following the *Mesolithic, which began in the Near East around 8000 BC and in Europe around 6000 BC, and was characterized by the adoption of *agriculture, village life, and the use of pottery.

neo-Malthusian A *pessimist view of the relationship between *population, *economic growth, and *resources, based on the ideas of Thomas *Malthus, who argued that population growth and economic growth would eventually be checked by absolute limits on resources such as food, energy, or water. This viewpoint grew in popularity particularly between the 1940s and the 1960s, when population growth and economic development were particularly strong in many countries. Many experts concluded that rapid population growth would eventually be checked by some absolute limit on resources (such as food, energy, or water). There was mounting evidence, too, that continued population growth and the environmental stresses associated with economic development could cause irreversible damage to the environmental systems that support life. This school of thinking was widely promoted through books such as *Limits to Growth.

neon (Ne) A natural, colourless, *inert *gas that comprises 0.0012% of the Earth's *atmosphere. When an electric current is discharged through it, neon produces a reddish-orange glow, and it is used in luminous tube signs and lights.

neotenin *See* JUVENILE HORMONE.

neoteny The retention of *juvenile body characters in the adult state.

Neotropical realm A *biogeographical realm that occupies Central and South America, and is dominated by *tropical forest, *savanna, and *desert.

neotropics The tropical parts of the New World (the Americas). *See also* PALAEOTROPICS.

NEP *See* NATIONAL ESTUARY PROGRAM.

NEPA *See* NATIONAL ENVIRONMENTAL POLICY ACT.

neptunianism A geological interpretation of features of the Earth's surface (including *rocks, *fossils, and *landforms) as the product of one or more major *floods, such as Noah's flood as described in the Old Testament of the Bible. *Contrast* UNIFORMITARIANISM. *See also* CATASTROPHISM.

neritic The relatively shallow waters within the *ocean over the *continental shelf, down to a depth of about 200 metres. This *ecological zone is generally richer in *nutrients than deep ocean waters.

NESHAPS *See* National Emission Standards for Hazardous Air Pollutants.

nest A structure in which animals (particularly *birds) lay eggs or give birth to their young.

nest box/structure An artificial box, platform, or other structure that enhances the reproductive habitat for target *species.

nest parasitism *See* brood parasitism.

nesting The nest-related activities of adult animals, including *nest building, egg laying, and departure of the young from the nest.

nesting cover Vegetation (including *grasses, low *shrubs, and *thickets) that protects nesting sites for quail, grouse, and many species of songbirds.

net economic welfare *Gross national product adjusted by subtracting the cost of problems such as *pollution and adding the value of beneficial, non-market activities such as leisure and *recreation.

net energy yield The total amount of useful *energy that is produced during the lifetime of an energy system, minus the energy that is used, lost, or wasted in making the useful energy available.

net migration The difference between the number of individuals who move in and the number who move out of a country or area within a specified period of time (normally one year). *Immigration minus *emigration.

net present value (NPV) The future stream of benefits and costs associated with a particular project or resource-using activity, converted into equivalent values today. This is estimated by assigning monetary values to benefits and costs, discounting future benefits and costs using an appropriate *discount rate, and subtracting the total discounted costs from the total discounted benefits. NPV is used in *cost–benefit analysis.

net primary production (NPP) The total amount of *energy or *biomass that is accumulated by *plants through *photosynthesis in a given area or habitat, over a specified period of time (usually a year). It is calculated from *gross primary production minus *respiration.

net primary productivity The amount of organic material that is produced by biological activity in an area or volume, in a given period of time, which is equivalent to *gross primary productivity minus the *respiration rate. It is conventionally expressed in grams per square metre per year (g m^{-2} year^{-1}), and varies from *ecosystem to ecosystem. The most productive ecosystems are *reefs and *estuaries, with natural *forests and freshwater a close second. The lowest productivities are found in the open *oceans and in *desert environments. Whilst mean net productivity levels on land are more than four times as high as those in the oceans, when these productivities are weighed according to the total area involved the oceans far exceed land in area and so total net primary productivity on land is only twice that in the oceans. It is estimated that over 70% of terrestrial net production is concentrated between latitudes 30°N and 30°S, which illustrates the ecological importance of the *tropical environment.

net radiation The balance between all incoming *energy and all outgoing energy carried by both *shortwave and *longwave radiation.

net radiative heating The net heating effect on the Earth of the *radiative forcing of climate, which can be positive or negative.

net reproduction rate A measure of *fertility, based on the number of daughters born to a woman given current birth rates, and her chances of living to the end of her child-bearing years.

net worth Net monetary value; total assets minus total liabilities.

netting In *greenhouse gas *emission control, the idea that all *emission sources in the same area that are owned or controlled by a single company can be treated as one large source, so an emission increase at one source can be offset with an emission reduction at another source in the same area. *See also* BUBBLE.

neurotoxin A *toxic substance, such as *lead or *mercury, that poisons nerve cells.

neustic community *See* NEUSTON.

neuston An *organism that rests or swims on the surface of a water body.

neutral Neither *acid nor *alkaline.

neutral soil A *soil solution with *pH of between 6.5 and 7.3.

neutral stability A stable atmospheric condition that exists in dry air when the *environmental lapse rate equals the *dry adiabatic rate, and in moist air when the environmental lapse rate equals the *saturated adiabatic rate.

neutralize To decrease the *acidity or *alkalinity of a substance by adding alkaline or acidic materials, respectively, in order to make it *neutral.

neutron A subatomic particle that is found in all *atoms except normal hydrogen, and which has no electric charge. *See also* ELECTRON, PROTON.

neutron-activation analysis A method of identifying the *isotopes of an *element by bombarding them with *neutrons and observing the *radioactive decay products that are emitted.

névé *See* GRANULAR SNOW.

new environmental paradigm (NEP) The view that humans represent only one among many *species on Earth, that human activities are determined by the *environment as well as by social and cultural factors, and that humans are strongly dependent upon the environment and its resources. *Contrast* DOMINANT SOCIAL PARADIGM.

New International Economic Order (NIEO) The emergence of a new global political economy in the latter part of the 20th century, in response to powerful economic, political, and technological forces that have transformed the world. It is reflected, for example, in new attitudes towards economic growth and development, international aid, regionalism, *multinational corporations, national policies, and *sustainable development.

new social movement Any of a series of social movements that have emerged in recent decades (including the *environmental movement, the women's movement, and the gay rights movement), which have different origins, strategies, and goals from more traditional social movements based on class and economics.

new town A planned *urban community that is developed in a *rural area, and is designed to be largely self-sufficient with its own housing, education, commerce, and recreation.

New Urbanism A planning movement to promote cities and towns with planned growth that minimizes damage to the environment, through the restoration of *urban centres and towns, the reconfiguration of sprawling *suburbs into communities of neighbourhoods and districts, the conservation of natural environments, and the preservation of the built legacy.

newly industrializing country Countries (such as South Korea, Taiwan, and Singapore) that have undergone major industrialization only in recent decades.

NFLRMP *See* NATIONAL FOREST LAND AND RESOURCE MANAGEMENT PLAN.

NFWF *See* NATIONAL FISH AND WILDLIFE FOUNDATION.

NGO *See* NON-GOVERNMENTAL ORGANIZATION.

niche overlap An overlap in resource requirements by two *species in the same *habitat or *ecosystem.

niche specialization The process by which a *species becomes better adapted, by *natural selection, to the specific characteristics of a particular *habitat.

nickel (Ni) A silver-white, hard, *malleable, *ductile *metal that can be highly polished, is resistant to corrosion, and is widely used in the manufacture of steel. It occurs naturally in all parts of the *environment, including plants and animals, is extremely persistent in soils, and is *carcinogenic to humans. *See also* OUTER CORE.

nicotine The main active ingredient of tobacco, which is addictive and extremely toxic to humans. It causes irritation of lung tissues, constriction of blood vessels, increased blood pressure and heart rate, and stimulation of the central nervous system. It is also used as an *insecticide.

NIEO *See* NEW INTERNATIONAL ECONOMIC ORDER.

night The time between *sunset and *sunrise, when it is dark outside. *Contrast* DAY.

nimbostratus (Ns) Mid-level or low *clouds that form shapeless, thick dark grey cloud layers and often bring rain or snow.

NIMBY *See* NOT IN MY BACKYARD.

NIREX An independent organization (the Nuclear Industry Radioactive Waste Executive) that was set up in the UK in 1982 to research, develop, and operate *radioactive waste disposal facilities on behalf of the *nuclear power industry. In 1985 it became a limited company whose mission is 'to provide the UK with safe, environmentally sound and publicly acceptable options for the long-term management of radioactive materials'. It advises the UK government and industry.

nitrate (NO_3) A naturally occurring salt that contains *nitrogen and *oxy-

NICHE

The functional role and position of a *species within a *community or *ecosystem, which is defined by such things as what the species does, how it feeds, what and when it eats, when and for how long it is active, how it reproduces, how it behaves, what particular part of the habitat it uses, and how it responds to temperature and moisture. In most ecosystems a number of species could perform a particular role, but at any one point in time only one species actually does. In *grassland ecosystems, for example, bison (in North America), antelopes and zebras (in Africa), and kangaroos (in Australia) play more or less the same role in different places. Two species cannot live in the same *habitat on a sustainable basis if they occupy identical niches. There are four important implications of niches. First, many species have overlapping niches. Second, the greater the diversity of niches within an ecosystem, the greater the variety of energy flows and thus the more stable the ecosystem. Third, the introduction of *exotic species into an ecosystem, either deliberately or by accident, can create problems because they sometimes compete with native species for niches and food supplies. Fourth, the removal of a species (or group of closely related species) can leave a niche empty and this can soon lead to reduced energy flow through the entire system, which can radically affect the structure and stability of the ecosystem overall. Also known as **ecological niche**. *See also* FUNDAMENTAL NICHE, POTENTIAL NICHE, REALIZED NICHE.

gen, can exist in the atmosphere or as a dissolved gas in water, is a *nutrient for plants and an inorganic *fertilizer, and can cause severe illness in humans and animals. It is produced in septic systems, animal feed lots, agricultural fertilizers, *manure, industrial wastewaters, *sanitary landfills, and *garbage dumps.

nitrate-forming bacteria *See* NITRIFYING BACTERIA.

nitric acid (HNO₃) A colourless or yellowish *acid that is made by distilling a *nitrate with *sulphuric acid. It is used especially in the production of *fertilizers and explosives and rocket fuels. It is caustic and corrosive, is a component of *acid rain, and acute exposure to it can cause irritation to the lungs, nose, throat, and skin.

nitric oxide (NO) A colourless, poisonous gas formed by the *oxidation of *nitrogen or *ammonia. It is the most common form of nitrogen emitted into the *atmosphere and is usually produced by fuel combustion. It is converted to

NITROGEN CYCLE

The processes by which *nitrogen circulates in both *organic and *inorganic phases between the *atmosphere and the *biosphere. Compared with the other major *biogeochemical cycles it is relatively fast but also very complex. The atmosphere is composed mostly of nitrogen gas (N_2). This is chemically unreactive, which means that it cannot be used directly by plants. In order to make it available to plants, some of the nitrogen is oxidized by energy released in *lightning, and it then dissolves in raindrops to form *nitric acid (HNO_3). Most (more than 90%) of the nitrogen in soils is fixed biologically, and is thus available to plants. The nitrogen gas is used by soil bacteria (particularly the bacteria living in nodules attached to the roots of *legumes such as peas and beans) which convert it into *ammonia (NH_3). Inorganic *nitrate (NO_3^-) and ammonia in the soil are absorbed by plants (via their roots) and turned into organic compounds (such as *proteins) in plant tissue. A proportion of this nitrogen is eaten by *herbivores, which use nitrogen in the form of amino acids from which they synthesize proteins. Some of the nitrogen is passed on in the form of proteins to the *carnivores that feed on the grazing animals. The nitrogen is ultimately returned to the soil as waste products (in urine and faeces) and when organisms die and decompose. Bacteria convert the organic nitrogen into ammonia or *ammonium (NH_4^+) compounds. Other bacteria convert these into *nitrites (NO_2^-) and then into nitrates (NO_3^-) which can be taken up again by plants. Some of the nitrate and ammonia that is not absorbed by plants is leached (washed) from the soil into groundwater and surface water (lake and rivers), where it provides nutrient supplies for aquatic plants. The oceans are also part of the nitrogen cycle. Some nitrogen accumulates as organic sediment which might through time be compacted and converted into sedimentary rock. The nitrogen can be released back into water by *weathering of the deposits and rocks. Ammonia (NH_3) evaporates quickly when it is released, thus recycling some nitrogen back into the atmosphere. From there it quickly dissolves in raindrops and is washed back to the ground. Special *denitrifying bacteria complete the cycle, converting nitrates to nitrite, nitrites to ammonia, and nitrates to gaseous nitrogen or nitrogen oxides. This effectively recycles all of the nitrogen compounds. The two most important ways in which human activities alter the natural nitrogen cycle are through the use in agriculture of *nitrogen fertilizers, and the release of *nitrogen oxides in *air pollution.

n

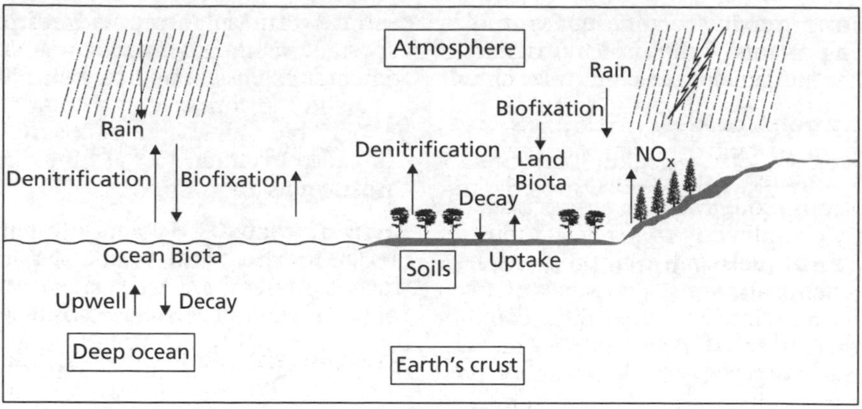

Fig 14 Nitrogen cycle

*nitrogen dioxide in the atmosphere, is a precursor of *ozone and *nitrate, and is a *toxic air pollutant.

nitrification The *oxidation of *ammonia to *nitrite and *nitrite to *nitrate by *micro-organisms. *Contrast* DENITRIFICATION.

nitrifying bacteria *bacteria that convert *nitrites into compounds that can be used by green plants to build *proteins. Also known as **nitrate-forming bacteria**. *See also* NITRITE-FORMING BACTERIA.

nitrite (NO₂) A *salt of nitrous acid that is commonly found in *soil, is produced by the chemical modification of *ammonium by specialized *bacteria, and is toxic to plants and animals at high concentrations. The second stage of the *nitrification process.

nitrite-forming bacteria *Bacteria that combine *ammonia with *oxygen to form *nitrites, as part of the *nitrification process.

nitrogen (N) A colourless, tasteless, odourless, and relatively unreactive gas that makes up nearly four-fifths (78% by volume) of the *atmosphere. It is a *macronutrient, an essential component of *proteins and *nucleic acids in living organisms, occurs in nature in a variety of forms (*ammonia, *ammo-nium, *nitrate, *nitrite), and is a key ingredient in many commercial *fertilizers. Nitrogen availability controls plant growth and thus rates of photosynthesis in many terrestrial ecosystems. It can be a *pollutant when nitrogen compounds are mobilized in the environment, for example by *leaching from manured or fertilized fields, discharges from septic tanks or feedlots, or emissions from vehicle engines. *Nitrogen oxides and ammonia also affect atmospheric chemistry through their involvement in *acid rain and their reactivity with *ozone. Nitrogen can also affect *water quality. *See also* EUTROPHICATION.

nitrogen dioxide (NO₂) A reddish-brown *gas that is produced by combustion, when *nitric oxide emitted from power plants and vehicles combines with *oxygen in the atmosphere. It is *toxic at high concentrations, reacts with *moisture in the air to form *nitric acid (which is highly corrosive to metals), and in the presence of sunlight and *volatile organic compounds it can contribute to the formation of ground-level *ozone, or *smog. It is also an important ingredient of *acid rain. *See also* PHOTOCHEMICAL SMOG.

nitrogen fixation A chemical process by which *nitrogen in the atmosphere is assimilated into *organic compounds, by *nitrogen-fixing bacteria and *nitrogen-fixing plants, and made available to other *organisms through *food chains.

nitrogen oxides (NO$_x$) Gas compounds of *nitrogen and *oxygen (*nitric oxide, *nitrogen dioxide, and *nitrous oxide) which are produced directly or indirectly from the combustion of *fossil fuels and from processes used in chemical plants. Emissions of nitrogen oxides into the atmosphere combine with oxygen to form nitrogen dioxide, and this in turn reacts with water vapour to form *nitric acid, which is an ingredient in *acid rain. Nitrogen dioxide is a *criteria pollutant. All of the nitrogen oxides are *greenhouse gases.

nitrogen-fixing bacteria *Bacteria that convert *nitrogen from the atmosphere or soil into *ammonia, which can then be converted into plant *nutrients by *nitrite- and *nitrate-forming bacteria.

nitrogen-fixing plant A plant that can assimilate and fix free *nitrogen from the atmosphere, through the work of *bacteria that live in its root nodules.

nitrogenous waste Any animal or vegetable residues that contain significant amounts of *nitrogen.

nitrous oxide (N$_2$O) A powerful and persistent *greenhouse gas, with a *global warming potential of 320. It is naturally created by *microbial activity in *soils, but it is increased by soil cultivation practices (particularly rice paddies), the use of commercial and organic *fertilizers, combustion of *fossil fuels, production of *nitric acid, and burning of *biomass.

NNR See NATIONAL NATURE RESERVE.

no longer threatened species *Species of plant and animal from which threats have declined, so their survival is not currently threatened.

no regrets policy An approach to *management that involves erring on the side of caution and planning well in advance. See also PRECAUTIONARY PRINCIPLE.

no till Planting a crop without preparing the ground in advance or working it afterwards. Also known as **zero till**.

NOAA See NATIONAL OCEANIC AND ATMOSPHERIC ADMINISTRATION.

noble gas See INERT GAS.

noble metal A chemically inactive *metal (such as *gold, *silver, or platinum) which is resistant to *corrosion or *oxidation. Also known as a **precious metal**.

noctilucent cloud A wavy, thin, bluish-white type of *cloud that forms at altitudes of 80–90 kilometres above the surface of the Earth, looks similar to *cirrus clouds, and is best seen at *twilight at high latitudes.

nocturnal Active only at night.

nocturnal inversion See RADIATION INVERSION.

nodule A small lump, such as a small rounded lump of mineral substance (for example, *chert) which is usually harder than the surrounding rock, or a small growth on the roots of most *legumes and some other plants, that plays a part in *nitrogen fixation.

NOEL See NO-OBSERVED-EFFECT LEVEL.

NOHSCP See NATIONAL OIL AND HAZARDOUS SUBSTANCES CONTINGENCY PLAN.

noise Unwanted or undesirable sound, particularly if it is loud and disturbs people's communication, sleep, study, or recreation.

noise abatement Reducing or eliminating unwanted *noise.

noise pollution *Pollution that is caused by *noise, for example from heavy traffic, busy factories, or large gatherings of people.

nomad A person who lives in no fixed place but has a wandering lifestyle, moving about in search of food or grazing land for their animals.

nomadism The practice of living by moving from place to place, having a *nomadic lifestyle. *Contrast* PASTORAL-ISM, SEDENTISM, TRANSHUMANCE.

nomenclature Names or terms forming a set or a system.

Non-Annex 1 Countries Countries that are not included in Annex I of the *United Nations Framework Convention on Climate Change, and that therefore do not currently have binding *emission reduction targets.

Non-Annex B Countries Countries that are not included in Annex B of the *Kyoto Protocol, and that therefore do not currently have binding *emission reduction targets.

non-aqueous phase liquid (NAPL) Any *contaminant that remains undiluted as a bulk liquid in *groundwater, such as spilled liquid *petroleum products (like *gasoline) that are not mixed into groundwater but float as a layer in the rock above it.

non-attainment area An area that does not meet *ambient *air quality standards for a *criteria air pollutant.

non-biodegradable Materials (such as glass, *heavy metals, and most types of plastics) that cannot be broken down by livings things into simpler chemicals.

non-combustible Not capable of catching fire or burning.

non-commercial cutting A tree-felling activity which does not yield a net income, usually because the *trees that are cut are too small, are of poor quality, or are not marketable.

non-commercial vegetative treatment The removal of *trees for reasons other than *timber production.

non-compliance Not obeying all the federal and state regulations that apply.

non-compliance coal A North American term for any coal that emits a relatively large amount of *sulphur dioxide (greater than 3.0 pounds per million *Btu) when burned. Also known as **high sulphur coal**.

non-consumptive use The use of *resources in ways that do not reduce supply. Examples include hiking, bird watching, and *nature study in a forest. *Contrast* CONSUMPTIVE USE.

non-consumptive value The value of natural resources which are not diminished by their use, that do not require the valuer to have access to them or make active use of them. Also known as **non-use value, passive use value**, or **existence value**. *Contrast* USE VALUE.

non-consumptive wildlife Any unharvested *wildlife that is valuable for observation, education, enjoyment, and as a natural member of an *ecological community.

non-consumptive wildlife use A general term for a variety of recreational activities related to wildlife. Primary non-consumptive uses include general wildlife observation, bird-watching, bird-feeding, and wildlife and bird photography. Secondary uses include nature walks, membership of wildlife organizations, ownership of wildlife pets, and visits to zoos.

non-conventional pollutant *See* NON-CRITERIA POLLUTANT.

non-criteria pollutant Any *pollutant (such as *asbestos, *benzene, *mercury, and *polychlorinated biphenyls) that is not a *criteria pollutant. Also known as **non-conventional pollutant** or **unconventional pollutant**.

non-ferrous metal Any *metal that isn't *iron or an *alloy that doesn't contain iron. Examples include *aluminium, *lead, and *copper.

non-forest land Land that is not used for *forestry or *timber production, but is used for other activities such as *farming, transport, industry, commerce, and housing.

non-game species *wildlife species that are not hunted for *sport.

non-governmental organization (NGO) A non-profit group or association that is separate from government, and whose purpose is to pursue particular social objectives (such as *environmental protection) or to serve particular constituencies (such as *indigenous peoples). Examples include the *World Wide Fund for Nature and the Red Cross.

non-hydric soil A *soil that has developed mainly under *aerobic conditions.

non-ionizing electromagnetic radiation A form of *radiation (for example from *microwaves, *radio waves, and low-frequency *electromagnetic fields from high-voltage transmission lines) that does not change the structure of atoms but does heat tissue and may cause harmful biological effects.

non-market cost See SOCIAL COST. See also TRUE COST.

non-market damage Financial damages that are generated by *environmental change (such as *climate change), which cannot be evaluated in terms of *market failure because of a lack of information and/or an inability to act on the available information.

non-market values Estimated values for goods and services that are not traded for money but are valued in terms of what reasonable people should be willing to pay rather than go without them. Examples include fish and wildlife values and scenic quality values. Contrast MARKET VALUES. See also CONTINGENT VALUATION METHOD.

non-metallic Not containing, resembling, or having the properties of a *metal.

non-native See EXOTIC.

non-native species See INTRODUCED SPECIES.

non-permeable surface Any surface material that will not allow water to penetrate, such as roads and concrete.

non-persistent pesticide A *pesticide which has only short-lasting harmful effects. Contrast PERSISTENT PESTICIDE.

non-point source Scattered sources of *pollutants, such as runoff from *agriculture, *forestry, an *urban area, *mining, *construction, *dams, land disposal, and *saltwater intrusion. Contrast POINT SOURCE.

non-potable water Water that is unsafe or unpalatable to drink because it contains *pollutants, *contaminants, *minerals, or infectious *agents. Contrast POTABLE WATER.

non-renewable energy *Energy (such as the *fossil fuels: *oil, *natural gas, and *coal) that comes from a *natural resource which is not replaced, or is replaced only very slowly, by natural processes.

non-renewable resource A *natural resource (such as *oil) which cannot be replaced when it is used up. Also known as **depletable resource, stock resource**. Contrast RENEWABLE RESOURCE.

non-silicate Rock forming minerals that do not contain *silicates.

non-statutory Not a legal requirement.

non-target species 1. A *species that is not intentionally targeted for control by a *pesticide or *herbicide, but which may suffer damage because of exposure to it.
2. That part of a catch (for example of fish) which excludes the *target species but includes the *bycatch and *byproduct.

non-transmissable disease A *disease (such as *cancer and heart disease) that cannot be transmitted by a *microorganism through human contact.

non-use value See NON-CONSUMPTIVE VALUE.

non-vascular plant A *bryophyte. Contrast VASCULAR PLANT.

no-observed-effect level (NOEL) The *exposure level at which no statistically or biological significant differences in the frequency or severity of any effect can be detected between an exposed and a control population.

noosphere That part of the *biosphere that is affected by human activities.

no-regrets policy An approach to *environmental policy that is based on the idea that the problem of global *climate change is linked to other critically important problems of *environment and *development, the combined risks of which are serious enough, and the eventual benefits of action great enough, to require bold initiatives without delay, even if they impose great immediate cost.

norm 1. Normal, average, or most common.

2. An expected standard of behaviour and belief that is established and enforced by a group.

normal distribution See GAUSSIAN CURVE.

normal fault The simplest type of vertical displacement *fault, in which one side falls relative to the other. The higher side survives as a cliff-like *escarpment. Normal faults can develop from either *compression pressure or from *tension pressure.

north The compass point that is at 0° or 360°.

north/northern General term that is used to describe the *industrialized countries which are located mainly in the *northern hemisphere and are sometimes referred to as the *First World. This includes Canada, the USA, Western European countries, the former Soviet Union, Japan, Australia, and New Zealand. *Contrast* SOUTH.

North Atlantic Drift See GULF STREAM.

North Pole The most northerly point on Earth, at a latitude of 90°N in the *northern hemisphere; the northern end of the Earth's rotational axis. *Contrast* SOUTH POLE.

northern hemisphere The half of the Earth that lies *north of the *equator. *Contrast* SOUTHERN HEMISPHERE.

northern lights See AURORA.

northern spotted owl A *threatened species of owl (*Strix occidentalis caurina*) which is nocturnal, feeds on small mammals and birds, and lives in *old growth forests of northern California and the Pacific Northwest of the USA and in southern parts of British Columbia, Canada. Its survival is under threat from the loss of old growth forest habitat, as a result of *logging and forest fragmentation, but also as a result of natural disasters such as fire, volcanic eruptions, and wind storms. The species is protected under the *Endangered Species Act and under *CITES.

North–South divide The gap, particularly in financial well-being, between richer *developed countries and poorer *developing countries. The term is also used to describe socio-economic differences between northern and southern England.

not at risk A general term for a *species that has been evaluated and found not to be at risk of becoming *extinct or *extirpated.

not in my backyard (NIMBY) The belief that certain types of hazardous activities and substances may be needed or acceptable, but not round here. *See also* LULU.

not threatened category of species A category for *species under threat, developed as *IUCN Conservation (Red Data Book) Categories, and widely used around the world. The only category within this group is *safe.

no-take reserve A special type of *nature reserve in which, by law, no one is allowed to fish or collect biological specimens. Rules could apply to one or all species.

notice of intent A formal notice, published in the US federal register, of intent to prepare an *environmental impact statement on a proposed action.

notification The process of official designation of a protected area, such as the notification of a *Site of Special Scientific Interest in the UK.

nowcast A short-term *weather *forecast, generally looking ahead six hours or less.

noxious Irritating, unpleasant, or harmful to living *organisms.

noxious species Plant species that can be harmful or even fatal when eaten by animals.

NPK *Nitrogen, *phosphorus, and *potassium, which are common ingredients in commercial *fertilizers.

NPL See NATIONAL PRIORITIES LIST.

NPP See NET PRIMARY PRODUCTION.

NPS See NATIONAL PARK SERVICE.

NPV See NET PRESENT VALUE.

N-rank See NATIONAL RANK.

NRAs See NATURAL RESOURCE ACCOUNTS.

NRC See NATIONAL RESPONSE CENTER.

nuclear energy *Energy or *power that is produced by *nuclear reactions. Nuclear power currently uses only *nuclear fission.

nuclear fission A *nuclear reaction process in which *isotopes of certain heavy elements (such as *uranium and *plutonium) are bombarded with *neutrons which splits them apart, creates two smaller *atoms, and releases a vast amount of *energy. Also known as **splitting the atom**.

nuclear fuel *Fuel (such as *uranium) that can be used to generate electricity in a *nuclear reactor in a *nuclear power plant.

nuclear fuel cycle The sequence of processes that are associated with the use of *nuclear fuel, from initial mining and milling of the *ore, through conversion and *enrichment of the nuclear fuel, its use in *nuclear power plants, and the management of *nuclear waste.

nuclear fusion A *nuclear reaction process in which several light *nuclei are combined to make a heavier one whose mass is slightly less than the combined mass of the lighter nuclei, which releases energy. This is the reaction that fuels the Sun, where hydrogen nuclei are fused to form helium; it is also one source of power in a *nuclear weapon.

nuclear medicine A branch of medicine which uses *radioactive substances for the diagnosis and treatment of disease (such as *cancer).

nuclear pile US name for a *nuclear reactor.

nuclear power See NUCLEAR ENERGY.

nuclear power plant/station An electric power plant that converts *nuclear energy into *electrical energy. Heat that is generated by a *nuclear reaction is used to make steam, which drives a *turbine, which in turn drives an electric generator.

nuclear reaction A change in the composition of a nucleus, caused by bombarding it with atomic or subatomic particles or very high energy radiation (for example in a *nuclear reactor). This process can form a new nucleus (*nuclear fusion) or split the original nucleus (*nuclear fission), and it releases a vast large amount of energy.

nuclear reactor A device in which *nuclear fission may be initiated, maintained, and controlled to produce energy, conduct research, or produce *fissile material for nuclear explosives. Known in the USA as a nuclear pile.

nuclear reactor and support facilities Facilities that are associated with the civilian use of *nuclear energy, including uranium mills, commercial power reactors, fuel reprocessing plants, and uranium *enrichment facilities.

nuclear test Military tests of *nuclear weapons that are carried out with government approval, in order to improve the design of nuclear weapons and to

n

study the impacts and effects of nuclear explosions.

nuclear war Warfare that uses *nuclear weapons. *See also* NUCLEAR WINTER.

nuclear waste *See* RADIOACTIVE WASTE.

nuclear weapon A device that releases *nuclear energy from a *nuclear reaction in an explosive manner, in order to damage structures and materials and to kill or injure large numbers of people. A *weapon of mass destruction. Also known as a **radiological weapon**.

nuclear winter A long period of darkness and extreme cold that scientists predict would follow a full-scale *nuclear war, caused by the blocking out of sunlight by a layer of dust and smoke in the atmosphere, which few organisms could survive. Worst case scenarios predict that nuclear war and nuclear winter would kill most plants and animals in the northern hemisphere and threaten the very survival of the human species.

nucleation The process by which substances change *phase between liquid, gas, and solid, as occurs for example during *condensation, *deposition, and *freezing.

nucleic acid A biological *molecule, composed of a long chain of *nucleotides, which controls activities in cells, helps in the synthesis of *proteins, and transmits *hereditary traits.

nucleotide The basis building block of *DNA, which consists of a sugar, a *phosphate, and an organic base that contains *nitrogen. Thousands of nucleotides are linked to form a *molecule of DNA or *RNA. *See also* NUCLEIC ACID.

nucleus 1. The centre of an *atom, which is occupied by *protons and *neutrons.
2. The part of a cell that contains the *chromosomes (*DNA).
3. A particle (for example of dust or salt) on which water molecules or ice accumulate to form *raindrops, *snow,

*hail, and other forms of *precipitation. *See also* CLOUD CONDENSATION NUCLEI.

nuclide Any species of *atom that exists for a measurable length of time, and is defined by the number of *protons and *neutrons in its *nucleus. A nuclide may be *radioactive.

nudation The creation of an area of bare land, either by natural events or by humans, which is the first stage in vegetation *succession.

nuée ardente A 'glowing cloud' that forms when a burning-hot ash, gas, steam, and rock emerges from a *volcano and flows downhill, burning everything in its path.

nuisance bloom A rapid increase of one or a few species of *phytoplankton in a water body, which discolours the surface water, often increases the concentration of *toxins, and reduces *water quality (particularly the concentration of *dissolved oxygen). *See also* ALGAL BLOOM.

nunatak A *mountain ridge that sticks up through an *ice sheet. They are usually angular with sharply defined ridges, caused by intense *freeze–thaw *weathering processes.

nuptial feeding Feeding of offspring by the male.

nurse tree A *tree that is grown to shelter and protect other young plants which are at a vulnerable stage of growth.

nursery A place where *plants are *propagated and grown to usable size, when they can be planted elsewhere.

nursery area An area (*range) that is used for *grazing during the summer months by a temporary social unit of *big game (cows and young calves).

nursery stock A *shrub or *tree species that is grown in a *nursery for subsequent planting out elsewhere.

nutrient Any chemical compound (such as *protein, fat, *carbohydrate, *vitamins, or *minerals) that provides the chemical energy (food) needed by living *organisms. *See also* MACRONUTRIENT, MICRONUTRIENT.

nutrient cycle The natural circulation of chemical elements (such as *carbon and *nitrogen) and compounds through specific pathways from the *abiotic parts of *ecosystems into the organic substances of the *biotic parts, and back again to the abiotic parts. A local example of the *biogeochemical cycle.

nutrient deficiency The lack of an adequate amount of a particular *nutrient which is important for the growth and health of plants and animals.

nutrient depletion A reduction in the level of *nutrients in a *habitat or *ecosystem, caused by uptake by plants, removal of plant residues, or *leaching, to a level at which they become unavailable for further uptake. The opposite of *nutrient enrichment.

nutrient enrichment An increase in the level of *nutrients in a *habitat or *ecosystem. The opposite of *nutrient depletion. *See also* EUTROPHICATION.

nutrient loading The quantity of *nutrients that is washed into a waterbody from its drainage basin, usually expressed as mass per unit area per unit time (kilograms per hectare per year, kg ha^{-1} year^{-1}).

nutrient pollution *contamination of *water resources by excessive inputs of *nutrients. *See also* EUTROPHICATION.

NWF *See* NATIONAL WILDLIFE FEDERATION.

NWS *See* NATIONAL WEATHER SERVICES.

objective A high-level statement of what is desired in any project or activity, often expressed as a specific statement of the measurable results that are to be achieved within a stated time period.

obligate Without option; of a *species, restricted to specific environmental conditions and thus unable to change its mode of feeding or ecological relationships.

obligate mutualist A beneficial *symbiont that lives exclusively in the *host and depends on the host for survival. *Contrast* FACULTATIVE MUTUALIST.

obligation A legal responsibility.

oblique-slip fault A geological *fault that involves both vertical (*dip-slope) and horizontal (*strike-slip) movements.

obliquity The angle of tilt of the Earth's axis of rotation.

obsidian A black or dark green *extrusive *volcanic glass that is formed from fast-cooling *lava and has a similar chemical composition to *granite. Also known as **volcanic glass**.

occidental From or relating to the western *hemisphere. *Contrast* ORIENTAL.

occluded front *See* OCCLUSION.

occlusion The final stage in the decay of a *depression, where the *cold front overtakes the *warm front of the depression, with cold air pushing under and lifting the warm sector away from the ground surface. Also known as **occluded front**. *See also* CYCLOGENESIS, FRONTOGENESIS.

occupational exposure Exposure (for example to *pollutants and *toxic *agents) that people incur in the course of their work.

Occupational Safety and Health Act (OSHA) A federal law in the USA that defines minimum health and safety standards for the workplace.

occurrence In *epidemiology, the frequency of a *disease or event in a *population. *See also* INCIDENCE, PREVALENCE.

ocean The body of saltwater that surrounds the continents and covers two-thirds of the surface of Earth. The average depth of the ocean floor is more than 3650 metres below sea level, and the total volume of the world's oceans is estimated at around 1370 million cubic kilometres.

ocean bed *See* OCEAN FLOOR.

ocean current The steady flow or circulation of surface water in a prevailing direction within the *ocean, as a non-tidal current. There are three types of oceanic currents: *drift currents, stream currents, and *upwelling currents. *See also* GULF STREAM.

ocean dumping The disposal of *hazardous wastes and other substances by dumping them in the deep waters of the *oceans. *See also* OFFSHORE DUMPING.

Ocean Dumping Act *See* MARINE PROTECTION, RESEARCH, AND SANCTUARIES ACT (1972).

ocean energy sources Existing *marine sources of *renewable energy include ocean *thermal energy, *tidal power, and *wave energy, and future possibilities include the construction of offshore *wind power stations, exploitation of submarine *geothermal energy and marine *biogas energy, and energy generated by exploiting *salinity gradients within the oceans.

ocean floor The *ecological zone that lies beyond the *continental shelf, *continental slope, and *continental

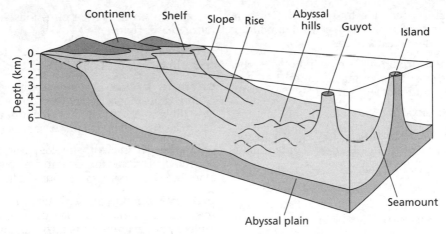

Fig 15 Ocean floor

rise; it underlies most of the major oceans, is often far from flat, and at depths greater than about 2000 metres is largely covered by fine-textured sediment (*ooze) composed mostly of organic matter. Also known as **abyss** or **ocean bed**.

ocean pollution *Pollution of ocean waters, both directly (such as by *oil spills) and indirectly via *rivers that flow into the sea.

ocean trench A deep, linear valley on the *ocean floor, which runs parallel to the *coast, and marks an active *subduction zone at the margins of one of the *continents, where one *crustal plate is pushed down beneath another at a *destructive plate boundary. The deepest known ocean trench is the Marianas Trench in the Pacific Ocean, which is 11 033 metres below sea level.

Oceanian realm A *biogeographical realm that covers the Pacific Ocean and is dominated by *tropical forest.

oceanic Relating to or occurring in the open sea, beyond the *continental shelf.

oceanic conveyor belt The idealized circulation pattern of the *ocean that is caused by *temperature and *salinity gradients which lead to the transport of heat in the North Atlantic.

oceanic crust That part of the Earth's *crust which is formed at *mid-ocean ridges, underlies the *oceans, and is generally between 5 and 10 kilometres thick. Oceanic crust is thinner, denser, and heavier than *continental crust. It consists mostly of *basaltic types of rocks, and geologists refer to this material as *sima. Because of the movements of *plate tectonics, the oceanic crust is nowhere older than about 200 million years. *Sea-floor spreading is continuously creating new oceanic crust, and *subduction is continuously destroying existing crust, like some huge non-stop geological conveyor belt.

oceanic front A boundary that separates masses of water within the *ocean that have different temperatures and densities.

oceanic island An *island in the ocean, which will often support a distinctive *community of plants and animals, and may be a *centre of endemism. Such islands are formed in various ways, including by breaking away from a continental landmass, volcanic action, coral formation, or a combination of processes.

oceanic ridge One of a number of extensive, sinuous, underwater *mountain chains which are found on the *ocean floor, and which mark the boundaries of some *crustal plates.

oceanography The study of the *oceans, their origin, composition, history, and ecology.

odour A smell.

odour threshold The lowest concentration of a chemical or *contaminant that can be smelled.

ODP *See* OZONE DEPLETION POTENTIAL.

OECD *See* ORGANISATION FOR ECONOMIC CO-OPERATION AND DEVELOPMENT.

offer In *greenhouse gas *emissions trading, the price at which the owner of an emission reduction, credit, or allowance is willing to sell. Also known as **ask**.

off-gas The *gas *effluent that is given off during any stage of an industrial process.

official development assistance Funding that is provided by the government of a *developed country to the governments of *developing countries, to pay for particular community, health, and commercial projects.

off-road vehicle (ORV) Any vehicle that is intended for use on unmade surfaces or rough terrain, including all-terrain vehicles, mini-bikes, trail bikes, dirt bikes, dune buggies, and golf carts, but excluding snowmobiles.

offset A form of credit-based *emissions trading. An offset is created when a source reduces its *emissions by more than it is required to, on a voluntary, permanent basis. The source can then trade that offset (the extra emissions) to new sources to allow growth or relocation, with regulators approving each trade. *See also* BUBBLE, NETTING.

offset ratio In *greenhouse gas *emissions trading, the ratio between the amount by which on-site emissions are allowed to increase and the size of *offset that is purchased. In US domestic emission trading programmes, new

sources must offset their emissions at a ratio greater than 1:1.

offshore All areas *seaward of the *coastline. *Contrast* INSHORE.

offshore breeze A *breeze or light *wind that blows from the land out over the sea. Also known as **land breeze**. *Contrast* ONSHORE BREEZE, SEA BREEZE.

offshore dumping The disposal or dumping of waste material off or away from the shore. *See also* OCEAN DUMPING.

offshore facility Any facility (other than a vessel) that is located in, on, or under the water *offshore, and is subject to the *jurisdiction of a particular country.

offshore waters Ocean waters beyond the *inshore, between about 60 and 250 *nautical miles from land.

off-site *See* EX SITU.

off-site facility A *facility for the treatment, storage, or disposal of *hazardous waste that is located away from the site where that material is created. *Contrast* ON-SITE FACILITY.

offspring Immediate *descendants.

offstream use Water that is withdrawn from surface or *groundwater sources for use at another place, for public water supply, industry, *irrigation, livestock, power generation, and other uses. *Contrast* INSTREAM USE.

ogive Alternating bands of dirty and clean ice below an *ice fall within a *glacier, formed by summer and winter movement over the ice fall. Also known as **Forbes band**.

O-horizon The layer rich in *organic matter within the *A-horizon of a *soil.

oil 1. In general, any viscous liquid, which includes natural plant and animals oils.
2. A thick, black, sticky *hydrocarbon substance that is used to produce *fuel (petroleum) and materials (plastics). Oil is one of the world's principal *fossil fuel

resources, and known reserves are fast being depleted. About two-thirds of the world's known *recoverable reserves are in the Middle East. Known world reserves of oil are expected to run out within 80 years using currently available technology. Since the 1970s technology has allowed the pumping of oil from offshore fields (including the North Sea) and from the frozen north of Alaska. Exploitation and use of oil resources has a number of *environmental impacts, including *oil pollution. Burning of petroleum is a major cause of *air pollution, and the transport of oil (particularly at sea in large *oil tankers) has caused some major environmental disasters, including *oil spills.

oil desulphurization A method of reducing *sulphur dioxide emissions from *power plants that burn *oil, in which the oil is treated with hydrogen (forming *hydrogen sulphide gas and reducing the sulphur content of the oil) before combustion.

oil fingerprinting A method of identifying the chemical characteristics of a particular *oil, which allows *oil spills to be traced to their source.

oil pipeline A pipe along which *oil is transported from the *refinery to other places.

oil pollution *Pollution of soil and water as the result of an *oil spill.

oil rig A structure that stands on the *sea-bed over a submarine *oilfield, which provides a stable base above water for drilling *oil wells.

oil shale A dark grey or black *shale that contains organic substances that can be distilled to yield liquid *hydrocarbons, but does not contain free *petroleum.

oil slick A layer of *oil floating on an area of water, usually as a result of an *oil spill.

oil spill An accidental leakage of *oil from a container such as a storage tank, *oil tanker, or *oil pipeline, which causes *pollution of the surrounding area. Oil spills at sea can causes extensive ecological and environmental damage. Toxic chemicals leach out from the oil and contaminate the surrounding water, poisoning sea life around and below the spill. Light oils can float freely on the ocean surface, and be dispersed over a wide area. Heavy oils sometimes form globules (small spherical lumps) which sink to the sea-bed and poison plants and animals that live there. When an oil slick drifts or is blown towards the coast it can seriously pollute the shore, fouling beaches and killing all shore life. The oil clogs up the feathers of birds, making it impossible for them to fly or even swim. Most other coastal creatures are suffocated by the oil. Procedures that are used to control and remove oil slicks include *containment, *absorption, and *dispersal. *See also* EXXON VALDEZ.

oil tanker A cargo ship that is designed to carry *crude oil in bulk.

oil well A vertical shaft that is drilled through rock to reach and extract the *oil that is stored naturally in an *oilfield beneath. At sea, the well is sunk from an *oil rig.

oilfield An area where *oil is trapped in *porous rocks beneath the ground or the *sea-bed, which can be extracted via an *oil well.

old age 1. The latter part of the *life cycle of an organism, after *maturity.
2. A late stage in the *cycle of erosion that is characterized by formation of a *peneplain near sea level. *Contrast* MATURITY, YOUTH.

old growth forest A North American term for an ecologically mature *forest that has not been significantly altered by humans for at least 120 years, and which contains old trees (more than 200 years old), fallen trees, trees with broken tops, and mature and dying trees. *See also* ANCIENT FOREST, OVERMATURE FOREST.

Oligocene An epoch of geological time within the *Tertiary subera, stretching from 35 to 23 million years ago.

oligohaline Relating to *brackish water, with a *salinity of between 0.5 and 5.0 parts per thousand (‰).

oligolectic A *pollinator which collects *pollen from a few species of plant. *Contrast* MONOLECTIC, POLYLECTIC.

oligotrophic Deficient in the *nutrients that are needed for plant growth. The term is usually applied to water or soil. *Contrast* EUTROPHIC.

oligoxenous A *parasite that lives within more than one host during its *life cycle. *Contrast* MONOXENOUS.

olivine An olive-green, greyish-green, or brown *silicate mineral that contains *magnesium and *iron, and is common in *basalt and in some *meteorites.

ombrotrophic An *ecosystem (such as a raised *mire) that is entirely dependent on atmospheric sources (mainly *precipitation) for its supply of water and *nutrients.

omnivore An *organism (such as a human) that eats both plants and animals. *Contrast* HERBIVORE, CARNIVORE.

omnivorous Capable of eating plant and animal material.

oncogenic Capable of producing tumours in animals, which can be either benign (non-cancerous) or malignant (cancerous).

oncology The study and treatment of cancer.

onion skin weathering *See* EXFOLIATION.

onshore On or towards the land.

onshore breeze *See* SEA BREEZE.

on-site facility A *facility for the treatment, storage, or disposal of *hazardous waste that is located on the site where that material is created. *Contrast* OFF-SITE FACILITY.

ontogenetic Of or relating to the origin and development of individual *organisms. *Contrast* PHYLOGENETIC.

ontogeny The history of the development and growth (*life cycle) of an individual.

oolitic limestone A type of *limestone formed as a *chemical precipitate, and consisting of small spherical grains of *calcium carbonate that have formed around tiny *nuclei.

ooze A fine-textured sediment that is formed from the remains of *pelagic organisms and is found on the *ocean floor.

opacity Lack of transparency, or the ability to keep light from passing through: for example, a glass window has almost 0% opacity, whereas a concrete wall has 100% opacity.

opaque Not allowing light to pass through.

opaque mineral A *mineral which transmits no light through a thin section under a microscope. Usually a *native metal, *sulphide, or metallic oxide mineral.

OPEC *See* ORGANIZATION OF PETROLEUM EXPORTING COUNTRIES.

open access system A *resource that is openly available to all users, without conditions or restrictions.

open burning The uncontrolled burning of *waste materials in the open, either in an outdoor *incinerator or an *open dump, either accidentally or on purpose.

open canopy forest *See* WOODLAND.

open cast mining The *mining of *minerals that occur near the surface, which are extracted through an open excavation. First the *overburden is removed, and then the mineral materials are broken up and removed. Also known as **opencut mining, open-pit mining, surface mining**. *See also* MOUNTAIN TOP REMOVAL.

open dump An uncovered *dump site in which *waste is left without environmental controls.

open forest *See* WOODLAND.

open range Unfenced, natural *grazing land, which includes *woodland as well as *grassland.

open sea That part of the ocean that lies beyond the *continental shelf. Similar to *high sea. *Contrast* COASTAL ZONE.

open space Public and private land that is undeveloped, which often retains its natural *vegetation, and is usually used for *recreation or as a reserve to protect natural areas.

open space zoning All types of land use *zoning (such as *large-lot zoning and *cluster zoning) which restrict land uses in order to preserve open space.

open system A *system that can exchange both matter and energy with its surrounding *environment. *Contrast* CLOSED SYSTEM.

open vegetation In *habitat evaluation, all *clearcuts, *meadows, and other openings.

opencut mining *See* OPEN CAST MINING.

open-pit mining *See* OPEN CAST MINING.

operational research The use of analytical methods, usually with a computer, adopted from mathematics for solving operational problems. Also known as **management science**.

opportunist species A plant or animal that takes advantage of whatever conditions exist in the *environment at the time.

opportunity cost *See* SOCIAL COST.

optimist, environmental A person who is inclined to take a favourable view of the *environmental crisis and what can be done about it, for example by believing that the development of new *technology, better use of *environmental economics, and improvement in *resource management will further expand the Earth's *carrying capacity. Also known as **technological optimist**.

Contrast PESSIMIST, ENVIRONMENTAL. *See also* SKEPTICAL ENVIRONMENTALIST.

optimization The process of making something as fully effective as possible, for example by selecting the most cost-effective combination of management practices.

optimum The most favourable condition under given circumstances.

optimum land use The *optimum use of available *land resources in a particular place or area, defined relative to stated management *objectives.

optimum yield The harvest level for a particular *species that achieves the greatest overall benefits, taking into account economic, social, and biological factors. This differs from *maximum sustainable yield, which considers only the biology of the *species.

option A right or agreement to buy or sell a specific commodity (including *emissions allowances) at a stipulated price and within a stated period of time. If the option is not exercised during that time, the money that has been paid for the option is forfeited.

option value The value that people place on having the option to enjoy something (such as the use of a particular *resource, or the ability to take part in a particular activity) in the future, even though they may not currently use it. The potential value of the resource for future (direct or indirect) use.

orchard A *plantation of fruit or nut trees that is managed to yield high quality produce on a *sustainable basis.

orchard heater Oil heaters that are placed in an *orchard in order to generate heat and promote convective circulations to protect fruit trees from damaging low temperatures. Also known as **smudge pot**.

order The fourth highest category (of seven) in the scientific system of classification for organisms (*taxonomy), below *class and above *family. Thus each class comprises more than one order, and each order comprises more than one family.

ordinance A regulation or statute (such as a local building code) that is enacted by a city government in the USA, under powers delegated to it by the state.

ordnance Military weapons and equipment.

Ordovician The second period of geological time in the *Palaeozoic era, dating from about 510 to 440 million years ago, during which the first species of fish and fungi appear.

ore Any naturally occurring *mineral or rock from which economically important constituents (such as *metals) can be extracted.

organelle A *cell structure that carries out a specialized function, such as *mitochondria.

organic Involving, related to, or derived from living matter or living organisms, including *compounds that contain *carbon.

Organic Administration Act (1897) US legislation that authorized the Secretary of Agriculture to manage the *National Forests to improve and protect the forests, to secure favourable conditions of water flow, and to furnish a continuous supply of timber.

organic agriculture The production of *crops without the use of *inorganic inputs (such as chemical *fertilizers and *pesticides). Fertilizer is derived from organic material such as animal *manure, *green manure, and *compost. Also known as **organic farming**. See also ALTERNATIVE AGRICULTURE.

organic chemistry The chemistry of substances that contain *carbon. Contrast INORGANIC CHEMISTRY.

organic compound Any compound that contains *carbon which is chemically bound to *hydrogen. Contrast INORGANIC COMPOUND.

organic farming See ORGANIC AGRICULTURE.

organic matter Material in *soil that contains *carbon and is derived from living plants and animals. The organic matter breaks down by *decomposition to produce *humus.

organic nitrogen *nitrogen that is bound to compounds (*amino acids and *proteins) which contain *carbon, and that must be subjected to *mineralization or *decomposition before it can be used by *plants. Contrast INORGANIC NITROGEN.

organic sedimentary rock A *sedimentary rock (such as *limestone, *chalk, *coal, and *oil shale) that is composed of the compacted decayed remains of plants and animals.

organic soil Any *soil that contains a high percentage (usually more than 25%) of *organic matter in the upper *horizons, where living roots are mostly found. Contrast MINERAL SOIL.

organic waste In general, any *waste that contains *carbon, which includes *paper, plastics, *wood, food waste, and *yard waste. The term is most often used to describe material that is directly derived from plant or animal sources, which can be decomposed by *micro-organisms.

Organisation for Economic Co-operation and Development (OECD) A forum that was founded in 1961 for monitoring economic trends in its member countries from the free-market democracies of North America, Western Europe, and the Pacific.

organism An individual animal, plant, or single-celled life form that is able to grow and reproduce. There are two main types of organism that derive energy from different sources—*autotrophs that produce energy, and *heterotrophs that get energy from consuming autotrophs.

Organization of Petroleum Exporting Countries (OPEC) An organization that was established 1960, consisting of 13 countries which control 60% of the world's *oil reserves. OPEC negotiates with oil companies on matters of oil production, prices, and future concession rights.

organochlorine A synthetic *organic compound that contains *chlorine, is highly *toxic, and has a variety of forms and uses including aerosol propellants, plasticisers, transformer coolants, and food packaging. Its greatest use was in *pesticides, in the form of *DDT, *aldrine, and *lindane. Organochlorines accumulate in the fatty tissue of animals, and can reach toxic levels in *predators through *biomagnification in the *food chain. Most countries have banned or severely restricted their use as pesticides. Also known as **chlorinated hydrocarbon**.

organophosphate A synthetic *organic compound that contains *phosphate, and is used as a *pesticide that works by interfering with an insect's nervous system. Organophosphates are generally considered safer than *organochlorines because they break down rapidly in the environment and do not *bioaccumulate, but are highly *toxic to mammals (including humans) and may be *carcinogenic.

organotin *See* TRIBUTYLTIN.

oriental From or relating to the eastern *hemisphere. *Contrast* OCCIDENTAL.

orientation Alignment, or relative direction.

ornamental A plant that is grown and prized for its beauty rather than for its produce (such as fruit). Such plants are often *exotic and planted deliberately.

ornithology The study of birds.

orogen A belt of deformed rocks that make up the *continental crust around a *craton. Many have been deformed into *fold mountains.

orogenesis The formation of *mountain ranges by *tectonic processes, particularly large-scale compression and intense upward displacement. Also known as **orogeny**.

orogenic belt A major range of *mountains on the continents that has been created by *folding and other deformation in a mountain-building episode.

orogeny *See* OROGENESIS.

orographic Relating to *mountains.

orographic precipitation *precipitation (usually *rain or *snow) that results from the lifting of a moisture-bearing *air mass over a topographic barrier such as a mountain range. The *leeward slope, beyond the barrier, is often dry and forms a *rain shadow.

orographic uplift The lifting of air over a topographic barrier such a *hill or *mountain.

orthodox seed *seed that can be dried and stored long-term, at low *temperature and *humidity, and remain viable.

ORV *See* OFF-ROAD VEHICLE.

OSHA *See* OCCUPATIONAL SAFETY AND HEALTH ACT.

Oslo Convention *See* CONVENTION FOR THE PREVENTION OF MARINE POLLUTION BY DUMPING FROM SHIPS AND AIRCRAFT.

osmosis The process by which a *solvent can pass through a *semipermeable *membrane, from a dilute solution into a more concentrated solution.

ostracod Tiny marine and freshwater *crustaceans that have a shrimp-like body which is enclosed in a bivalve shell. Also known as **seed shrimp**.

Our Common Future The title of the 1987 report of the *World Commission on Environment and Development, more commonly known as the *Brundtland Report, which defined *sustainable development.

Our Stolen Future A controversial book by Theo *Colborn, Dianne Dumanoski, and John Peterson Myers (1997), which outlined how *toxic pollutants such as chemical *pesticides cause birth defects, sexual abnormalities, and reproductive failures in wildlife. It built on Rachel *Carson's *Silent Spring.

outbreeding The *mating of individuals which are genetically unrelated. Also known as **outcross**. See also BREEDING, INBREEDING.

outcrop An exposure of *bedrock at the Earth's surface.

outcross See OUTBREEDING.

Outdoor Recreation and Wilderness System A system, managed by the US Department of Agriculture *Forest Service, that provides outdoor *recreation opportunities and *wilderness experiences within the USA.

outer core See CORE.

outfall The place where a *sewer, *drain, or *stream discharges.

outflow The flow of air out of a *thunderstorm, or of water out of a *lake or *reservoir.

outgassing The release of gas (such as *water vapour, *nitrogen, or *argon) from cooling *molten rock or the interior of the Earth.

output The energy, matter, or information that leaves a *system. Contrast INPUT.

outwash Sand and gravel that is deposited by *meltwater streams in the *proglacial environment, or beyond the margin of an active *glacier.

outwash plain The relatively flat area beyond the margins of a retreating *glacier, where *meltwater streams draining from the ice deposit *sand, *gravel, and *mud washed out from the glacier. Most outwash plains are wide and extensive, but some (*valley trains) are confined within a narrow valley.

overbank deposition The deposition of *alluvial sediment on the surface of a *floodplain by a river under *flood (over the bank) conditions.

overbrowsing The *overgrazing of shrubs and trees by a large population of animals that *browse.

overburden The layer of *soil, *sediment, and *rock that covers a *mineral deposit (such as *coal), and must be removed to allow *surface mining.

overcast The sky condition when more than nine-tenths of the sky is covered by *clouds.

over-consumption The use of *renewable resources faster than natural processes can replace them, which is not *sustainable.

overdraft See AQUIFER DEPLETION.

overexploitation The use or extraction of a *resource to the point of *depletion (for *inorganic resources) or *extinction (for *organic resources), or the reduction of a population to a level below the minimum needed for *sustainable yield. Activity that exceeds the *carrying capacity. Also known as **overharvesting**.

overfishing *fishing beyond a *sustainable level, caused by harvesting so many fish (particularly immature individuals) of a particular species that the *breeding stock left is not large enough to replenish the species. *Overexploitation of fish populations.

overflow channel See MELTWATER CHANNEL.

overgrazing The damage or destruction of natural vegetation (particularly *grasses and *forbs) that is caused when too many *herbivores are allowed to *graze on it before it has a chance to recover from previous grazing. A form of *overexploitation.

overharvesting See OVEREXPLOITATION.

overland flow The flow of water over the ground surface towards *stream channels. Also known as **sheet flow**.

overmature forest An *old growth forest in which tree growth has almost ceased, and decay and deterioration are increasing.

overpopulation A situation in which an existing *population is too large (perhaps because of a *population explosion) to be adequately supported by available *resources on a *sustainable basis, at current levels of *consumption. A population size that exceeds the *carrying capacity of the *environment, and is likely to lead to a *population crash. Also known as **overshoot**. *See also* MALTHUS, THOMAS.

overshoot *See* OVERPOPULATION.

overstocked A managed *forest in which the trees are so closely spaced that they compete for *resources, and grow more slowly as a result.

overstocking Allowing too many animals to *graze a particular area, usually for a short period of time. Continued overstocking will lead to *overgrazing.

overstorey The *crown or upper *canopy of branches and leaves in a stand of *stratified vegetation, which decreases the amount of sunlight that reaches the ground below. *Contrast* UNDERSTOREY.

overt release An announced release of a biological *agent, by terrorists or others. *Contrast* COVERT RELEASE.

overtopped *See* SUPPRESSED.

overturn *See* TURNOVER.

overturned fold A type of *fold in rock, which is created where compressional forces are asymmetrical, so that one limb of the fold is pushed over the other limb and over-rides it. *See also* NAPPE.

overwinter To survive over the *winter period.

oviparous Egg laying; producing fertilized eggs that develop and hatch outside the female's body

ox-bow lake A *lake that is formed on the *floodplain of an *alluvial river when the river cuts through the narrow

neck of a large *meander. Also known as a **cut-off**.

oxidant A substance that causes *oxidation in other substances by underoing *reduction itself. Also known as an **oxidizer** or **oxidizing agent**.

oxidation 1. A chemical process in which an element reacts with *oxygen, where oxygen is gained, or electrons are lost.
   2. A common form of *chemical weathering of *rocks, which involves the combination of rock materials (such as *silicates or *carbonates of *iron or *manganese) with oxygen and water.

oxidation pond An artificial *pond or *lagoon that contains water in which *waste is consumed by *bacteria as part of a waste treatment process. Also known as **sewage lagoon**.

oxide A compound of *oxygen with another element or group. Examples include *sulphur dioxide and *nitrogen oxides.

oxidize To add *oxygen to or combine with oxygen, or to lose electrons.

oxidizer *See* OXIDANT.

oxidizing agent *See* OXIDANT.

oxisol A type (*order) of deeply weathered soil that contain *clays of *iron and *aluminium oxides, has low *fertility, and is *acidic.

oxyfuel *See* OXYGENATED FUEL.

oxygen (O) An invisible, colourless, tasteless, odourless *gas that exists naturally in air (21% by volume), accounts for 46.6% of the Earth's *crust, and is the most abundant element on Earth. It is produced by plants during *photosynthesis, and it is essential for *aerobic *respiration (it is breathed by humans and other animals). The Earth's early atmosphere probably contained little oxygen, but after the *evolution of the first oxygen-producing plants (around 1900 million years ago) the proportion of oxygen in the air increased steadily. Oxygen forms an average of 70% of the atoms in living matter and it is a basic building block of carbohydrates, fats, and proteins. It is chemically very active and combines with most elements to form oxides, such as *water, *carbon dioxide, and *silicon dioxide (*quartz). *Ozone (O_3) is an *allotrope of oxygen. The process of *oxidation speeds up *metabolic processes, and oxygen is essential for *combustion.

oxygen demand *See* BIOLOGICAL OXYGEN DEMAND, CHEMICAL OXYGEN DEMAND.

oxygen depletion A decrease in the concentration of *dissolved oxygen in a *waterbody downstream from a *point source of pollutants, which is caused by the uptake of *oxygen by *bacteria, *fungi, and *invertebrates in the water as they break down the pollutants by *aerobic *decomposition. Also known as **oxygen sag**.

oxygen isotope Any of the oxygen *atoms that have the same atomic number (number of *protons) but different mass numbers (different numbers of *neutrons). The two common stable *isotopes of oxygen are ^{16}O and ^{18}O, where ^{16}O is the more abundant of the two.

oxygen isotope ratio The ratio between the two stable *isotopes of oxygen (^{16}O and ^{18}O) in a sample such as old layers of ice or shells from ocean sediments. The ratio is temperature dependent and provides a reliable indicator (*proxy) of past climatic conditions.

oxygen minimum layer The level in a *waterbody, usually below the *thermocline, below which there is very little *dissolved oxygen.

oxygen sag *See* OXYGEN DEPLETION.

oxygen solubility The ability of oxygen *gas to dissolve into water.

oxygenated fuel A special type of *gasoline which has been blended with *alcohols or ethers (which contain oxygen), in order to reduce emissions of *carbon monoxide and other *greenhouse gases. Also known as **oxyfuel**.

ozonation The application of *ozone to water for purification purposes (as a *disinfectant, and to improve taste or control odour).

ozonator A mechanical device that creates *ozone, which is used to *oxidize water in order to eliminate *organic wastes.

ozone (O_3) A highly reactive *greenhouse gas, made of three atoms of *oxygen. Ozone comprises 0.00006% of the Earth's *atmosphere. It is formed when the molecule of the stable form of oxygen (O_2) is split by *ultraviolet *radiation or electrical discharge, for example when *lightning flashes through a cloud. There are two distinctly different ozone problems: depletion of *stratospheric ozone, and increasing concentrations of *low-level ozone in the *troposphere. The latter comes partly from natural sources, but most is produced by *pollution (particularly *photochemical reactions involving vehicle exhaust fumes). Ozone is a constituent of *smog, is at least partly responsible for some of the damage attributed to *acid rain, can damage plants, and causes respiratory problems (including asthma attacks) in people. It is *toxic to humans at a concentration as low as 1 part per million (ppm) in air. The US *Environmental Protection Agency recommends that people should not be exposed for more than 1 hour a day to ozone levels of 120 parts per billion (ppb), while the *World Health Organization recommends a lower limit of 76–100 ppb.

ozone depleter Any of a group of chemicals that destroy *ozone and attack the *ozone layer in the *stratosphere. Most are chemically stable compounds that contain *chlorine or *bromine, which remain unchanged long enough to drift up to the stratosphere. The best known are *chlorofluorocarbons (CFCs), but others include *halons (which are used in some fire extinguishers), *methyl chloroform and *carbon tetrachloride (solvents), some

CFC substitutes, and the *pesticide *methyl bromide.

ozone depletion potential (ODP) A relative index of the extent to which a particular chemical may cause *ozone depletion, compared with CFC-11 and CFC-12 which have an OPD of 1.0 Other *CFCs and *HCFCs have ODPs that range between 0.016 and 1.0, *halons have ODPs ranging up to 10, and *HFCs have an ODP of 0 because they do not contain *chlorine.

ozone hole General name for a sharp seasonal decrease in the concentration of ozone in the stratosphere that occurs over *Antarctica in the spring, was first detected during the 1970s, and is growing larger (a larger decrease, over a larger area) through time. *See also* OZONE DEPLETION.

ozone layer The natural protective layer of *ozone in the *stratosphere that is formed naturally by a *photochemical reaction with solar *ultraviolet *radiation. Ozone is constantly being formed, broken down, and re-formed above about 40 kilometres, and it sinks and accumulates at the 20–25 kilometres level. The ozone layer is not uniform around the world; it varies in density, being least dense over the equator and

OZONE DEPLETION

The destruction of the stratospheric *ozone layer, which shields the Earth from *ultraviolet (UV) radiation, caused by the breakdown of certain compounds (*chlorofluorocarbons or *halons) that contain *chlorine and/or *bromine, which destroy *ozone molecules in the stratosphere. In 1985 British scientists discovered a seasonal thinning (usually referred to as a hole), the size of the USA, in the ozone layer over the *Antarctic. Field monitoring showed a 40 % thinning in the ozone layer. Their evidence suggested that the hole was growing bigger through time, and *satellite surveillance confirmed the initial reports. The main cause of the thinning was believed to be a three-fold increase in atmospheric *CFCs within 10 years. Annual measurements indicate that the ozone hole over the Antarctic appears to be getting bigger through time. More recently, reports that the ozone layer above the Arctic and parts of Western Europe is thinning have caused concern. Whilst much of the thinning of the ozone layer can be blamed on air pollution by CFCs, some natural pollutants—including aerosols from the 1991 Mount Pinatubo volcanic eruption—aggravate the problem. From the mid-1980s onwards the issue of ozone depletion has been taken seriously. In 1985 the *Vienna Convention for the Protection of the Ozone Layer was initially signed by 22 countries. Not all countries agreed on the urgency of the problem. Since 1987 the *Montreal Protocol was strengthened to work towards a faster phase-out of CFCs by 1996. By 1990 many European countries had made great progress in phasing out the use of CFCs. By the end of 1994 some *industrialized countries had already halted production of CFCs and others were well on target to meet the 1996 deadline. *Developing countries were given a 10-year extension. If the ozone layer is damaged, more UV radiation would pass straight through the atmosphere. This would have a variety of possible impacts, depending on the extent and location of the damage. Agricultural crops would be scorched and yields would fall. Marine *plankton would be seriously affected. *Human health would suffer – there would be more eye cataracts, more skin *cancer, and more problems arising from damage to people's immune systems. The problems are likely to affect many if not all countries. Also known as **ozone thinning**. *See also* OZONE HOLE.

most dense over high latitudes (beyond 50°N and 50°S). It also changes through the year, and is best developed over polar regions in early spring because ozone is neither formed nor destroyed during the polar night, when ozone transported from lower latitudes is stored over the poles. The ozone layer absorbs some of the ultraviolet rays from the Sun, thereby reducing the amount of potentially harmful radiation that reaches the Earth's surface. *See also* MONTREAL PROTOCOL.

ozone monitoring Measurements of background levels of ozone that have been collected over 20 years at sites a long way from pollution sources (including the South Pole, Barrow in Alaska, and *Mauna Loa in Hawaii). These *baseline surveys show that ozone concentrations vary through the day and from season to season. Efforts are being made through the *International Geosphere-Biosphere Programme to collect more information on changing ozone levels in the atmosphere and on the ways these might promote other changes in atmospheric chemistry.

ozone thinning *See* OZONE DEPLETION.

PA *See* PRELIMINARY ASSESSMENT.

Pacific Ocean The largest *ocean in the world, which covers a third of the Earth's surface, contains about 25 000 *islands (more than half of the global total), and is regularly affected by *typhoons and *hurricanes, *volcanoes, *earthquakes, and *tsunamis.

Pacific Rim General term for the 34 countries and 23 island states that are situated in and around the *Pacific Ocean and share similar political, economic, and environmental interests.

Pacific Ring of Fire A *volcanic zone that runs right around the edges of the *Pacific Ocean, along a series of *subduction zones of *convergent plate boundaries, where the moving plates of the Pacific plunge beneath the confining plates to the east, north, and west. It extends through the Andes in South America, runs through Central America and Mexico, along the west coast of the USA into Alaska, across the Aleutian Island chain, through the Kamchatka Peninsula, through Japan and the Philippines, and into New Zealand and the margins of Antarctica.

pack ice Seawater that is frozen into thick blocks of *ice, under permanently cold conditions, such as those found around the *polar ice caps. *See also* SEA ICE, ICE FLOE.

packaging Materials that are used for the containment, protection, handling, delivery, and presentation of goods.

packed tower *See* TRICKLING FILTER.

paddock An enclosed area of *grazing land for animals, such as horses.

paedogenesis A form of reproduction in *insects in which the larval stage reproduces without maturing first. *See also* LARVAL PAEDOGENESIS, PUPAL PAEDOGENESIS.

PAH *See* POLYCYCLIC AROMATIC HYDROCARBON.

pahoehoe Volcanic *lava that solidifies into ropy or corded shapes with a smooth surface. The name is a Hawaiian word meaning satin-like.

palaearctic A *biogeographical region that covers Europe, Asia, and North Africa.

Palaearctic realm A *biogeographical realm that covers much of Europe and Asia, and contains a variety of *biomes including *tundra, *grassland, *deciduous forest, *coniferous forest, *chaparral, and *desert.

palaeo- A prefix meaning old or ancient, particularly *prehistoric. Spelled paleo in North America.

palaeobiology The study of the origin, growth, and structure of *fossil animals and plants as living *organisms.

palaeobotany The study of *fossil plants.

Palaeocene The first epoch of geological time within the *Palaeogene, between 65 and 56 million years ago.

palaeoclimate The *climate of a particular period in the geological past, before historical records or instrumental observations.

palaeoclimatology The study and reconstruction of past *climates and of *climate change.

palaeoecology The study of ancient *ecosystems.

palaeoenvironment A previous or ancient *environment.

palaeoenvironmental proxy Any of a number of remnants of past *environments (including *pollen grains, *tree rings, lake sediments, *ice cores, and *coral skeletons) that can be used to how *climate and environment have changed through time.

palaeoenvironmental reconstruction The study of ancient *environments and of *environmental change.

Palaeogene A period of *geological time within the *Tertiary subera, dating from about 65 to 24 million years ago; the early age of *mammals and a period of major *extinctions of species.

palaeogeographical map A map that shows the surface *landforms and *coastline of an area at some time in the geological past.

palaeogeography The study of the geography (particularly the *environment and distribution of physical features) of particular periods in the geological past.

palaeohydrology The study of the structure and dynamics of the *water cycle at particular periods in the geological past.

Palaeo-Indian The earliest, well-documented *archaeological cultures in North America, dating back some 6000 to 12 000 years before the present.

palaeolimnology The study of past *freshwater, *saline, and *brackish environments, based on analysis of sediment cores which preserve evidence of water chemistry and biology in the past.

Palaeolithic The second part of the *Stone Age, following the *eolithic, which began in Europe between about 750 000 and 500 000 BC and lasted until the end of the last *ice age about 8500 BC.

palaeomagnetism The study of variations in the intensity and direction of the Earth's *magnetic field in the geological past, as recorded in *rocks. Also

known as **magnetostratigraphy**. *See also* MAGNETIC REVERSAL.

palaeontological area An area in the USA which has been designated as containing significant (usually *fossil) remains of flora and (non-human) fauna dating from *prehistoric times.

palaeontology The study of ancient *fossil remains of plants and animals that are found in *sedimentary rocks, and of their geological contexts.

palaeosol An ancient buried *soil that preserves evidence of past *environmental conditions and processes.

palaeotropics The tropical parts of the Old World (Africa and South East Asia). *See also* NEOTROPICS.

palaeowind The prevailing *wind direction in a particular area at some time in the geological past, inferred from the structure of ancient *sand dunes or the distribution of *volcanic ash.

Palaeozoic The period of geological time between the *Precambrian and the *Mesozoic, about 570 to 245 million years ago, during which all classes of *invertebrates except insects emerged, and seed-bearing plants, *amphibians, and *reptiles appeared.

palatable Pleasing to the taste.

palatable species Plants that are preferred by *grazing and *browsing animals, usually because they are sweeter or softer than other plants.

palatable water Water that is free from most colour, cloudiness, unpleasant tastes, and odour.

paleo- 1. North American spelling of *palaeo-.
  2. In North American *archaeology, a cultural period between 40 000 and 12 000 years ago.

palm A member of the Aracaceae (previously Palmae) family. This contains more than 200 genera and more than 2700 species, many of which are economically important for food, fibre, canes, waxes, wood, thatch, and other uses. A common tree in *agroforestry.

Palmer drought severity index An indicator of long-term deficits or surpluses of *soil moisture, based on temperature, precipitation, and soil type, which is used to gauge the severity of *drought conditions in a particular area.

Palouse prairie A North American *prairie *grassland and meadow-*steppe vegetation dominated by grasses, which covers a large area of northwestern Idaho, southeastern Washington, and eastern Oregon.

paludification The spread of boggy conditions across an area as a result of the gradual rising of the *water table, as the accumulation of *peat impedes the drainage of water.

palustrine Relating to *wetland or *marsh *habitats.

palustrine wetland Any non-tidal *wetland that is dominated by trees, shrubs, persistent emergent plants, emergent *mosses or *lichens, and small, shallow open water ponds. Examples include *swamps, *marshes, *bogs, and *fens.

palynology *See* POLLEN ANALYSIS.

pampas The extensive grassy *prairie plains of *temperate South America.

PAN *See* PEROXYACETYL NITRATE.

pan A compact soil horizon that has a high *clay content.

pandemic A *epidemic that occurs over a very wide area, and affects a large proportion of the population.

Pangaea The original landmass that existed probably between 200 and 250 million years ago, and was surrounded by the *Panthalassa Sea. It broke apart by *continental drift into two portions, *Laurasia and *Gondwanaland, with the *Tethys Sea between them.

panorama A broad view, usually of an attractive natural *landscape or *scenery.

Panthalassa The ancient *sea that covered most of the Earth's surface when all of the land that existed was in one large landmass (*Pangaea), probably between 200 and 250 million years ago.

pantropical Throughout the *tropics as a whole, between 23°30'S and 23°30'N latitude. *Contrast* NEOTROPICS, PALAEOTROPICS.

paper A thin sheet of material made of *cellulose pulp which is derived mainly from *wood, but also from rags and certain grasses. The pulp is processed into flexible leaves or rolls, which have many uses including for writing, printing, drawing, wrapping, and covering walls.

paper mill Mills (factories) that produce *paper from wood pulp.

parabiosis 1. The joining of two organisms which share one blood circulation. 2. The condition of living together.

paradigm A worldview, shared set of assumptions, or widely accepted idea of how the world works, which forms the basis for *hypotheses and explanations.

paraffin 1. A *liquid fuel refined from *petroleum or shale. 2. A flammable waxy solid that is made in the same way as the liquid fuel, and is used for sealing and waterproofing and in candles.

parallel drainage pattern A *drainage pattern in which small rivers flow generally parallel to one another over a sloping surface of uniform rock resistance.

parapatric speciation A form of *speciation in which the new *species evolves from *contiguous *populations.

paraquat A *defoliant *herbicide that is commonly used to kill particular types of crops (such as marijuana) and is *toxic to humans.

parasite A plant or animal that lives in or on a *host (an animal or plant of a different species) for at least part of its life, from which it obtains *nutrients. The relationship does not usually kill or

p

benefit the host organism. *See also* ENDO-PARASITE, ECTOPARASITE.

parasitism A win–lose form of *symbiosis between two different species in which one (the *parasite) benefits and the other (the *host) is harmed. *See also* BROOD PARASITISM.

parasitoid A *parasite that eventually kills its host.

parataxonomist A specialist in the monitoring of local *biodiversity, who is recruited locally and trained in the field.

parent element An *isotope (parent) that is transformed into a different (*daughter product) isotope by *radioactive decay.

parent material The *mineral material from which a *soil develops, which is usually unconsolidated *weathered *bedrock or *sediment (*alluvium, *colluvium, or wind-blown deposits).

Pareto optimum In welfare economics, the optimum allocation of products and services to people in a particular setting (such as a country) that produces maximum benefit to the *population as a whole, and when it is impossible to make one or more individuals better off without making one or more other people worse off.

Parícutin An explosive *volcano that grew suddenly and rapidly in central Mexico in 1943. Over a 2-week period early that year, the area around Parícutin had been shaken by *earthquakes. On 20 February the ground cracked and a *vent opened up in a cornfield. First steam, then *volcanic dust and then hot fragments started to pour out. Soon molten rock was pouring out, and it started to build up a cone around the central vent. The eruption continued for eight months, and the cone grew to a height of about 450 metres. The accompanying *lava flows buried the village of Parícutin and nearby settlements. By 1952 the volcano had become *dormant.

Paris green *See* COPPER ACETOARSE-NITE.

park-like structure A stand of *vegetation that is dominated by large, scattered *trees with open spaces between them, and is usually maintained by natural ground *fires.

parthenogenesis Development of an egg without fertilization.

partial duration series Flow records that are used in *flood frequency analysis, which include all peak flows with a discharge greater than a chosen threshold (such as *bankfull discharge). *Contrast* ANNUAL MAXIMUM SERIES.

partial pressure The *pressure that is exerted by one *gas in a mixture of gases.

partial retention In *forest management, a method of tree harvesting in which only part of the *stand is removed, and the rest is left standing.

participation Engagement with and taking part in an activity. The process through which *stakeholders influence and share control over priority setting, policy-making, resource allocations, and access to public goods and services.

participation rate The portion of a *population that participates in a particular activity, such as a *recycling programme.

participatory action research (PRA) A type of *action research in which active collaboration between *stakeholders, practitioners, and researchers defines the content, process, and results of the research, and in which the activities are often designed to identify or monitor solutions to particular problems.

participatory monitoring and evaluation (PME) A process through which key *stakeholders actively engage in monitoring or evaluating a particular project, programme, or policy.

participatory rural appraisal (PRA) A group of participatory approaches and methods that emphasize local knowledge and enable local people to make their own appraisal, analysis, and plans, which helps development practitioners, government officials, and local people to work together to plan appropriate programmes. Also known as **rapid rural appraisal**.

participatory technology development (PTD) An approach to *rural development that is based on involving farmers in developing agricultural technologies that are appropriate to their particular situation, based on their own practical experience.

particle A tiny mass of solid or liquid material.

particle size The size of individual particles or grains of material, such as *soil or *sediment, expressed either in millimetres diameter or using the *phi (ϕ) scale. Size classes for sediments range from very fine *clays (less than 0.002 millimetres in diameter) and *silts (between 0.002 and 0.05 millimetres), to coarse *sands (range from 0.05 to 2.0 millimetres), and *gravels (larger than 2 millimetres).

particle-size analysis Determination of the size distribution of particles in a sample of *soil or *sediment, which is usually determined by sieving or *sedimentation (settling in liquid).

particulate Composed of *particles of solid or liquid matter, such as *soot, *dust, *aerosols, fumes, and *mists.

particulate filtration A procedure used in *air cleaning, in which *particles are removed by passing the air through fine filters.

particulate loading The concentration of particles in a sample of air or water, usually expressed as mass per unit volume (such as milligrams per litre, mg L^{-1}).

particulate pollution *Pollution in the form of small liquid or solid *particles suspended in the atmosphere or in water.

particulates Fine solid or liquid *particles in the air or in an emission. A *conventional or *criteria pollutant.

parts per billion (ppb) The amount of one substance in parts that is in a total sample of one billion (10^9) parts.

parts per million (ppm) The amount of one substance in parts that is in a total sample of one million (10^6) parts.

party A group of people who are engaged in a particular activity, such as a state (or regional economic organization such as the *EU) that agrees to be bound by a particular *treaty.

pascal (Pa) The SI unit of *pressure, which is equal to one hundredth of a *millibar, or one newton per square metre.

passenger pigeon A migratory pigeon (*Ectopistes migratorius*) that as late as the mid-19th century was one of the most abundant birds in North America. By 1900 it had become *extinct in the wild as a result of wholesale hunting for restaurants and other markets. The last known specimen died in captivity in 1914.

passive detector A measuring device that functions on its own, without any energy input or ongoing attention from the user.

passive heat absorption Using natural materials and structures without any moving parts to collect and store heat.

passive solar design Designing buildings in order to maximize the use of *solar radiation to warm and light the interior of a building, through such measures as careful siting and orientation, use of energy efficient windows, and provision of appropriate levels of insulation and heat storage.

passive solar heating A form of *solar heating that uses *solar radiation to heat buildings directly, by trapping it within the structure of the building and then releasing it slowly. It relies on natural air flow rather than pumps to distribute the heat within the building. Glass greenhouses and conservatories are good examples. *Contrast* ACTIVE SOLAR HEATING.

passive solar power Using or capturing *solar energy (usually to heat water) without any external power.

passive use value See NON-CONSUMPTIVE VALUE.

pastoral Relating to shepherds or herdsmen, devoted to raising sheep or cattle, or more generally a romantic or idealized image of *rural life. See also BUCOLIC.

pastoral agriculture Farming based on rearing *livestock, such as beef cattle, dairying, and sheep. Contrast ARABLE.

pastoral nomad A *nomad who herds animals, and has no permanent place of residence.

pastoralism A form of *nomadism in which people move animals from place to place with no permanent human settlement. See also TRANSHUMANCE.

pasture A field covered with grass or herbs and suitable for grazing by livestock. See also ANNUAL PASTURE, NATIVE PASTURE, PERMANENT PASTURE, ROTATION PASTURE.

pasture management *Land management that is designed to ensure the sustainable growth of *pasture plants, in order to provide high quality feed and encourage the growth of desirable *grasses and *legumes while crowding out weeds, *brush, and inferior grasses.

pastureland Land that is used primarily for *pasture.

patch An area of *vegetation that has a uniform structure and composition, and differs from the surrounding vegetation.

patch cut A *clearcut *logging operation that creates small openings in a *forest in patches of between about 10 to 100 acres, which are separated by living forest.

patch dynamics Changes in the distribution of *patches of *habitat within a landscape that are caused by *disturbance and regrowth.

patch size The size of a *patch of *vegetation or *habitat.

patent An *intellectual property right relating to inventions (innovative processes or products).

paternoster lake A lake that is formed on a *pro-glacial *floodplain after the glacier ice has melted. Such lakes are often found as a series, or chain.

pathogen An *organism (such as a *bacterium, *fungus, or *virus) that can cause *disease in another organism.

pathogenic Capable of causing *disease.

pathological Related to or caused by *disease.

pathway The physical route that a *chemical or *pollutant takes from its source to the exposed organism.

patterned ground A range of ground surface features that are found in *periglacial environments, and are the result of *freeze–thaw action (*physical weathering) at the top of the *permafrost. The most common features are sorted *stone polygons (on low-sloping ground) and *stone stripes (on higher slopes). Also known as **sorted ground**.

pavement Any rock that is exposed at the Earth's surface in the form of a more or less horizontal surface.

PCA See POLYCYCLIC AROMATIC.

PCB See POLYCHLORINATED BIPHENYL.

PE See POTENTIAL EVAPOTRANSPIRATION.

peak The highest point (summit) of a *mountain.

peak discharge The maximum volume of flow at a particular point in a stream during a *runoff event.

peat Partially decomposed vegetable matter that forms in boggy ground and can be cut and dried for use as a fuel and in gardening. It forms in the initial stages of the *coal series.

peat bog See BOG.

peatland Any *wetland ecosystem that accumulates partially decayed plant matter. Also known as **bog** or **fen**.

pebble bed reactor An advanced design of *nuclear reactor which uses an inert or semi-inert gas (such as *helium, *nitrogen, or *carbon dioxide) rather than water as a coolant, at very high temperature, to drive a *turbine directly.

ped A small unit (*aggregate) of *soil that is composed of individual particles of sand, silt, clay, and other soil material that stick together into a specific structure (such as a crumb or granule), and is formed by natural processes. *See also* SOIL STRUCTURE.

pedalfer A *soil in which *aluminium and *iron accumulate. It is produced by *laterization and *podsolization, and occurs mainly in humid climates.

pedigree The line of descent (list of ancestors) of an individual or a *purebred animal or plant.

pedigree breeding A system of *breeding plants or animals in which individuals are selected from a *cross on the basis of their desirability, judged individually and from their *pedigree record.

pediment A gently sloping surface, usually covered with *gravel, that has been created by *erosion in front of a mountain range in an *arid region.

pedocal A *soil in which *calcium accumulates. It is produced by *calcification and occurs mainly in subhumid, semi-arid, and arid climates.

pedogenesis The process of *soil formation.

pedology The study of the formation, characteristics, and distribution of *soils.

pedon The smallest three-dimensional unit of *soil, which defines a block (usually between 1 and 10 square metres in area) with relatively uniform properties.

pedosphere The thin outer layer of the Earth's surface, which is made up of *soil.

PEL *See* PERMISSIBLE EXPOSURE LIMIT.

pelagic Of or relating to the *open sea or *ocean, free from any direct influence of the *shore or *sea-bed.

pelagic organism A marine organism that can move vertically upwards and downwards within a water body such as a sea or *fjord, between the surface and the bed. Pelagic organisms are generally free-swimming (*nekton) or floating (*plankton).

pelagic sediment Deep-sea sediments that are made up of fine-grained material (such as *clay, *radiolarian ooze, and *foraminiferal ooze) that slowly settles from surface waters.

pelt The fur coat of a *mammal.

peneplain A large flat or gently undulating area close to *sea level, formed by a long period of *erosion, which represents the end product of the ideal *cycle of erosion.

penetrometer An instrument that is used to determine the resistance to penetration of a *soil, which reflects soil strength.

peptide A substance formed by two or more *amino acids; *proteins are made of multiple peptides.

per capita Literally 'each head', meaning each individual.

per capita consumption The amount of a particular commodity that is used by each individual, on average.

perched aquifer *Groundwater that is separated from the underlying main body of groundwater (*aquifer) by unsaturated rock (*aquiclude). Also known as **perched groundwater, perched water table**.

perched groundwater *See* PERCHED AQUIFER.

perched water table *See* PERCHED AQUIFER.

percolate To drain or seep slowly through a *porous substance.

percolation The downward movement of water (*soil moisture) within a soil, through the *soil horizons, under the force of *gravity. This water may eventually pass into the underlying bedrock and become part of the *aquifer. *Contrast* INFILTRATION.

percoline A hollow at the base of a *hillslope where water (*soil moisture) seeps out onto the ground surface and becomes *overland flow. Also known as a **seepage line**.

perennial Persisting from year to year.

perennial flow River flow that continues through the year, and only dries up during prolonged *drought. *Contrast* EPHEMERAL FLOW, INTERMITTENT FLOW.

perennial plant A plant that lives for three or more *growing seasons. Also known simply as a **perennial**. *Contrast* ANNUAL.

perennial stream A *stream or river that flows throughout the year, from source to mouth, and that dries up only during prolonged *drought and starts to flow again when regular *precipitation is restored.

perfluorocarbon (PFC) One of a group of human-made chemicals that are composed of *carbon and *fluorine only (CF_4 and C_2F_6). They were introduced as alternatives to *ozone-depleting substances, are emitted as *by-products of industrial processes, and are used in manufacturing. They do not harm the *ozone layer in the stratosphere, but are powerful *greenhouse gases.

performance standard zoning Land use *zoning regulations that define broad criteria for determining the acceptability of certain industries, land uses, and buildings, rather than specification standards or detailed requirements.

performance-based resource estimate An estimate of the amount of a particular *mineral deposit that is available underground, based mainly on the ability of existing technology to extract the mineral under existing and probable future economic conditions.

periglacial The cold climate region adjacent to a *glacier or *ice sheet, in which the ground is largely permanently frozen (*permafrost) but may thaw during the summer.

perihelion The point in the orbit of any planet when it is closest to the Sun. For the Earth this occurs on 3 or 4 January each year. The opposite is *aphelion.

period A unit of time; in geological terms a unit that is shorter than an *era but longer than an *epoch.

periodic table A chart listing all of the known *elements of the universe, arranged in columns with similar chemical properties, in order of increasing *atomic number.

periphery The edge or outer part of an object, away from the centre.

periphyton *algae that grow attached to rocks, stems, twigs, and bottom sediments in a freshwater lake or river.

permaculture Permanent *agriculture: a *sustainable form of *agriculture that is designed to enhance local *ecosystems and increase local *biodiversity, for example by providing *fuel, materials for shelter and home, and *habitat for livestock, as well as food.

permafrost Permanently frozen soil or ground, which is a common in *periglacial environments such as the arctic *tundra. *See also* PATTERNED GROUND.

permanent pasture A *pasture, usually created rather than natural, which consists mainly of *introduced *perennial plants that are permitted to remain for a number of years. *Contrast* ANNUAL PASTURE.

permanent retrievable storage Placing waste storage containers in a secure place (such as a special building, salt mine, or cavern in bedrock), where they can be inspected periodically and retrieved if necessary.

permanent water A watering place that supplies water to *wildlife and people at all times during the year or during a *grazing season.

permanent wildlife opening An area of land that is managed in order to create and maintain a *wildlife *habitat of grass, low shrub, and/or herbaceous ground cover.

permeability A measure of the rate at which water can *percolate through a soil or rock, usually expressed in cubic metres per second $(m^3 s^{-1})$. Also known as **hydraulic conductivity**.

permeable Material (such as a *soil or *rock) that permits fluids (such as *water) to pass through it in both directions. *Contrast* IMPERMEABLE. *See also* PERVIOUS.

Permian The last *geological period in the *Palaeozoic era, between about 290 and 245 million years ago, during which an estimated 96% of all species died out in a *mass extinction (including many *corals, *brachiopods, and *trilobites), and *reptiles diversified and grew in dominance.

permissible dose The maximum *dose of a *chemical that an individual may receive in a given period of time, without it causing significant harm.

permissible exposure limit (PEL) The permissible concentration of a harmful physical *agent (such as an air *contaminant) to which most workers can be safely exposed for 8 hours a day, 40 hours a week, over a working lifetime (40 years), without adverse effects.

permit A licence or legal document that gives the holder official permission to do something.

peroxyacetyl nitrate (PAN) An *air pollutant that is created by the action of sunlight on *hydrocarbons and *nitrogen oxides in the air, and is an ingredient of *smog.

perpetual resource Any *resource that is inexhaustible on a human time-scale.

persistence Continuity, such as the continued presence of a species at a particular location or of a *non-degradable chemical (such as a *pesticide) in the environment.

persistence time The time it takes for a *pesticide to become *inert.

persistent Long lasting or never ceasing.

persistent organic pollutant (POP) A chemical substance that is *persistent in the environment, *bioaccumulates through the *food web, and can damage human health and the environment.

persistent pesticide Any *pesticide (such as *DDT) that does not break down chemically or only breaks down very slowly, so it remains in the *environment after *a growing season. *Contrast* NON-PERSISTENT PESTICIDE.

person-rem A measure of the exposure of a *population to *radiation, calculated as the average dose per individual (in *rem) multiplied by the number of people exposed.

PERT *See* PROGRAMME EVALUATION AND REVIEW TECHNIQUE.

perturbation *See* RADIATIVE FORCING.

pervious A material (such as a *soil or *rock) that is able to let a liquid (such as water) penetrate it or infiltrate in. *Contrast* IMPERVIOUS. *See also* PERMEABLE.

pessimist, environmental A person who is inclined to take an unfavourable view of the *environmental crisis and what can be done about it. For example they may believe that the human *population already exceeds the Earth's *carrying capacity, and this is reflected in global environmental problems such as the growing rate of *extinction of species, *global warming, and increased *poverty. Also known as an **ecodoom pessimist**. *Contrast* OPTIMIST, ENVIRONMENTAL.

pest Any *organism (such as birds, rodents, flies, and larvae) that directly or indirectly interferes with human activities and causes annoyance, economic damage, or health problems. Like *weeds, pests are usually *organisms that are simply in the wrong place as far as human comfort is concerned.

pest control Any activities that are undertaken in order to restrict, reduce, or eliminate particular *pests in a given area.

pesticide A chemical (such as an *insecticide, *fungicide, *rodenticide, *herbicide, or *germicide) that is used to kill or control *pests, such as *insects, *weeds, or *micro-organisms.

pesticide persistence The *persistence of a *pesticide, that is usually expressed in terms of *half-life (the length of time required for one-half of the original quantity to break down) and often divided into three categories: non-persistent (with a typical soil half-life of less than 30 days), moderately persistent (30 to 100 days), or persistent (half-life of more than 100 days).

pesticide resistance A situation in which a *population of *pests can become *resistant to particular *pesticides so they are no longer affected by them. This can happen through behavioural change (which helps them to avoid the pesticide), biochemical change (which allows them to *detoxify the pesticide), or some other genetic characteristic that reduces their *susceptibility to the pesticide.

pesticide tolerance The amount of *pesticide residue that is allowed by law to remain in or on a *crop that has been harvested, which is set well below the point where the chemicals might be harmful to consumers.

pesticide treadmill The tendency of *pests to become resistant to the effects of particular *pesticides, as a natural part of the evolutionary process. New and more *toxic pesticides then have to be used, to which pests may eventually become resistant, and the spiral continues. *See also* PESTICIDE RESISTANCE.

pet A *domesticated *animal or *bird that is kept for companionship or amusement.

petrified forest The *fossilized remains of a former *forest, in the form of large logs and stumps, usually *in situ*. *See also* SILICIFICATION.

petrochemicals Chemicals that are derived from *crude oil or *natural gas and are used in the manufacture of many industrial chemicals, *fertilizers, *pesticides, plastics, synthetic fibres, paints, and medicines.

petroglyph A *prehistoric carving or drawing on a natural rock surface.

petrol *See* GASOLINE.

petroleum A naturally occurring liquid that is found in certain *sedimentary rocks and is composed of *hydrocarbons formed by the *anaerobic decay of organic matter. Petroleum can be *refined to produce *gasoline, *paraffin, and *diesel oil. Also known as **crude oil**. *See also* MINERAL OIL, NATURAL GAS, OIL SHALE.

petroleum derivative A chemical that is formed when *petroleum breaks down in contact with *groundwater.

petroleum geology The study of the occurrence and exploitation of *oil and *gas fields.

petroleum product Any product that is derived from *petroleum or *natural gas.

petroleum refinery An installation (*refinery) that makes *petroleum products from *crude oil, unfinished oils, natural gas liquids, other *hydrocarbons, and alcohol.

PFC *See* PERFLUOROCARBON.

pH A measure (short for potential hydrogen) of the concentration of hydrogen ions and thus the level of *acidity or *alkalinity of a solution. It is based on a (logarithmic) scale from 1.0 to 14.0, in which 7.0 represents a neutral state (such as distilled water), 1.0 is the most *acidic, and 14.0 is the most *alkaline or *basic.

phalaenophily *Pollination by moths.

Phanerozoic The most recent period of geological time, since the end of the *Proterozoic, which covers roughly the last 570 million years, during which life on Earth has become more varied, complex, and abundant.

pharmaceuticals Medicinal drugs which are prescribed by a doctor in order to cure, prevent, or treat diseases and relieve pains.

phase One of the three forms in which any matter can exist, as a solid (for water, in the form of tiny ice crystals), a liquid (as droplets of *water), or a gas (*water vapour). As the matter is converted from one phase to another it undergoes *phase changes.

phase change A change in the form (*phase) in which matter is present. For water, the six phase changes are *melting, *freezing, *evaporation, *condensation, *melting and evaporation, and *sublimation.

phenetic classification An approach to the classification of *organisms that is based on observable similarities and differences between taxa, with no assumptions about evolution. Also known as **phenetics**. Largely replaced in recent years by *cladistics.

phenetic species concept The definition of *species as groups of individuals that look similar to each other and distinct from other groups. See BIOLOGICAL SPECIES CONCEPT, CLADISTIC SPECIES CONCEPT, ECOLOGICAL SPECIES CONCEPT, RECOGNITION SPECIES CONCEPT.

phenetics See PHENETIC CLASSIFICATION.

phenocryst A large *crystal in an *igneous rock, that is surrounded by a finer-grained *matrix.

phenol A white, crystalline compound (C_6H_5OH) that is derived from *benzene and used in the manufacture of some resins, weed killers, plastics, and disinfectants, and in the *petroleum refining process. It is *toxic to humans, and may be *carcinogenic.

phenology The study of periodic biological phenomena, such as flowering, breeding, and migration, especially in relation to climate.

phenotype The outward appearance, physical attributes, or behaviour of an *organism that develop through the interaction of the *environment and genetic makeup (*genotype) during growth and development. Individuals with the same genotype can have different phenotypes in different environments. See also ACQUIRED TRAIT, PHENOTYPIC ADAPTATION, TRAIT.

phenotypic adaptation A non-genetic form of biological *adaptation in *organisms, which occurs through behavioural changes in individuals.

pheromone Chemical messages that are relayed between individuals of the same or different animal and insect species, in order to attract mates, promote social cohesion, mark trails and territory, and as warnings.

pheromone trap A trap which uses either a natural or synthetic insect sex attractant *pheromone.

phi (φ) scale A scale that geologists use to measure the *particle size of *sediments, based on the logarithm of size rather than size in millimetres. The scale has zero in the fine sand range, with positive values for finer material (down to *clay) and negative values for coarser material (up to large *boulders).

philosophy A statement or system of beliefs, which influence thought, action, and knowledge. See also ENVIRONMENTAL PHILOSOPHY.

phloem The tissue in a plant that transports dissolved *nutrients from the leaves to the other parts of the plant. See also VASCULAR PLANT.

p

phosphate A *non-metallic salt of *phosphorus and *oxygen which occurs naturally. Phosphates are a constituent of *nucleic acids and are essential for life. Plants use phosphates during *photosynthesis, and they are generally a *limiting factor. Phosphates are also released into the environment by *fertilizers and *detergents, act as a *nutrient pollutant in water, and contribute to *eutrophication.

phosphorescence See BIOLUMINESCENCE.

phosphorus (P) A *non-metallic element and one of the elements (*macronutrient) essential for the growth of *organisms. Phosphorus compounds are major constituents in the tissues of both plants and animals. Increased phosphorus levels result from the discharge of phosphorus-containing materials (such as *fertilizers) into surface waters, and this can contribute to *eutrophication.

phosphorus cycle The natural *biogeochemical cycle through which *phosphorus is moved through *environmental systems, from rocks through the *biosphere and *hydrosphere and back to rocks.

phosphorus plant Any *facility that uses electric furnaces to produce *phosphorus for commercial use, such as high grade phosphoric acid, *phosphate-based detergent, and organic chemicals.

photic Of, relating to, or caused by light. Contrast APHOTIC.

photic zone The zone of a waterbody through which light penetrates, in which *photosynthetic organisms live. Also known as the **epipelagic zone**. Contrast BATHYPELAGIC ZONE.

photoautotroph An *autotroph that uses light energy from the Sun to manufacture food, via *photosynthesis. Compare CHEMOAUTOTROPH.

photochemical oxidant A *conventional or *criteria pollutant that is cre-

ated by the action of *sunlight on *nitrogen oxides and *hydrocarbons. See also PHOTOCHEMICAL SMOG.

photochemical pollutant Any chemical that reacts photochemically (in the presence of *sunlight) to destroy *ozone in the *stratosphere.

photochemical reaction A chemical reaction that is triggered by *sunlight.

photochemical smog A brownish atmospheric *haze that is often found above large urban or industrial centres during hot dry weather, and is caused by reactions between *air pollutants (particularly *oxygen, *nitrogen oxides, and unburned *fuel from vehicle exhausts) in the presence of *sunlight. Such smogs were first observed in California in the 1940s, but they are becoming increasingly common in many heavily built-up areas. Many plants wilt and sometimes die when exposed to photochemical smogs, and people and animals suffer from irritation of the eyes and lungs. One way of controlling photochemical smogs is to cut down on the emissions of pollutant gases (particularly nitrogen oxides) from vehicle exhausts. During the 1990s new vehicle exhaust emission controls were introduced in North America and in many European countries (except Britain) to try to tackle this problem. New *catalytic converters on vehicle exhausts are designed to reduce exhaust gas emissions.

photochemistry The branch of *chemistry that deals with the chemical action of light.

photodegradable Capable of being broken down by *ultraviolet light, for example in *sunlight.

photodissociation Chemical decomposition involving *sunlight, in which molecules are split into their constituent atoms. Also known as **photolysis**.

photoelectric Relating to the conversion of light (*radiant energy) to electricity.

photolysis *See* PHOTODISSOCIATION.

photon A unit of *electromagnetic energy which has no electric charge and is generally regarded as a particle with zero mass.

photo-oxidant An air pollutant (such as *ozone, *nitrogen dioxide, or *PAN) that is created by *photochemical reaction.

photoperiod The duration of the *daylight period.

photoperiodism The response of *organisms to changes in *day length, for example through changes in growth rate, feeding pattern, migration, and reproduction.

photophilous Tolerant of strong light. *Contrast* PHOTOPHOBIA, SCIOPHILOUS.

photophobia Very sensitive to light. *Contrast* PHOTOPHILOUS.

photorespiration *Respiration in plants that is triggered by O_2 rather than by CO_2, which occurs in cool-season plants during the light period, and which produces energy that is not in a useful form for plant growth.

photosphere The bright visible surface of the Sun.

photosynthate The *carbohydrate (including starch and sugar) that is synthesized during the process of *photosynthesis.

photosynthesis A complex process occurring within the cells of green plants where sunlight is utilized in combination with *carbon dioxide and *water, in the presence of *chlorophyll, to produce oxygen and simple carbohydrates, which are forms of energy that can be directly used by plants. *Contrast* CHEMOSYNTHESIS.

photosynthesizer A plant that carries out *photosynthesis.

photosynthetic efficiency The percentage of available light that is captured by *plants and used in *photosynthesis to make useful products.

photosynthetic rate The rate at which plants convert dissolved *carbon dioxide and *bicarbonate into useful products by *photosynthesis.

phototactic Moving in response to light.

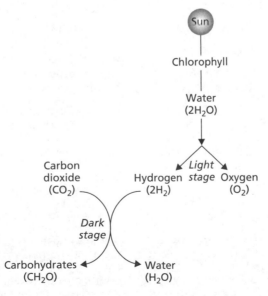

Fig 16 Photosynthesis

phototroph Any *autotrophic *organism (including mosses, ferns, conifers, flowering plants, *algae, and photosynthetic *bacteria) that obtains energy from light by photosynthesis.

phototropism The growth of plants towards *sunlight or other light source.

photovoltaic Technology for converting *sunlight directly into *electricity, usually with *photovoltaic cells.

photovoltaic cell An electronic device, consisting of layers of semiconductor materials, that is capable of converting incident light directly into an electric current. Also known as **solar cell**.

photovoltaic energy The electrical energy produced when sunlight incident on a *photovoltaic cell is converted into electrical current.

phragmites Reeds that grow in *wetland habitats such as *marshes, lake margins, or river banks.

phreatic Of or relating to *groundwater.

phreatic surface The upper surface of *groundwater within an *aquifer.

phreatic zone See SATURATED ZONE.

phreatophyte See HYDROPHYTE.

phyletic evolution *Genetic changes that occur within the line of descent of a *species.

phyletic gradualism The belief that *evolution and *speciation occur over a long period of time through slow, progressive change. Contrast PUNCTUATED EQUILIBRIUM.

phylogenetic Relating to evolutionary development. Contrast ONTOGENETIC.

phylogenetic species concept The idea that a *species is the smallest recognizable group of individual organisms within which there is a parental pattern of ancestry and descent. Contrast BIOLOGICAL SPECIES CONCEPT.

phylogenetic systematics See CLADISTICS.

phylogenetics The study of the relationships between *organisms; a branch of *systematics.

phylogeny The evolutionary history and line of descent of a *species.

phylum The second highest category (of seven) in the scientific system of classification for organisms (*taxonomy), below *kingdom and above *class. Each kingdom comprises more than one phylum, and each phylum comprises more than one class. For example, the phylum Mollusca includes slugs, snails, clams, squid. The phylum Chordata includes all of the 'higher' animals (including fish, *amphibians, *reptiles, *birds, and *mammals), and the phylum Arthropoda (*arthropods) covers *insects, spiders, and *crustaceans.

physical and chemical treatment Processes that are used to clean up pollution (including toxic materials in surface and *groundwater, *oil spills, and hazardous materials on or in the ground) and in large-scale *wastewater treatment facilities. Physical processes may include *air stripping or *filtration. Chemical treatment includes *coagulation, *chlorination, or *ozonation.

physical environment The *inorganic surroundings that are necessary for life, such as air, water, soil, rock, and sunlight.

physical factor Any non-living (*abiotic) factor, such as temperature, light, water, minerals, and climate, that influence an *organism.

physical oceanography The study of the world's *oceans, particularly in terms of water masses and their motions.

physical planning A form of *land use planning, particularly in *urban areas, which seeks to achieve a physically attractive environment by regulating the relationships between structures and their sites and surroundings.

physical quality of life A measure of economic welfare that is more sophisticated than *gross national product, and is based on three variables: percentage of the population which is literate, infant *mortality rate, and average *life expectancy after the age of one.

physical weathering See MECHANICAL WEATHERING.

physics The study of matter and energy, and their interactions.

physiognomic Based on physical features.

physiognomy External appearance: an old fashioned term to describe the *topography and other physical characteristics of a *landform and its *vegetation.

physiographic climax See CLIMAX COMMUNITY.

physiographic province An area or region across which *elevation, *relief, *lithology, geological structure and geological history are similar, and different from those in surrounding regions.

physiographic region An area with similar *landforms.

physiography The natural physical features of the land surface.

physiological drought A condition caused by high salt levels in the *root zone of a soil, in which water in the soil is attracted to the salt, leaving less available for uptake from the soil by plants. This causes similar symptoms of stress in plants to those caused by normal *drought.

physiological race A geographically defined population of a *species that is physiologically distinct from other populations.

physiology The study of functions and processes in living things.

phyto- As a prefix (such as *phytoplankton), relating to plants.

phytochemical Relating to the *chemistry of plants, plant products, and plant processes.

phytology See BOTANY.

phytomedicine The use of medicinal products that are based on plants.

phytonutrient Any *nutrient that is derived from plants.

phytoplankton Microscopic floating aquatic plants, such as *algae and *diatoms.

phytoremediation *Remediation that uses plants to remove *contaminants from *soils, *sediments, and *groundwater. See PHYTOTREATMENT.

phytotoxic Harmful or poisonous to plants.

phytotoxin A substance that causes reduced growth in or kills plants.

phytotreatment The cultivation of specialized plants that absorb particular *contaminants from the soil through their roots or foliage. A form of *phytoremediation.

picocurie (pCi) A unit for measuring radioactivity, often expressed as picocuries per litre of air. one picocurie = one million millionth of a *curie = 0.037 *becquerel.

picoplankton *Planktonic organisms that range in size from 0.2 to 2.0 *micrometres.

piedmont An area of gently rolling hills at the base of a *mountain or mountain range.

piedmont glacier A type of *glaciation in which large *valley glaciers converge and coalesce to form an almost stagnant *ice sheet.

piezometer An instrument that is used to measure *hydraulic pressure.

piezometric surface The surface of the static water level in a *confined aquifer.

pile 1. The *fuel element in a *nuclear reactor.
2. A heap of waste.
3. A column of wood, steel, or concrete

that is driven into the ground to support a structure.

pillow lava Volcanic *lava that has solidified under water and resembles heaps of pillows.

pilot balloon A small *meteorological balloon that is filled with helium and tracked as it rises through the *atmosphere, in order to determine how wind speed and direction change with altitude.

pilot test A small study that is carried out prior to a large-scale study, in order to test the suitability of a technique or procedure.

Pinchot, Gifford A US politician and conservationist (1865–1946) who served as the first Chief of the *US Forest Service (1905–10), was Republican Governor of Pennsylvania (1923–27 and 1931–35), and is remembered for coining the term conservation and advocating scientific approaches to the planned use and renewal of US forest reserves.

pinnacle A tall peak on a mountain, or an isolated column of rock.

pinyon–juniper woodland A distinctive North American woodland vegetation with scattered trees of medium height, often mixed with sagebrush and other scrub, which is common in the dry arid mountain regions in the west.

pioneer species Species of plants, *lichens, and *micro-organisms that dominate the vegetation community in the early stages of *succession, because they can colonize bare sites.

pipe *See* SOIL PIPE.

piridaben A type of *insecticide that is used primarily in controlling *mites in tree fruit crops.

pisceculture *See* AQUACULTURE.

pit mining The extraction of material of value from below the surrounding land surface, by excavation of an *open cast mine or a *quarry.

pitchblende The most common *ore of *uranium, which is mainly composed of the mineral *uraninite. *See also* POLONIUM.

placer A deposit of a valuable *ore (such as mineral-bearing *gravel or *sand) that has formed by a concentration of heavy minerals in flowing water, for example by a stream or waves.

placer deposit A mass of *sand, *gravel, or similar *alluvial material, that can be extracted through *placer mining.

placer mining The extraction of valuable heavy *minerals from a *placer deposit by washing sand and dirt away with a powerful jet of water, leaving behind the desired *mineral (such as *gold). Also known as **hydraulic mining**.

plagioclimax A form of *arrested development where human activities (such as *pollution, *habitat change, and *overgrazing) arrest the natural *succession.

plain A large area of level, open land.

planet A large body made of rock and metal, or predominantly liquid and gas, which orbits around a *star. *See also* MOON.

planetary albedo The fraction (approximately 30%) of incident *solar radiation that is reflected by the Earth-atmosphere system and returned to space, mostly by backscatter from *clouds in the atmosphere. *See also* ALBEDO.

planetary boundary layer The lowest level of the *atmosphere, where *friction is an important force and vertical mixing is common.

planetary scale *See* GLOBAL SCALE.

planktivore An animal that eats *plankton, usually a fish that feeds on *zooplankton.

plankton *Aquatic organisms (such as *zooplankton and *phytoplankton) that are unable to swim because they have no motor power of their own, so they drift around with the *tide and ocean currents. *contrast* NEKTON.

planktonic Of or relating to the *plankton.

planktonic foraminifera Single-celled *protozoan *micro-organisms that live near the surface of the ocean and have skeletons made of *calcium carbonate. *See also* FORAMINIFERA.

planned grazing system A system of *grazing in which two or more grazing units are alternately rested from grazing, in a planned sequence, over a period of years.

planned ignition A fire that is started by a deliberate management action.

planned unit development (PUD) zoning *See* CLUSTER ZONING.

planning The act of drawing up plans and establishing a broad outline for goals, policies, and procedures that will accomplish agreed objectives. *See also* MANAGEMENT.

planning horizon The period of time which will be considered in the planning process to which a particular plan relates.

plant Any living thing that is not an *animal; *organisms that belong to the kingdom Plantae. Plants cannot move, lack a nervous system, have cellulose cell walls, and produce food from *sunlight and water via *photosynthesis.

plant breeding The deliberate improvement of *crops that are grown for food and fibre, based on selecting *genes that produce beneficial traits (such as increased production or decreased costs) and transferring them from one plant to another.

plant community A group of plants (such as *broadleaved *woodland) that grows together in a particular area, under given environmental conditions (soil, climate, etc.).

plant kingdom The taxonomic *kingdom (*Plantae) that comprises all living or extinct *plants.

plant retrogression The reverse of *succession, in which a particular area is successively occupied by less-developed vegetation. In grassland management, for example, continuous *overgrazing of *pasture can cause the plant community to change from desirable, highly productive grasses to low quality, unproductive grasses.

Plantae In *taxonomy, the *kingdom of plants. Also known as the ***plant kingdom**.

plantation An area of *forest that has been deliberately sown or planted to produce a cash crop of trees.

plantation agriculture A commercial *agricultural system found in tropical countries, in which specialized crops such as bananas, coffee, flowers, and cacao are grown primarily for export to *developed countries.

planula The tiny *larval swimming stage of *corals.

plasma 1. A hot, electrically neutral gas of ions and free electrons.
2. The liquid portion of blood, in which cells are suspended.

plasma membrane The outer membrane of a living cell.

plasma-arc reactor An *incinerator that operates at extremely high temperatures and destroys highly *toxic wastes that do not readily burn.

Plasmodium A *parasitic *protozoan that causes malaria in humans.

plastic A property of any material that can flow.

plastic limit The moisture content at which a *soil changes from solid to *plastic.

plastics Durable, flexible, synthetic-based products that can be shaped by the application of heat or pressure, some of which (such as *PVC) are difficult to *recycle and pose problems because of their *toxic properties.

plastron The bottom part of the shell of a *crustacean, such as a turtle or tortoise. *Contrast* CARAPACE.

plate *See* CRUSTAL PLATE.

plate boundary A zone of *seismic and *tectonic activity along the edges of *crustal plates, which indicate relative motion between plates.

plate tectonics A widely accepted theory that the Earth's surface is divided into rigid *crustal plates, which can move relative to each other (at an average speed of about 70 kilometres per million years; 7 centimetres a year) because of the *plastic nature of the underlying upper *mantle. Material in the mantle behaves like a plastic, and it is believed that huge slow-moving *convection currents within it can move sections of the overlying crustal rocks with them. This drives the movements of the plates. Six major plates (Eurasian, American, African, Pacific, Indian, and Antarctic) are recognized, together with a number of smaller ones. The plate margins coincide with zones of *seismic and *volcanic activity, and plate movements are driven by *sea-floor spreading and the development of *mid-oceanic ridges. These large-scale crustal movements shape the Earth's surface and create all of its major natural features including *mountains, ocean basins, deep *ocean trenches, *volcanoes, and *earthquakes. *Subduction at destructive plate margins creates a series of large-scale landforms including mountain belts, volcanoes, and ocean trenches. *See also* CONTINENTAL DRIFT.

platy Soil *aggregates that are long and flat, rather than round or spherical.

playa A dry, flat-floored lake bed in a *desert, which may *flood from time to time, during *ephemeral flows after *storms. Many playas have a white, crusty layer of salts, formed by *salinization.

Pleistocene The first *epoch within the *Quaternary period, which began about 1.5 million years ago and lasted until about 10 000 years ago. During the Pleistocene climate switched between *glacial and *interglacial conditions through a number of cycles. Geological evidence from many locations throughout North America supports a four-phase model of glacial advances (called the *Kansan, *Nebraskan, *Illinoian, and *Wisconsinan), although there is evidence that many parts of Europe experienced five glacial periods. Ice covered much of the *northern hemisphere, largely removing all traces of earlier landscapes across much of Europe and North America and replacing them with distinctive glacial landforms, including the *Great Lakes. *Moraines indicate the outer limits of ice advance, which correlate closely with the distribution of glacial deposits and landforms. Most of the Pleistocene ice had melted by the start of the *Holocene, about 10 000 years ago, but valley glaciers remain in some high altitude mountain areas in both hemispheres and ice caps survive in Greenland and other parts of the *Arctic, and in *Antarctica.

Pleistogene *See* QUATERNARY.

plesiobiosis A rudimentary form of social *symbiosis, which involves the close proximity of nests of different species of social insects (such as ants).

pleuston *Organisms that float on the sea surface.

Pliocene The final *epoch of the *Tertiary period, between about 5.2 and 1.6 million years ago, when climate was relatively warm with little *glaciation, and more and larger species of *mammal evolved.

plot A small piece of land.

plough A farm tool with one or more heavy blades that break the soil surface, cut a furrow, and turn the soil over in preparation for sowing and growing crops. Spelled plow in North America.

plow *See* PLOUGH.

plucking *See* QUARRYING.

plugging **1.** The process of stopping the flow of water, oil, or gas into or out of a rock formation through a *borehole or *well.

2. The process of filling and abandoning a well.

plume A visible or measurable discharge of a *contaminant that is released from a particular source into the air, *groundwater, or surface water.

pluton *See* INTRUSION.

plutonic *See* INTRUSIVE.

plutonium (Pu) A heavy, *radioactive, metallic element (atomic number 94) that is similar to *uranium. It undergoes *fission when bombarded with *neutrons and is used in the production of *nuclear energy and the explosion of nuclear weapons. It occurs in nature in trace amounts, but is also produced as a *byproduct of the fission reaction in a uranium-fuelled nuclear reactor, and can be recovered for future use. *See also* BREEDER REACTOR.

pluvial A period during which the *climate is relatively wet. A pluvial period lasts for decades or longer, between *interpluvials.

pluvial lake A *lake that is formed during a *pluvial period.

pluvial period *See* PLUVIAL.

PM Particulate material.

PM₁₀ Particulate material in the air which is less than 10 *micrometres in diameter and includes soot, dust, smoke, fumes, and *aerosols. The particles small enough to be inhaled by humans and can damage health. PM₁₀ is a *criteria air pollutant that can cause reduced visibility.

PME *See* PARTICIPATORY MONITORING AND EVALUATION.

PNA *See* POLYNUCLEAR AROMATIC HYDROCARBON.

PNC *See* POTENTIAL NATURAL COMMUNITY.

pneumoconiosis A lung disease that is caused by the long-term inhalation of fine particulates.

PNV *See* PRESENT NET VALUE, POTENTIAL NATURAL VEGETATION.

poach To hunt, trap or fish illegally.

poaching Hunting, trapping, or fishing illegally.

pocosin A swamp or marsh vegetation underlain by poorly drained, mostly organic soil, which is found on the southeastern coastal plain of the USA, particularly North Carolina.

podsol A shallow, greyish-white *acidic soil with relatively little organic content and few soil organisms, that develops under *coniferous or heath vegetation in *temperate to cold moist climates. Podsols are usually infertile and unsuitable for agriculture.

podsolization A *soil-forming process by which the upper *horizon of a *soil becomes *acidic through the *leaching of bases which are deposited in the lower horizons (for example, in a *pedalfer). It usually occurs under forests in areas with cool, moist climates and quite severe winters, where *decomposition of *organic matter is inhibited and a peaty mat of acidic organic material builds up on the soil surface. Mixing of organic matter and mineral matter within the soil is very limited because there are relatively few organisms.

poikilotherm A cold-blooded animal (such as fish and amphibians, but excluding birds and mammals) whose body temperature varies with the temperature of its surrounding environment. *Contrast* HOMEOTHERM.

point bar A low ridge of *sand and *gravel that is deposited along the inner bank of a *meandering *alluvial stream, where flow velocity is relatively low.

point of exposure The point at which an individual or *population may come into contact with a chemical of concern that originates from a particular site.

p

point source *Pollution that originates from a stationary source or a fixed facility, such as a pipeline or a factory. *Contrast* NON-POINT SOURCE.

poison A chemical that adversely affects health by causing injury, illness, or death.

polar At, near, or relating to the North or South Pole, or to the ends (poles) of a magnet.

polar air mass A cold *air mass that forms in a *high-latitude source region.

polar climate A *climate zone that has no warm season, receives limited precipitation, and experiences penetrating cold with temperatures below 10°C throughout the year.

polar continental air mass An *air mass that develops over large land masses at high latitude, contains cold, dry, and stable air, and brings cool conditions in summer and cold conditions in winter.

polar easterlies Easterly winds that develop at high latitudes *poleward of the *subpolar low.

polar front A belt of *low pressure in the *atmosphere at about 60° latitude, where the *westerlies meet air that flows from the *poles towards the *equator.

polar front jet stream A strong, generally westerly *jet stream wind that is concentrated in a relatively narrow and shallow current in the upper *troposphere above the *polar front.

polar high A major *wind belt that circles the Earth, in which the strong, dry winds blow from the northeast in the northern hemisphere, and from the southeast in the southern hemisphere.

polar ice cap An area of permanent ice at high latitudes, beyond the *tundra, where average temperature in each month is below freezing point, and annual *precipitation is low (usually less than 250 millimetres a year). *Antarctica and the interior of Greenland are the only large land masses in this climate zone.

polar maritime air mass An *air mass that develops over the oceans in higher latitudes, contains cool, moist air, and brings heavy cloud cover and dull conditions in winter and mild fair weather in summer.

polar region The area around the *poles, in which the climate is permanently cold and inhospitable.

polar wandering The natural long-term cycle of movement of the Earth's *magnetic field. Evidence of polar wandering is preserved in the pattern of magnetic changes (*palaeomagnetism) in submarine rocks adjacent to *mid-ocean ridges and created by *sea-floor spreading. These show that the Earth's *magnetic field has changed as the poles have moved due to *continental drift at roughly half-million-year intervals, with shorter reversal periods in between the major ones.

polarity Having a positive or negative charge.

polder A low-lying area of land that has been reclaimed from the sea by being artificially drained and protected from flooding by building *dykes, and is kept dry by continuous pumping.

pole 1. A point on the Earth where the axis of rotation intersects the surface or a magnetic pole. *See also* NORTH POLE, SOUTH POLE.

2. A young tree. *See also* SAPLING.

poleward Towards the *poles.

policy A written plan of action that is adopted by an individual or group in order to accomplish some particular social or economic goal.

policy cycle The process by which problems are identified and acted upon in the public arena, for example by local and national government.

political ecology An umbrella term for a variety of projects that involve politics and the environment, including attempts to study politics using the language and methods of *ecology, the study of political struggles for control over *natural resources, and research on *biodiversity and *natural resource exploitation that is intended to inform public policy.

politics Social relations that involve authority or power; the science and methods of government. *See also* ENVIRONMENTAL POLITICS.

pollard A traditional form of *woodland management in which mature trees are felled at about two metres above ground level in order to promote the regrowth of a crown of poles above the reach of *grazing animals. *See also* COPPICE.

pollen Tiny, powdery grains derived from seed plants that contain the male reproductive cells. The pollen of each plant has a distinctive size and shape, which helps in *pollen analysis

pollen analysis A technique that is used to reconstruct past *climate. Variations in the frequency of *pollen grains at different layers within a soil, sediment, or deposit can be analysed and the results plotted on a *pollen diagram. Also known as **palynology**.

pollen basket Part of the back (posterior) legs of the honeybee, in which it stores *pollen.

pollen diagram A graph of the results of a *pollen analysis for a particular site, which shows changes in the relative proportions of the *pollen of different plants over a period of time, and indicates changes in vegetation and climate.

pollination The transfer of *pollen from the male part of a flower (*stamen) to the female part of a flower (the style and *stigma), by a *pollinator. *See also* MONOLECTIC, OLIGOLECTIC, POLYLECTIC.

pollinator An agent that carries *pollen to the female part of a flower, including beetles (*cantharophily), flies (*myophily), wasps (*specophily), ants (*myrmecophily), bees (*melittophily), butterflies (*psychophily), and moths (*phalaenophily).

pollutant A substance that *pollutes or causes *pollution. *See also* AIR POLLUTANT, WATER POLLUTANT.

pollutant pathway The route along which a particular *pollutant is distributed, for example within a building or through an *environmental system.

pollute To make impure or dirty with harmful or poisonous substances.

polluted Contaminated by *pollution.

polluter A person or organization that causes pollution of the environment.

polluter pays principle (PPP) The principle that the *polluter should pay the costs of controlling the *pollution they generate, and of cleaning up any environmental damage that is caused by that pollution.

pollution The process of contaminating or polluting. *See also* AIR POLLUTION, NOISE POLLUTION, WATER POLLUTION.

pollution charge The fee or charge that is levied per unit of pollution, based on the *polluter pays principle.

pollution control Any activities that are carried out in order to reduce or eliminate *pollution.

pollution episode A serious air pollution or water pollution incident in a particular area at a point in time. Also known as an **episode**.

pollution indicator Any *indicator that is used to show the presence of specific environmental conditions or *pollutants.

pollution permit An approach to the control of pollution emissions in which *polluters can bid for a *permit that allows them to create a given amount of *pollution. Permits can be resold, and the government can gradually reduce the number of permits that are available so that total pollution emissions can be controlled.

Pollution Prevention and Control (PPC) An approach to the control of *pollution in the UK, introduced under the EU

p

*Integrated Pollution Prevention and Control Directive (IPPC Directive), that requires any operator of a factory or plant to obtain a *permit from the *environmental regulator and comply with the conditions in that permit. This replaced *integrated pollution control in 2000.

pollution prevention pays programme An approach to *environmental management that is based on preventing pollution (by elimination, reduction, or substitution) rather than dealing with it once it has occurred.

Pollution Standards Index (PSI) A system developed by the US *Environmental Protection Agency for measuring pollution levels for the major *air pollutants that are regulated under the *Clean Air Act, based on the health effects of several pollutants. It ranges from 0 (healthy) to 500 (extremely unhealthy). In 1999 the PSI was replaced by the *Air Quality Index (AQI).

polonium (Po) A *radioactive element that occurs in *pitchblende and other ores that contain *uranium.

polychlorinated biphenyl (PCB) A synthetic, *organic chemical that was once widely used in electrical equipment, specialized hydraulic systems, heat transfer systems, and other industrial products. It is highly *toxic and a potent *carcinogen, and the management of PCB wastes is tightly regulated.

polyculture *Crop cultivation in which a variety of different plants are grown together in the same *field. *Contrast* MONOCULTURE.

polycyclic aromatic (PCA) *See* POLYCYCLIC AROMATIC HYDROCARBON.

polycyclic aromatic hydrocarbon (PAH) A class of *organic compounds that contains hydrogen and carbon, and which are found in products such as *petroleum and creosote. Some are *carcinogens. Also known as **polynuclear aromatics** (PNAs) or polycyclic aromatics (PCAs).

polycyclic organic matter (POM) A broad class of compounds that are formed primarily from *combustion. POM is present in the air in *particulate form, and comes from diverse sources including vehicle exhaust, fires, and hazardous waste sites.

polyembryony Producing two or more young from the same egg.

polygamous Having more than one mate at a time.

polygenetic Caused by multiple processes, or created in a number of phases during which different processes were dominant.

polyhaline Relating to *brackish water which has a *salinity between 18 and 30 parts per thousand.

polylectic A *pollinator which collects *pollen from many species of plant. *Contrast* MONOLECTIC, OLIGOLECTIC.

polymer A substance that is made of many repeating chemical units or *molecules. Natural polymers include *proteins (polymers of amino acids) and synthetic polymers include *PVC (a polymer of vinyl chloride).

polymictic A *freshwater lake that mixes completely from time to time.

polymorphic Something (like a *species) that has more than two distinct forms.

polymorphism The occurrence of more than one distinct form of individual in a *population.

polynuclear aromatic (PNA) *See* POLYCYCLIC AROMATIC HYDROCARBON.

polyp A small tube-like *marine animal which lives in warm, clear seas and grows with one end attached to the seabed, to rocks, or to other polyps. On the other end is a mouth surrounded by finger-like, stirring tentacles. Live *coral is made of polyps.

polyp bailout The process by which some stony *corals release themselves from their skeleton and drift to a new location, usually induced by stress.

polyphagous Feeding on or using many kinds of plant or animal, usually plant species from several families. *Contrast* MONOPHAGOUS.

polyphenism The occurrence of several *phenotypes in a *population that may be caused by environmental influences but is not due to different *genetic types.

polyphyletic A group of *species that do not have one common *ancestor species.

polyploid An organism that contains two or more sets of *genes or *chromosomes.

polyvinyl chloride (PVC) A tough *polymer plastic that has many everyday uses (including footwear, electrical cable, packaging, and toys), is not *biodegradable, and releases *hydrochloric acid when burned.

polyxenous In *symbiosis, having two or more *host families.

POM *See* POLYCYCLIC ORGANIC MATTER.

pond A pool of still water, often artificially created. A type of freshwater *wetland.

pool A shallow body of freshwater.

pool and riffle A common sequence of *bedforms in *alluvial channels, which tend to have narrow, deep sections (the pools) with fine *sediment on the bed, alternating with wider, shallower sections with *gravel beds (the riffles). This pool and riffle sequence appears to play an important role in the development of *meandering.

POP *See* PERSISTENT ORGANIC POLLUTANT.

population 1. A group of individuals of the same *species who are living in the same area at the same time and share a common *gene pool, which makes it possible for them to interbreed.

2. In statistics, the complete set of individuals, alive or not. *Contrast* SAMPLE.

population at risk A subgroup of a *population, which is more likely to be exposed to a particular agent, or is more sensitive to that agent, than is the general population.

Population Bomb An influential book by Paul and Anne *Ehrlich (1968), in which they predicted imminent famine and disaster on a scale unprecedented in world history.

Population Connection A campaigning organization in the USA, originally called *Zero Population Growth, which was founded by Paul *Ehrlich in 1968 and adopted its new name in 2002. Its main focus is on the relationships between population, environment, and society.

population crash A sudden sharp reduction in the size of a *population that can be caused by *disease, environmental stress (such as *pollution), or when its numbers exceed the *carrying capacity of its *habitat. Also known as **dieback**. *Contrast* POPULATION EXPLOSION. *See also* MALTHUSIAN GROWTH.

population density The number of individuals of a *species that is found in a particular area at a given point in time.

population dose The total *dose of *radiation that is received by individuals in a population that is exposed to a particular source or event.

population dynamics The study of the factors that affect the growth, stability, and decline of *populations, and of the interactions between those factors.

population explosion A sudden rapid rise in the number of individuals in a population in which the *birth rate exceeds *mortality rate. It generally brings increased environmental damage, through increased *resource use, *waste production, *pollution, and *land use change. A population explosion is also widely associated with social impacts such as low living standards, low education standards, unemployment and *poverty, *malnutrition and *starvation, *civil unrest, large-scale *migration (both internal and international), the growth of huge *megacities and associated *shanty towns, and unsustainable pressures on government institutions and national economies. *Contrast* POPULATION CRASH. *See also* MALTHUSIAN GROWTH, OVERPOPULATION.

population genetics The study of the patterns and causes of variation in *genes among a group of individuals within a *population.

population momentum The tendency for *population growth to continue beyond *replacement level, because of a relatively large number of individuals of child-bearing age.

population policy A *policy or set of policies, usually agreed at national level, designed to manage population growth and improve the quality of life of individuals. *See also* INTERNATIONAL CONFERENCE ON POPULATION AND DEVELOPMENT.

population projection A *projection of the likely rate and pattern of growth of a population under particular conditions, particularly assumptions about future levels of *fertility, *mortality, and *migration. Every two years the *United Nations Population Division produces a set of population projections for every country, based on updated population data and revisions in projection methodology. The 1998 projections, for example, were and they based on three *scenarios, namely low growth, medium growth, and high growth. By 2050 the UN expects total world population to grow to between 7.3 billion and 10.7 billion. Under the high growth scenario world population will continue to increase after 2050; under the low growth scenario it will have slowly begun to decline. The medium growth projection of 8.9 billion people by 2050 is considered the most likely. Even under the low growth forecast, the world will have to meet the needs of an extra 1.3 billion people because *fertility in *developing countries is twice as high as that in *developed countries; developing countries have a younger age structure, which creates a momentum for continued *population growth for several decades ahead; and continued reductions in *mortality will add extra growth, particularly in countries with a relatively low *life expectancy.

population pyramid A special type of bar chart that shows the distribution of the *population (within a country, region, or the world) by gender and age, usually with the younger ages at the bottom and with males on the left and females on the right. Slowly growing populations (such as the USA) typically have a relatively large elderly population, so they have a population pyramid of reasonably uniform width from bottom (young) to top (old). Rapidly growing populations (such as Kenya) have a disproportionately large number of young people and thus more of a true pyramid shape, becoming narrower towards the top. The pyramid for a population that is not growing, or is decreasing, often has a bulge in the middle-age range, with relatively few young and old people. Population pyramids for all countries can be seen on the website of the US Census Bureau (http://www.census.gov/ipc/www/idbpyr.html).

population viability analysis An analysis that estimates the minimum numbers that are required for a *population to be viable or self-sustaining.

pore space Spaces within a body of soil, *sediment, or rock.

porifera The *phylum that comprises sponges.

porosity A measure of the amount of *pore space within a soil, sediment, or rock, which determines its capacity for holding water, and is expressed as the amount of pore space compared to the total volume of soil or rock, as a percentage.

porous Able to absorb fluids.

porphyritic *igneous rock that contains relatively large crystals set within a finer grained *matrix.

positive charge An electrical charge that is created by having fewer *electrons than *protons.

positive feedback *Feedback that amplifies the response of the system. *Contrast* NEGATIVE FEEDBACK.

possibly extinct A *species which is known only from historical evidence, but for which there are hopes that it may be rediscovered.

post A pole or timber stake that is set up to mark something, or to support other structures for fencing.

post-closure The period of time after a waste management or manufacturing facility has been closed, which for monitoring purposes is often taken to be 30 years.

post-consumer material *See* SECONDARY MATERIAL.

post-consumer recycling The reuse of waste materials that have been generated from residential and consumer waste, such as converting wastepaper from offices into corrugated boxes or newsprint.

post-consumer waste Waste materials (such as office paper and *aluminium cans) that have been used by consumers and recovered from the *waste stream, which are then used as raw materials to make new products.

post-glacial After the *ice age, or after a *glacier has melted from a particular area.

post-industrial society A modern economy that is dominated by services and information, rather than by industry.

post-materialism A philosophy or worldview that emphasizes quality of life over the acquisition of material goods.

post-modern A late 20th-century view (in art, thought, and society) that what we take to be reality is in effect a human (social) construction rather than a physical entity, which promotes a distrust of objectivity, authority, universality, and moral and ideological absolutes.

post-modernism A belief that individuals are constructs of social forces, and that there is no absolute transcendent truth that can be known.

postulate **1.** A basic principle.
   2. To make an assumption that is used as the basis for reasoning or hypothesizing.

potable water Water that safe to drink. Also known as **drinking water**.

potash alum *See* ALUM.

potassium (K) A silver-white, soft, light, low-melting metallic element of the alkali metal group that occurs abundantly in nature and accounts for 2.59% of the Earth's *crust. It is a *macronutrient, one of three main *fertilizers (with *nitrogen and *phosphorus) that are essential for plant growth, and is generally not considered toxic.

potential dose *See* ADMINISTERED DOSE.

potential energy *Energy which is possessed by a body because of its position or state, such as water stored behind a *dam which can be used to produce *hydroelectricity.

potential evaporation The amount of water that could be evaporated if it was available, which depends on air temperature, *insolation, and wind speed. *Contrast* ACTUAL EVAPORATION.

potential evapotranspiration (PE) The amount of water that could be evaporated or transpired if it was available. *Contrast* ACTUAL EVAPOTRANSPIRATION.

potential instability *See* CONVECTIVE INSTABILITY.

potential natural community (PNC) The *biotic community that would become established on a site if all stages in the *succession were completed without interference from humans, under the present environmental conditions.

potential natural vegetation (PNV)
The plant community that would develop on a site if all stages in the *succession were completed without interference from humans, under the present environmental conditions.

potential niche The maximum possible distribution of a species in the environment. *Contrast* REALIZED NICHE.

potential receptor Any organism or environmental medium that is in the pathway of a particular *contamination from a discharge.

potential resource Any part of the natural or human environment that is currently not thought to have *value, but which one day may become valuable as a result of developments in technology and know-how, and/or changes in market conditions.

potential yield The maximum annual *yield of a *renewable resource (such as timber from a forest or fish from the sea) that can be obtained on a *sustainable basis by using intensive management practices.

potentially responsible party (PRP)
An individual or company that is potentially responsible for, or contributing to, contamination at a *Superfund site, and therefore potentially liable for *cleanup costs.

potentiation The ability of one chemical to increase the effect of another chemical.

potentiometric surface An imaginary surface that represents the level to which water rises in wells in a *confined aquifer: similar to the *water table of an *unconfined aquifer.

pothole 1. A small, rounded depression in the *bedrock of a stream bed, formed by *abrasion of small pebbles and cobbles in a strong current.
2. A system of caves and underground rivers formed by water over a prolonged period of time.

POTW *See* PUBLICALLY OWNED TREATMENT WORKS.

powder snow Dry, loose, unconsolidated *snow.

power The rate at which *energy is supplied, usually measured in *watts.

power plant A North American term for a place where *electricity is generated, either by burning *fossil fuels, controlling *nuclear reactions, tapping hot subsurface rocks (*geothermal), or diverting water to turn turbines (*hydroelectric). Also known as **power station**.

power station The UK term for *power plant.

ppb Parts per billion.

PPC *See* POLLUTION PREVENTION AND CONTROL.

ppm Parts per million.

ppmv Parts per million by volume.

PPP *See* POLLUTER PAYS PRINCIPLE.

PRA *See* PARTICIPATORY ACTION RESEARCH.

prairie A large area of flat or rolling *grassland, dominated by grasses and wildflowers, with no or few trees, that often grows on fertile soil. *See also* DRY PRAIRIE, HILL PRAIRIE, MESIC PRAIRIE, MIXED GRASS PRAIRIE, SAND PRAIRIE, SHORTGRASS PRAIRIE, TALLGRASS PRAIRIE, WET PRAIRIE.

prairie pothole A marsh type of vegetation dominated by small seasonally or permanently flooded depressions and emergent *hydrophyte vegetation, that is found in North America on the prairies of the upper Midwest and the Canadian plains, and provides a habitat for the majority of North American ducks during migratory nesting and the breeding season.

Precambrian The earliest unit of geological time, which ended with the beginning of the *Palaeozoic about 570 million years ago, and accounts for about 90% of geological time. The first

POVERTY

The state of having little or no money and few or no material possessions. More than a fifth of the global population today live in poverty, subsisting on less than US $1 a day. There are clear links between poverty, population growth, and environmental problems. These are most evident at the level of individual households or communities, but they can also be detected at the national scale. Poverty is often accompanied by illiteracy, *malnutrition and poor *health, low status of women, and exposure to *environmental hazards. Poverty and lack of economic opportunities can force people to exploit marginal resources (e.g. by *overgrazing land or *overharvesting forests); this sets in motion a repeating cycle of *environmental deterioration, *marginalization, hardship, and poverty. Poverty is associated with a wide variety of health risks and problems, including inadequate *sanitation, unsafe *drinking water, *air pollution, and crowding. Poverty is linked to *fertility, and in many places women from low-income families have more children than women from richer families in the same society. Problems of rapid population growth are compounded by poverty and *inequality, and efforts are being made in many countries to reduce poverty as a means of improving the health of children and mothers, reducing the problems of rapid *urban development, and ensuring adequate nutrition for everyone. *See also* ABSOLUTE POVERTY, RELATIVE POVERTY.

living *organisms appeared during the Precambrian.

precautionary principle A proactive method of dealing with the environment based on the idea that if the costs of current activities are uncertain but are potentially both high and irreversible then society should take action before the uncertainty is resolved. The *Convention for the Prevention of Marine Pollution by Dumping from Ships and Aircraft (Oslo Convention) (1972) is an early example of the use of the precautionary principle since this convention, and subsequent ones, used a system of 'black' and 'grey' lists covering substances that could not be disposed of at sea and those for which a licence was required. Substances could be placed on the black list even if scientific evidence of harm was not fully available, so that discharges did not continue just because there was no proof of environmental damage. The *Vienna Convention for the Protection of the Ozone Layer (1985) and *Montreal Protocol (1987) were at the time quite unique amongst international environmental initiatives, because it was probably the first time

that serious attempts were made to avert a major environmental problem *before* the more serious consequences and side-effects were obvious. Previously governments would have insisted on waiting for conclusive scientific proof of cause–effect links before introducing appropriate preventive legislation and embarking on costly behaviour-changing strategies. Also known as the **do-no-harm principle**. *Contrast* WAIT-AND-SEE PRINCIPLE. *See also* NO REGRETS POLICY.

precedent An act or decision that can be used as an example in dealing with subsequent similar situations.

precedent law *See* COMMON LAW.

precession The variation in the Earth's axis of rotation that traces out the path of a cone over a period of about 23 000 years.

precious metal *See* NOBLE METAL.

precipitate A *solid that separates from a *solution.

precipitation 1. The process by which moisture (*water vapour) in the *atmosphere condenses and falls as *rain, *snow, *sleet, or *hail.

2. In chemistry, the action of causing a substance to be deposited in solid form from a solution.

precipitation scavenging The process by which rain or snow removes *particulates from the *atmosphere and deposits them on the ground surface. *See also* RAINOUT.

precipitator An *air pollution control devices that collect particles from an *emission by mechanical or electrical means.

pre-commercial thinning In commercial *forestry, removing some of the trees from a stand that are too small to be sold (for lumber or house logs), so that the remaining trees will grow faster.

pre-commitment period In *air pollution *emissions reduction, the period of time before the first *Commitment Period under the *Kyoto Protocol, when the operational framework is established and the involvement of signatories to the protocol is determined.

precursor A substance from which another substance is formed.

pre-cycling Making purchasing decisions that will reduce waste. This includes buying goods with less packaging, choosing products that will last longer, and avoiding single-use or disposable products, the most commonly cited example of disposable products being nappies.

predation An interaction between organisms in which one (the *predator) eats the other (the *prey). *See also* PARASITISM.

predator A *carnivore; an animal (such as a shark) that hunts and kills other animals (the *prey) for food.

predator control An interaction between *predator and *prey in which the predator controls the size of the prey population.

prediction An indication in advance based on observation, experience, or scientific reason. *Contrast* PROJECTION.

pre-existing use *Land use that existed before particular planning constraints were introduced.

preferable species *Plant species which are preferred by animals and are grazed out of choice.

pre-glacial Before *glaciation or before an *ice age.

prehistoric Dating back to before written historical records begin. In Europe this includes the *Stone Age, the *Bronze Age, and the *Iron Age. In North America prehistory is usually taken to refer any time before AD 1540.

Preliminary Assessment (PA) The initial step of a *site assessment of *Superfund sites, based on collecting and reviewing available records and information. *See also* SITE INSPECTION.

premature death A death that occurs before the average age within that population and which is usually attributable to a specific cause (such as an accident or *disease).

preparation The treatment of *waste materials prior to composting, which includes grinding, shredding, sorting, and adding sewage sludge.

preparatory cut In *forest management, the removal of trees towards the end of a *rotation in order to open the *canopy so the crowns of seed-bearing trees can enlarge, which improves seed production and encourages natural regeneration.

preparedness All measures that are taken before an event occurs, that allow for prevention, mitigation, and readiness. *See also* CONTINGENCY PLAN, DISASTER MANAGEMENT.

prescribed burn A planned fire in a predetermined area, in order to meet particular *resource management objectives such as to improve *silviculture, *wildlife management, *grazing, or reduction of fire hazard. Also known as **controlled burn, prescribed fire.**

prescribed fire *See* PRESCRIBED BURN.

prescription Written instructions or a written statement that defines management objectives.

present net value (PNV) A way of comparing the value of something now with the value of it in the future, which is used to compare alternative projects that have different cost and revenue flows. Also known as **present net worth**. *See also* COST–BENEFIT ANALYSIS, DISCOUNT RATE.

present net worth *See* PRESENT NET VALUE.

preservation The activity of protecting something (such as old and historic buildings, sites, structures, and objects) from loss or danger. *Contrast* CONSERVATION.

preservative A chemical substance used to protect organic material (such as food or wood) from *decomposition or *fermentation.

preserve 1. To keep or maintain in an unaltered condition.
2. A reservation in which animals are protected. Also known as a **reserve** or **nature reserve**.

pressed wood product Material that is made from wood veneers, particles, or fibres that are bonded together with an adhesive under heat and pressure, and is used in making buildings and furniture.

pressure The force that is applied per unit area. *See also* ATMOSPHERIC PRESSURE.

pressure gradient The rate of change in *atmospheric pressure between two points, which reflects the distance between *isobars on a *weather map, and affects wind speed (a steep pressure gradient produces strong winds). Also known as **barometric gradient**.

pressure group Any group of people who share special interests, are politically active, and advocate particular policies or approaches. Examples include *Friends of the Earth (FOE) and *Green-peace. Also known as an **interest group** or **special interest group**. *See also* ENVIRONMENTALISM.

pressure-tube anemometer An instrument that measures *wind speed from the pressure exerted through a tube aligned to point into the wind.

pressurized water reactor (PWR) A type of *light-water *nuclear reactor in which water is used as a coolant, but it is kept under enough pressure so that it does not boil. Electricity is generated by turbines which are turned by steam that is formed in a secondary cooling system. *Contrast* BOILING WATER REACTOR.

pre-treatment A range of processes that are used to reduce, eliminate, or alter the nature of *wastewater pollutants from non-domestic sources before they are discharged into *publicly-owned treatment works.

prevailing westerlies The dominant *westerly winds that blow in middle latitudes on the *poleward side of the *subtropical high-pressure areas. Also known as **westerlies**.

prevailing wind The direction from which the wind blows most often.

prevalence The number of individuals in a *population who have a particular *disease or condition at a given time. *Contrast* INCIDENCE. *See also* OCCURRENCE.

Prevention of Significant Deterioration (PSD) A programme, specified in the US *Clean Air Act, whose goal is to prevent *air quality from deteriorating significantly in areas of the country that currently comply with *ambient air quality standards.

prey An animal that is killed and eaten by other animals (the *predators).

price elasticity A measure of the sensitivity of *supply and *demand to changes in price.

Price-Anderson Act (1957) US legislation that established an insurance scheme for *nuclear power that is backed by taxpayers. It limits the amount of insurance that *nuclear power plant owners must carry, and shields nuclear power from free market forces by capping the liability of nuclear power plant owners in the event of a catastrophic accident or attack at levels which are significantly lower than likely costs.

primary ambient air quality standards *Air quality standards that are designed to protect human health. Contrast SECONDARY AMBIENT AIR QUALITY STANDARDS.

primary consumer A *herbivore, which eats *primary producers and so sits above them in a *food chain. Contrast SECONDARY CONSUMER. See also HETEROTROPH.

Primary Drinking Water Regulation A regulation that applies in the US to public water systems and specifies a contaminant level which it is judged will not adversely affect human health.

primary economic activities Agriculture and mining. Contrast SECONDARY ECONOMIC ACTIVITIES, TERTIARY ECONOMIC ACTIVITIES.

primary effect See DIRECT EFFECT.

primary energy Energy that exists in natural resources (such as *coal, *crude oil, *sunlight, *uranium) that has not been converted or transformed by humans.

primary forest A *forest that is largely undisturbed by human activities. Contrast SECONDARY FOREST.

primary market The *marketplace into which shares are sold when they are first issued. Under the *Kyoto Protocol, the exchange of *emission reductions, *offsets, or *allowances between buyer and seller where the seller is the originator of the supply. Contrast SECONDARY MARKET

primary metabolite Any compounds which occurs in all living organisms and are essential for life, such as *carbohyd-

rates, the essential *amino acids, and *polymers derived from them.

primary mineral 1. A *mineral such as *feldspar or *mica which occurs or occurred originally in an *igneous rock.
2. Any *mineral which occurs in the *parent material of the *soil. Contrast SECONDARY MINERAL.

primary nutrient The elements *phosphorus, *potassium, and *sodium, which must be taken up and used by plants in sufficient quantities for them to complete their *life cycles. Contrast SECONDARY NUTRIENT.

primary pollutant Chemicals (such as *carbon monoxide, *hydrocarbons, *nitrogen compounds, *particulate matter, and *sulphur dioxide) that are released directly into the air in a harmful form. Contrast SECONDARY POLLUTANT.

primary producer An *autotroph that captures energy from the environment and turns it into *biomass through *photosynthesis or *chemosynthesis, and forms the base (lowest *trophic level) of a *food chain. Examples include green plants on land and *phytoplankton in oceans. See also CHEMOTROPH.

primary production See PRIMARY PRODUCTIVITY.

primary productivity The amount of new plant *biomass that is formed by *photosynthesis or *chemosynthesis in a given period of time. Primary productivity varies a great deal from one type of *ecosystem to another. The estimated average net primary productivity for the world (in grams per square metre per year, $g\ m^{-2}\ year^{-1}$) is 303; the average for major world biomes is reefs and estuaries 2000, forest 1290, freshwater 1250, cultivated land 650, woodland 600, grassland 600, tundra 140, open ocean 125, desert three. For cultivated ecosystems the estimated average net primary productivity (in $g\ m^{-2}\ year^{-1}$) is sugar cane 1726, rice 496, potatoes 402, wheat 34. Also known as **primary production**. Contrast SECONDARY PROD-

UCTIVITY, TERTIARY PRODUCTIVITY. *See also* GROSS PRIMARY PRODUCTION, NET PRIMARY PRODUCTION.

primary range A *grazing area (*range) which livestock prefers to use because it contains vegetation they like. When there is limited or no *range management animals are likely to over-graze these areas.

primary rock *See* IGNEOUS ROCK.

primary standard The pollution limit for *criteria air pollutants that is based on health effects. *Contrast* SECONDARY STANDARD. *See also* NATIONAL AMBIENT AIR QUALITY STANDARDS.

primary succession An ecological *succession that begins on a newly formed soil or on a new surface that has been exposed for the first time (for example as the result of a landslide or volcanic eruption). *Contrast* SECONDARY SUCCESSION.

primary treatment The initial treatment that is provided at a *wastewater treatment plant, in which solids and liquids are separated from *domestic sewage by screening and sedimentation. Also known as **primary waste treatment**. *See also* ADVANCED WASTEWATER TREATMENT, PRIMARY WASTE TREATMENT, SECONDARY TREATMENT.

primary waste treatment *See* PRIMARY TREATMENT.

primary wave *See* P-WAVE.

primate Any of the 195 species of *mammal that belongs to the order Primates, which includes monkeys, apes, and humans.

prime agricultural land The best available land which is capable of producing acceptable yields of *crops with acceptable inputs and minimal *environmental damage. Also known as **prime farmland**.

prime farmland *See* PRIME AGRICULTURAL LAND.

primeval Having existed from the earliest times. *See also* PRIMORDIAL.

Primitive Area A large tract of land within a US *National Forest, that has been set aside for *preservation in its natural condition with no alteration or development permitted except for measures to protect the spread of fire. Such areas were reclassified as *Wilderness Areas under the 1964 *Wilderness Act.

primitive area A category of land defined within the *recreation opportunity spectrum as an 'essentially unmodified natural environment of fairly large size', with few interactions between users. Motor vehicles are not permitted, and the area is managed to be largely free from evidence of human control.

primitive character In *cladistics, a character that is shared among members of a large group (*clade) and is thought to have arisen early in the *evolution of the group.

primitive cultivar A crop that has been developed from a *landrace.

primitive recreation Those types of *recreation activities that are associated with land having no roads, which includes hiking, backpacking, and cross-country travel.

primitive ROS A classification of *wilderness and *recreation opportunity on the ROS (*recreation opportunity spectrum) scale, based on an unmodified environment, where trails may be present but structures are rare, and where contact with other people is unlikely.

primordial First created or developed. *See also* PRIMEVAL.

principle of competitive exclusion *See* COMPETITIVE EXCLUSION PRINCIPLE.

principle of limiting factors The principle that the factor (such as a particular *nutrient, *water, or *sunlight) that is in shortest supply (the *limiting factor) will limit the growth and development of an *organism or a *community.

prion A small abnormal *protein that is found in brain cell membranes, which is responsible for mad cow disease (*Bovine Spongiform Eucaphalopathy) and causes new variant *Creutzfeld–Jakob disease in humans.

prior appropriation A doctrine of water law that allocates the rights to use a particular body of water on a first-come first-served basis.

prior informed consent In the context of the international trade in *hazardous chemicals, a procedure that seeks to share responsibility between exporting and importing countries in protecting *human health and the environment by sharing information about the hazards associated with those chemicals, and encouraging importing countries not to proceed without agreement.

priority habitat A category in the *EU Habitats Directive that is applied to *habitats that are deemed to be of conservation importance within the *European Union.

priority pollutant A group of around 130 chemicals (of which about 110 are *organics) that appear on a US *Environmental Protection Agency list because they are toxic and relatively common in industrial discharges.

priority species Any species that is targeted in the *UK *Biodiversity Action Plan (BAP) for national conservation effort, and is the subject of a *Species Action Plan or *Species Statement in the UK BAP.

prisere The whole sequence of *sere *communities in *succession.

pristine Natural, unspoiled, or undeveloped.

private forest *Forest land that belongs to individuals and corporations.

private good Any good or service whose consumption by one person excludes consumption by others. *Contrast* PUBLIC GOOD.

private land Land that belongs to individuals and corporations.

private opportunity cost The *opportunity cost of using a resource that is experienced by an individual, excluding any *externalities.

private property Things (*property or goods) that belong to individuals as opposed to public ownership, which gives those individuals the exclusive right to possess, use, or dispose of those things.

private value The value to an individual of using or not using a particular resource.

privatization The process of changing a government-owned resource (such as a *National Forest) to private control and ownership.

process waste Any *toxic pollutant or combination of pollutants that is an unavoidable product from any manufacturing process.

producer An *organism that synthesizes food molecules from inorganic compounds using an external energy source, usually by *photosynthesis. *Contrast* CONSUMER. *See also* PRIMARY PRODUCER.

product stewardship A product-centred approach to *environmental protection, which expects those in the product *life cycle (manufacturers, retailers, users, and disposers) to share responsibility for reducing the *environmental impacts of products.

product water *See* FINISHED WATER.

production well A *well that has a high enough *yield that it can be used for public use, either for water supply or for industrial purposes.

productive Describing an area that is to provide goods and services and to sustain *ecological values.

productivity In general, the rate of output per unit of input, a measure of efficiency. In ecology, the rate of production of *biomass. *See also* GROSS PRIMARY PRODUCTIVITY, NET PRIMARY PRODUCTIV-

ITY, PRIMARY PRODUCTIVITY, SECONDARY PRODUCTIVITY, TERTIARY PRODUCTIVITY.

profile 1. A cross-section through *topography or a *landscape.
  2. A vertical section through a *soil from the surface into the relatively unaltered material. *See also* SOIL PROFILE.
  3. Variations in the physical and chemical properties of water at different depths within a *waterbody.

profundal zone The bed of a body of deep water, where insufficient light penetrates for rooted plants to become established.

progeny Children or offspring, the immediate *descendants of a plant or animal (including humans).

pro-glacial The area adjacent to (usually in front of) a *glacier.

pro-glacial lake A lake in front of a retreating *glacier or *ice sheet, which is usually fed by *meltwater. *See also* GLACIAL LAKE.

programme evaluation and review technique (PERT) *See* CRITICAL PATH METHOD.

programmed harvest The part of the *potential yield of a *crop that is planned for *harvest in any particular year.

prohibition zoning Any land use *zoning regulations which only allow particular types of use in certain kinds of areas, such as *floodplain zoning that only allows uses which do not reduce the flood water storage capacity.

project management The deliberate planning, control, and coordination of all aspects of a project (initiating, planning, executing, controlling, and closing), in order to achieve the agreed objectives.

project scenario The emission forecast made in an *emission reduction project, which is compared with the *business-as-usual (*baseline) scenario in order to determine the emission reductions that have been achieved.

projection A *forecast that is based on historic and current information. *Contrast* PREDICTION.

prokaryote An *organism that does not have a nucleus containing genetic information. *See also* EUKARYOTE.

proleg The fleshy, unsegmented leg of a caterpillar.

promontory A narrow strip of land that extends out into the sea. Also known as **headland**.

propagate 1. To increase the number of plants of a particular population by means of cuttings.
  2. The breeding of new lifestock through natural methods.

propagation Increasing or spreading of a plant or plants by natural reproduction.

propagule A part of a plant that can produce another plant which includes *seeds, *roots, and *rhizomes.

propane (C_3H_8) A colourless, heavy gaseous *hydrocarbon that is found in *crude oil and *natural gas, and is used as *fuel and in *petrochemicals.

propellant Something that propels, such as a gas (such as *butane, *propane, and *nitrogen oxides) that has a high *vapour pressure and is used to force material out of *aerosol spray cans.

proper resource pricing The pricing of *natural resources at levels which reflect their combined *economic values and *environmental values.

property A place or an object that is owned.

property rights The rights of an owner of *private property, which usually include the right to use the property as they see fit (within limits, such as *zoning) and the right to sell it when and to whom they decide to.

proportionality principle The concept that control measures or a response should generally be proportional to the risk.

proposed species In the USA, any species of fish, wildlife, or plant that is proposed to be listed under the *Endangered Species Act.

prospect 1. To search for something useful or valuable.
2. The possibility of future success.

prospecting The search for *mineral deposits, for example by removing *overburden, or by drilling cores to obtain subsurface samples.

prospective study A study that follows a group (*cohort) of individuals over time, in order to determine whether exposure to particular factors produces adverse effects on the group or on different subgroups. Also known as **cohort study**.

prospector A person who explores an area for useful or valuable *resources, such as *mineral deposits.

protandry A state in *hermaphroditic organisms that is characterized by the development of the male stage before the appearance of the corresponding female feature, thus ensuring against self-fertilization. *Contrast* PROTOGYNY.

protected area Land that is set aside, usually for *conservation purposes, where development is banned or seriously restricted. There are many different types of protected areas, ranging from *national parks (where conservation is one of a number of priorities) to *nature reserves and *biosphere reserves (where it is the main objective).

Protected Area Category Ia An *IUCN category of *protected area, a strict *Nature Reserve, which is a protected area managed mainly for scientific research or monitoring.

Protected Area Category Ib An *IUCN category of *protected area, a *Wilderness Area, which is a large area in an unmodified or slightly modified state which is protected and managed so as to preserve its natural condition.

Protected Area Category II An *IUCN category of *protected area, a *National Park, which is a protected area managed mainly for *ecosystem conservation and *recreation, which provides a foundation for scientific, educational, spiritual, recreational, and visitor opportunities, all of which must be environmentally and culturally compatible.

protected area system A system that legally establishes *protected areas and sites which are managed for *conservation objectives.

Protected Landscape or Seascape An *IUCN Management Category (V) for protected areas, which is designed 'to maintain nationally significant natural landscapes that are characteristic of the harmonious interaction of man and land while providing opportunities for public enjoyment through recreation and tourism within normal life style and economic activity of these areas'.

protection Defence against harm or danger: any activity (including insect and disease control, fire protection, and law enforcement) that reduces losses or risks, tends to maintain basic conditions and values, and reduces damage or injury to people and property.

protective plant Plants that are grown to protect crops, soils, or land from environmental damage.

protein A large organic molecule composed of one or more chains of *amino acids in a specific order, which is required for the structure, function, and regulation of cells, tissues, and organs. Each protein has unique functions. Examples include *hormones, *enzymes, and *antibodies.

Proterozoic An *aeon of geological time that followed the *Archaean, extending from 2.5 billion years ago to about 570 million years ago (the start of the *Phanerozoic). During this time period the first bacteria and fungi and primitive multicelled organisms appeared and developed. Proterozoic means 'early life'.

protese inhibitor A drug that inhibits the ability of a *virus to make copies of itself.

protist Any organism that belongs to the kingdom *Protoctista.

Protista See PROTOCTISTA.

protocol A code of correct conduct, or series of formal steps for conducting a test. See also MONTREAL PROTOCOL.

Protocol on Preparedness, Response and Co-operation to Pollution Incidents by Hazardous and Noxious Substances (2000) An international *protocol that follows the principles of the *International Convention on Oil Pollution Preparedness, Response and Co-operation (1990), and seeks to ensure that ships carrying *hazardous and noxious liquid substances are covered, or will be covered, by regimes similar to those already in existence for oil incidents. Also known as the **HNS Protocol**.

Protoctista In *taxonomy, the *kingdom that includes all of the single-celled *eukaryotes (the *protozoa, *slime moulds, and *eukaryotic algae), as distinct from the multicelled *Plantae, *Fungi, and *Animalia. Previously known as Protista.

protogyny A state in *hermaphroditic systems that is characterized by the development of female reproductive organs before the appearance of the corresponding male feature, thus ensuring against self-fertilization. Contrast PROTANDRY

proton A subatomic particle that is found in all *atoms and has a positive electric charge. See also ELECTRON, NEUTRON.

protoplasm A term previously used to describe the components of living tissues. See also CYTOPLASM.

protoplast A plant, bacterial, or fungal cell that has had its cell wall removed.

protozoa Microscopic, usually single-celled *micro-organisms that are larger and more complex than *bacteria, live in the soil or freshwater, and are non-photosynthetic and feed upon dead or live bacteria and fragments of organic matter. Some protozoa cause human disease, including *malaria and *sleeping sickness. A member of the kingdom *Protoctista.

proven reserve Deposits of a mineral resource (such as a *fossil fuel) whose location and extent are known, as opposed to potential but unproved (*discovered) deposits, and which can be extracted economically.

provenance Source area.

province A political or administrative region within a country.

provincial rank (S-rank) The conservation status rank of a species, subspecies, or variety of plant or animal within a country, based largely on the total number of known sites within that country and the degree to which they are potentially or actively threatened with destruction. See also GLOBAL RANK.

provinciality effect Increased *diversity of species caused by geographical isolation.

proximate cause The special or effective cause of a particular change, such as the combustion of *fossil fuel, which causes increased levels of *carbon dioxide in the atmosphere, which in turn causes *global warming.

proximate factor The general cause of a particular change, such as variations in day length, which affects the behaviour or organisms in various ways. Contrast ULTIMATE FACTOR.

proximity principle The principle that *wastes should be treated close to where they arise (ideally within the boundary of the plant or community in which they are generated), in order to avoid exporting them to other places.

proxy Substitute or surrogate.

proxy data Data that are used to represent a situation, phenomenon, or condition for which no direct information (such as instrumental records) is available.

proxy indicator An *indicator that is used to provide *proxy data. There are four main types of proxy climate indicator, for example, which are historical (oral or written records), biological (records of faunal and floral growth and distribution), geological (terrestrial deposits and features, and marine ocean sediment cores), and glaciological (*ice cores).

PRP *See* POTENTIALLY RESPONSIBLE PARTY.

pruning Cutting live or dead branches from plants (particularly the branches of trees, or the sides and tops of *hedges), in order to improve the quality of the growth. *See also* THINNING OUT.

psammosere The vegetation *succession that occurs on moving sand such as also *sand dunes.

PSD *See* PREVENTION OF SIGNIFICANT DETERIORATION.

PSI *See* POLLUTION STANDARDS INDEX.

psychophily *Pollination by butterflies.

psychrometer An instrument consisting of a wet-bulb and a dry-bulb thermometer that is used to measure the amount of *water vapour in the air.

PTD *See* PARTICIPATORY TECHNOLOGY DEVELOPMENT.

pterygote An insect with wings or a wingless insect that is believed to have evolved from winged ancestors. *Contrast* APTERYGOTE.

public estate *See* PUBLIC LAND.

public forest *Forest land in the USA that belongs to municipal, state, and federal governments.

public good Any good or commodity whose benefits are available to everyone, from which no one can be excluded, and from which no single individual can enjoy all of the benefits.

public health The health or physical well-being of a whole community.

public land Land that is owned in common by all and is managed by the government (town, county, state, or federal). Also known as **public estate**.

Public Rangelands Improvement Act (1978) US legislation that amended the *Federal Land Policy Management Act (1976) to provide the *US Bureau of Land Management and the *US Forest Service with additional direction and authorities in managing public rangelands.

Public Utilities Regulatory Policy Act (PURPA) (1978) US legislation that requires utilities to buy power from eligible *cogeneration sources, small hydro, or waste-fuelled facilities, under contracts, at an agreed cost. It also requires utilities to provide a backup supply of *electricity to customers who choose self-generation.

publicly owned treatment works (POTW) A *municipal *wastewater treatment facility.

puddle 1. A small shallow area of water on the ground surface.
2. To destroy the structure of the surface soil by physical methods such as the impact of raindrops, poor cultivation with implements, and trampling by animals.

puddled soil A dense *soil that is almost *impervious to air and water, having been *puddled when wet.

pulp Raw material (cellulose *fibres) that is made from *trees and used in producing paper products.

pulpwood Low quality *softwood that is harvested for use as a source of *fibre in the manufacture of *pulp, for making products such as paper and textiles.

p

pulse *See* LEGUME.

pumice A light-coloured *volcanic rock that contains many trapped gas bubbles formed by the explosive eruption of *magma, is light enough to float on water, and can develop as layers to form *tuff when hardened.

pump and treat technology A treatment method in which *contaminated water is pumped out of the ground and then treated before being discharged back into the environment.

pumped storage scheme A *hydroelectric scheme in which the water that flows though the *turbines is recycled by pumping it back to the upper storage reservoir during off-peak periods, where it can be used again to produce *electricity during peak periods.

pumping level The level of the water surface in a *well when pumping is in progress.

pumping test A test that is conducted in a *well in order to determine the characteristics of the *aquifer or well.

punctuated equilibrium A model of *evolution in which change occurs in short, relatively rapid phases, followed by longer periods of stability (*stasis). This model helps to explain some important discontinuities in the *fossil record, including *mass extinctions (in which many different species appear to have died out at roughly the same time), and the lack of intermediate forms (fossil evidence of the existence of transitional forms between species, which would indicate stages in the long-term evolution of new species). *Contrast* PHYLETIC GRADUALISM.

pupa In an *insect, the stage of metamorphosis between being a larva and an adult, during which the body is transformed.

pupal paedogenesis The production of young by *pupae.

purebred An *animal (usually a horse) that has been bred for many generations

from members of the same *breed or strain.

purification The process of making something pure; free from anything that debases, pollutes, or contaminates.

PURPA *See* PUBLIC UTILITIES REGULATORY POLICY ACT.

purse-seining A long fishing net used at sea to encircle and trap fish, but which also often catches dolphins and other marine species.

putrefaction The biological *decomposition of organic matter under *anaerobic conditions, which produces offensive smells. *See also* BUTYRIC FERMENTATION.

putrescible Material (usually *waste) that is able to rot quickly enough to cause offensive smells and attract flies.

PVC *See* POLYVINYL CHLORIDE.

P-wave A type of *seismic wave, caused by an *earthquake, which is refracted (bent) as it passes through the liquid outer *core of the Earth and can be detected and measured at *seismograph stations. Also known as **primary wave**. *Contrast* S-WAVE.

PWR *See* PRESSURIZED WATER REACTOR.

pycnocline A layer in the *ocean where the water density increases rapidly with depth, which acts as a strong barrier to vertical mixing of seawater.

pyramid of biomass An *ecological pyramid that shows the relationship between total *biomass (weight of organic matter) and *trophic level within a *food web: biomass usually decreases at higher levels, because whilst the individual organisms are larger, there are fewer of them.

pyramid of energy An *ecological pyramid that shows the relationship between *energy (in calories per unit of area (square metres) over a given period of time, usually a year) and *trophic level within a *food web, as energy is lost through respiration so that relatively little of the original energy is ultimately available to the top *carnivores.

pyramid of numbers An *ecological pyramid which shows the relationship between the number of individuals and their *trophic level within a *food web. The number of individuals usually decreases at each level; a large number of *primary producers supports a small number of larger *consumers.

pyranometer An instrument which is used to measure *diffuse and direct *solar radiation.

pyrethrin A form of *organic *insecticide which is derived from certain species of chrysanthemums and breaks down in sunlight in a few days. In high *doses it kills many garden insect pests, but many insects can recover from exposure to low doses.

pyroclast Broken fragments of rock that are thrown out during a *volcanic eruption, including *lapilli, *pumice, *volcanic ash, and *volcanic bombs.

pyroclastic flow A hot, dense mixture of *ash, *pumice, *rock fragments, and *gas that is formed during an *explosive volcanic eruption and flows downhill at great speed.

pyroclastic rock A rock that is formed by the accumulation of fragments of volcanic rock scattered by a *volcanic explosion.

pyrolysis The chemical *decomposition of a substance by heat in the absence of oxygen, which produces various *hydrocarbon gases and a carbon-like residue.

p

QELRC *See* QUANTIFIED EMISSIONS LIMITATION AND REDUCTION COMMITMENT.

quadrat A sampling frame or square that is marked out for quantifying the number or percentage cover of a particular species within a given area. *See also* TRANSECT.

quadruped An animal that walks on four legs.

qualitative Relating to or expressed in terms of quality, rather than quantity. *Contrast* QUANTITATIVE.

quality assurance A programme or set of procedures and activities that are designed to ensure an acceptable level of quality.

quality of life The level of well-being of life style, and the physical conditions in which people live. *Contrast* STANDARD OF LIVING.

Quantified Emissions Limitation and Reduction Commitment (QELRC) Legally binding targets and timetables defined under the *Kyoto Protocol for the limitation or reduction of *greenhouse gas emissions for *developed countries. Also known as **Quantified Emissions Limitation and Reduction Objectives** (QELROS).

Quantified Emissions Limitation and Reduction Objectives (QELROS) *See* QUANTIFIED EMISSIONS LIMITATION AND REDUCTION COMMITMENT (QELRC).

quantitative Relating to or expressed in terms of quantity (numbers), rather than quality. *Contrast* QUALITATIVE.

quarantine Enforced isolation or restriction of free movement of individuals, usually imposed to prevent the spread of a *contagious disease.

quarry 1. A surface excavation from which clay, stone, or minerals are extracted. *See also* MINE.
2. An animal that is hunted or caught for food.

quarrying 1. A process of *physical/mechanical weathering that involves repeated freeze–thaw cycles at the base of a *glacier or *ice cap, weakening the underlying rock, fragments of which are plucked out by the ice as it flows over it. Also known as **plucking**.
2. Industrial scale surface extraction, using machinery, of minerals such as limestone for aggregates in road building or china clay for the pottery industry.

quartz The most common rock-forming *mineral, composed of crystals of silicon dioxide (SiO_2), which may be glassy or opaque.

quartzite A resistant *metamorphic rock in which individual sand grains are cemented together by *quartz, formed by applying great heat and pressure to *sandstone.

Quaternary The most recent period of *geological time, covering the last 2 million years, which includes the *Pleistocene and the *Holocene. Also known as **Pleistogene**.

quiescence Dormancy, at rest.

race *See* SUBSPECIES.

RACM *See* REASONABLY AVAILABLE CONTROL MEASURES.

RACT *See* REASONABLY AVAILABLE CONTROL TECHNOLOGY.

rad Short for 'radiation absorbed dose'. A unit of measurement that was used for the *absorbed dose of *radiation. It has been replaced by the *gray (1 rad = 0.01 *joules per kilogram = 0.01 gray).

radar Short for radio detection and ranging, an instrument that is used to detect and determine the range to distant objects (including rainstorms) by measuring the strength of the *electromagnetic signal reflected back from it.

radial drainage A *drainage pattern in which the rivers radiate outwards from a central high point, such as the dome of a uniform rock outcrop or the centre of a *volcano.

radiant energy *See* RADIATION.

radiant heat transfer The movement of heat between two surfaces that are exposed to each other but are not touching, and that have very different temperatures.

radiation The transfer of heat and other energy by means of *electromagnetic waves. The Earth is warmed by shortwave radiant energy from the Sun, and it warms the overlying atmosphere by longwave *radiation. This heating process operates in much the same way as a domestic central heating radiator heats a room. Also known as **radiant energy**.

radiation balance The balance on Earth between incoming shortwave solar *radiation and outgoing longwave radiation. Also known as **radiative balance**.

radiation budget A measure of all the inputs and outputs of *radiation relative to a particular system, such as the *Earth.

radiation dose The amount of *radiation that is absorbed by a material, system, or tissue in a given amount of time, expressed in *roentgens. *See also* ABSORBED DOSE.

radiation exposure Most (82%) of the *radiation to which people are exposed comes from natural sources, including radioactive elements in the Earth and in our bodies, cosmic radiation from space, and radon gas. Of the 18% that comes from human activities, most is from medical *X-rays and *nuclear medicine.

radiation fog *See* GROUND FOG.

radiation inversion An increase in temperature with height in the atmosphere that is caused by radiational cooling of the ground surface. Also known as **nocturnal inversion**.

radiation sickness The syndrome associated with intense acute exposure to *ionizing radiation, symptoms of which include nausea, diarrhoea, headache, lethargy, and fever.

radiation standard Regulations that define maximum exposure limits for protection of the public from *radioactive materials.

radiative balance *See* RADIATION BALANCE.

radiative cooling The process by which the surface of the Earth, and the air in contact with it, cools by the emission of longwave *radiation, which usually happens at night.

radiative forcing The increase or decrease in the *radiation balance of the Earth, which is caused by the presence of a particular atmospheric constituent, such as *carbon dioxide or *aerosol. An increase leads to heating of the Earth's surface, while a decrease leads to cooling. Also known as **forcing** or **external forcing**.

radiative transfer The movement of heat in the form of long-wavelength *infrared *radiation.

radiatively active gas Any gas (such as *water vapour, *carbon dioxide, *methane, *nitrous oxide, *chlorofluorocarbons, and *ozone) that absorbs incoming *solar radiation or outgoing *infrared radiation.

radical ecology The view that environmental problems can only be solved by a radical revision of attitudes and values, rather than through economic and political reform. See also DEEP ECOLOGY, ECOFEMINISM, SOCIAL ECOLOGY.

radio telemetry The science and technology of gathering and transmitting data automatically over great distances, for example from an artificial *satellite to a monitoring station on Earth.

radio wave A type of *radiant energy, with a wavelength of 3 to 30 GHz within the *electromagnetic spectrum, that is used to transmit radio and television signals.

radioactive Giving off or capable of giving off radiant energy (*radiation) in the form of particles or rays, as in *alpha particles, *beta particles, and *gamma rays.

radioactive decay The natural, spontaneous decay of the *nucleus of an atom to form another nucleus, which releases *alpha or *beta particles and/or *gamma rays at a fixed rate.

radioactive isotope See RADIOISOTOPE.

radioactive pollution *Contamination by a *radioactive material.

radioactive substance A substance that emits *ionizing radiation, and is thus *radioactive.

radioactive waste Waste material that is *radioactive, is produced as a *byproduct of *nuclear reactions through power generation and nuclear research, and is classified into *low-level waste, *intermediate-level waste, and *high-level waste. Also known as **nuclear waste**.

radioactivity The spontaneous decay or disintegration of an unstable atomic nucleus, which is accompanied by the emission of *radiation.

radiobiology The study of the principles, mechanisms, and effects of *radiation on living things. See also RADIOECOLOGY.

radiocarbon (^{14}C) A *radioactive *isotope of *carbon that is generated naturally in the atmosphere, has a * half-life of nearly 6000 years, and is used in *carbon dating.

radiocarbon dating See CARBON DATING.

radiochemical species Individual *radioactive elements that produce *radioactivity, such as radium-226, cobalt-60, strontium-90, and tritium.

radioecology The study of the effects of *radiation on plants and animals in natural communities. See also RADIOBIOLOGY.

radioisotope A *radioactive *isotope. Also known as **radionuclide**.

radioisotope dating See RADIOMETRIC DATING.

radiolaria Marine *protozoa that have skeletons made of *silica.

radiolarian ooze A silica-rich deep-sea sediment that is composed largely of the skeletons of *radiolaria. See also PELAGIC SEDIMENT.

radiolarite A *lithified sedimentary rock formed from *radiolarian ooze.

radiological Relating to *radiology.

radiological monitoring Locating and measuring *radiation using instruments that can detect and measure ionizing radiation.

radiological weapon See NUCLEAR WEAPON.

radiology The diagnostic and therapeutic use of *radiation.

radiometric dating A precise method of determining the *absolute age of rocks by measuring the different isotopes of elements within the rocks (such as *carbon-14, uranium-238, and uranium-235), and comparing them with known rates of decay of *radioisotopes. Also known as **radioisotope dating**.

radionuclide See RADIOISOTOPE.

radiosonde An instrument that is attached to a *weather balloon and transmits measurements of pressure, humidity, temperature, and winds as it ascends up to about 30 000 metres. See also RAWINSONDE.

radium (Ra) A rare radioactive *isotope (radium-266) which occurs in *uranium ores, such as *pitchblende, and has a *half-life of 1622 years.

radius of influence The maximum distance from a *well over which vacuum or pressure occurs, which influences the movement of soil gas or *groundwater. The distance between a water well and the limit of the *cone of depression.

radius of vulnerability zone The maximum distance from the point of release of a *hazardous substance, over which the concentration in the air could reach the *level of concern under specified weather conditions.

radon (Rn) A naturally occurring, colourless, odourless *radioactive gas (radon-222) which is about eight times denser than air, is formed by the *radioactive decay of *radium, and has a *half-life of 3.82 days. It is one of the more important natural sources of *radioactiv-ity in the environment, and causes health problems (including *cancer) in humans. Radon is released naturally from rocks, soils, and minerals that contain radium, and the gas is normally trapped in underground rocks, but when the rocks are weathered or crack the gas escapes. Some of it is dissolved in *groundwater, and can cause a health risk by contamination of water supplies. Most of the radon escapes directly into the lower atmosphere, where it is mixed, dispersed, and diluted. It normally disappears quickly, because of its short half-life and rapid dispersion by winds, preventing the build-up of dangerously high concentrations. The gas and its *radon daughter products emit *alpha particles, and may cause lung cancer.

radon daughter A short-lived radioactive decay product of *radon that decays into lead *isotopes, which can be spread through the air and cause health problems (particularly lung damage if inhaled) in humans. Also known as **radon decay product** or **radon progeny**.

radon decay product See RADON DAUGHTER.

radon progeny See RADON DAUGHTER.

RAFU See REASONABLY ANTICIPATED FUTURE USE.

rain Liquid *precipitation that falls as droplets of water. See also CONVECTION RAIN, FRONTAL RAIN, OROGRAPHIC PRECIPITATION, RAINFALL.

rain forest See RAINFOREST.

rain gauge An instrument that is used to measure the amount of *rainfall over a given period of time.

rain making The processes by which *rain is formed in the atmosphere.

rain shadow The dry *leeward slope beyond a mountain barrier that gives rise to *orographic precipitation.

rainbow An arc of coloured light that is formed in the sky when the Sun's rays are reflected and refracted (bent) by drops of water in rain or mist, and display the colours of the *visible light part of the *electromagnetic spectrum in sequence: violet (0.390–0.455 micrometres), blue (0.455–0.492 micrometres), green (0.492–0.577 micrometres), yellow (0.577–0.597 micrometres), orange (0.597–0.622 micrometres), red (0.622–0.770 micrometres).

Rainbow Warrior A ship that is operated as a campaigning and research vessel by *Greenpeace; the first *Rainbow Warrior* was sunk by the French secret service in Auckland harbour, New Zealand, on 10 July 1985.

raindrop A large *cloud droplet.

raindrop erosion *Erosion of exposed *soil on a bare, sloping surface that is caused by the impact of *raindrops, which loosens and splashes particles and moves them downslope. Also known as **rainsplash** or **splash erosion**.

rainfall 1. A form of *precipitation in which *rain falls as *raindrops.
  2. The quantity of *precipitation that falls as *rain in a given period of time, usually expressed as depth of water (millimetres).

rainfall intensity The rate at which rain is falling at any given point in time, usually expressed in millimetres per hour (mm h^{-1}).

rainfall interception *See* INTERCEPTION.

rainfed farming Growing crops or animals under conditions of natural *rainfall, without the use of *irrigation.

rainforest A large, dense, evergreen forest that grows in a climate which has high humidity, constant temperature, and abundant rainfall (generally over 380 centimetres per year). There are both *tropical rainforests (such as the Amazon) and *temperate rainforests (such as in the Pacific Northwest of North America).

Rainforest Action Network (RAN) An international *environmental group

that works to protect the Earth's *rainforests and support the rights of their inhabitants.

Rainforest Alliance An international *environmental group whose objective is to protect *rainforest *ecosystems and the people and wildlife that depend on them by transforming land use practices, business practices, and consumer behaviour.

rainout The removal of a *pollutant within clouds by *precipitation scavenging.

rainsplash *See* RAINDROP EROSION.

rainstorm A *storm with *rain.

rainwash *See* WASH.

raised beach A former *beach that is now situated above the present *sealevel as a result of earth movement or changes in global sea level.

raised bog A type of *bog that has a shallow dome-shaped cross-section, so that the bog surface (at least in the centre) is raised above the normal level of the *groundwater.

RAM *See* RELATIVE ATOMIC MASS.

Ramsar Convention An international convention on wetlands of international importance (1971), which was designed to promote the *conservation of internationally important *wetland *habitats, particularly in order to protect *waterfowl and other migratory water birds.

Ramsar site A *wetland that is of international importance and has been designated under the *Ramsar Convention.

RAN *See* RAINFOREST ACTION NETWORK.

random Lacking any predictable order or plan.

range 1. *See* RANGELAND.
  2. *See* NATURAL RANGE.
  3. *See* STATISTICAL RANGE.

range allotment *See* ALLOTMENT.

range condition The status of the vegetation and soil of a particular area of *rangeland, compared with the natural potential (*climax) for that site.

range management The management of *rangeland in order to sustain maximum livestock production and *conserve the *natural resources.

range of variability The observed limits of change in the composition, structure, and function of an *ecosystem over a given period of time, which results from both natural and human factors.

rangeland A large area of *grassland or open *woodland on which livestock can *graze and *browse. *See also* OPEN RANGE, PRIMARY RANGE, SECONDARY RANGE.

ranger A person who is responsible for managing and protecting a forest or park (such as a *national park), enforcing the laws that apply there, and offering information to visitors.

rank vegetation *Grassland or *marsh vegetation that has grown abundantly without being cut or grazed for some time, and as a result has become tall, tussocky, and dominated by coarse species of grass.

rapid needs assessment A group of techniques that are designed to document community needs (such as for food, shelter, healthcare, electricity, radio, and telephones) following a natural *disaster, usually within 48 hours.

rapid rural appraisal *See* PARTICIPATORY RURAL APPRAISAL.

rapids A section of a *river where *slope and flow *velocity both increase, so there is much *white water and many exposed rocks.

raptor A bird of prey (such as a falcon, hawk, eagle, or owl) that hunts and eats meat.

raptorial Adapted for catching and holding prey, often by having sharp claws and spines or bristles.

RARE *See* ROADLESS AREA REVIEW AND EVALUATION.

rare A *threatened category of species defined by the *IUCN (*Red Data Book) as 'taxa with small world populations that are not at present *endangered or *vulnerable, but are at risk. These taxa are usually localized within restricted geographical areas or habitats or are thinly scattered over a more extensive range'.

RARE II *See* ROADLESS AREA REVIEW AND EVALUATION II.

rare and endangered species *Species that have small *populations, either naturally or as a result of human impacts.

rare gas A gas (such as helium or neon) that is *inert or unreactive, does not readily take part in chemical reactions, and is thus very stable. *See also* INERT GAS, NOBLE GAS.

rarity In *conservation assessment, seldom found or occurring.

rarity rank A rank (*G-rank, *N-rank, or *S-rank) that is assigned to a *species or ecological *community to indicate its degree of *rarity at the global, national, or subnational level, respectively.

ratification Making something valid by formally ratifying or confirming it, for example after a country has signed a particular *convention, *protocol, or *treaty.

rattan A climbing *palm that has very long tough stems and is used for making baskets and furniture.

ravine A deep, narrow, steep-sided valley that has been eroded by running water.

raw humus *Humus that consists mainly of well preserved but fragmented plant remains.

raw land A North American term for undeveloped land or land that has not been subdivided in preparation for development. Also known as **undeveloped land**.

raw material Natural, unprocessed material or resources that are converted by a manufacturer into a final product.

raw sewage Untreated domestic or commercial *wastewater.

raw water Untreated surface water or *groundwater.

rawinsonde A *radiosonde that is tracked to measure variations in *temperature, *pressure, *relative humidity, and wind *speed and direction at different levels within the *atmosphere.

RAWS *See* REMOTE AUTOMATIC WEATHER STATION.

Rayleigh scattering Scattering within the *atmosphere that is caused by gas molecules and fine particles of *dust, which reflects *radiation in all directions and makes the sky appear blue.

RCRA *See* RESOURCE CONSERVATION AND RECOVERY ACT.

reach A stretch of stream or river that lies between inflowing *tributaries.

reaction 1. A chemical change: the interaction of two or more substances to form new substances.
2. The fourth stage in plant *succession.

reactive waste *Hazardous waste which is normally unstable, reacts violently with water, can give off toxic fumes, and can explode if heated under pressure.

reactivity The ability of a material to undergo a chemical *reaction with the release of *energy, for example as a result of mixing or reacting with other materials, application of heat, or physical shock. One of the four *characteristics that is used to define *hazardous waste in North America.

reactor core The innermost part of a *nuclear reactor that consists of the fuel, the *moderator (in a thermal reactor), and a coolant, where the *fission reaction takes place and the level of radiation is highest. Also known as the **core**.

readiness In *disaster management, the phase of preparations that is designed to deal with an accident or incident.

reaeration The absorption of *oxygen into water from the *atmosphere.

reafforestation *See* REFORESTATION.

reagent A substance that is used to cause a *reaction, especially to detect another substance.

realized niche The actual distribution of an *organism in the environment. *Contrast* POTENTIAL NICHE.

real-time monitoring Monitoring and measuring events, processes, or changes as they are happening.

reasonable maximum exposure (RME) The maximum exposure to a particular *pollutant that is reasonably expected to occur at a particular site, such as a *Superfund site, combining all relevant pathways.

reasonably anticipated future use (RAFU) The likely future use of a site or facility given current use, local government planning, and *zoning.

reasonably available control measures (RACM) A broad term that is used in the USA to refer to technological and other measures for *pollution control that are reasonably available for use, technologically feasible, and economically viable.

reasonably available control technology (RACT) Pollution control technology that is reasonably available for use, technologically feasible, and economically viable.

rebound A recoil or movement back from an impact, particularly the *upwarping of the Earth's crust after additional weight (for example during *glaciation) is removed from it. *Contrast* SUBSIDENCE.

REC *See* RENEWABLE ENERGY CERTIFICATE.

recalcitrant A pollutant or substance that is not *biodegradable or is only biodegradable with difficulty, and that therefore accumulates in water, soil, and vegetation.

recalcitrant seed Seed that does not survive drying and freezing for long-term storage.

recarbonization A process in which *carbon dioxide is added to water that is being treated to lower the *pH.

receiving water Any body of water into which *wastewater or treated *effluent are discharged.

receptor A species, population, community, habitat, or ecosystem that may be exposed to one or more particular *contaminants or *agents.

receptor site The site to which individuals of an *introduced species are taken in order to be introduced from another place (the *donor site).

recessional moraine See END MORAINE.

recharge To refill a *waterbody or an *aquifer with water by natural or artificial processes. See also GROUNDWATER RECHARGE.

recharge area An area across which water enters (*recharges) an *aquifer via *infiltration. Also known as **recharge zone**.

recharge rate The quantity of water per unit of time that refills an *aquifer.

recharge zone See RECHARGE AREA.

reclaim To return something (such as land) to its original condition.

reclaimed land Land that has been brought into new productive use through *reclamation.

reclamation 1. See LAND RECLAMATION. 2. See WASTE RECLAMATION.

recognition species concept The definition of *species as a set of organisms that recognize one another as potential mates, through a shared mate recognition system (such as a particular bird song). See also BIOLOGICAL SPECIES CONCEPT, CLADISTIC SPECIES CONCEPT, ECOLOGICAL SPECIES CONCEPT, PHENETIC SPECIES CONCEPT.

recombinant bacteria *Micro-organisms whose genetic makeup has been altered by the deliberate introduction of new *genetic elements, so that their offspring also contain these new genetic elements.

recombinant DNA *DNA that has been genetically engineered, such as new DNA produced by joining together pieces of DNA from different sources.

recombinant DNA technology See GENETIC ENGINEERING.

recombination The formation of new combinations of *genes, which occurs naturally in plants and animals during the production of sex cells (sperm, eggs, *pollen) and their subsequent joining in fertilization. It also occurs artificially, through *genetic engineering.

recommended maximum contaminant level (RMCL) The maximum level of a *contaminant in drinking water at which no known or anticipated damage to human health would occur.

reconnaissance survey A preliminary inspection or survey of an area (such as a *forest, *range, *watershed, or wildlife area), usually executed rapidly and at a relatively low cost, in order to gain general information that will be useful for future management.

reconstruction Rebuilding structures that have been damaged or destroyed, for example by a *natural hazard or *disaster.

record of decision (ROD) A document, associated with an *environmental impact statement (EIS), that publicly and officially discloses the decision of the responsible official concerning which alternative assessed in the EIS is to be implemented.

recoverability The ability of a *habitat, *community, or individual to recover from damage that is caused by an external factor.

recoverable Capable of being recovered, regained, or restored.

recoverable resource A *resource that can be recovered with current *technology, but whose recovery is not economical under current conditions.

recovered material Waste materials and *byproducts which have been recovered or diverted from *solid waste.

recovery 1. Return to an original state, such as the natural capacity of an open system for self-repair.
2. In *conservation, the process by which the decline of an *endangered or *threatened species is arrested or reversed, and threats neutralized, so that its survival in the wild can be ensured.
3. In *mineral processing, the amount of mineral that is separated and recovered in a mill, as a percentage of that calculated to be the original ore.
4. In *waste management, the process of obtaining materials or energy resources from solid or liquid waste.
5. In *disaster management, the coordinated process of supporting communities that have been affected by disaster in *reconstruction of the physical infrastructure and restoration of emotional, social, economic, and physical well-being.

recovery plan 1. A *plan to restore areas that have been affected by a particular *disaster.
2. A plan for the conservation of an *endangered or *threatened species.

recovery rate The percentage of usable materials in a *solid waste stream that is recovered and *recycled, as a percentage of the total waste.

recreation Any activity that refreshes, satisfies, and brings enjoyment to people, in which they engage on a voluntary basis during *leisure time. See also CONCENTRATED RECREATION, DISPERSED RECREATION, INTENSIVE RECREATION, PRIMITIVE RECREATION.

recreation land An area of land and water that is used primarily for *recreation.

recreation opportunity spectrum (ROS) A continuum of land types that is used in the USA to characterize *recreation opportunities in terms of setting, activity, and experience opportunities. The spectrum contains six classes: *primitive area, *semi-primitive non-motorized area, *semi-primitive motorized area, *roaded natural area, *rural area, and *urban area.

recreation resource A scenic or *wilderness feature or setting that has significance or value for *recreation, or a recreation facility.

recreation visitor day See VISITOR DAY.

recreational area An area of land or water that has been set aside for use by the public for *recreation activities and enjoyment.

recruitment The addition of new individuals to a *population by reproduction, which is commonly measured as the proportion of young in the population just before the *breeding season.

rectangular drainage pattern A *drainage pattern that is similar to the *trellis pattern, but with the pattern largely controlled by *faults or *joints that intersect at high angles.

recurrence interval (RI) The average period of time, usually measured in years, between two successive *floods of a given size (*discharge) at a particular location within a river system, as calculated by *flood frequency analysis. Also known as **return period**.

recurrent selection A *breeding method that is designed to increase the frequency of favourable *genes within a species, via repeated cycles of *selection and selected crossing of individuals.

recyclable Any material (such as glass, paper, *aluminium, and some plastics) that can be collected and sold for *recycling at a net cost no greater than the cost of collection and disposal.

recycle To reclaim or reuse old materials in order to make new products.

recycling The reprocessing of discarded *waste materials for reuse, which involves collection, sorting, processing, and conversion into raw materials which can be used in the production of new products.

recycling mill A facility where recovered (*recycled) materials are reprocessed into new products.

recycling scheme An organized scheme for *recycling solid waste materials. *See also* KERBSIDE RECYCLING SCHEME.

red bag waste *See* INFECTIOUS WASTE.

Red Data Book A global list (*Red List) of *endangered and *threatened species, which is produced and regularly updated by the *World Conservation Union (IUCN). It contains colour-coded information sheets arranged by species: red for species that are *endangered, amber for *vulnerable, white for *rare, green for out of danger, and grey for species that are indicated to be endangered, vulnerable, or rare but with insufficient information to be properly classified. The three main groups of categories are *threatened, *unknown, and *not threatened.

Red Flag warning A term used by *weather forecasters to alert land managers to weather conditions that could lead to extensive *wildland *fires.

Red List The list of *endangered species that are included within a *Red Data Book.

red tide A reddish-brown discoloration of surface sea waters, usually along the *coast, caused by *nutrient pollution that encourages abnormally rapid growth and thus high concentrations of certain *dinoflagellates that produce *toxins which can be fatal to fish and other organisms. Such a tide can be red, green, or brown, depending on the coloration of the plankton; it is often transported by real *tides.

redox Short for reduction–oxidation potential, which is a measure of how easy it is for organic reactions to take place and is used as an indication of *water quality.

redox potential A measurement of the state of *oxidation of a *system.

reduced tillage system A *tillage system (such as *minimum tillage) that preserves soil, saves energy and water, and increases crop yield.

reducer organism *See* DECOMPOSER.

reduction A chemical process that involves the addition of *hydrogen, removal of *oxygen, or the addition of *electrons to an element or compound. *Contrast* OXIDATION.

reductionism The idea, which underpins modern science, that it is possible to explain phenomena in terms of their component parts, so that complex systems can be understood in terms of their constituents. *Contrast* HOLISM.

redwood The soft red wood of either of two species of huge coniferous *sequoia trees.

reed A tall, woody, *perennial *grass that has a thin hollow stem.

reedbed *Wetland dominated by stands of common reed (*Phragmites australis*).

reef A ridge of rocks, often made of *coral, that lies submerged near the surface of the sea and can pose a hazard to passing ships.

reef ball An artificial *reef that is built to restore damaged *coral reefs, and to create new fishing and scuba diving sites.

reef knoll A dome-like mass of *limestone that has grown upwards from an active *coral reef.

reference site A sampling site that is selected for its relatively undisturbed conditions, which provides a *benchmark in comparative studies.

reference year The *benchmark year on which *environmental quality targets are established. For example, the *Kyoto Protocol uses 1990 as the reference year against which *Annex I nations are required to control their emissions of *greenhouse gases.

refine To remove unwanted substances from something such as *petroleum, which is refined to produce *petrol, *paraffin, and *diesel oil.

refinery A factory where a substance such as *petroleum is refined. *See also* PETROLEUM REFINERY.

reflection The return of light or sound waves from a surface.

reflectivity *See* ALBEDO.

reforestation Replacing forest. Also known as **reafforestation**. *See also* AFFORESTATION.

reformer A device that extracts pure *hydrogen from *hydrocarbon fuels (such as *natural gas, *methanol, *ammonia, *gasoline, or vegetable oil) so it can be used in a *fuel cell.

reformulated gasoline A specially refined cleaner-burning blend of *gasoline that reduces emissions from vehicle exhausts, particularly *volatile organic compounds (VOCs) and *hazardous air pollutants.

refraction The deflection or bending of the path of a wave, such as the bending of light as it passes through different parts of the *atmosphere, or the deflection of a sea *wave as it approaches the *shore.

refrigerant A chemical that cools air as it evaporates. Many refrigerants contain *CFCs and are harmful to the *ozone layer in the Earth's atmosphere.

refuge A safe place that offers protection or shelter, such as an area (for example a *nature reserve) that has been set aside for the purpose of conserving *species and their *habitats.

refugee A person who is forced to leave his/her home and seek safety in another region or country as an exile, usually as a result of persecution.

refugium An area that remains unchanged while the areas that surround it change a great deal, and which therefore provides a *refuge for *relict species or populations that require specific habitats.

refuse All forms of *solid waste or *rubbish. Also known as **trash** or **garbage**.

refuse-derived fuel Fuel that is made from the *organic residue of *municipal solid waste, which is shredded, formed into pellets or small briquettes, then dried to make fuel for *power plants.

reg A *desert surface that is covered by loose stones and gravel, but with no *sand.

regelation The process by which a substance (for example the ice in a *glacier) melts under pressure and then refreezes after the pressure is released.

regeneration The regrowth, by either natural or artificial means, of vegetation and *ecological communities that have been damaged or destroyed.

regeneration cut In *forest management, a cutting strategy in which old trees are removed while favourable environmental conditions are maintained in order to promote and enhance the natural establishment of new trees.

regeneration harvest In *forest management, a *clearcut that removes an existing stand of trees in order to prepare the site for *regeneration.

regenerative farming Farming techniques and methods of *land management that are designed to restore soil productivity by measures such as *crop rotation, planting *ground cover, protecting the surface with *mulch, and reducing the input of synthetic chemicals and mechanical compaction.

regime Any system of control, or more specifically a system of government. *See also* ENVIRONMENTAL REGIME.

regimen A treatment plan, formalized schedule, or strict, regulated plan that is designed to reach particular goals.

r

region A geographical area of land and/ or water that is distinguished by certain natural features, climatic conditions, or fauna or flora. *See also* MORPHOGENETIC REGION, NATURAL REGION, PHYSIOGRAPHIC REGION.

regional climate *See* MACROCLIMATE.

regional diversity *See* GAMMA DIVERSITY.

regional haze Atmospheric *haze that extends over a large area and is produced by many different sources.

regional metamorphism *Metamorphism of large areas of the Earth's *crust, usually during mountain building at *convergent plate margins. *Contrast* CONTACT METAMORPHISM.

regional plan The plan for a region, for example to assist regional government or authority to manage *natural resources on a *sustainable basis or meet particular social or economic goals.

regional planning The process of developing regional plans.

Regionally Important Geological/ Geomorphological Site (RIGS) In the UK, sites (other than *SSSIs) that are considered worth protecting for their educational, research, historical, or aesthetic importance, because of their geology.

regolith The layer of *unconsolidated material (*soil and mineral *subsoil) that lies between the soil surface and the solid *bedrock below, composed mainly of weathered bedrock or *sediment. Also known as **regosolic soil**.

regosolic soil *See* REGOLITH.

regression 1. A statistical method for studying and expressing the nature of relationships between two or more variables.
  2. A slow lowering of *sea level that exposes land along the *shoreline.

regulated forest Forest land that is managed to produce a *sustained yield of timber.

regulated river A river in which the natural processes and/or the *river channel itself have been deliberately altered for management purposes, such as the construction of a large *dam for *water supply, to generate *hydroelectric power, or to reduce downstream *flooding.

regulation An authoritative rule; a principle or condition that normally governs behaviour. *See also* DIRECTIVE.

regulator An official or agency responsible for the control and supervision of a particular activity or area of public interest, or which oversees and applies legislation.

rehabilitate To restore to a good condition or bring back to an original condition.

rehabilitation The process of returning a disturbed site, *habitat, or *ecosystem to its original condition.

reintroduction The renewed introduction of a species to an area from which it has almost or completely disappeared within historical times.

rejuvenation The process of restoring an *environmental system to a more youthful condition, such as the renewed downcutting of a river system as a result of *tectonic *uplift. *See also* LONG PROFILE.

relative abundance The number (*abundance) of *organisms of a particular kind which is present in a *sample or *community, relative to the total number of organisms in that sample or community.

relative age Age that is expressed relative to other ages ('younger than' or 'older than'), rather than in number of years. *Contrast* ABSOLUTE AGE.

relative atomic mass (RAM) The mass of an *atom of a particular chemical *element, relative to one atom of carbon (which has a RAM of 12). Formerly known as **atomic weight**.

relative chronology A *chronology that determines the age of a feature or event relative to the age of other features or events (for example, younger than or older than). *Contrast* ABSOLUTE CHRONOLOGY.

relative dating A range of methods for establishing the *relative age of an object or material, for example based on the relative position of individual *strata within *sedimentary rocks.

relative frequency *See* FREQUENCY.

relative humidity The amount of *water vapour that is present in the air at a particular point in time, as a percentage of the amount that the air could hold if it was totally saturated. *Contrast* ABSOLUTE HUMIDITY. *See also* HUMIDITY.

relative poverty A state of *poverty that is characterized by scarcity rather than lack of economic necessities, which does not immediately threaten life or health. *Contrast* ABSOLUTE POVERTY.

relative relief The difference between the highest and lowest points in a particular area. *Contrast* ABSOLUTE RELIEF.

relative risk A measure of comparative ˄risk of developing a particular *disease or condition, within a defined period of time.

relative scarcity A condition that exists when a particular *resource is in short supply in one or more areas, because of inadequate or disrupted distribution. *Contrast* ABSOLUTE SCARCITY. *See also* SCARCITY.

relay cropping A *cropping system in which two or more *crops are grown one after the other in the same field in the same year.

release 1. In general, to discharge or liberate something.
2. In *pollution control, any process (including spilling, leaking, pumping, pouring, emitting, emptying, discharging, injecting, escaping, *leaching, dumping, or disposing) by which a *hazardous or *toxic chemical, or extremely hazardous substance, is discharged into the *environment.

release cutting The removal of competing vegetation in order to allow desired tree species to grow.

release rate The quantity of a *pollutant that is released from a *source into the *environment, over a given period of time.

relict An *organism or *species that survives as a remnant of an otherwise *extinct flora or fauna, in an *environment that has changed much from the one in which it originated.

relict distribution The distribution area of a *relict species, which is a remnant of a previously wider *range.

relief 1. The difference in *elevation between the high and low points of a land surface.
2. Assistance in time of difficulty, such as the provision of immediate shelter, life support, and human needs to people who are affected by a *disaster.

rem The abbreviation for roentgen equivalent man, a measurement of *radiation in terms of biological effect on human tissue. *See also* DOSE EQUIVALENT.

remedial action Long-term *cleanup activities, including the cleanup, removal, containment, isolation, treatment, or monitoring of *hazardous substances that have been released into the *environment, in order to reduce or eliminate the long-term risks to *human health or the environment from exposure to *contaminants.

remediation Action that is taken to correct or treat a *pollution problem, which usually involves the *cleanup or other methods that are used to remove or contain a *toxic spill or hazardous materials from a *hazardous waste site (such as a *Superfund site). Also known as **amelioration**.

remedy A method, technique, or process that is designed to treat or clean up contaminated air, soil, sediments, or water. *See also* REMEDIATION.

remote automatic weather station (RAWS) A self-contained weather station that automatically acquires, processes, and stores local weather data which are then transmitted via radio, a landline, or through a *satellite (*GOES) to a receiving station elsewhere.

remote sensing Collecting information about an object or event from a distance, without being in physical contact with it, such as through *air photography or *satellite imaging. For example, satellite monitoring of the Earth's *radiation balance, atmospheric dynamics, the *oceans and ocean/atmosphere interactions, land resources, *glaciers, and *ice caps is proving to be particularly useful in building up a detailed picture of how the environment works and how it is being affected by human activities. Satellites are also being increasingly used in monitoring *natural hazards and *natural disasters. One promising development is the *World Environment and Disaster Observation System (Wedos).

remotely operated vehicle (ROV) An unmanned submersible vehicle that is controlled from the sea surface, the uses of which include photographing and repairing underwater structures and collecting samples of sediment from the deep ocean floor.

removal action *See* SHORT-TERM CLEANUP.

removal cut In *forest management, the removal of the last seed-bearing trees or shelter trees after *regeneration is established. *See also* SHELTERWOOD CUT.

renewable Something that can be used again because it can be replaced or replenished: a *natural resource that is capable of being replaced by natural *ecological cycles or sound management practices.

renewable energy *Energy that is obtained from sources that are for all practical purposes *inexhaustible, which includes moving water (*hydroelectric power, *tidal power, and *wave power), *thermal gradients in ocean water, *biomass, *geothermal energy, *solar energy, and *wind energy. Contrast this with energy from sources such as *fossil fuels, of which there is a finite supply which is exhaustible. Also known as **renewables**.

Renewable Energy Certificate (REC) Under the *Kyoto Protocol, a tradable certificate of proof of the amount of *renewable energy that is used in the manufacture of a particular good or commodity. In many countries 1 REC = the equivalent of one megawatt hour (Mwh) of electricity from a renewable generation source. Also known as **Renewable Energy Credit**, and known in the UK as **Renewable Obligations Certificate**. *See* RENEWABLE PORTFOLIO STANDARD.

Renewable Energy Credit *See* RENEWABLE ENERGY CERTIFICATE.

Renewable Obligations Certificate (ROC) *See* RENEWABLE ENERGY CERTIFICATE.

Renewable Portfolio Standard (RPS) A benchmark that defines the proportion of total annual energy sales that individual energy retail suppliers must obtain from *renewable energy sources, under the *Renewable Energy Certificate scheme.

renewable resource A *natural resource (such as fresh water, a forest, or *renewable energy) that is replaced at a rate which is at least as fast as it is used, which has the ability to renew itself and be harvested indefinitely under the right conditions, but which can be converted into a *non-renewable resource if subject to *overexploitation. Also known as **flow resource, replenishable resource**. *See also* SUSTAINABLE YIELD.

renewables *See* RENEWABLE ENERGY.

replacement level A *population growth rate of zero, when the number of births equal the number of deaths in a population over a year. *See also* ZERO POPULATION GROWTH.

replacement-level fertility *See* ZERO POPULATION GROWTH.

replenishable resource *See* RENEWABLE RESOURCE. *Contrast* DEPLETABLE RESOURCE.

reportable quantity (RQ) The minimum amount of a *hazardous substance that, when spilled, must be reported to the proper authorities in the USA under *CERCLA.

repository A long-term storage *facility or archive. One preferred approach to the safe and sustainable management of *nuclear waste is long-term underground storage of the waste in specially developed and managed sites, but efforts to set up such repositories have been hampered by public opposition. Key site requirements for nuclear waste repositories include the following: the site must be away from major centres of population, the site must be accessible by road and rail, geological conditions must be suitable (it must be possible to tunnel, for example), prospects of minimizing *groundwater contamination must be good, it must be possible to store radioactive waste safely there for at least 10 000 years, and the site must be geologically stable (away from *fault lines and *earthquake zones). The controversy surrounding the proposed repository at *Yucca Mountain in Nevada illustrates some of the complexity of the issues involved.

representative area An area that is typical of its surroundings in terms of physical features and ecological patterns.

representative sample A *sample whose characteristics are the same as those of the original *population or reference population.

representativeness In *conservation assessment, typical of a feature, habitat, or group of species. Also known as **typicalness**.

reprocessing The chemical treatment of spent *nuclear fuel in order to separate potentially reusable constituents (such as

*uranium and *plutonium) from unwanted *nuclear waste products.

reproduction The act of making copies; in biology, the process of generating offspring.

reproductive health hazard Any *agent that damages the reproductive system of an adult male or female, or the developing foetus or child.

reproductive isolation The isolation from each other of two parts of a *population, the individuals of which would otherwise be capable of interbreeding.

reproductive potential *See* BIOTIC POTENTIAL.

reptile Any *cold-blooded *vertebrate of the class Reptilia, which includes tortoises, turtles, snakes, lizards, alligators, and crocodiles. *See also* HERPETOFAUNA.

research Systematic investigation that is designed to establish facts.

Research Natural Area An area that contains *natural resource values of scientific interest, which is managed primarily for research and educational purposes. *See also* NATURAL AREA PRESERVE.

reservation *See* RESERVE.

Reservation and Other Tribal Lands A North American term for all lands that lie within the boundaries of any Indian reservation, including rights of way and all lands that are held in trust for or supervised by any Indian tribe.

reserve An area of land that is set aside and managed for a particular purpose, such as *nature conservation. Also known as a **reservation**. *See also* BIOSPHERE RESERVE, GAME RESERVE, NATURE RESERVE.

reserves That part of an identifiable supply of *resources (such as a *mineral deposit) that can be extracted profitably, at present prices, with current technology.

reservoir A natural or artificial pond, lake, or basin that is used for the storage, regulation, and control of water. *See also* IMPOUNDMENT.

residence time The period of time during which a substance remains in a particular area or in a particular part of an *environmental system, such as water retained in a *reservoir, or a *greenhouse gas retained in the *atmosphere. Typical residence time are: water vapour in the lower atmosphere, 10 days; carbon dioxide in the atmosphere, 5–10 days; aerosol particles in the lower atmosphere, one week to several weeks; aerosol particles in the upper atmosphere, several months to several years; water in the biosphere, two million years; oxygen in the biosphere, 2000 years; carbon dioxide in the biosphere, 300 years; surface water in the Atlantic Ocean, 10 years; deep water in the Atlantic Ocean, 600 years.

resident A permanent inhabitant or non-migratory animal.

residential land Land which is primarily used for permanent dwellings, such as houses and flats, including the associated infrastructure (such as roads, drains, and sewers).

residential sector A North American term for all private residences (occupied or vacant, owned or rented), including mobile homes as well as secondary homes such as summer homes. *See also* COMMERCIAL SECTOR, INDUSTRIAL SECTOR, TRANSPORTATION SECTOR.

residential waste *See* HOUSEHOLD WASTE.

residual 1. The difference between an observed and an expected or predicted value.
2. The amount of a *pollutant that remains in the *environment after a natural or technological process has taken place, such as the *sludge that remains after initial *wastewater treatment, or the *particulates that remain in air after it has passed through *scrubbing or other processes.

residual herbicide A *herbicide that persists in the soil after it has been ap-plied, and injures or kills germinating weed seedlings over a short period of time. *Contrast* CONTACT HERBICIDE.

residual risk The risk that remains after the application of selected safeguards, particularly the health risk from air pollutants that remains after the application of the *maximum achievable control technology.

residual stand In *forest management, the trees that are left standing after any cutting operation.

residuals *See* SLUDGE.

residue Matter that remains after other materials have been removed, such as the *fly ash that is left after *incineration, or the dry solids that remain after a sample of water or *sludge has been evaporated.

resilience The rate at which a system regains structure and function following a stress or perturbation. *Contrast* RESISTANCE. *See also* SENSITIVITY.

resin A thick, sticky liquid that is produced by certain plants and becomes hard when exposed to the air. Fossilized resin is known as *amber.

resistance 1. Any mechanical force that slows or opposes motion.
2. A measure of the degree to which a body opposes the passage of an electric current.
3. The ability of a system to withstand being disturbed by an external factor. *Contrast* RESILIENCE. *See also* INERTIA, STABILITY.

resistor A device that is used to control the flow of electric current by providing *resistance.

resolution 1. The accuracy with which something can be measured, or the scale at which it is measured.
2. Finding a solution to a problem.
3. A decision to do something or to behave in a certain manner.

resource An available supply of something that is valued because it can be used for a particular purpose, usually to satisfy particular human wants or desires. *See also* AESTHETIC RESOURCE, CULTURAL RESOURCE, NATURAL RESOURCE.

resource allocation The deliberate distribution and use of *resources in an organized and efficient way, in order to meet defined goals or objectives, or to complete particular tasks in a timely and efficient manner.

resource conservation An approach to the *conservation and *sustainable use of *natural resources that is based on reducing the amount of *solid waste that is generated, reducing overall resource consumption, and using recovered resources.

Resource Conservation and Recovery Act (RCRA) (1976) US legislation that amended the *Solid Waste Disposal Act (1965) by establishing a regulatory system for the *Environmental Protection Agency to regulate and manage *solid waste, including *hazardous waste, requiring safe and secure procedures to be used in treating, transporting, storing, and disposing of hazardous wastes, and preventing new *Superfund sites from arising.

resource depletion Using a *non-renewable resource and/or using a *renewable resource faster than it can naturally be replaced.

resource management 1. The deliberate *management of *natural resources.
2. A task within *project management that ensures that adequate resources are allocated to a particular activity (for example after a *disaster).

resource management system (RMS) A system that integrates *conservation practices and *resource management techniques in order to maintain or improve soil, water, plant, and related resources.

resource partitioning The sharing of resources by different *species within a biological community, through the de-

velopment of specialized *niches, thus reducing direct competition.

resource plan A *plan that is prepared by government and used to guide public policy and action on the use, treatment, and management of soil and water resources.

resource recovery The recovery of materials, fuel, or energy, particularly from *solid waste.

resource recovery facility Any *facility at which *solid waste is processed in order to extract, convert to energy, or otherwise separate and prepare solid waste for *reuse.

resource recovery system A solid *waste management system that is designed for the collection, separation, recycling, and recovery of solid wastes, including the disposal of non-recoverable waste residues.

Resource Reserve An *IUCN Management Category (VI) for *protected areas, which is designed 'to protect resources of the area for future use and prevent or contain development activities that could affect the resource pending the establishment of objectives that are based upon appropriate knowledge and planning'.

resource scarcity A shortage or deficit of a particular *resource. *See also* ECONOMIC SCARCITY.

resource values A broad term used in North America to describe products or services that depend upon *ecological processes, including *water quality and quantity, *forage, *fish, *wildlife, *timber, *recreation, *energy, *minerals, and cultural and heritage resources.

resource war A military conflict that is driven by the attempt by one country to dominate and appropriate the *natural resources (such as the oil reserves) of another country.

Resources for the Future An independent environmental organization that is based in the USA, and whose mission is 'to improve environmental and natural resource policymaking worldwide through objective social science research of the highest caliber'.

respiration Breathing; a *biochemical process by which living *organisms take up *oxygen from the environment and consume *organic matter, releasing both *carbon dioxide and heat.

respiration rate The rate at which the concentration of *oxygen is reduced by *respiration.

response action Action that is taken by an agency in order to address the risks posed by the release or threatened release into the environment of *hazardous substances.

rest To remain inactive. In land management, to leave an area of *grazing land ungrazed or unharvested for a particular period of time such as a year or a *growing season.

restitution The act of making good, or of giving the equivalent, for any loss, damage, or injury. *See also* RESTORATION.

restoration The act of restoring, renovating, or re-establishing something to close to its original condition, such as the structure and function of a *damaged habitat or *ecosystem.

restoration cost The estimated likely total cost of cleaning up a site and restoring it to a natural condition after resource-using activities (such as *mining) have ceased there.

restoration ecology The study of the structure and *regeneration of plant and animal *communities, particularly the deliberate *colonization and revegetation of *derelict land, for example after activities such as *mining and *waste disposal and after land has been released from *agricultural use.

restricted use area (RUA) In North America, an area of land that is closed to particular activities in order to protect damaged or fragile *habitats and *ecosystems, and allow them to recover naturally.

retardation The extent to which something is delayed or held back, such as the retention of *contaminants in soils and deposits by one or more physical, chemical, or biological factors.

retention The process of holding back or retaining, such as the portion of an original stand of trees that is retained after felling, or the amount of *precipitation on a drainage area that does not escape as *runoff.

retention time The average period of time for which *water resides in a *lake or *reservoir. Also known as **flushing rate, turnover rate**.

retrofit To add machinery or equipment to an existing piece of equipment or system in order to correct a defect or add capability, such as the addition of a *pollution control device on an existing facility without making major changes to the generating plant. Also known as **backfit**.

return period *See* RECURRENCE INTERVAL.

return stroke An electrical discharge that moves upward along a *lightning channel from the ground to the cloud.

reuse The repeated use of a product or material in its same form, for either the same or a different purpose. *Contrast* RECYCLING.

revegetation Re-establishing and developing plant cover, by either natural or artificial means, such as reseeding. *See also* REFORESTATION.

revenue–cost analysis A method of comparing alternatives which is based on analysing the monetary income that each alternative would generate in relation to its cost. *See also* COST–BENEFIT ANALYSIS.

reverse fault An overhanging type of vertical displacement *fault, in which one side is pushed upwards over the other at an angle (the *dip of the *fault line), as a result of strong *compression pressure.

reverse osmosis A method for the *desalination of water, which involves passing the *saline water through a *permeable membrane (film), normally under pressure. Relatively pure water passes through the membrane, leaving much of the dissolved salt behind.

reversible change A change that can be reversed to give a return to the original state without a major alteration in the surroundings.

reversible effect An effect that is not permanent, such as an adverse effect that decreases when exposure to a *toxic chemical stops.

reversible reaction A *chemical reaction that can proceed in either direction, depending on the concentration of reacting materials.

review To reappraise or evaluate, particularly to reconsider objectives and policies, or to evaluate progress against objectives and policies.

revolving door 1. In general, an organization or institution that has a high rate of turnover of personnel or membership.
2. A US term for the movement of people in and out of government in a manner that implies that, while the individual is not working for the government, he or she is working for an interest group that benefited from the policies that the individual pursued while working for the government.

Reynolds number (*Re*) A number associated with fluid flow, which represents the relative importance of *viscous forces and *inertial forces in a fluid. As *Re* increases, inertial forces become more important.

rhizome A horizontal, underground stem system comprising the *roots and leafy stems of plants.

rhizosphere The soil zone that immediately surrounds plant roots, which is modified by the increased number of *micro-organisms that live there in association with plant roots.

R-horizon *See* D-HORIZON.

rhyolite A fine-grained *extrusive *volcanic rock, similar in composition to *granite.

RI *See* RECURRENCE INTERVAL.

ria A drowned river valley.

ribbon development A narrow strip of continuous building development, for example along a road or along the *coast.

ribonucleic acid (RNA) A molecule with a similar structure to *DNA that is involved in a number of cell activities, especially *protein synthesis (because it decodes the instructions that *genes carry for protein synthesis).

richness, species The number of species in a community, habitat, or sample. *See also* DIVERSITY.

Richter scale A scale for describing the magnitude of an *earthquake, devised by Charles Richter in 1935 and based on the amount of energy generated. It is a logarithmic scale that ranges along a progression from zero (minor event) to nine (major event): severe earthquakes have magnitudes greater than seven. The scale is given in Appendix 7. *Compare* MERCALLI SCALE.

ride An area (often a strip) of *woodland that has been cleared, in order to provide access, fire breaks, and open areas for game and wildlife.

ridge 1. A long area of relatively *high pressure in the atmosphere.
2. A long section of high ground between adjacent *valleys.

ridge planting A *conservation farming method in which seeds are planted in ridges, which creates warmer soil temperatures and traps rainwater in the furrows between the ridges.

ridging The building or intensification of a *ridge of *high pressure in the atmosphere.

riffle A rocky shoal or *sandbar that lies just below the surface of a *stream. *See also* POOL AND RIFFLE.

r

rift valley A *valley that has formed along a *rift zone where tensional forces have pulled the Earth's *crust apart.

rift zone A system of crustal fractures arising from tensional forces, similar to that which occurs at *mid-oceanic ridges.

rifting The geological process by which a long trough bounded by normal faults is formed by fractures of the Earth's *crust where adjacent *crustal plates have been pushed apart, similar to that which occurs at *mid-oceanic ridges.

right of way The legal right to cross land or *property that is owned by someone else, including the strip of land that is used for a road, railroad, or power line, to which the operating company has special right of access.

right to roam A UK campaign by the Rambers Association and others to be allowed to walk across private land and land owned by the Ministry of Defence.

rights Entitlements that are assured by custom, law, or property.

RIGS *See* REGIONALLY IMPORTANT GEOLOGICAL/GEOMORPHOLOGICAL SITE.

rill A small *gully or channel that is formed by *soil erosion on a hillslope or a gently sloping surface, such as a bare *spoil heap or recently ploughed field.

rill erosion The *erosion of exposed *soil on a sloping surface that is caused by the cutting of small water channels by concentrated *surface runoff. It is intermediate between *sheet erosion and *gully erosion.

rime A form of *frost that is created by the freezing onto solid objects of supercooled water droplets in *fog, and appears as a mass of ice crystals on surfaces such as grass.

ring net *See* SEINE NET.

Ringelmann chart A series of shaded illustrations that are used to measure the *opacity of *air pollution emissions, ranging from light grey (number one) to black (number five). The chart is used to set and enforce *emission standards.

Rio Declaration on Environment and Development A general statement of intent which was agreed in 1992 at the *United Nations Conference on Environment and Development and centres on the well-being of people, on the right of states to control their own natural resources, and on their obligation not to damage the environment of other countries. Amongst other things it recommends the *precautionary principle, internalization of *environmental costs, use of *environmental impact statements, and the *polluter pays principle. *See also* SUSTAINABLE DEVELOPMENT.

Rio Earth Summit *See* UNITED NATIONS CONFERENCE ON ENVIRONMENT AND DEVELOPMENT.

rip current A current that flows strongly away from the *shore through gaps in the *surf zone at intervals along the *shoreline.

riparian Relating to the banks of a *stream, *lake, or *waterbody.

riparian forest Trees and shrubs that grow alongside a river or lake. Also known as **gallery forest**.

riparian rights The *rights of an owner of land which borders a *waterbody to use or control all or part of that waterbody, including the right to prevent diversion or misuse of upstream waters.

riparian vegetation Plants that are adapted to grow in the moist conditions found along the banks of streams, lake lakes, and waterbodies.

riparian zone A stream and all of the vegetation on its banks.

ripple A very small-scale *bedform, like a miniature sand dune, which is created at low *flow velocity in the fine sediment of an *alluvial channel. Ripples are always found in a group.

rip-rap A layer of medium to large rocks or logs that is used to stabilize banks along ponds, lakes, rivers, and reservoirs.

rising tide *See* FLOOD TIDE.

risk A measure of the likelihood or *probability that damage to life, health, property, and/or the environment will occur as a result of a particular *hazard. *See also* ACCEPTABLE RISK, AVOIDABLE RISK, UNACCEPTABLE RISK.

risk analysis The process of identifying where *risks might arise, estimating the likelihood of them happening, and analysing the trade-off of risk against costs and benefits.

risk assessment Methods that are used to estimate short- and long-term harmful impacts on human health or the environment which arise from exposure to *hazards associated with a particular product or activity.

risk avoidance Management that is designed to help ensure that risks will not occur. *Contrast* RISK MITIGATION.

risk characterization A summary of what is known about the adverse health, safety, and environmental impacts that are likely to arise from exposure to a particular *hazard.

risk communication A process of providing the public with information about the *risks associated with particular products, substances, activities, and technologies.

risk estimate An assessment of the likelihood that *organisms that are exposed to a particular *pollutant will develop an adverse response (such as *cancer).

risk evaluation The process of determining the acceptability of a particular *risk, based on perceived costs and benefits.

risk factor Any factor that increases the chance of an individual developing or aggravating a particular medical condition, such as heart disease or *cancer.

risk management The systematic application of *management policies, procedures, and practices to the tasks of identifying, analysing, evaluating, treating, and monitoring *risk.

risk mitigation The management of undesirable events as they occur, when the associated *risk could not be avoided. *Contrast* RISK AVOIDANCE.

risk perception The perception by people of the health, safety, or environmental *risks that are or may be associated with particular activities or in particular settings.

risk reduction Any action that is taken to reduce the likelihood and impact of a particular *risk, which includes education, *regulation, and *remediation.

risk-based targeting The direction of *resources (for example in *disaster management) to those areas that have been identified as having the highest potential or actual adverse impact on *human health and/or the *environment.

risk–benefit analysis A comparison of the relative *risks and benefits associated with a particular *hazard, in order to determine its acceptability.

risk-specific dose The *dose that is associated with a specified level of *risk.

Riss The third phase of *glaciation in Europe during the *Pleistocene; equivalent to the *Illinoian in North America.

river A large natural *stream of water that flows along a *channel.

river bank *See* RIPARIAN, STREAM BANK.

river basin *See* DRAINAGE BASIN.

river bed *See* STREAM BED.

river blindness A disease in humans (onchocerciasis) which is caused by an *infestation with slender threadlike roundworms that are deposited under the skin by the bite of blackflies. The disease can cause blindness and is common in Africa and tropical America.

river capture *See* RIVER PIRACY.

river channel *See* STREAM CHANNEL.

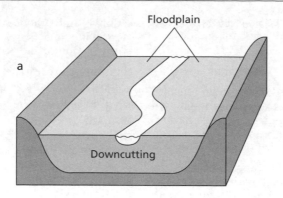

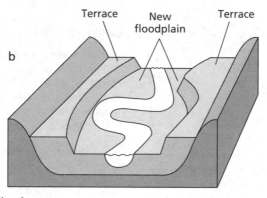

Fig 17 River terrace development

river diversion scheme An engineering scheme that is designed to redirect *flood water on a *floodplain and make it flow away from built-up areas, normally by constructing artificial channels.

river equilibrium An *equilibrium channel form that is maintained by adjustments of *hydraulic geometry, *channel pattern, and *slope, in order to preserve a balance of energy moving through the system. Equilibrium exists when the energy provided to do work (transport the sediment and shape the river landscape) is just enough to perform the work that has to be done. *See also* DYNAMIC EQUILIBRIUM.

river flow The water that flows along a river, ultimately to the sea, as part of the *water cycle. It includes water from *overland flow, *throughflow, *inter-flow, *soil pipes, *percolation, and *groundwater, as well as *precipitation that falls directly into rivers and lakes. *See also* EPHEMERAL FLOW, INTERMITTENT FLOW, PERENNIAL FLOW.

river form The *morphology of a river channel, which differs between *alluvial channels and *bedrock channels, which display different forms of *river equilibrium.

river network *See* DRAINAGE NETWORK.

river piracy The *erosion of large rivers in which one actively eroding river captures the *headwater of an adjacent river, altering the pattern of its *drainage network.

river regulation *See* REGULATED RIVER.

river terrace A flat platform of land that lies above the level of the valley floor, which may be a surviving remnant of an abandoned *floodplain. *Contrast* TERRACE.

river valley A *valley that has been eroded by a *river.

riverine Of or related to a *river. *See also* ALLUVIAL, FLUVIAL, RIPARIAN.

riverscape The *landscape of or surrounding a *river.

RMCL *See* RECOMMENDED MAXIMUM CONTAMINANT LEVEL.

RME *See* REASONABLE MAXIMUM EXPOSURE.

RMS *See* RESOURCE MANAGEMENT SYSTEM.

RNA *See* RIBONUCLEIC ACID.

road An open way, generally public rather than private, which is used for travel or transportation.

roaded natural area A category of land defined within the *recreation opportunity spectrum as 'predominantly natural appearing environments with moderate evidence of the sights and sounds of man', with moderate to high interaction between users, conventional motorized use allowed and catered for in the design of the facilities, and resource use practices that are evident but harmonized with the natural environment.

Roadless and Undeveloped Area A large area of a *National Forest in the USA which contains no roads and is suitable for designation as a *wilderness area.

Roadless Area Review and Evaluation (RARE) The US national inventory of roadless and undeveloped areas that lie within the *National Forests and *National Grasslands, which was carried out during the 1960s.

Roadless Area Review and Evaluation II (RARE II) An updated version of the *Roadless Area Review and Evaluation that was conducted by the *US Forest Service in 1979, which included consideration of potential *wilderness designations for these areas.

roadway Any surface that is designed, improved, or normally used for vehicle travel.

ROC *See* RENEWABLE OBLIGATIONS CERTIFICATE.

roche moutonnée An outcrop of bare, resistant rock which has been carved by *glacier ice moving over it. The upstream (*stoss) side is shaped by *abrasion, and is usually smoothed, rounded, and streamlined and often has *striations. The downstream (*lee) side is usually steep and angular because it is eroded by *quarrying processes. Also known as a **rock knob**.

rock A solid, cohesive, aggregate of one or more crystalline minerals. *Oxygen makes up about 47% of exposed rocks, *silicon makes up about 28%, and a further 24% is made of *aluminium, *iron, *potassium, *calcium, *sodium, and *magnesium. *Contrast* IGNEOUS ROCK, METAMORPHIC ROCK, SEDIMENTARY ROCK. *See also* ROCK CYCLE.

rock cycle The never-ending series of natural geological processes that continuously refashions, redistributes, and recycles material from and on the Earth's *crust, including *weathering, *erosion, *sediment transport, *deposition, and *lithification of *sedimentary rocks, *metamorphism, and all the processes that create *igneous rocks. The rock cycle operates in much the same way as the major *biogeochemical cycles and the global *water cycle, but much more slowly over the full span of *geological time. Also known as the **geological cycle**.

rock flour A *glacial sediment composed of finely ground rock particles, which is formed by *abrasion of rocks underneath a moving *glacier.

rock glacier A mass of rock (*talus) that is held together by *ice that moves downslope, like a *glacier.

rock knob *See* RÔCHE MOUTONNÉE.

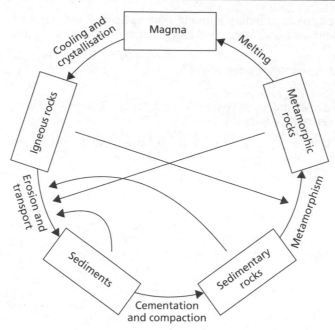

Fig 18 Rock cycle

rock salt Common salt, or *sodium chloride, a white, crystalline, evaporate mineral. Also known as **halite**.

rock type A division of *rocks into three major groups (*igneous rocks, *sedimentary rocks, and *metamorphic rocks), depending on how they were formed.

rock varnish A thin, dark coating of manganese that forms on the surface of *rocks in *arid climates.

rockfall A very sudden, rapid *mass movement *erosion process, involving the fall of weathered rocks down very steep slopes or cliffs. These rocks are often deposited as a *scree slope.

rockslide A *landslide of rocks.

Rocky Flats The site near Denver, Colorado, USA of a weapons production facility run by the US Atomic Energy Commission, which operated from 1952 to 1989. Cleanup of the site began in 1992 in preparation for turning the site into a *national wildlife refuge.

ROD See RECORD OF DECISION.

rodent Any of an order (Rodentia) of relatively small gnawing mammals (such as a mouse, a squirrel, or a beaver) that have in each jaw a single pair of teeth with a chisel-shaped edge.

rodenticide A chemical *pesticide that is used to control or kill *rodent pests (such as rats and mice) and prevent them from damaging food and crops.

roentgen (R) A unit of exposure to ionizing *radiation from *gamma rays or *X-rays. See also REM.

roguing The removal of undesirable plants from *hybrid seed production plots, in order to produce a better quality of seed.

Romanticism An early 19th century movement which emphasized the personal, emotional, and dramatic through the use of exotic, literary, or historical subject matter. The movement affected political, philosophical, and artistic thought throughout Western Europe. See also TRANSCENDENTALISM.

rookery A colony of breeding animals, usually birds or seals.

roost A perch or place where birds or bats rest or sleep, often in groups.

root The lower (usually underground) part of a *plant, which anchors it in the soil and absorbs *water and mineral *nutrients from the soil by means of the *root hairs.

root climber A plant (such as ivy) that raises itself up by means of *adventitious roots, which adhere to the plants that support it. *See also* SCRAMBLER, TWINER.

root hair The thin hair-like growth from the *root, through which a plant absorbs *nutrients from the *soil.

root zone The upper part of a *soil that is penetrated by *plant roots.

ROS *See* RECREATION OPPORTUNITY SPECTRUM.

rosette A short, circular cluster of *leaves on a plant.

Rossby wave A horizontal wave-like motion that is found in the *mid-latitude westerly circumpolar air stream. It occurs in the upper *troposphere, forms and decays over a period of one or two months, and is followed by *geostrophic winds as they flow around the world.

rotary kiln incinerator An *incinerator that has a rotating combustion chamber that is designed to keep waste moving, thus allowing it to vaporize and burn more easily.

rotating-cup anemometer An instrument that measures *wind speed from the speed of rotation of small cups on its top.

rotation 1. The spinning of a body (such as the Earth) about its axis.
2. A planned recurrent sequence, for example of *crops. *See also* CROP ROTATION.

rotation cycle The planned number of years between successive planting of crops within a *rotation.

rotation pasture A *pasture composed of *introduced *perennial plants and/or self-seeding *annual species, which is one unit in a *crop rotation of five years' duration or less.

rotational cropping *See* CROP ROTATION.

rotational set-aside A form of *crop rotation in which the location of *set-aside land is periodically switched from place to place within a farm holding.

rotenone An organic *insecticide that is derived from the roots of some *legumes and is used to control beetles and similar insects.

rotor cloud A turbulent *altocumulus or *cumulus cloud formation that is found in the *lee of some *mountain barriers when winds cross the barrier at high speed.

rough fish Any species of *fish that is considered to be useless for food or sport, or even as bait. *Contrast* GAME FISH.

roughage Indigestible parts of food, that are high in *fibre but low in *nutrients.

roundwood *See* INDUSTRIAL TIMBER.

route of exposure The way in which an organism might come into contact with a chemical substance, such as by drinking (*ingestion), breathing (*inhalation), or bathing (skin contact).

ROV *See* REMOTELY OPERATED VEHICLE.

row crop Cultivated land on which the *crops are grown in rows.

row intercropping Growing two or more *crops simultaneously, in rows.

RPS *See* RENEWABLE PORTFOLIO STANDARD.

RQ *See* REPORTABLE QUANTITY.

r-selection *Natural selection that favours species with high reproductive rates, at the expense of species which live a long time. *Contrast* K-SELECTION.

r-strategy A life strategy in which *species are best adapted to live in a variable *habitat by having a high rate of reproduction. *Contrast* K-STRATEGY.

RUA *See* RESTRICTED USE AREA.

rubber *See* LATEX.

rubbish *See* SOLID WASTE.

rudaceous rocks Coarse-grained *clastic *sedimentary rocks (such as *conglomerates) which have *particles larger than two millimetres across.

ruminant An animal (such as a cow or goat) that chews the cud, having a multi-chambered stomach that allows it to digest *cellulose.

runoff That part of *precipitation, *snow or ice melt, or *irrigation water which flows across the land to *streams or other *waterbodies.

rural Living in or characteristic of farming or country life; beyond the limits of a *city, *town, *village, *hamlet, or any other designated residential or commercial area.

rural area A category of land defined within the *recreation opportunity spectrum as 'a natural environment that has been substantially modified by development of structures, vegetative manipulation, or pastoral agricultural development'. In a rural area resource use practices may be used to enhance specific recreation activities and to maintain vegetative cover and soil, sights and sounds of humans are readily evident, and interaction between users is often moderate to high. Many facilities are designed for use by a large number of people and facilities are often provided for special activities; facilities for intensive motorized use and parking are available.

rural development The development of the *countryside, in order to improve living conditions, increase employment, and enrich cultural life.

rural diversification Increasing the number and variety of ways of making a living in the countryside, beyond traditional activities such as *farming, and including new activities such as *recreation and *tourism, and craft industries.

rural settlement A group of houses in the *countryside, which can take the form of a *dispersed settlement, a *hamlet, or a *village. *See also* URBAN SETTLEMENT.

rural–urban fringe Where a *town or *city meets the *countryside.

rural–wildland interface A North American term for where *rural development and *wildland meet and intermix, with no clearly defined zone of separation.

rustic Simple, characteristic of *rural life.

rusting The *corrosion of *iron.

ryolite A relatively viscous, *extrusive type of *granite; a typical *felsic rock.

r

SAC *See* SPECIAL AREA OF CONSERVATION.

sac Any membranous bag or cavity.

sacred grove A stand (*grove) of trees that is regarded as having spiritual value or meaning.

sacrifice area/site A patch of land in which overuse is allowed or even encouraged, in order to protect other more fragile areas or to optimize the overall benefits of *natural resource use across an area.

sacrificial seed crop *Seed that would normally be sown for a commercial harvest (such as *cereal) but which is sown instead to be consumed by seed-eating animals and birds.

safe 1. Free from danger or *risk.
2. A *not threatened category of species defined by the *IUCN (*Red Data Book) as 'neither rare nor threatened'.

Safe Drinking Water Act (SDWA) (1974) US legislation that sets standards for drinking water quality and oversees the states, localities, and water suppliers who implement those standards. It is used with the *Resource Conservation and Recovery Act and the *Comprehensive Environmental Response, Compensation, and Liability Act to protect and clean up *groundwater by setting *water quality standards.

safe water *Water that is considered safe for drinking because it does not contain harmful *bacteria, *toxic materials, or chemicals, although it may have taste, odour, colour, and certain mineral problems.

safe yield The amount (*yield) of water that can be taken from a source of supply each year, over a period of years, without depleting the source beyond its ability for natural replenishment.

safety Freedom from danger or *risk.

Saffir–Simpson scale A scale for measuring the strength of *hurricanes, which runs from one (minimal damage likely) to five (catastrophic damage potential), based on the maximum sustained wind speed generated by the hurricane. The scale is given in Appendix 4.

Sagebrush Rebellion An organized resistance movement which gathered momentum in the 1960s in the western USA, and especially in the state of Nevada, calling for a return of the control of federal lands to individual states. Also known as the **Wilderness Lands Sagebrush Rebellion**.

Sahara A *sand desert which stretches across much of northern Africa, and is the world's largest *desert (nine million square kilometres). As recently as 6000 years ago the Sahara was covered with grasses and shrubs, and climate modelling of why it suddenly turned into the driest region on Earth shows that the transition from grass to sand could have occurred in just 300 years. The model starts with a slow reduction in solar heating of the atmosphere (to simulate the effect of changes in the Earth's tilt and orbit), which gradually weakened the monsoons over India and North Africa, thinning the vegetation cover. After a few thousand years of gradual change, the impoverished vegetation could no longer preserve soil moisture and maintain the cycle of evaporation, atmospheric circulation, and precipitation that drove the African monsoon. It is suggested that this triggered an abrupt

SAHEL

A narrow zone of land about 500 kilometres wide that runs across much of northern Africa, south of the *Sahara and north of the *equatorial zone. Annual rainfall is normally between 100 millimetres and 200 millimetres, and most of the rain falls between June and September. The natural vegetation is mainly sparse savanna grassland and shrubland, and traditional forms of agriculture include nomadic herding and (in some places) the cultivation of peanuts and millet. The area has always been marginal because of the limited rainfall and naturally arid conditions. Water is always in short supply in the Sahel region because of the low total rainfall, the concentration of rain into the wet season, and the persistently high temperatures (and thus rapid rates of evaporation). But water shortage became particularly serious between 1968 and 1974, when rainfall in the Sahel was significantly below the long-term average. This triggered the worst drought in North Africa in about 150 years and was probably associated with an increasingly arid regional climate. Between 1969 and 1990 rainfall in the Sahel was below the long-term average in every year. During the drought of 1973–4, for example, an estimated 100 000 people died, mostly through starvation. Millions of cattle died too. The situation grew worse in the 1980s as the continued lack of rainfall caused widespread famine. Initially many people died through lack of drinking water. But as the drought continued, lack of water caused crop failures and livestock had nothing to graze. The land became completely barren and the soil was baked by the sun to form a hard surface which is impossible to plough.

switch to a desert climate about 5400 year ago, producing the initial Sahara Desert that has since changed in response to climate change and human disturbance.

saline Salty, a solution of salts (*sulphates and *chlorides of *sodium and *calcium).

saline deposit A deposit of salt (*sulphates and *chlorides of *sodium and *calcium) in or on a soil, which reduces soil *fertility and inhibits the growth of *vegetation or *crops.

saline intrusion *See* SALTWATER INTRUSION.

saline soil A soil that contains enough soluble salts (*sulphates and *chlorides of *sodium and *calcium) to reduce its fertility. Saline soils are common in hot climates, and usually have fairly uniform *profile and contain little *humus.

salinization A *soil-forming process that involves the accumulation of un-

usually high concentrations of dissolved salts (*sulphates and *chlorides of *sodium and *calcium), often as a result of large-scale *irrigation schemes in *semi-arid areas, where much of the soil water is *evaporated off leaving behind the salt residue. Salts can accumulate over time to threshold concentrations at which they become toxic and inhibit commercial agriculture. Salt damages the root hairs on plants (through which they take in water and nutrients from soil moisture) and reduces yields. Through time *salinity in the upper soil increases and plants start to show symptoms of stress. This type of salinization is widely associated with irrigation agriculture in *drylands, particularly in Australia, China, Egypt, India, Pakistan, the former Soviet Union, and the USA.

salinity A measure of the concentration of dissolved salts (*sulphates and *chlorides of *sodium and *calcium) in water, normally expressed as parts per thousand (‰). *Freshwater has a salinity of less than 0.5‰ (it is described as *lim-netic), *seawater has a salinity greater than 30‰ (it is *euhaline), and *brackish water an intermediate salinity. Salinity is not a fixed property of water, because it can be increased (usually inadvertently, through *salinization) or decreased (on purpose, by *desalination). Seawater is saline, because it contains dissolved salts—the most important of which are sodium chloride (77.8‰), magnesium chloride (9.7‰), magnesium sulphate (5.7‰) and calcium sulphate (3.7‰). These are derived from weathering and erosion of rocks on the continents, as well as material derived from natural forest fires, volcanic eruptions, and air pollution. Seawater in the open ocean away from obvious sources of pollution has an average salinity of about 35‰, varying between about 34 and 36‰. Run-off in rivers normally has a salinity of more or less zero, whereas salinity in the Great Salt Lake in the USA is more than 150‰ because of the high evapor-ation loss in the hot, dry climate.

salinity gradient The change in *sal-inity with depth in a *waterbody, such as the *ocean.

salmonid A member of the *fish fam-ily *Salmonides*, which includes salmon, trout, and char.

SALR *See* SATURATED ADRABATIC LAPSE RATE.

salt The common name for the chem-ical compound *sodium chloride (NaCl).

salt flat A site with poor drainage in an *arid climate, where the water evapor-ates, leaving behind salts (*sulphates and *chlorides of *sodium and *calcium) in and on the soil.

salt spray *See* SEA SPRAY.

salt weathering A type of *physical weathering of rocks that is most com-mon in *arid climates, and involves the growth and expansion of salt crystals within cracks in rocks. Also known as **salt wedging**.

salt wedging *See* SALT WEATHERING.

saltation A process of *bedload move-ment in a river, that involves the hop-ping or bouncing of small *sand-sized particles along the river bed in the *tur-bulent water flow.

saltatorial Adapted for hopping or moving by leaping.

saltmarsh An area of low-lying ground along a *coast (usually within a sheltered *bay), which is regularly flooded by *sea-water at average *high tide during the *growing season, and where salt-toler-ant plants grow.

saltwater The water of the *sea and *ocean, which has a much higher *salin-ity than *freshwater.

saltwater intrusion The *salinization of *groundwater in coastal areas, which occurs mainly where large quantities of fresh *groundwater are abstracted for supply purposes (usually by pumping it out) and more is taken out than naturally flows in to replenish it. It is a principal cause of salinization in many coastal areas because saltwater, which is per-manently stored in *sediments beneath the *sea-bed, can percolate inland be-neath the freshwater *aquifer. Saltwater intrusion has affected coastal aquifers in many areas, particularly in arid climates. The United Nations Environment Pro-gramme estimates that by the early 1990s it had already affected 70 000 square kilometres in China, 200 000 square kilometres in India, 32 000 square kilometres in Pakistan and the Near East, and 52 000 square kilometres in the USA. It also affects parts of southern Europe. Also know as **saline intrusion**.

salvage To save from ruin, harm, or destruction, including by the *reuse of waste materials.

salvage logging In *forest manage-ment, the logging of dead, damaged, or diseased trees in order to improve the overall health of a forest.

sample In statistics, the particular set of individuals, alive or not, about which measurements are taken. *See also* REPRESENTATIVE SAMPLE. *Contrast* POPULATION.

sampling The process of selecting a representative set of specimens from the full *population, so that the subset can be used to make inferences about the population as a whole.

San Andreas fault A major *fault line system that runs for 1040 kilometres up the west coast of California, defined by the boundary between the Pacific *crustal plate and the North American plate. The Pacific plate is sliding northwest at a speed of a few centimetres a year. Recent *global positioning system measurements put the rate of movement along the San Andreas fault between 1990 and 1993 at about 46 millimetres a year. The San Andreas fault is the source of a number of major destructive *earthquakes along western California, including the 1906 San Francisco earthquake, a similarly large earthquake at Fort Tejon in 1857, and the smaller Loma Prieta earthquake of 1989.

sanction The granting of formal and explicit approval or authorization, or a penalty that is imposed for failure to observe a command or agreed policy.

sand Mineral rock fragments (*sediment) which have a *particle size between 0.06 millimetres and 2.0 millimetres, which is between −1.0 and 4.0 on the *phi (ϕ) scale.

Sand County Almanac A book by Aldo *Leopold (1948), which describes the changing nature, seasons, and his reflections through the year, during which he lived with his family on a run-down farm in rural Wisconsin.

sand desert A type of hot *desert (such as the *Sahara), which is often covered by *sand dunes and dominated by wind action.

sand dune A ridge (*dune) of *sand that is formed when sand grains are blown by the wind, and which migrates forward as the individual sand grains are pushed downwind.

sand mining The deliberate extraction of *sand from beaches and other deposits for use elsewhere (for example by the construction industry). *Contrast* BEACH REPLENISHMENT.

sand prairie A type of *prairie that grows on extensive *sand deposits, which are usually well-drained and leached, and have nutrient-poor soils.

sand storm *See* HABOOB.

sandbar A ridge (*bar) of *sand that has been built up by water *currents, especially in a *river or in *coastal waters.

sandblasting 1. A *physical weathering process in which rock is eroded by the impact of sand grains carried by the wind. *See also* VENTIFACT.
2. A mechanical process that uses sand at high pressure to remove surface material from an object, for example to remove rust from metal objects, or to clean buildings from discoloration.

sandflat A flat area of *sand on a *coast or in an *estuary.

sandstone A *clastic *sedimentary rock composed of compacted and cemented grains of sand (mostly *quartz); it varies in colour, depending on what *mineral it contains and what material cements it together.

sanitary Clean and virtually free from germs (*micro-organisms), thus posing little if any danger of infecting something that comes into contact with it.

sanitary landfill A traditional approach to *solid waste management, in which material is simply dumped in holes in the ground (*landfill) and then covered with soil

sanitary sewer An underground pipe (*sewer) that carries off domestic or industrial *wastewater but not storm water.

sanitary waste Waste material (such as *refuse) that is generated by normal housekeeping activities and is not hazardous or *radioactive.

sanitary water Domestic wastewater that comes from kitchen, bathroom, and laundry sinks, baths, and washing machines. Also known in the United States as **gray water**.

sanitation The control of physical factors in the environment that could harm human health.

sanitation salvage In *forest management, the removal of dead, damaged, or susceptible trees, mainly in order to prevent the spread of *pests or *disease and thus to promote forest health.

sanitization The process of cleaning to levels judged safe by public health standards.

Santa Anna The name given to a warm, dry *adiabatic wind which blows downslope on the *lee side of the mountains in the Los Angeles basin (USA). *See also* CHINOOK, FOEHN.

sap A watery solution of sugars, salts, and minerals (such as *nitrates) that circulates through the *vessels of a *vascular plant, from the roots to the leaves. Also known as **xylem sap**.

saphrophyte A type of *heterotrophic *organism that feeds on soluble organic compounds from dead plants and animals. Some absorb compounds that are already dissolved, others (including some *fungi and many *bacteria) break down undissolved foods by secreting digestive *enzymes onto them.

sapling A young tree, usually less than 10 centimetres in diameter. Also known as a **pole**.

saprolite A soft, fully decomposed *rock that is formed *in situ in tropical areas by the *chemical weathering of *igneous or *metamorphic rocks.

saprophagous Feeding on dead or decaying organic matter.

saprophyte An *organism (such as a *fungus or *bacterium) that lacks *chlorophyll and derives its food from dead or dying organic matter, allowing

*nutrients to be recycled within an *ecosystem.

sapwood The outer zone of wood in a *tree, that lies between the *bark and the *heartwood.

SARA *See* SUPERFUND AMENDMENTS AND REAUTHORIZATION ACT.

satellite **1.** A natural object that orbits a *planet.
2. An artificial object that is placed in orbit around the Earth or another planet in order to collect information or for communication. *See also* REMOTE SENSING.

saturated Filled to capacity (for example, with water).

saturated adiabatic lapse rate (SALR) The rate at which the *temperature of a parcel of *saturated air decreases as it is lifted in the *atmosphere. The rate depends on temperature and *pressure; typical SALR rates are between about 3 and 6°C (the higher the moisture content, the lower the lapse rate). *Contrast* DRY ADIABATIC LAPSE RATE. *See also* LAPSE RATE.

saturated air Air that contains as much *water vapour as it can possibly hold under existing conditions (of temperature and pressure), so that *condensation will occur if any more water vapour is added.

saturated flow The movement of water in a soil that is completely filled with water. *Contrast* UNSATURATED FLOW.

saturated zone That portion of a soil or an *aquifer in which all of the pore space is filled with water. Also known as the **phreatic zone**. *Contrast* UNSATURATED ZONE, ZONE OF SATURATION.

saturation The act of making something totally *saturated.

saturation point The maximum concentration of water vapour that a parcel of air can hold at a given temperature and pressure.

saturation vapour pressure The maximum amount of *water vapour that is needed to keep moist air in *equilibrium with a surface of pure water, which is the maximum amount of water vapour that the air can hold at a given temperature and pressure. *See also* VAPOUR PRESSURE, RELATIVE HUMIDITY.

savanna Extensive open tropical *grassland, with scattered *trees and *shrubs, which covers large areas of Africa, North and South America, and northern Australia.

sawlog A log that is large enough to be cut into sawn *timber.

sawtimber Trees that are large enough to be cut into *timber.

SBS *See* SICK BUILDING SYNDROME.

scarcity A shortage or lack of something (such as a particular *natural resource), so that *supply is insufficient to meet *demand. *See also* ABSOLUTE SCARCITY, RELATIVE SCARCITY.

scarification Loosening *topsoil by mechanical action (for example by scratching the surface of it) in order to prepare the ground for regrowth by direct seeding or natural seed fall.

scarp *See* ESCARPMENT.

scattered clouds The sky condition when between one-tenth and five-tenths is covered by *clouds.

scattered showers *Showers that cover between a quarter and a half of an area.

scattering Dispersing or spreading wide, such as the process in which a beam of light is diffused by collisions with *particles that are suspended in the *atmosphere.

scavenge 1. In ecology, to feed on *carrion or refuse.
2. In meteorology, a process by which *raindrops are formed.

scavenger An *organism (including an animal or bird) that feeds on the dead or rotting bodies of other organisms. *See also* CARRION.

scavenger fish *See* ROUGH FISH.

scavenging *See* PRECIPITATION SCAVENGING.

scenario A description of how a situation might develop, based on a particular set of assumptions and factors, which is used to evaluate available options and select ways of dealing with uncertainty. *See also* WORST CASE SCENARIO.

scenery The appearance of the natural and human-made features of a *landscape.

scenic Having attractive (usually natural) *scenery.

Scenic and Recreational River Part of the *National Wild and Scenic Rivers System in the USA, which includes rivers or sections of rivers that are readily accessible by road or railway, may have some development along their shorelines, and may have undergone some impoundment or diversion in the past.

scenic area Any place that is designated as containing particularly attractive *scenery, and which is specially managed in order to preserve these qualities.

scenic corridor A North American term for the view seen from a road. *See also* VISTA.

schist A medium-grained *metamorphic rock in which the *minerals are compressed into a plate-like arrangement, formed by the applying great heat and pressure to banded *gneiss.

schistosoma Blood fluke; a flatworm *parasite in the blood vessels of mammals.

schistosomiasis *See* BILHARZIA.

schlerophyll Plants with hard, woody, short, and often spiky *evergreen leaves.

schlerophyll forest A forest *biome that is found in *mediterranean climates and is dominated by scattered woody *scrub and widely spaced trees, most of which are less than six metres high.

science The systematic study of the natural world and its physical and biological processes, through observation, identification, description, experimental investigation, and theoretical explanations.

Science and Survival A book by Barry *Commoner (1963), in which he writes that 'the separation of the laws of nature among the different sciences is a human conceit; nature itself is an integrated whole', and 'technology has not only built the magnificent material base of modern society, but also confronts us with threats to survival which cannot be corrected unless we solve very grave economic, social, and political problems'.

Scientific Committee on Problems of the Environment (SCOPE) A global, interdisciplinary group of scientists and scientific institutions that assembles, reviews, and assesses the information available on induced environmental changes and the effects of these changes on people.

scientific method The approach that *science uses to gain knowledge, based on making observations, formulating laws and theories, and testing theories or hypotheses by experimentation.

scientific name *See* LATIN NAME.

scientific notation A method of writing numbers in terms of powers of ten. For example the number 0.000118 would be represented as 1.18×10^{-4} or 1.18E−04 where E stands for exponent, as in the exponent that ten is raised by.

Scientific Reserve/Strict Nature Reserve An *IUCN Management Category (I) for protected areas that are designed 'to protect nature and maintain natural processes in an undisturbed state in order to have ecologically representative examples of the natural environment available for scientific study, environmental monitoring, education, and for the maintenance of genetic resources in a dynamic and evolutionary state'.

sciophilous A term describing a shade-loving plant which thrives in habitats of low light intensity. *Contrast* PHOTOPHILOUS, PHOTOPHOBIA.

sclerite A general term for any single plate of the *exoskeleton of an *organism.

SCOPE *See* SCIENTIFIC COMMITTEE ON PROBLEMS OF THE ENVIRONMENT.

scoping A process for identifying which potential environmental impacts, alternatives, and other issues will be addressed in an *environmental impact statement.

scour The effect of *abrasion, usually by sand or gravel, on the bed of a *river or the *sea.

scrambler A *root climber plant that raises itself above other vegetation with the aid of thorns or hooks. *Contrast* TWINER.

scrap 1. A small fragment of something.
2. Waste material that is discarded from manufacturing operations but may be suitable for *reprocessing.

scrap metal Discarded metal that is suitable for *reprocessing.

scree A slope that is covered with loose rock fragments that have broken off a cliff or steep slope above as a result of *frost action and *rockfall. Also known as **talus** or **talus cone**.

screening 1. The sifting of *sewage through a screen in order to remove coarse *solids that are floating and suspended in it.
2. Testing a large number of individuals from a population in order to identify which ones have a particular genetic trait, characteristic, or biological condition.

scrub Dense vegetation consisting of stunted *trees or bushes which colonize open ground, particularly *grassland. *See also* MAQUIS.

scrub oak–mountain mahogany shrubland A North American *shrubland vegetation dominated by mountain mahogany, serviceberry, and/or Gambel oak, accompanied by numerous *grass and *forb species, which is found in temperate areas such as the Central Rockies.

scrubber An *air pollution control device that uses a spray of water to trap *pollutants and cool *emissions.

scrubbing The process of purifying a gas by washing it with a liquid in a tower. *See* FLUE GAS SCRUBBING, LIMESTONE SCRUBBING.

S-curve A curve that describes *logistic growth in a *population, so-called because of its shape. *Contrast* J-CURVE.

scyphozoa The true jellyfish; a member of the phylum Cnidaria.

SDWA *See* SAFE DRINKING WATER ACT.

SEA *See* STRATEGIC ENVIRONMENTAL ASSESSMENT.

sea Part of an *ocean or a large body of *saltwater that is at least partially enclosed by *land.

sea breeze A coastal *breeze that blows from sea to land. Also known as **onshore breeze**. *Contrast* LAND BREEZE.

sea defence wall An artificial structure that is built along a *coastline in order to protect cliffs or property from *erosion by *waves and *tides.

sea ice *Ice that is formed by the freezing of seawater and can extend out from the land as *pack ice, for example around the *Antarctic.

sea level The level of the surface of the *ocean, based on the average level between high and low *tides, which is used as a *datum when calculating the height of features on land or depth in the ocean. It is used as a standard base for measuring heights and depths on the Earth, and it is the baseline for plotting the global *hypsometric curve, and for calculating the heights of *mountains and the depth of *ocean trenches. *Atmospheric pressure readings are usually expressed as sea level equivalents in order to eliminate the effects of *altitude.

sea level change A long-term change in the relative level of land and sea, which can be caused by *isostatic adjustment of the land (as a result of *tectonic forces), and/or by *eustatic change (change in the volume of water that is stored in the *oceans). Geological evidence suggests that during the height of the last *glaciation (about 18 000 years ago) sea level was between 110 metres and 140 metres lower than it is today, and large areas of the *continental shelves were dry land. Sea level has risen through the post-glacial period (the so-called *Flandrian *transgression), rapidly at first, but levelling off about 5000 to 6000 years ago. It has remained relatively close to its present position over the last 5000 to 6000 years. Global warming has caused a global rise in sea level of between 15 and 20 centimetres over the last century; roughly 2–5 centimetres of the rise has been caused by melting of *glaciers and *ice caps, and another 2–7 centimetres has resulted from the expansion of ocean waters as they warm.

sea spray The spray that blows from the tops of *ocean waves. Also known as **salt spray**.

sea-bed The bottom of a *sea or *ocean.

sea-bed mining Extracting *natural resources (such as nodules of *manganese) from the *sea-bed.

sea-floor spreading The process by which *crustal plates beneath the *oceans are pushed apart as new *magma is pushed upwards from within the Earth's crust to the surface of the ocean floor, in the central trough of the *mid-ocean ridge, where it cools. The new rock mirrors and preserves the *magnetic polarity of the Earth at the time of formation. *See also* CONSTRUCTIVE PLATE BOUNDARY, PLATE TECTONICS.

seamount A flat-topped submarine *mountain. Also known as a **guyot**.

search and rescue In *disaster management, the process of locating and recovering victims and applying first aid and basic medical assistance.

seascape A coastal or maritime *landscape. *See also* PROTECTED LANDSCAPE OR SEASCAPE.

season A natural division of the year, defined by the *equinoxes and *solstices (the four seasons: *winter, *spring, *summer and *autumn/fall) or by atmospheric conditions (for example, the *monsoon season).

seasonal Occurring during or dependent upon a particular *season.

seasonal plant A plant that flowers and completes its life cycle within a single wet–dry season in equatorial regions.

seasonal polyphenism A type of *polyphenism in which different forms of a species are produced at different times of the year. It is a common form of *phenotypic plasticity among insects.

seaward In the direction of the *sea. *Contrast* LANDWARD.

seaweed Any of the large plants that grow in *the sea, particularly marine *algae such as kelp.

Secchi disc A disc that is lowered into a waterbody in order to determine the clarity or transparency (*opacity) of the water, based on the depth at which its black and white markings can no longer be seen from the surface.

second growth *See* SECONDARY FOREST.

Second World Centrally planned economies, such as those of Eastern Europe between 1948 and 1990. *Contrast* FIRST WORLD, THIRD WORLD.

secondary ambient air quality standards *air quality standards that are designed to protect human welfare, including the effects on vegetation and fauna, visibility, and structures. *Contrast* PRIMARY AMBIENT AIR QUALITY STANDARDS.

secondary consumer A *carnivore, which sits above (at a higher *trophic level than) the *primary consumers in a *food chain because it eats them. Examples include: on land, lions that eat wildebeest, hawks that eat field mice; in the soil, soil organisms that decompose herbivore remains; in the oceans, herring that eat copepods.

secondary economic activities Manufacturing activities. *Contrast* PRIMARY ECONOMIC ACTIVITIES, TERTIARY ECONOMIC ACTIVITIES.

secondary effect *See* INDIRECT EFFECT.

secondary forest A *forest that grows on land after the original mature *old growth forest has been *harvested, or died because of *fire or insect attack. Also known as **second growth**. *Contrast* PRIMARY FOREST.

secondary market The exchange (under the *Kyoto Protocol) of *greenhouse gas *emission reductions, *offsets, or *allowances between buyer and seller where the seller is not the originator of the supply. *Contrast* PRIMARY MARKET.

secondary material A material that is recovered from *post-consumer wastes and is reused in the manufacture of a product. Also known as **post-consumer material**.

secondary mineral *Minerals (such as *clays and *oxides) that are formed from the material released by *weathering. *Contrast* PRIMARY MINERAL.

secondary nutrient The elements *sodium, *calcium, and *magnesium that must be taken up and used in sufficient quantities for plants to complete their life cycles. *Contrast* PRIMARY NUTRIENT. *See also* ESSENTIAL ELEMENT.

secondary pollutant A *pollutant that is created when *primary pollutants combine with each other and with other substances, as happens for example when *sulphur dioxide reacts with *oxygen and moisture to form *sulphuric acid and give rise to *acid rain. *Contrast* PRIMARY POLLUTANT.

secondary production *See* SECOND-
ARY PRODUCTIVITY.

secondary productivity The amount
of new biomass (weight) that is produced
by plant-eating animals (*herbivores) in
a given period of time. Also known as
secondary production. *Contrast* PRIMARY
PRODUCTIVITY, TERTIARY PRODUCTIVITY.

secondary range A *grazing area
(*range) that is lightly used or unused
by livestock when there is limited or no
*range management. It will not nor-
mally be fully used until the *primary
range has been overused.

secondary recovery Extracting more
*oil or *gas from a reservoir than can
normally be recovered by normal flowing
and pumping operations, for example by
pumping pressurized gas, steam, or containing chemicals water into a *well.

secondary rock *See* SEDIMENTARY
ROCK.

secondary standard The pollution
limit for *criteria air pollutants that is
based on environmental effects (such as
damage to property, plants, or visibility).
Contrast PRIMARY STANDARD.

secondary succession The re-estab-
lishment of *vegetation on land that has
had a vegetation cover at some time in the
past but which might have been
destroyed by natural processes (including
*lightning fires, severe *storms and *vol-
canic eruptions) or by human interference (including *forest clearance,
building construction, and fires). *Contrast*
PRIMARY SUCCESSION. *See also* SUCCESSION.

secondary treatment The second
treatment that is provided at a *wastewater treatment plant, in which *bacteria consume suspended organic matter
in the waste that remains from the *primary treatment of sewage. This process
removes only limited amounts of *nitrogen and *phosphorus, which must
be removed by *advanced wastewater
treatment. Also known as **secondary wastewater treatment**. *See also* PRIMARY
WASTE TREATMENT.

secondary use Use that is incidental to
the main purpose of an area or building.

secondary wastewater treatment
See SECONDARY TREATMENT.

secondary wave *See* S-WAVE.

secondary woodland *Woodland on
a site that has not been continuously
wooded since 1600 AD. *Contrast* ANCIENT
WOODLAND.

second-growth forest A *forest that
has grown back after being logged, as a
result of *secondary succession.

sector A segment or part of the *economy of a country in which particular
activities take place. *See also* COMMERCIAL
SECTOR, INDUSTRIAL SECTOR, RESIDENTIAL
SECTOR, TRANSPORTATION SECTOR.

secular **1.** Long term change that takes
place slowly, as opposed to *seasonal or
*cyclical changes.
   2. Not sacred or religious, relating to
the worldly.

secure landfill A solid waste disposal
site (*landfill) that is lined and capped
with an *impermeable barrier in order
to prevent leakage or *leaching.

sedentary Not moving, staying in one
place. *See* SESSILE.

sedentism Living in one main place all
year round; having a *sedentary lifestyle.
Contrast NOMADISM.

sedge A plant of the family Cyperacae,
which resemble *grasses but have solid
stems, and often grow in *wetlands.

sediment Grains of solid (usually
*mineral) material that have been depos-
ited by some natural process.

sediment delivery ratio The proportion of the *soil that is eroded from upland sources that actually reaches a
*stream channel or storage *reservoir.

sediment discharge *See* SEDIMENT
LOAD.

sediment load The amount of *sediment that is transported through a

*stream cross section over one year, which includes chemicals carried in solution (the *dissolved or *solution load), small particles carried in suspension (the *suspended load), and larger particles that are rolled along the river bed (the *bedload). Also known as **sediment discharge** or **sediment yield**.

sediment pond A natural or artificial ^pond that is used for the recovery (by settling) of solids from *effluent or *runoff.

sediment yield See SEDIMENT LOAD.

sedimentary rock A type of *rock that is formed by deposition, either of cemented fragments of rocks that have been eroded from the land (*clastic sedimentary rocks), organic materials (*organic sedimentary rocks), or chemicals that are deposited from water (*chemical precipitates). Also known as **secondary rock**. See also ROCK CYCLE, DIAGENESIS.

sedimentation The deposition of *sediment from a state of *suspension in water or air. Also known as **siltation**.

sedimentation basin See CLARIFIER.

sedimentation tank A *wastewater tank in which floating wastes are skimmed off and settled *solids are removed for disposal elsewhere.

sedimentology The study of the nature and character of *sediments, and of how they form.

seed The fertilized part of a *plant that contains food products for *germination.

seed bank A special type of *gene bank that is designed for the *ex situ conservation of individual plant varieties through the preservation and storage of *seeds.

seed productivity The amount of *viable *seed that is produced by an individual plant.

seed shrimp See OSTRACOD.

seed tree In *forest management, a tree that is left standing in order to provide *seed for the natural *regeneration of the area that has been harvested.

seedbed The *soil in which seeds are deposited and *germinate.

seeding Putting seeds in or on soil to allow it to *germinate and grow. See also BROADCAST SEEDING, DRILL SEEDING.

seedling The young plant that grows from a seed that has *germinated.

seep A small *spring, *pool, or other place where water trickles naturally from the ground.

seepage The slow movement or *percolation of water through *soil or *rock.

seepage lake A *lake that has either an inlet or an outlet but not both and which receives water mainly from *precipitation and *groundwater rather than from inflow along *tributary streams.

seepage line See PERCOLINE.

seiche A *wave on the surface of a *lake or landlocked *bay that is caused by local atmospheric changes, tidal currents, or *earthquakes.

seif dune A type of *sand dune that is common in hot *deserts, in which long ridges of wind-blown sand are aligned parallel to the dominant wind direction.

seine net A traditional form of *ocean fishing involving the use of curtain nets that are drawn out and pulled along, in order to enclose a shoal of fish. Also known as a **ring net**.

seismic Relating to *earthquakes, particularly the energy waves that are generated by *earthquake activity.

seismic activity *earthquake activity.

seismic seawave See TSUNAMI.

seismic survey A technique for determining the detailed structure of the *rocks beneath a particular area, which is based on sending acoustic *shock waves into the rocks and measuring the signals that are reflected back.

seismic wave A *shock wave that is generated by an *earthquake and measured using a *seismograph. The waves travel at different speeds, depending on the density of the material they pass through. There are two main types of seismic wave, the *S-wave (secondary wave) and the *P-wave (primary wave).

seismicity The geographical and historical distribution of *earthquakes, or the number of earthquakes in a particular period of time.

seismograph An instrument for measuring and recording Earth tremors and *earthquakes, based on recording the passage of *seismic waves through the ground.

seismology The study of *earthquakes and *seismic activity.

selection Choice between available options. See also ARTIFICIAL SELECTION, NATURAL SELECTION.

selection pressure A measure of the effectiveness of *natural selection in altering the *genetic composition of a *population.

selective breeding See ARTIFICIAL SELECTION, CAPTIVE BREEDING.

selective cutting See SELECTIVE THINNING.

selective pesticide A chemical *pesticide that is designed to affect only particular types of *pests, leaving other plants and animals unharmed.

selective thinning In *forest management, the periodic removal (*thinning) of individual trees or groups of trees in order to improve or regenerate a *stand. Also known as **selective cutting**. See also LINE THINNING, MECHANICAL THINNING.

selenium (Se) A relatively rare *non-metallic element that occurs naturally in rock and soils in certain regions. It is released from *copper and *lead refineries and municipal wastewater. It is essential to humans and animals in low concentrations and is toxic at very high concentrations. See also ACID MINE DRAINAGE.

self-pollinated See INBRED.

self-sufficient Able to look after yourself, and provide for all of your needs yourself without help from others.

seller In *greenhouse gas *emissions trading, under the *Kyoto Protocol, a legally recognized entity (such as an individual, corporation, not-for-profit organization, or government) that sells reductions, *credits, or *allowances to another legally recognized entity through a sale, lease, trade, or other means of transfer.

selva See TROPICAL RAINFOREST.

semi-arid A moderately dry climate, with amounts of rainfall between 250 and 500 millimetres a year, which is usually enough to support the growth of short, sparse *grass. See also STEPPE.

semi-confined aquifer An *aquifer that is partially overlain by a rock formation which has low *permeability, through which water can pass only slowly to *recharge the aquifer. Contrast CONFINED AQUIFER.

semi-improved grassland *Grassland that has been modified by the application of *fertilizers and/or *herbicides, and by intensive *grazing or *drainage. Contrast IMPROVED GRASSLAND.

semi-natural habitat A habitat which has been affected directly or indirectly by human activity. Contrast NATURAL HABITAT.

semiochemical A naturally occurring message-bearing *biochemical that organisms use to communicate with each other, and to interrogate their environment. See also ALLOMONE, KAIROMONE, PHEROMONE.

semi-permeable A material that will allow some but not all substances to pass through it. Contrast IMPERMEABLE, NONPERMEABLE, PERMEABLE.

semi-primitive motorized area A category of land defined within the *recreation opportunity spectrum as 'a predominantly natural or natural appearing environment of moderate to large size', with a low concentration of users, motorized recreation use permitted only on unmade local primitive roads, and the area managed to have subtle minimum on-site controls and restrictions.

semi-primitive non-motorized area A category of land defined within the *recreation opportunity spectrum as 'a predominantly natural or natural-appearing environment of moderate to large size', with low interaction between users, motorized recreation use not permitted, local roads possibly used on a limited basis for other resource management activities, and the area managed to have minimal but subtle on-site controls and restrictions.

semi-volatile organic compound (SVOC) A group of *organic compounds that are composed primarily of *carbon and *hydrogen, and have a tendency to *evaporate (volatilize) into the air from water or soil.

senescence The loss of function that accompanies aging, for example in a plant or animal or in a *waterbody that is subject to *eutrophication.

sensible heat Heat that causes a change in temperature (for example of air in the atmosphere) by changing the speed at which the *molecules move.

sensible heat flux The transfer of heat that is associated with thermal heating of parcels of air.

sensible temperature The sensation of temperature that a human body feels, rather than the actual temperature of the *environment as measured with a *thermometer.

sensitive species Plant or animal *species which are susceptible to changes in *habitat or to impacts from human activities.

sensitivity The degree of change in a system that is associated with a given degree of stress or perturbation. See also CLIMATE SENSITIVITY, SUSCEPTIBILITY.

septic system A domestic *wastewater treatment system (consisting of a *septic tank and a soil absorption system) into which wastes are piped directly from the home. *Bacteria *decompose the waste, *sludge settles to the bottom of the tank, and the treated *effluent flows out into the ground through drainage pipes.

septic tank An underground tank for storing *wastewater from a home. See also SEPTIC SYSTEM.

sequential cropping A pattern of *multiple cropping in which one *crop follows another on the same land, either with or without a break, in successive seasons.

sequester To separate or lock up. See also SEQUESTRATION.

sequestration The removal (*sequestering) of *greenhouse gases (particularly *carbon dioxide) from the *atmosphere by *biomass (such as trees, soils, and plants), via *photosynthesis, in order to help reduce the *greenhouse effect and *control global warming. See also CARBON SEQUESTRATION, FOREST SEQUESTRATION.

sequoia Either of two species of huge *coniferous trees, that reach heights of up to 100 metres. See also REDWOOD.

serac Jagged pinnacles of ice on the surface of a *glacier or *ice sheet.

seral Of or relating to the series of stages (*seres) in ecological or plant *succession.

seral stage A recognizable step or stage in the development of an ecological *community through *succession.

sere A transitional stage in the development of a plant community as it changes, through *succession, from bare ground to *climax. The full sequence of seral communities is a *prisere.

sericulture The culture of silkworms (*Bombyx mori*) for the production of silk.

sessile 1. An organism that is *sedentary, fixed, or attached and unable to move at will. *Contrast* MOTILE.
2. A leaf that has no stalk, and is attached to the plant directly at the stem.

seston *Particulate matter that is suspended in *seawater.

set-aside Agricultural land that is removed from food production as part of the *Common Agricultural Policy, which allows farmers to claim special payments and creates an opportunity to adopt land management practices that benefit *wildlife.

settleable solids Suspended material that is heavy enough to sink to the bottom of a *wastewater treatment tank.

settlement 1. An area where one or more groups of families live together, either in the *countryside (*See also* RURAL SETTLEMENT) or in towns and cities (*See also* URBAN SETTLEMENT).
2. The act of establishing a *colony.
3. A legal agreement that is reached between two or more individuals or parties.

settling A gradual sinking to a lower level by *gravity, such as the deposition of suspended solids on the bed of a *waterbody.

settling basin *See* CLARIFIER.

settling tank A storage tank for the treatment of drinking water and *wastewater, in which heavy particles sink by *gravity to the bottom from where they can be removed and disposed of. *See also* SEPTIC TANK.

severe storm A major *storm that is associated with a *depression. Such storms are given different names in different parts of the world, including *hurricane (throughout most of the western hemisphere), *typhoon (through most of

Asia and across the Pacific), *cyclone (in the Indian Ocean and Australia), and *willy-willy (in some South Pacific islands).

severe thunderstorm A *thunderstorm that produces heavy *precipitation, frequent *lightning, strong, gusty surface winds, or *hail, and can cause *flash floods and damage structures.

Seveso The site in northern Italy of a major *air pollution incident on 10 July 1976, when a cloud of chemicals containing *dioxin was released from a chemical factory, to which an estimated 37 000 people were exposed between Milan and Lake Como, raising concern about the prospects of long-term health problems.

Seveso II Directive A directive of the *European Union on the control of major accident hazards involving dangerous substances, that is designed to help prevent such hazards in the wake of the *Seveso accident by making information more widely available about the distribution of hazardous substances, and by the adoption of major accident prevention policies and strategies.

Seveso site A site where hazardous substances are located, as defined under the *Seveso II Directive. Such sites are referred to as either upper tier or lower tier, depending on the level of *risk.

sewage The *organic waste and *wastewater that is produced by residential and commercial establishments, and is discharged into *sewers.

sewage contamination The introduction of untreated *sewage into a *waterbody.

sewage lagoon *See* OXIDATION POND.

sewage outfall The point at which *sewage is discharged, often from a pipe, into a *waterbody.

sewage sludge The solid portion of *sewage that contains *organic matter along with the *algae, *fungi, *bacteria, and *protozoans that consume it. Also known as **biosolids** and **sludge**.

sewage system The system of pipes (*sewers) that carries *sewage. Also known as **sewerage**.

sewage treatment See WATER TREATMENT.

sewage treatment plant A facility that is designed to receive the *wastewater from domestic sources, and to remove from it any materials that reduce *water quality and threaten public health and safety when they are discharged into receiving streams or *waterbodies.

sewer A channel or network of pipes that carries *wastewater and *stormwater runoff from the source to a treatment plant or receiving stream. See also SANITARY SEWER, STORM SEWER.

sewerage See SEWAGE SYSTEM.

sex ratio The number of females per thousand males within a *population.

sexual reproduction The production of new individuals through *fertilization of male and female *gametes.

sexual selection A type of *natural selection that acts differently on males and females of the same species.

shade intolerant Plants (particularly trees) that need a lot of *sunlight to regenerate and grow.

shade tolerant Plants (particularly trees) that can regenerate and grow in the shade of surrounding vegetation.

shadow price The *social value of a resource (such as habitat or wilderness), for which a meaningful economic price is difficult to calculate.

shale A fine-grained, grey or black *clastic *sedimentary rock that is composed of compacted and hardened *silt and *clay, with clear *laminations.

shale oil A *petroleum-like substance that is found in high concentrations in some *shale rocks. See also KEROGEN.

shallow ecology A *worldview or set of beliefs which reflects a utilitarian and *anthropo centric attitude to nature, based on materialism and *consumerism. It seeks technological solutions to major environmental problems, rather than a change in human behaviour and valves. For example, shallow ecology promotes the recycling of waste rather than preventing waste in the first place. Contrast DEEP ECOLOGY.

Shannon–Wiener index A measure of *biodiversity, based on the number of *species (species richness) and the way in which the individuals are apportioned into those species (evenness) within a given area. The index number is large when the species diversity is large and/or apportioned evenly amongst the species, and it is small when the species diversity is small and/ or apportioned unevenly amongst the species. This method is based on information theory, and it is used when there are too many individuals within the sample area for each one to be identified and counted. The index (H) is calculated using the formula $H = -\sum_{i=1}^{N_0} [p_i \times \log p_i]$, where i is the total number of individuals in one species, p_i is the number of individuals of one species in relation to the number of individuals in the population, and N_0 is the total number of species in the sample. See also SPECIES DIVERSITY INDEX.

shanty town A peripheral area of a town or city that is inhabited by the very poor. It usually consists of simple houses with little or no infrastructure or services and is located in marginal environments with low *environmental quality and high environmental *risk. See also SQUATTER SETTLEMENT.

shear See WIND SHEAR.

shear strength A *force that acts on a *particle, which serves to resist movement and is thus an opposing force to *shear stress.

shear stress A *force that acts on a *particle, which exists because gravity pulls objects downhill so it opposes

S

*shear strength. It promotes downslope movement, and its strength is determined mainly by slope angle (steeper slopes have higher shear stress, all else being equal).

shear wave *See* S-WAVE.

shearing 1. To cut the fleece off an animal using electric or hand-held shears.
2. To trim back and shape tree branches.
3. The deformation a body by moving one part of it relative to another.

sheep dip A liquid *disinfectant and *insecticide in which sheep are immersed to protect them from pests and diseases.

sheet erosion The *erosion of a layer of exposed *soil from a sloping surface by *runoff water or wind.

sheet flow *See* OVERLAND FLOW.

sheet lightning A bright flash of *lightning from a distant *thunderstorm that lights up part of the *cloud.

shelf break The sharp change in slope on the ocean floor between the *continental shelf and the *continental slope.

shelter *See* COVER.

shelterbelt A natural barrier composed of rows of *trees and *shrubs that are planted to block wind flow and drifting snow, reduce *soil erosion, and protect buildings, gardens, orchards, or sensitive crops from high winds. Also known as a **windbreak** or **windstrip**.

sheltered In terms of *wave exposure, coasts with a restricted *fetch which are protected from danger or bad weather.

shelterwood cut In *forest management, the gradual removal of the entire *stand of trees over a period of partial cuttings, which allows an *even-aged stand to regenerate in the partial shade. *See also* REMOVAL CUT.

shepherd A person who herds sheep on an open *range.

shield In a nuclear power plant, a wall to protect people from exposure to harmful *radiation.

shield area *See* CRATON.

shield volcano A *volcano that is built from repeated non-explosive flows of *lava, covers a large area, and has relatively shallow slopes, like many of the volcanoes in Hawaii.

shifting cultivation *See* SLASH AND BURN.

shingle 1. A general term for coarse *beach material, which is often well-rounded and composed of hard resistant rock, and a mixture of *gravel, pebbles, and larger material.
2. Wood or slate building material that is used as roofing or siding on a building.

shoal 1. A *sandbank in a river or stream, that is visible under low flow.
2. A large group of *fish.

shock load The arrival at a *water treatment plant of water that contains high amounts of *algae, *suspended solids, or other *pollutants.

shock wave A large *compression wave (such as a *seismic wave or *sonic boom) that is caused by a shock to the medium through which the wave travels.

shoot The above-ground portion of a *plant, on which the leaves grow. *Contrast* ROOT.

shooting Killing wildlife, often *game as a *field sport, by gunfire.

shopping arcade *See* SHOPPING MALL.

shopping centre *See* SHOPPING MALL.

shopping mall A North American term for a building or set of buildings that contain stores and have interconnecting walkways that may or may not be enclosed, and which make it easy for people to walk from store to store. In the United Kingdom and Australia these are called **shopping centres** or **shopping arcades**. *Contrast* STRIP MALL.

shore The strip of land that borders a relatively large *waterbody (such as a *lake or *ocean) and is above the *high water mark. *See also* LITTORAL.

shoreline *See* COASTLINE.

short term A brief period of time, in environmental terms often defined as less than about three to five years. *Contrast* LONG TERM, MEDIUM TERM.

short-distance migration A pattern of *migration in which species move within, rather than between, *temperate or *tropical zones.

shortgrass A North American *prairie *grassland vegetation dominated by grasses that are short and can tolerate drought conditions, which is common on the dry upland plains east of the Rocky Mountains.

shortgrass prairie A type of *prairie that grows in areas that receive between 30 and 40 centimetres of precipitation per year and have high *evaporation rates.

short-range planning Planning ahead up to five years. *Contrast* LONG-RANGE PLANNING. *See also* SHORT TERM.

short-term cleanup Cleanup of a *contaminated site that seeks to minimize immediate threats to public health and the environment, for example after an emergency fire, spill, or release. Also known as **removal action**.

shortwave radiation *Radiant energy in the visible and near-ultraviolet *wavelengths (between about 0.1 and 2.0 micrometres), such as that emitted by the Sun. *Contrast* LONGWAVE RADIATION.

shower A sudden downpour or short period of heavy *precipitation.

shredding 1. Reducing the size of a particular material (such as waste paper for *recycling) by tearing or chopping it. **2.** Mixing and grinding *soil in order to make it more homogeneous and increase *permeability.

shrink–swell clay *Clay that expands (swells) when wet and shrinks when dry.

shrub A low, woody, *perennial *plant that has no trunk but several major branches at or near the base giving a bushy appearance, usually less than six metres tall.

shrubbery A collection of *shrubs that are growing together, which may have been planted.

shrubland An open or closed *stand of *shrubs up to about two metres tall.

SI *See* SYSTEME INTERNATIONALE.

sial Part of the Earth's *crust, which is named after its two main chemical constituents, silica (Si) and alumina (Al). It has a lower density than *sima, and forms the *continental crust.

sibling species Species that look so alike that human observers find it hard to distinguish between them.

sick building syndrome (SBS) A set of symptoms that affect some people while they are inside a particular building but which disappear when they leave the building, and which cannot be traced to specific *pollutants or sources within the building.

side effect An effect (often undesired or negative) other than the one intended.

Sierra Club A US environmental organization founded by John *Muir in 1892 that promotes public education, litigation, and outings and conferences.

significant effect A serious adverse impact in terms of such things as size, geographical extent, duration, frequency, or degree of reversibility.

silage *Fodder that is made by storing and fermenting green *forage plants in a *silo.

Silent Spring A book written by Rachel *Carson, published in 1962, which described the loss of birds killed by *pesticide poisoning from agricultural

S

*insecticides, and which had a significant impact on *attitudes towards the *environment, and on the development of modern *environmentalism.

silica A *mineral composed of *silicon dioxide that is found in all *volcanic rocks and is an important ingredient in the Earth's *crust.

silicate A mineral or rock whose composition includes *silicon and *oxygen.

siliceous Mainly or entirely composed of *silicon.

silicification The process in which *silica replaces the original material of a substance, such as wood. See also PETRIFIED FOREST.

silicon (Si) A naturally occurring, *non-metallic *element, which is a major component of many types of rocks and minerals and accounts for 27.2% of the Earth's *crust

silicon dioxide (SiO₂) A white or colourless solid that becomes soluble at high temperatures. Also known as **sand**. See also SILICA.

sill A sheet-like horizontal intrusion of *magma between existing rock strata.

silo A tall, cylindrical structure (or pit) in which *fodder is stored to make *silage.

silt Fine *sediment with a *particle size between 0.002 and 0.06 millimetres, which is between 4.0 and 9.0 on the *phi (ϕ) scale; coarser than *clay but finer than *sand.

siltation See SEDIMENTATION.

siltstone A fine-grained *clastic *sedimentary rock composed of compacted and hardened *silt-size particles.

Silurian The third period of the *Palaeozoic era, covering the period from about 440 to 408 million years ago, during which the first species of plants and insects appeared.

silver (Ag) A naturally occurring, soft, white, *precious metallic element that is used in coins and jewellery, and has many other uses.

silviculture The care and cultivation of *forest trees. See also FORESTRY.

silvopastoral system An *agroforestry land use system in which *trees or *shrubs are grown and animals graze or browse. See also FOREST GRAZING

silvopisciculture Growing trees as part of a fish-farming enterprise.

sima The *oceanic crust of the Earth, which is named after its two main chemical constituents (*silica (Si) and *magnesium (Ma)), is denser than *sial that covers it in places, and where it is not covered by sial it comprises the ocean bed.

Simon, Julian US academic and environmental *optimist (1932–98) who is best known for his work on population and natural resources, including the controversial book The *Ultimate Resource.

simulation An experiment run as a model of reality, often using a computer, to predict the behaviour of a *system under certain conditions.

single occupancy vehicle (SOV) A privately operated vehicle whose only occupant is the driver, who uses it mainly for personal travel and daily commuting to work. Contrast HIGH OCCUPANCY VEHICLE.

sink 1. A place where a substance can be stored naturally, such as plants which are a sink for *carbon dioxide (because they transform it by *photosynthesis into *organic matter, which either stays in the plant or is stored in the soil). Contrast SOURCE. See also CARBON SINK.
   2. See SINKHOLE.

sink habitat A *habitat that is a net importer of individuals, because local reproduction is not sufficient to balance local *mortality. Contrast SOURCE HABITAT.

sink population A population that occupies a *sink habitat. Contrast SOURCE POPULATION.

sinkhole A steep-sided depression that is found in some *limestone (*karst) areas and is caused by the collapse of an underground channel or cavern as a result of *solution weathering. Also known as **sink**.

sinking A method of controlling an *oil spill, using a *sinking agent to trap the oil and sink it to the bottom of the *waterbody where the mixture of oil and agent eventually undergoes natural *biodegradation.

sinking agent A chemical additive that is added to a floating *oil spill in order to sink it below the water surface.

sinter A chemical *precipitate that is formed from the minerals (mainly *silica) in *geothermal water, especially from *geysers.

sinuosity A measure of the degree of *meandering within a river, defined as the ratio of stream length to valley length. Tightly meandering rivers travel much further over a given length of valley and so they have high sinuosity.

sinuous Winding, with many curves or turns.

SIP *See* STATE IMPLEMENTATION PLAN.

sirocco A dry *desert wind that blows from the Sahara in north Africa across the Mediterranean Sea (where it picks up moisture) towards Sicily and southern Italy, where it is experienced as a hot, humid wind.

SITE *See* SUPERFUND INNOVATIVE TECHNOLOGY EVALUATION.

site A piece of land on which something is located.

site assessment The formal process by which the US *Environmental Protection Agency determines whether a potential site should be placed on the *National Priorities List (NPL). It can consist of a *Preliminary Assessment (PA) or a combination of a PA and a *Site Inspection (SI).

site capacity The *carrying capacity of a particular *site for a particular use.

site class A form of *land classification based on ecological factors and the potential production capacity (such as *crop or *forest yield) of a particular site.

Site Inspection (SI) A stage in *site assessment in the USA, following the *Preliminary Assessment, during which the *Environmental Protection Agency gathers information from a *Superfund site in order to use the *Hazard Ranking System to determine whether the site should be placed on the *National Priorities List.

Site of Special Scientific Interest (SSSI) A *conservation site of national importance in the UK that has been designated for special protection under the Wildlife and Countryside Act 1981 because of its plants, animals, or geological or physiographical features.

site preparation The treatment of a site (for example with mechanical clearing, burning, or *herbicides) to prepare it for planting, such as *reforestation.

site remediation The process of cleaning up a *hazardous waste disposal site. *See also* REMEDIATION.

siting The process of choosing a location for a *facility.

size class In *forest management, one of three categories of tree stem diameters that is used to classify timber, which are seedling/sapling (less than five inches (12.7 centimetres) in diameter), pole timber (five to seven inches (12.7 to 17.8 centimetres)), and sawtimber (greater than seven inches (17.8 centimetres)).

Skeptical Environmentalist, The A book by Bjorn Lomborg, published in 2001, which challenges the belief that the global environment is progressively getting worse, using statistical information from reputable international research institutes. It is one of the most controversial books on the environment in recent years. *See also* OPTIMIST, ENVIRONMENTAL.

skidding Hauling logs by sliding them from where they are cut to a collection point.

skimming Using a machine to remove a physical *pollutant (such as *oil) from the surface of a *waterbody.

skin absorption The ability of some *hazardous chemicals to enter the human body by passing directly through the skin and entering the bloodstream.

sky The *atmosphere and outer space beyond it, viewed from the ground.

skyline logging A method of *logging in which timber is transported from where it is cut to a collecting point by means of an overhead trolley that runs along a cable system. Also known as **skyline yarding**.

skyline yarding See SKYLINE LOGGING.

slack A low-lying area of ground in an area of coastal *sand dunes, which is often damper than the surrounding sand and may even contain temporary *pools.

slack water The minimum velocity of a *tidal stream, at the point of cross-over between successive *flood currents and *ebb currents, when water is relatively calm.

SLAPP See STRATEGIC LAWSUIT AGAINST PUBLIC PARTICIPATION.

slash See LOGGING RESIDUE.

slash and burn A method of clearing a patch of *forest land for *cultivation, in which the trees are first felled and then burned to add *nutrients to the soil. Also known as **milpa, swidden agriculture** or **shifting cultivation**.

slash disposal The removal of *logging residue, often in order to reduce the risk of *fire spreading.

slashing In *forest management, cutting back the less tough, competing vegetation (such as *ground cover like bracken) to improve tree growth.

slate A black or grey, fine-grained *metamorphic rock that is formed by the applying great heat and pressure to *shale and *mudstone, and is useful as a building and roofing material.

sleeping sickness An *encephalitis that is caused by a *virus, which was *epidemic between 1915 and 1926, and whose symptoms include paralysis of the eye muscles and extreme muscular weakness.

sleet *Precipitation in the form of *rain that contains some *ice or *snow that melts as it falls. Also known as **ice pellet**.

slickenside A polished rock surface that is created when one mass of rock slides over another in a *fault plane.

slide See LANDSLIDE.

slime mould A mass of *protoplasm that has characteristics of both plants and animals. A member of the kingdom *Protoctista.

sling psychrometer An instrument that is used to measure the amount of *water vapour in the air, using wet and dry bulb *thermometers mounted on a frame.

slip See LANDSLIP.

slope The angle at which something is inclined, which is normally expressed as fall (drop in height) (metres) per unit distance (kilometres), or metres per kilometre. Also known as **gradient**.

slope stability The resistance of an inclined surface to failure by sliding or collapsing.

slough A *stagnant *swamp type of vegetation that is dominated by grasses, sedges, and rushes and is commonly found on the muddy backwater of a river, marsh, or tidal flat.

slow sand filtration A *water treatment process in which water is drained slowly through a bed of *sand in order to remove particles from it.

sludge Solid matter that is removed during the treatment of *water or *wastewater. Also known as **residuals**. *See also* SEWAGE SLUDGE.

sludge digester A tank in which organic *sewage sludge is decomposed by *micro-organisms, which releases energy and converts much of the sewage to *methane, *carbon dioxide, and *water.

sludge processing *See* AEROBIC DIGESTION.

slum An area in a *city that has poor-quality living conditions and marked poverty.

slump A *landslide in which the rock mass tilts back as it slide down from a *cliff or *escarpment.

slurry A *suspension of fine solid *particles in a liquid, such as liquified animal *manure.

small game *Birds and small *animals (*game) that are hunted or trapped.

smallholding A small plot of land on which fruit and vegetables are cultivated, usually for sale. *See also* HORTICULTURE, MARKET GARDEN.

smart growth Comprehensive planning and environmentally sensitive development of land that is designed to restore sense of community, enhance natural and cultural resources, expand choice in transportation, employment, and housing, reduce air pollution, promote public health and healthy communities, and increase the efficiency of investments in infrastructure.

Smart Growth Network A US non-profit *anti-environmental *front group that promotes the principles and practice of *smart growth as a way of reconciling the tensions between *conservation and *development. *Compare* WISE USE MOVEMENT.

smelter An industrial plant for *smelting.

smelting The process of extracting *metal from *ore by heating and melting, which separates out the impurities.

Smith-Lever Act (1914) US legislation that established a federal–state co-operative extension programme to provide education for the public about agricultural and *natural resources.

smog A form of *air pollution that is caused by the interaction of *pollutants and *sunlight, often restricts *visibility, and is sometimes hazardous to health. *See also* PHOTOCHEMICAL SMOG.

smoke A type of *air pollution in the form of a visible cloud of fine particles suspended in the air, which is produced by the burning of *fossil fuels, particularly *coal.

smokestack *See* STACK.

Smokey the Bear A mascot of the *US Forest Service, which was adopted in 1944 to educate the public on the dangers of forest fires. Royalties from licensed use of the mascot are used to fund education programmes on prevention of forest fire.

SMR *See* STANDARDIZED MORTALITY RATIO.

SMSA *See* STANDARD METROPOLITAN STATISTICAL AREA.

smudge pot *See* ORCHARD HEATER.

SMZ *See* STREAMSIDE MANAGEMENT ZONE.

snag A standing dead or decaying *tree, which provides nesting sites, feeding sites, and *habitats for *wildlife.

snow Frozen *water vapour in the atmosphere that falls in light white flakes (*snowflakes) and settles on the ground as a white layer. A form of *precipitation.

snow advisory A statement (*advisory) that is issued in the USA when *snow is expected to create hazardous travel conditions.

snow avalanche The rapid downslope movement of large quantities of *snow.

snow cornice A mass of *snow or *ice that projects over a mountain ridge.

snow density The mass of *snow per unit volume, which is equal to the water content of the snow divided by its depth.

snow flurry An intermittent light shower of *snow.

snow line The lower limit of permanent *snow cover, below which snow doesn't accumulate.

snow melt The conversion of *snow into *runoff and *groundwater, as *temperature rises.

snow pellet A form of *precipitation that consists of a *snow crystal and a *raindrop frozen together. Also known as **graupel**.

snow shower See SNOW SQUALL.

snow squall An intermittent heavy shower of snow that greatly reduces visibility. Also known as a **snow shower**.

snowbank A mound or heap of *snow.

snowdrift A mass of *snow that has been heaped up by the wind, often in the *lee of an obstruction.

snowfall The rate at which *snow falls or accumulates on the ground, which is usually expressed in depth of snow (in centimetres) over a six-hour period.

snowfield A permanent large area of *snow on the ground, most commonly found at high *altitude or *latitude.

snowflake A six-pointed cluster of ice crystals which falls from a *cloud.

snowline The *altitude above which *snow remains on the ground throughout the year, which decreases from the *equator (around 5000 metres) towards the *poles (close to *sea level).

snowpack The total amount of *snow and* ice on the ground, including fresh new snow and older snow and ice which has not melted.

snowstorm A *storm accompanied by heavy *snowfall.

soakaway A place where water soaks into the ground.

soap A natural cleaning agent that is made by the reaction of a fat or oil and an *alkali. See also INSECTICIDAL SOAP.

social capital The degree of social cohesion which exists in a community, which is reflected in social networks, norms, and trust.

social class A collection of people with similar position or social rank.

social construction of reality A process in which people's experience of reality is determined by the meaning they attach to that reality. See POST-MODERNISM.

social cost The value of the best alternative use of a particular scarce *resource, or the potential benefit that is foregone from not following the best alternative course of action. Also known as **non-market** cost or **opportunity cost**.

social Darwinism The belief that social structure is determined by how well people are suited to particular living conditions. See also NATURAL SELECTION.

social ecology The view that environmental problems arise from fundamental social problems, and that they cannot be understood or solved without dealing with problems within society, including economic, ethnic, cultural, and gender conflicts. See also RADICAL ECOLOGY.

social forestry Planting and looking after trees or shrubs in order to maintain or improve the quality of life of local communities. See also COMMUNITY FORESTRY.

social impact The changes in social and cultural conditions, which can be positive or negative, which directly or indirectly result from an activity, project, or programme. Contrast ENVIRONMENTAL IMPACT.

social inequality A condition in which different members of a society have unequal amounts of income, prestige, and social power.

social justice The objective of creating a fair and equal society in which each individual matters, their rights are recognized and protected, and decisions are made in ways that are fair and honest.

social movement A large number of people who act together in order to pursue some shared objective, such as *environmentalism.

social sciences Those disciplines that are concerned with the study of human behaviour, which include anthropology, economics, psychology, geography, political science, and sociology.

social value The non-economic value that society puts on a resource and that is recognized by most, if not all, people, such as the benefits to human health of clean air and water. *See also* SHADOW PRICE.

socialism An *economic system in which the means of production are controlled by the state. Also known as **socialist economy**. *Contrast* CAPITALISM. *See also* MARXISM.

socialist economy *See* SOCIALISM. *Contrast* CAPITALIST ECONOMY, MARKET ECONOMY.

society A broad grouping of people (including a community or nation) that shares common traditions, institutions, collective activities, and interests.

sociobiology The study of behaviour in social animals (including humans), based on the assumption that such behaviour is biologically based, genetically encoded, and evolves through the process of *natural selection.

socioeconomics The social and economic conditions of a particular area or population.

sociology The study and classification of social groups and human societies.

SOCs *See* SYNTHETIC ORGANIC CHEMICALS.

sod 1. Established grass, *turf, or *sward.
2. A strip of turf that is cut and transplanted to establish grass cover in another place.

sodar A device that is used to detect and determine the range to distant objects or changes in temperature or humidity in the atmosphere, based on the scatter and reflection of sound energy. Short for 'sound detection and ranging'. *See also* SONAR.

sodgrass A North American *grassland vegetation dominated by grasses (such as big bluestem, *Andropogon gerardii*) that spread horizontally, like the grass in a lawn, and are characteristic of *tallgrass prairies.

sodicity The accumulation of *sodium in *soil, which adversely affects soil structure and increases *toxicity to plants.

sodium (Na) A naturally occurring, silver-white, soft, waxy, ductile element of the *alkali metal group, that can form various salts with *halogens and *metals, is abundant in nature, and is not generally considered to be *toxic. When dissolved in water it can be used to indicate *salinity.

sodium chloride (NaCl) Common salt.

soft detergent A cleaning agent (*detergent) that breaks down naturally in the environment.

soft energy path An approach to *economic development that relies on renewable sources of *energy, including *solar power, *wind power, and *water power. *Contrast* HARD ENERGY PATH.

soft mast soft fleshy fruits potentially eaten by wildlife (examples include persimmon, wild grapes, blackberries, blueberries, huckleberries, mulberries, plums, autumn olive, and crab apples).

soft water Water that does not contain a significant amount of dissolved minerals such as salts of *calcium or *magnesium. *Contrast* HARD WATER.

softened water Water that is treated in order to reduce its hardness (due to *calcium and *magnesium).

S

softwood A general term for a *coniferous tree and for the wood from such trees. *Contrast* HARDWOOD.

soil The thin layer of disintegrated rock particles, organic matter, water, and air that covers most of the land surface to an average depth of about 20 centimetres. Soil is a typical *open system with inputs, throughputs and outputs of matter and energy, and it is intimately linked with other *environmental systems, particularly climate and vegetation. Through *weathering and *erosion, soils are linked to the lithosphere system and have a place (albeit a short-term one) within the *rock cycle. *See also* PEDOGENESIS.

soil acidification Soil that becomes *acid, due to the production of nitric acid and sulphuric acid by *bacteria followed by *leaching. *See also* ACIDIFICATION.

soil aeration The process by which air in the soil is replenished by air from the atmosphere. The degree of *aeration depends mainly on the *porosity of the soil, and poorly aerated soils contain more *carbon dioxide and less *oxygen.

soil aggregate A soil aggregate is made of individual particles of *soil stuck together, and determines *soil texture. Thus, for example, a soil with a granular structure is composed of relatively small, rounded aggregates, and a soil with a blocky structure is made up of aggregates with sharp corners and irregular shapes.

soil amendment *See* SOIL ENHANCEMENT.

soil and water conservation practices Control measures that are adopted in order to reduce the loss of soil and water from a particular area.

Soil and Water Resource Conservation Act (1977) US legislation that provides for a continuing appraisal of soil, water, and related resources, including fish and wildlife habitats, and a soil and water conservation programme to assist landowners and land users in furthering soil and water conservation.

Soil Association A UK environmental campaigning and certification organization, whose mission is to promote *sustainable *organic food and farming.

soil association A group of named *soil taxonomic units that occur together in a characteristic pattern over a geographical region.

soil bulk density The mass of *soil per unit of volume, which is a measure of *compaction.

soil class A group of soils that have a definite range in a particular property such as *acidity, *slope, *texture, or *structure.

soil classification The systematic arrangement of *soils into groups or categories on the basis of their characteristics. Also known as **soil taxonomy**. *See also* COMPREHENSIVE SOIL CLASSIFICATION SYSTEM.

soil climate The moisture and temperature conditions that exist within a particular soil.

soil compaction The reduction of soil volume as a result of applied load, vibration, or pressure, which leads to a decrease in *soil bulk density and *soil porosity.

soil complex A mapping unit (smaller than a *soil association) that is used in detailed *soil surveys where two or more defined *taxonomic units are so completely intermixed geographically that separating them is impossible.

soil concretion *See* CONCRETION.

soil conditioner Any substance (such as *compost) that enriches the physical condition of *soil and increases its *organic content as part of *soil enhancement.

soil conservation Methods that are used to reduce *soil erosion, prevent depletion of soil nutrients and soil moisture, and enrich the *nutrient status of a soil.

Soil Conservation Act (1935) US legislation that provided for control and prevention of *soil erosion, delegated all activities relating to soil erosion to the Secretary of Agriculture, and established the Soil Conservation Service.

Soil Conservation and Domestic Allotment Act (1936) US legislation that established the Agricultural Conservation Program, which provides cost-sharing funds to landowners for the protection of soil and water resources, including practices to improve tree planting and timber stands.

soil creep See CREEP.

soil drainage See DRAINAGE.

soil enhancement Adding material (including *fertilizers, *organic materials, *sand, or *lime) to a *soil in order to improve its fertility, long-term chemical or physical status, or stability. Also known as **soil amendment** or **soil improvement**. See also SOIL CONDITIONER.

soil erodibility A measure of the inherent susceptibility of a *soil to *erosion, regardless of *topography, *vegetation cover, *management, or *weather conditions.

soil erosion The removal of *topsoil by *raindrop erosion and *sheet erosion, as a result of human activities, particularly if the protective vegetation has been removed. A number of factors promote soil erosion, including *deforestation, *overgrazing, *compaction (which destroys soil structure), farming on steep slopes without *terraces, and the practice of leaving ground barren for long periods in winter. The erosion of soil by running water can be tackled in a number of ways, including: maintaining a protective plant cover on vulnerable soils, perhaps by planting crops in alternate strips (grass and grains); reducing the amount of time for which bare soil is exposed, by

using *mulch or reintroducing *crop rotations; *contour ploughing; building terraces that intercept and stop the downslope movement of soil; and netting, a temporary form of erosion control in which netting is placed over bare soil surfaces after they have been seeded. The erosion of soil by wind can be reduced by maintaining a plant cover to reduce wind speed (thus energy), contour ploughing, planting strips of row crops in intervals across a field of cereals, and using trees and shrubs as shelter belts or windbreaks. See also ACCELERATED EROSION.

soil fabric The arrangement, size, shape, and frequency of the particles in a *soil.

soil factor See EDAPHIC FACTOR.

soil fertility The ability of a *soil to support plant growth by providing water, *nutrients, and a medium in which growth can occur.

soil formation The process by which *soil is formed as a result of interactions over time between parent material (*rock), *climate, *topography, and *organisms. Also known as **pedogenesis**.

soil gas Gas (such as *radon) that is contained in the pore spaces between *soil particles.

soil horizon A relatively uniform layer within a *soil profile that has distinct characteristics. See also A-HORIZON, B-HORIZON, C-HORIZON, D-HORIZON.

soil improvement See SOIL ENHANCEMENT.

soil loss equation See UNIVERSAL SOIL LOSS EQUATION.

soil map A map that shows the location and extent of the different *soil associations and *soil complexes in a particular area.

soil moisture See SOIL WATER.

soil moisture deficit The difference between the current amount of *soil moisture and the moisture which would be in that soil at *field capacity.

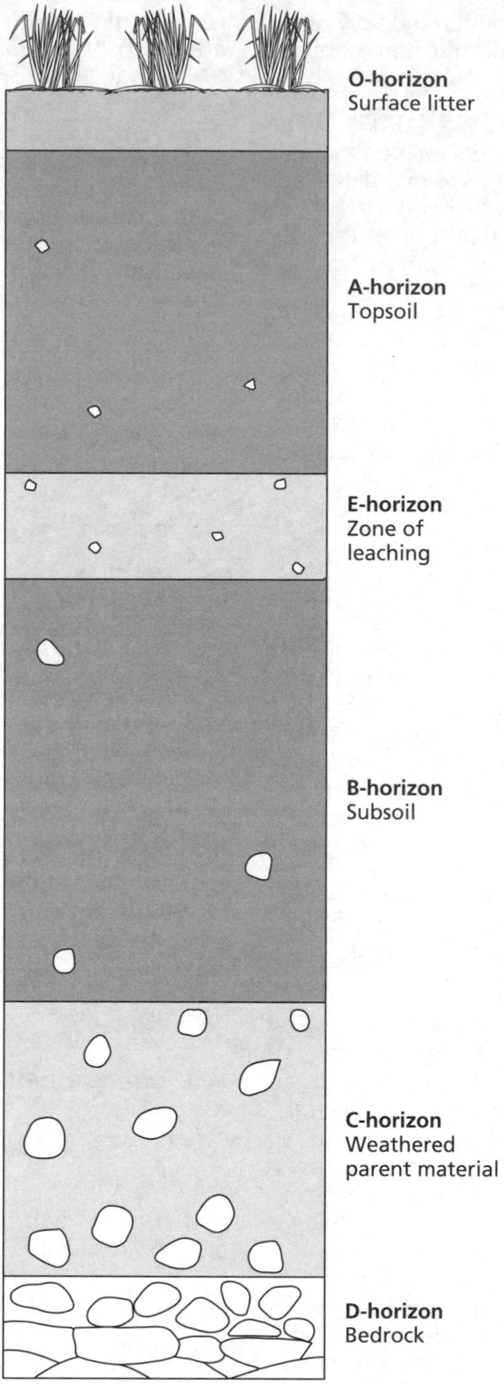

O-horizon
Surface litter

A-horizon
Topsoil

E-horizon
Zone of
leaching

B-horizon
Subsoil

C-horizon
Weathered
parent material

D-horizon
Bedrock

Fig 19 Soil horizon

soil monolith A vertical section through a *soil, which is often preserved with resin and mounted for display.

soil order The highest level within a *soil classification. There are 11 soil orders in the *Comprehensive Soil Classification System, which are *alfisols, *andisols, *aridisols, *entisols, *histosols, *inceptisols, *mollisols, *oxisols, spodsols, *ultisols, and *vertisols.

soil organic matter See HUMUS.

soil phase A subdivision of a *soil series, based on characteristics (such as slope) that affect the use and management of the soil but not enough to create a separate soil series.

soil pipe 1. A drain that conveys liquid waste from toilets.
2. A natural small-scale tunnel within a *soil, usually part of a network, that plays a part in *soil erosion and *gully development. Also known as **pipe**.

soil productivity The capacity of a *soil to produce a particular *yield of crops or other plants, under a given specified system of *management.

soil profile The vertical arrangement of layers or horizons in a *soil. See also SOIL HORIZON.

soil reaction A measure of the degree of *acidity or *alkalinity of a *soil. See also pH

soil region An area across which *soils are relatively uniform. There are 50 soil regions in the USA, which vary with climate and vegetation.

soil resource inventory See SOIL SURVEY.

soil salinization A *soil that has become damaged by large quantities of *salts, usually from poor quality *irrigation water, under conditions of poor or impaired drainage. See also SALINIZATION.

soil series A group of *soils that have similar *soil profile characteristics and are derived from similar *parent materials.

soil solution The liquid phase of the *soil, which comprises the water and the *solutes that are dissolved within it.

soil structure The arrangement of *soil particles into larger particles or clumps.

soil survey The systematic examination, description, classification, and mapping of *soils within an area. Also known as a **soil resource inventory**.

soil taxon A named group *of soils which have a similar *soil taxonomy.

soil taxonomic unit See SOIL TAXON.

soil taxonomy See SOIL CLASSIFICATION.

soil texture The relative proportion of the various size groups of individual particles (*sand, *silt, and *clay) in a *soil.

soil type A subdivision of a *soil series defined in terms of surface *soil texture.

soil water *Water that is contained in the *pore spaces of a *soil, as *gravitational water, *capillary water, and *hygroscopic water. Also known as **soil moisture**.

soil-forming factor Any natural factor that influences *soil formation, the most important of which are *parent material, climate, natural vegetation, living organisms, slope, and time.

soil-forming process The processes by which *soil formation occurs, which include *leaching, *eluviation, *illuviation, *laterization, *podsolization, *calcification, *acidification, and *salinization.

soiling index A measure of the darkening potential of smoke and soot suspended in one cubic metre of air.

solar activity Activity of the *Sun.

solar architecture The design and building of structures that are energy efficient in terms of *solar energy, involving a combination of technologies including *solar panels and *photovoltaic systems.

S

solar cell A means of indirectly harnessing *solar energy, using panels made from a semiconductor material (usually *silicon) that generate electricity when illuminated by *sunlight. Solar cells can be used at any *latitude, and large installations are required because of the low efficiency of the technology (currently about 15%). The energy they produce is relatively expensive. Also known as a **photovoltaic cell**.

solar collector A device that collects *solar energy, concentrates it, and converts it to useful forms of energy. Typical systems are based on heat-absorbing (usually black) panels containing pipes through which air or water is circulated, either by thermal convection or by a pump. The air or water is heated as it passes through the panel, and the heat is usually circulated around a building through pipes. Such collectors require a lot of sunshine to work efficiently, and are not really suited to high latitudes. The energy produced is relatively expensive because of the high capital cost of installing the heater systems. A form of *active solar heating.

solar constant The rate at which *solar energy is received just outside the Earth's atmosphere, which is about 1360 *watts per square metre (W m^{-2}). The exact rate varies a little through time, depending on variations in the total amount of energy released by the Sun, variations in the Earth's *rotation around its axis, and variations in the Earth's *orbit around the Sun.

solar cycle The natural regular cycle of variation in the frequency or number of *sunspots, which has a length of roughly 11 years. Also known as the **sunspot cycle**. *See also* MAUNDER MINIMUM.

solar energy Energy that is derived from the *radiation of the *Sun, which is a *sustainable, *renewable, nonpolluting, and relatively reliable energy source. Nearly all the energy we use is ultimately solar, except· perhaps for *tidal energy (which is generated mainly by the gravitational pull of the Moon) and *nuclear energy. *Fossil fuels and *biomass fuels are ultimately derived from solar energy. *Hydropower and *wind power are generated from air and water circulations, which are ultimately driven by solar heating. Solar radiation (*insolation) arrives at the Earth's surface at a fairly low temperature, so sunshine is often thought of as low-grade energy that is not suitable for most uses. It can be made much more useful by concentrating the energy, using *solar collectors, or by *using solar cells. Also known as **solar power**.

solar flare A bright eruption of hot gas in the Sun's outer atmosphere, usually associated with active groups of *sunspots.

solar heating The use of solar *energy or *sunlight to heat buildings. Traditional approaches to heating buildings have relied heavily on *passive solar heating, particularly in sunny places. More recently, technologies have been developed to pump heat around buildings, using *active solar heating or passive solar heating. There is great scope for expanding the use of solar heating systems because the energy supply is inexhaustible and the technology is already fairly well developed and is likely to become cheaper and more efficient in the future. This type of energy use produces less *pollution than burning *fossil fuels or running *nuclear reactors and most places receive enough *sunlight to make this approach possible, even if it supplements other types of heating systems that can be switched on when there is insufficient sunlight.

solar luminosity A measure of the brightness of the *Sun, which is determined by the amount of *solar radiation it is emitting.

solar power *See* SOLAR ENERGY.

solar radiation Energy that comes from the *Sun in the form of *electromagnetic radiation, mostly at *wavelengths of between 1.0 and 0.1

micrometres. Very shortwave (high-energy) *gamma and *X-rays are absorbed in the upper *atmosphere and none reaches the surface. Much of the radiation within the *ultraviolet (UV) range (0.2–0.4 micrometres) is filtered by atmospheric gases, particularly at wavelengths below 0.29 micrometres where most of the UV (which causes sunburn) is absorbed by stratospheric *oxygen and *ozone. The atmosphere is transparent to wavelengths longer than 0.29 micrometres, but *water vapour absorbs energy in a series of narrow bands between 0.9 and 2.1 micrometres. As well as its composition, the amount of solar radiation also changes as it passes through the Earth's atmosphere. Less than half of the radiation that arrives at the edge of the atmosphere (the *solar constant) eventually reaches the Earth's surface. Most of what is lost is reflected back into space, and has no further impact on the Earth or its atmosphere. About 10% of what is lost is absorbed or scattered by ozone, water vapour, and particles of dust in the atmosphere. *See also* INSOLATION, RAYLEIGH SCATTERING.

solar system The collection of *celestial bodies that orbit around the *Sun.

solar thermal electric technology A range of technologies that are similar in concept to *solar heating technologies in using *sunlight to generate heat, but they create enough heat to power a *generator, which is then used to produce *electricity.

solar wind The steady stream of charged particles (mostly *hydrogen and *helium) that flows away from the *Sun out into the *solar system at all times. The solar wind creates *auroras and *geomagnetic storms when it interacts with the Earth's upper atmosphere.

solar year The time it takes the Earth to make one orbit around the *Sun, which is roughly 365.24 days.

solder A *metallic compound of *tin and *lead that is used to seal joints between pipes.

solfatara A *volcano that no longer erupts *lava and *volcanic ash, but releases steam and gases from small *vents. The steam might be exploited for *geothermal energy.

solid waste Solid materials which are discarded from industrial, commercial, mining, or agricultural operations, and from community activities. Also known as **garbage, refuse**, or **rubbish**.

solid waste disposal The placement of *solid waste that is not *salvaged or *recycled into its permanent, final location, such as a *landfill.

Solid Waste Disposal Act (1965) US legislation on *waste disposal that provided states with money for research on the disposal of solid wastes. It was replaced by the *Resource Conservation and Recovery Act (RCRA) (1976).

solid waste management The systematic collection, source separation, storage, transportation, transfer, processing, treatment, and disposal of *solid waste.

solidification The conversion of a liquid into a solid, either naturally (such as the freezing of water into ice) or by adding material (like cement) to a liquid *waste stream to make it more solid and less soluble. Also known as **chemical fixation**.

solifluction A type of *soil creep that occurs in areas underlain by *permafrost, under *periglacial conditions.

solitary Living alone, not in *colonies. *Contrast* GREGARIOUS.

solonetz A type of *grassland *soil that is found in a *subhumid or *semi-arid climate, under grass and shrub vegetation.

solstice The day when the Sun at midday is either highest or lowest in the sky, which determines the *seasons. In the *northern hemisphere the Sun is highest in the sky when it is over the *Tropic of Cancer on 22 June, and this defines the *summer solstice; the Sun is lowest in the sky when it is over the *Tropic of

S

Capricorn on 22 December, which defines the *winter solstice. The solstices, and thus the seasons, are reversed in the *southern hemisphere.

solubility The ability of a substance to dissolve into another. *See also* GAS SOLUBILITY.

soluble Capable of being dissolved. *Contrast* INSOLUBLE.

solum That part of the *soil that is capable of supporting life, which is mainly the *A-horizon and *B-horizon within a *soil profile.

solute A substance that can be dissolved into another substance (a *solvent).

solute load *See* SOLUTION LOAD.

solution A mixture of one or more *solutes dissolved in a *solvent.

solution load The *sediment that is carried in *solution by a *river. Also known as dissolved load, **dissolved solids**, or **solute load**.

solvent A substance that has the ability to dissolve another (a *solute). *See also* UNIVERSAL SOLVENT.

soma The entire body of an *organism, excluding the *germ cells.

somatic Relating to or characteristic of the *body (*soma).

somatic cell Any cell other than a *germ cell.

somatic damage Damage that is caused to the *chromosomes in individual body tissue cells (such as skin cells or lung cells) by environmental factors (for example, *UV light in the case of skin cells and tobacco smoke/air pollution or radiation from *radon decay products in lung cells), which can accumulate over successive cell divisions and lead to *cancer. *Contrast* GENETIC DAMAGE.

sonar A device for detecting and locating objects by means of sound waves which are sent out and reflected back by the objects. Short for 'sound navigation and ranging'. *See also* SODAR.

sonde *See* RADIOSOND, RAWINSONDE.

sonic boom A loud explosion-like sound in the air that is caused by a shock wave from an aircraft (or any object) travelling at or above the speed of sound.

Sonoran Desert One of the largest and hottest *deserts in North America, which straddles part of the border between the USA and Mexico and covers large parts of the states of Arizona, California, and Sonora. It has a dry, clear climate and contains some unique plants and animals, such as the saguaro cactus.

soot Fine *carbon dust that is formed by the incomplete *combustion of *fossil fuels, and gives *smoke its colour.

sorption The action of soaking up (*absorption) or attracting (*adsorption) substances, which is used in many *pollution control systems.

sorted ground *See* PATTERNED GROUND.

sound wood *Timber that is in a good condition, free from damage, decay, or defects.

sounding 1. A measure of the depth of water.
2. A plot of variation in height of one or more atmospheric parameters (such as temperature, humidity, pressure, or wind speed) at a particular point in time, usually based on data from a *radiosonde.

soundscape The full range of noises that are heard in a particular place or area.

sour water *Wastewater that contains containing foul-smelling material, usually *sulphur compounds.

source Any area (such as a *city), place (such as a *power plant), or object (such as a *chimney stack) from which particular *pollutants are released, or at which *waste materials are produced. *Contrast* SINK.

source habitat A *habitat that is a net exporter of individuals. *Contrast* SINK HABITAT.

source population A population that occupies a *source habitat. *Contrast* SINK POPULATION.

source reduction The elimination or reduction of *waste material at the *source, by modifying the process which produces the waste. *See also* BACK-YARD COMPOSTING, INTEGRATED WASTE MANAGEMENT.

source region A region where *air masses originate and acquire their properties (particularly temperature and moisture).

source separation Segregating *waste materials where they are generated, such as separating paper, metal, and glass from other wastes in order to make *recycling simpler, more efficient, and more cost-effective.

south 1. The compass point that is at 180° from *north.
2. A general term ('the South') for the less developed countries that are located in the *southern hemisphere and are often referred to as the *Third World.

South Pole The most southerly point on Earth, at 90°S latitude in the *southern hemisphere; the southern end of the Earth's rotational axis. *Contrast* NORTH POLE. *See also* ANTARCTICA.

southern hemisphere The half of the Earth that lies *south of the *equator. *Contrast* NORTHERN HEMISPHERE.

southern lights *See* AURORA.

Southern Ocean The area of *ocean that lies within the *Antarctic Convergence.

Southern Oscillation A periodic reversal of the pattern of *atmospheric pressure across the tropical parts of the Pacific Ocean during *El Niño events, which occurs on average every 2.3 years, and is accompanied by variations in wind speed, ocean currents, sea-surface temperatures, and precipitation in the surrounding areas. *See also* ENSO.

southern pine forest A *coniferous forest *biome in the USA that is characterized by a warm, moist climate.

SOV *See* SINGLE OCCUPANCY VEHICLE.

sovereignty The exercise by a state of absolute power over its territory, system of government, and population.

SPA *See* SPECIAL PROTECTION AREA.

space 1. An empty area or the distance between two or more objects.
2. The area in which the solar system, stars, and galaxies exist.

Spaceship Earth The metaphor of the *Earth as a spaceship, travelling through *space as a self-contained system, dependent on its own vulnerable supplies of natural resources (including air, water, and soil). There are some interesting implications of the Spaceship Earth metaphor. For example: spaceships are self-contained, and as such must carry all the resources they require; spaceships are small relative to the vastness of space through which they travel; spaceships are complex systems (involving many computer control systems, air conditioning, wiring, and so on) with many components fitted together into a stable, working whole; no spaceship would be able to operate for long without a constant supply of energy to power all of its systems; in many spaceships some of the energy comes from within (from batteries and other non-renewable fuel sources) and some from outside (they use solar energy by means of solar panels, for example); conditions in spaceships are carefully controlled to suit human life; if conditions in a spaceship change too much or too fast they can become unsuitable for human survival; spaceships don't last for ever!

spatial distribution The arrangement of individuals in a particular area or *space.

spawn The eggs of certain *aquatic *organisms (particularly fish and *amphibians), or the act of producing such eggs or egg masses.

SPCC *See* SPILL PREVENTION, CONTROL AND COUNTERMEASURE.

Special Area of Conservation (SAC) An area within the *European Union that is designated for protection under the *Habitats Directive, because of its *natural habitats and/or populations of *species. *See also* NATURA 2000.

special interest group A large group of people who share a common cause, interest, or issue, which uses its organizational power to promote change which is beneficial to its members. *See also* PRESSURE GROUP.

Special Protection Area (SPA) An area within the *European Union that has been designated for protection because it contain important populations of *birds and habitats for birds. *See also* NATURA 2000.

special risk group A group of people (such as children or the elderly) which is at high risk because of its *sensitivity or *exposure to one or more particular *hazards.

special waste Items of *solid waste, such as household hazardous waste, bulky wastes (including refrigerators and furniture), tyres, and used oil.

specialist A species that has a very narrow *range in terms of *habitat or food requirements. *Contrast* GENERALIST.

speciation The process by which new species originate through *mutation, *natural selection, and *evolution, dividing one species into two, which at least in theory are unable to interbreed. Rapid speciation is known as *adaptive radiation. *Contrast* EXTINCTION.

species A population of organisms that reproduce with one another but not with other populations. This preserves the identity of the individual species through time, and it raises questions about how new species can evolve from existing ones. Species are defined in various ways, including the *biological species concept, *cladistic species concept, *ecological species concept, *phenetic species concept, and *recognition species concept. *See also* BIODIVERSITY, BIOLOGICAL CLASSIFICATION, LATIN NAME.

Species 2000 An international project that was set up to index all of the known *species of organisms (*animals, *plants, *fungi, and *micro-organisms) on Earth, in order to provide a baseline dataset for comparative studies of global *biodiversity.

species abundance A measure of the total number of individuals of a particular *species in a defined area, population, or community. *Contrast* SPECIES RICHNESS.

Species Action Plan A special part of a *biodiversity action plan, which sets objectives and targets for the maintenance or enhancement of the population and range of particular species, and the actions necessary to achieve them, over a period of 10–15 years. *See also* LOCAL BIODIVERSITY ACTION PLAN.

species at risk *Species whose future is at risk, either globally or within a particular region. The *IUCN classification defines species in terms of the degree of threat they face, as *extinct, *possibly extinct, *endangered, *vulnerable, *rare, *no longer threatened, and *status unknown.

species composition The types and number of *species that occupy a particular area.

species concept The basis on which a *species is defined. *See also* BIOLOGICAL SPECIES CONCEPT, CLADISTIC SPECIES CONCEPT, ECOLOGICAL SPECIES CONCEPT, PHENETIC SPECIES CONCEPT, RECOGNITION SPECIES CONCEPT.

species diversity The number and relative abundance of *species that are present in a particular community. *See also* BIODIVERSITY.

species diversity index The ratio between the total number of *species in a particular *community, and some measure of the relative importance of individual species (for example, in terms of number, *biomass, or *productivity). *See also* SHANNON–WIENER INDEX.

species extinction *See* EXTINCTION.

species name *See* LATIN NAME.

species of concern A North American term for a *native species that is not listed as *endangered or *threatened but whose population is declining fast enough to cause concern among biologists.

species recovery plan A plan that is designed to halt the decline of an *endangered species, and increase its population, through protection, *habitat management, *captive breeding, disease control, or other techniques.

species richness The number of *species that are present in a *habitat or *ecosystem. Also known as **alpha diversity**.

Species Statement Within the UK *Biodiversity Action Plan, an overview of the status of *species and broad policies developed to conserve them.

species survival plan A *captive breeding programme, usually within a *zoo or *aquarium, that is designed to aid the recovery of a particular species by breeding a healthy, genetically diverse, and self-sustaining captive population.

species turnover *See* BETA DIVERSITY.

species–area relationship In *island biogeography, the relationship between the area of the island and the number of species that occur within it; as area increases, the number of species present (*diversity) usually also increases.

specific conductance A measure of the *conductivity of a sample of water (its ability to conduct an electrical current), which reflects its content of *dissolved solids.

specific gravity The relative weight of a *mineral compared with the weight of an equal volume of pure water, which has a specific gravity of one.

specific heat The amount of heat that is required to raise the temperature of one gram of a substance by 1°C. *See also* HEAT CAPACITY.

specific humidity The mass of *water vapour in a given mass of air, usually expressed as grams per kilogram (g kg^{-1}): a measure of *humidity.

specific yield The amount of water that a unit volume of *saturated *permeable rock will yield when drained by *gravity.

specimen An individual, item, or part that is considered typical of a group, class, or whole.

specophily *Pollination by wasps.

spectrum The distribution of *electromagnetic radiation in order of *wavelength. *See also* ELECTROMAGNETIC SPECTRUM.

spent fuel *Nuclear fuel from a *commercial reactor that is no longer capable of sustaining the fission process. *See also* NUCLEAR FUEL CYCLE, REPROCESSING.

spill The deliberate release of water from a *dam, through the *spillway. *See also* OIL SPILL.

Spill Prevention, Control and Countermeasure (SPCC) A North American plan that outlines how a facility will prevent *oil spills, and how it plans to control and contain an oil spill to keep it from reaching surface water.

spillway The overflow structure of a *dam, through which excess or flood flows are released.

spit A low ridge or *bar of sand and gravel that projects from the shore into the sea, and is formed by the *deposition of *sediment moved along the *coast by *longshore drift.

splash erosion *See* RAINDROP EROSION.

splitting the atom *See* NUCLEAR FISSION.

spodic horizon The B-horizon in a *spodosol, in which *aluminium and *organic matter accumulate.

spodosol A soil *order in the *Comprehensive Soil Classification System, comprising sandy soils that develop under forests (particularly *coniferous forest), are *acidic, and have accumulations of *organic matter and *iron oxides and *aluminium oxides in the *B-horizon.

spoil The *overburden of rock and soil that is removed to gain access to the coal or mineral material in *surface mining.

spoil bank A mound of mine refuse that is created by the deposit of *overburden or waste rock. Also known as spoil heap.

spoil heap See SPOIL BANK.

spore The microscopic seed-like cells of *fungi that grow directly into a new plant.

sport fish See GAME FISH.

sports utility vehicle (SUV) A vehicle that has the load-hauling and passenger-carrying capacity of a large station wagon or minivan, which may or may not be suitable for off-road driving.

SPOT Système Probatoire d'Observation de la Terre, a *satellite *remote sensing system which was launched in 1986 and provides information on a commercial basis for environmental research and monitoring, ecology management, and for use by the media, environmentalists, and legislators.

spot forward See FORWARD CONTRACT.

spot zoning A special type of land use *zoning that allows particular activities in an isolated plot of land which would not be allowed in the surrounding area.

sprawl Low density development on the edge of cities and towns, which is usually poorly planned, occupies much land, and depends heavily on vehicles. See also SUBURBAN SPRAWL.

spray A jet of *vapour or small droplets.

spraying The application of a liquid (such as a *fertilizer, *herbicide, or *insecticide) in the form of small particles which are ejected from a sprayer.

spread To extend, scatter, or disperse, for example to expand the geographical distribution of an *organism within a particular area.

spreading zone The area around a *mid-ocean ridge, under the sea, where newly formed *crust is moving away from the central axis of the ridge, at rates of 1–10 centimetres per year.

spring 1. A place where *groundwater flows naturally from the *water table in rock onto the land surface.
2. The *season of growth, between *winter and *summer, when temperatures at mid-latitudes increase as the Sun approaches the *summer solstice. Astronomically this is the period between the *vernal equinox and the *summer solstice, which covers the months of March, April, and May in the northern hemisphere, and September, October, and November in the southern hemisphere.

spring line A line of *springs, usually defined by changes in rock type.

spring tide A *tide which occurs about every two weeks and coincides with the new and full moon, which has a large tidal range because the gravitational forces of the Moon and Sun complement each other. Contrast NEAP TIDE.

springwood See EARLYWOOD.

sprinkler irrigation A method of *irrigation that uses mechanical devices that sprinkle water over crops to simulate rain.

squall A sudden, violent *wind that is often accompanied by *rain.

squall line A line of *thunderstorms or *squalls that may extend over several hundred kilometres.

squatter Someone who settles upon land without permission or authority.

squatter settlement A *shanty town that occupies land without the owner's permission.

S-rank *See* PROVINCIAL RANK.

SSSI *See* SITE OF SPECIAL SCIENTIFIC INTEREST.

St Elmo's fire A bright electric discharge that is projected from objects when they are in a strong electric field, such as during a *thunderstorm.

stability 1. A condition of the *atmosphere that exists when the *dry adiabatic lapse rate is greater than the normal *lapse rate, so that a parcel of air that is caused to rise will tend to sink back to its original height without cooling to the *dew point, so no *condensation occurs.
2. The propensity of a system to attain a steady state or stable oscillation. *See also* PERSISTENCE, RESILIENCE, SENSITIVITY.

stabilization The process that is used to convert harmful bacteria and odours in sludge into inert, harmless material, which usually involves *aerobic or *anaerobic *digestion.

stabilization pond *See* LAGOON.

stable air Air that has little or no tendency to rise, which is usually accompanied by clear dry *weather. *See also* STABILITY.

stable equilibrium A *system that tends to return to the same *equilibrium after disturbance.

stable isotope An *atom whose *nucleus is not *radioactive and therefore does not alter (*decay) through time.

stack 1. A large, tall *chimney through which combustion gases and smoke are released into the air. Also known as a **smokestack**.
2. An isolated pillar of rock in the sea, the remnant of a former *cliff that has been eroded by *wave attack.

stack effect An upward flow of air in a building that is caused by warm air rising, which creates positive pressure at the top of a building and negative pressure at the bottom. It can disrupt *ventilation and air circulation within the building.

stack gas *See* FLUE GAS.

stade A brief period of glacial advance (*glaciation) within a glacial stage. *Contrast* INTERSTADE.

stadial A cold period, within a generally warmer climate, which is too short-lived for an *ice sheet or *glacier to fully form.

stage The level of the water surface relative to a given *datum, at a given location, such as in a *stream channel.

stagnant Not in a current or stream, and thus still rather than moving or flowing.

stagnation A state of inactivity.

stakeholder A person or group who have a vested interest in a particular project, activity, or issue because they are involved in it or affected by it.

stakeholder analysis An analysis of the interests of different *stakeholders in a project, activity, or issue, such as a conservation project.

stalk The main *stem of a *herbaceous plant.

stand A group of similar plants in a particular area. *See also* TIMBER STAND.

stand density The degree of crowding of trees in a *stand or a *forest, often expressed as the number of trees per unit area.

standard A basis for comparison, such as the maximum concentration of a particular *air pollutant or *effluent that is adopted by a regulatory authority as enforceable.

standard atmosphere A hypothetical vertical distribution of temperature, pressure, and density of air in the *atmosphere which, by international consent, is regarded as representative of the overall atmosphere, for example for use in designing missiles and air-craft.

Standard Metropolitan Statistical Area (SMSA) An *urban region in the USA that has at least 100 000 inhabitants, and has strong economic and social ties to a central *city of at least 50 000 people.

standard of living The level of material well-being of an individual or group, in terms of goods and services available to them. A measure of *quality of life.

standardized mortality ratio (SMR) The ratio of the number of deaths in a population to the expected number of deaths in a standard population adjusted for age and possibly other factors such as gender or race.

standing crop The total amount of *biomass in an *ecosystem at a particular point in time, usually expressed as the weight (in grams) of living material per unit area or volume.

standing wave *See* HYDRAULIC JUMP.

starch A complex *carbohydrate that is found mainly in the seeds, fruits, roots, and other parts of plants, particularly in corn, potatoes, wheat, and rice.

starter castle *See* McMANSION.

starting delivery date The first specified future date when the *forward contract *emissions reductions will be delivered under the *Kyoto Protocol.

starting price The price of the *emissions reductions at the *starting delivery date in the bid or offer for a *forward contract, under the *Kyoto Protocol.

startling coloration The presence of very strong colours in a species. *Contrast* CAMOUFLAGE, DISRUPTIVE COLORATION.

starvation A state of extreme hunger that results from lack of *essential nutrients over a prolonged period of time.

stasis A period of little or no detectable change, for example in the *evolution of a *species.

State Implementation Plan (SIP) A detailed description of the programmes a state in the USA will use to carry out its responsibilities under the *Clean Air Act, which must be approved by the *Environmental Protection Agency.

state land Land that is owned and administered by the state in which it is located.

state of matter One of the three normal states of matter, which are *solid, *liquid, and *gas.

state of the environment report A report that provides up-to-date information relating to *environmental quality and *natural resources, which usually includes an inventory of what is there, an assessment of their state and quality, a baseline against which to compare changes, and the prospect of monitoring of changes through time.

state park A park or recreation area in the USA that is owned and administered by the state in which it is located.

Statement of Forest Principles The informal name of a non-legally binding framework for international action in dealing with the causes and impacts of *deforestation, particularly within the tropics, that was agreed at the *United Nations Conference on Environment and Development in 1992. Its formal title is Non-Legally Binding Authoritative Statement of Principles for a Global Consensus on the Management, Conservation and Sustainable Development of all Types of Forests.

static equilibrium A type of *equilibrium that occurs where force and reaction are balanced, and the properties of the system remain unchanged over time. *Contrast* STEADY-STATE EQUILIBRIUM.

stationary front A *front between warm and cold air masses, which is moving very slowly or not at all.

stationary source A source of *pollution (such as a *smokestack) that does not move. *Contrast* MOBILE SOURCE.

statistic A measure that is calculated from a *sample of data.

statistical range The difference between the upper and lower values in a set of data.

statistics A branch of mathematics that deals with the collection, analysis, interpretation, and presentation of large amounts of numerical data.

status unknown An *unknown category of species that is defined by the *IUCN *Red Data Book as 'no information is available with which to assign a conservation category'.

statute An *act that is passed by a legislative branch of government, which declares, commands, or prohibits something, and which describes *law and defines the time within which parties must take action to comply.

statutory A legal requirement, authorized by *statute.

statutory agency A government body which has powers defined in law, such as the *Environment Agency in the UK and the *Environmental Protection Agency in the USA.

statutory law Law which is enacted by the legislative branch of government, in contrast to case law or common law

steady state A state in which the properties of a *system remain constant and there are no net exchanges of mass or energy with the surrounding *environment over time. Also known as **equilibrium state**. *See also* HOMEOSTASIS.

steady-state economy An *economy in which birth and death rates are relatively low, and the use of renewable energy sources and recycling of materials is relatively constant over time.

steady-state equilibrium A type of *equilibrium in which the average condition of the system remains unchanged over time. *Contrast* STATIC EQUILIBRIUM.

steam fog *See* EVAPORATION FOG.

Stefan–Boltzmann equation An equation that can be used to calculate the *black body temperature of a body (including a *planet, such as the *Earth).

stem The upright structure (*stalk) that supports a *plant or *fungus, usually above the ground and in a direction opposite to the *roots.

stemflow The water (*precipitation after *interception) that trickles down the *stems of plants and eventually reaches the ground below the vegetation as part of the *throughfall.

steno- A prefix denoting the ability of an *organism to tolerate only a narrow range of changes in environmental conditions. *Contrast* EURY-.

stenohaline Tolerant of a narrow range of *salinity.

stenothermal Tolerant of a narrow range of *temperature.

stenotopic Found in only one or a few *habitats.

steppe An extensive, mid-latitude, *temperate, grassland *biome that covers much of Eurasia, particularly Russia.

sterile Incapable of reproducing.

sterile male release An approach to fly pest control, which is based on releasing sterile males with which females mate only once, producing eggs that do not hatch.

sterilization The removal or destruction of all *micro-organisms, using heat, irradiation, gas, or chemicals. Also known as **disinfection**.

stewardship Responsibility for the sustainable management and use of a particular *resource or place. *See also* LAND ETHIC.

stewardship management *See* TOTAL RESOURCE MANAGEMENT.

stochastic A process or event that involves a *random variable or behaviour,

S

or a pattern that results from random effects, but with some direction. *Contrast* DETERMINISTIC.

stock **1.** Farm animals that are bred and kept for meat or milk. *See also* LIVESTOCK.
2. A plant or stem onto which a *graft is made.
3. A *discordant *intrusive igneous feature that is smaller than a *batholith, with no distinct shape. Also known as **boss**.

stock resource *See* NON-RENEWABLE RESOURCE.

Stockholm Conference *See* UNITED NATIONS CONFERENCE ON THE HUMAN ENVIRONMENT.

Stockholm Declaration A statement of 'common principles to inspire and guide the peoples of the world in the preservation and enhancement of the human environment' which was agreed at the 1972 *United Nations Conference on the Human Environment.

stocking density The number of *livestock per unit area of land.

stockpile A reserve, held in store for future use or a special purpose.

stomata The small openings in leaves, herbaceous stems, and fruits, through which gases and *water vapour pass from the surrounding air.

Stone Age The *prehistoric period of human culture in Europe that lasted up to about 2000 BC (up to the dawn of the Bronze Age), during which stone was the main material used for making tools and weapons. Archaeologists usually divide the Stone Age into four periods: the *eolithic, the *Palaeolithic, the *Mesolithic, and the *Neolithic.

stone polygon A type of *patterned ground on low slopes, in which stones are moved across the ground surface by *freeze-thaw action and are aligned in the shape of polygons. *Contrast* STONE STRIPE.

stone stripe A type of *patterned ground on steep slopes, in which stones

are moved across the ground surface by *freeze-thaw action and are aligned in stripes oriented downhill. *Contrast* STONE POLYGON.

storage General term for the temporary holding (for example in containers, tanks, waste piles, and surface impoundments) of *waste material while it awaits treatment or disposal.

storm A violent weather episode that often includes extreme conditions, such as strong *winds (force 11 on the *Beaufort scale), heavy rain, hail, or snow, often accompanied by *thunder and *lightning.

storm flow The rapid runoff on a river *hydrograph that is created directly by the input of *precipitation from a *storm.

storm hydrograph *See* HYDROGRAPH.

storm sewer A *sewer system that collects and carries rain and snow runoff to a point where it can soak back into *groundwater or flow into surface waters.

storm surge An abnormally high wall of seawater that is pushed ashore by high winds, and is normally associated with the approach of a *hurricane.

storm warning A *warning that is issued about an impending *storm, which involves sustained surface winds of greater than about 90 kilometres an hour.

stormwater The portion of *precipitation which flows over land into storm sewers or surface waters.

stormwater collection system The equipment (which includes pipes, pumps, and conduits) that is used to collect and transport surface water from *precipitation, and/or *wastewater, to and from storage or treatment areas.

stoss The side of a slope that faces the direction of flow of ice, wind, or water. *Contrast* LEE.

strain **1.** A genetic variant, or group of organisms within a *species that differ very little from similar groups. *See also* CULTIVAR.

2. Deformation of a material as a result of applied force. *See also* ISOSTATIC ADJUSTMENT.

strangler An *epiphytic plant (such as a vine) whose aerial roots extend down the trunk of a supporting tree and grow around it, eventually strangling the tree.

strata Multiple layers (*beds) of rock (plural *strata).

strategic environmental assessment (SEA) The systematic process of evaluating the *environmental impacts of a policy, plan, or programme and its alternatives, including the preparation of a report on the evaluation and the use of the findings in publicly accountable decision-making. *See also* ENVIRONMENTAL IMPACT ASSESSMENT.

strategic lawsuit against public participation (SLAPP) Lawsuits that have no merit in themselves but are brought against private citizens who act in the public interest, usually by a large corporation of a wealthy individual, in order to intimidate them and silence them by burdening them with huge costs of legal defence.

strategic mineral Any *mineral that a country uses (for example for defence purposes) but which it cannot produce itself.

strategic plan A plan that an organization develops in order to help align its structure and resources with defined priorities, missions, and objectives.

strategy A statement or framework that describes how a particular programme will achieve its desired goals and objectives.

stratification To separate into layers (*strata), such as the layers (*lamination) within a *sedimentary rock, or the layers of vegetation within a *tropical rainforest. *See also* THERMAL STRATIFICATION.

stratified Layered. *See also* STRATIFICATION.

stratified drift Well-sorted layers of sand and gravel that have been deposited by glacial *meltwater.

stratified estuary An *estuary in which *freshwater from the land lies on top of denser *saltwater from the sea.

stratified seed *Seed that has been stored in a cool, moist environment in order to develop to the point where it can *germinate.

stratified vegetation The layers (*strata) of vegetation within a forest, which are defined by species, age, or size of plant into the tree layer, shrub layer, and herb layer. *See also* CANOPY, OVERSTOREY, UNDERSTOREY.

stratiform A *cloud that develops horizontally rather than vertically.

stratigraphic Relating to or determined by the layers (*strata) within a sediment or *sedimentary rock.

stratigraphic sequence The series of layers (*strata) within a sediment or *sedimentary rock, which can be interpreted by *stratigraphy.

stratigraphic unit A group of closely related layers (*strata) within a *sedimentary rock.

stratigraphy The study of the origin, composition, distribution, and succession of *strata in sedimentary rocks, on the basis of fossil content (*biostratigraphy) and lithology (*lithostratigraphy).

stratocumulus (Sc) A soft, grey, rolling low *cloud which is usually found below about 2400 metres, has a flat base and rounded top, is thicker than *stratus, and is often accompanied by weather changes and *precipitation.

stratopause The boundary in the *atmosphere between the *stratosphere and the *mesosphere, at a height of about 50 kilometres, where the air temperature is about 0°C.

stratosphere The layer of the *atmosphere between the *troposphere (below) and the *mesosphere (above), the lower level of which varies with latitude from about nine kilometres at the poles to 16 kilometres near the equator, and which extends to an altitude of around 50 kilometres. In this layer there is little *water vapour but much *ozone, and temperature remains stable or rises slightly with altitude. The stratosphere is not uniform, and conditions change between the lower and upper parts of it. The lower part of the stratosphere, up to an altitude of about 35 kilometres is a transitional zone above the *tropopause in which air temperature is fairly constant in the region of –55°C. Temperature rises with height in the upper stratosphere, between about 35 and 60 kilometres, reaching about 0°C at the stratopause. *See also* OZONE DEPLETION, OZONE LAYER.

stratospheric ozone *Ozone that is found in the *stratosphere. It blocks out harmful *ultraviolet rays from the Sun, including those that may cause damage to human health and the environment. *Contrast* TROPOSPHERIC OZONE. *See also* OZONE LAYER.

stratotype A sequence of sediment or rock at a particular place, in which a *stratigraphic unit or boundary is defined, which is used as a global standard for comparing all other stratigraphic units of its kind (defined by age and *lithology).

stratovolcano A *volcanic cone that consists of both *lava and *pyroclastic rocks.

stratum A horizontal layer of *sedimentary rock or vegetation. (plural *strata). *See also* STRATIFICATION.

stratus (St) A thick, grey, flat, low-level type of *cloud, the most common of all low clouds, which rarely extends higher than about 1500 metres, can produce *drizzle or *snow but rarely produces heavy *precipitation, and creates *fog when it touches the ground. Stratus clouds are commonly found during the winter in humid climates in middle and high latitudes, and can persist for days or weeks.

straw The plant residue that remains after *seeds are removed in threshing.

streak The colour of a *mineral when it is in powdered form.

stream A natural body of running water that flows along a *channel. *See also* CREEK, RIVER.

stream bank The sides of a *stream channel. Also known as **river bank**.

stream based standard *See* RECEIVING WATER QUALITY STANDARDS.

stream bed The bottom of a *stream channel. Also known as **river bed**.

stream channel A *watercourse, or natural *channel along which water flows. Also known as **river channel**.

stream discharge The rate of flow of water along a stream over a given period of time, which depends on volume and velocity of flow, and is usually expressed in cubic metres per second ($m^3 \ s^{-1}$).

stream frequency The number of *stream segments per unit drainage area (usually per square kilometre). Dense networks have high stream frequencies. *Contrast* DRAINAGE DENSITY.

stream order The relative position or rank of a *stream segment in a drainage network, with the smallest permanent streams defined as first order, second order defined as where two first-order streams join, third order as two second-order streams join, and so on.

stream segment A reach of a *stream channel, usually defined as between two successive tributaries.

stream use classification A system for classifying *streams according to the intended use of the water, such as for *recreation or *irrigation.

streamflow Discharge that occurs in a natural *stream channel, which includes *runoff and water that is added or removed through *diversion or flow in a *regulated river.

streamline The curved path of any medium (such as a parcel of air or a flow of water or ice) that flows steadily over or around an obstacle.

streamside buffer A *buffer zone or strip of permanent undisturbed vegetation (usually forest or grass) that is left between a *stream and an adjacent area of more intensive land use (such as agriculture or urban development). It is designed to reduce *soil erosion and protect water quality by filtering out pollution before it reaches the water. *See also* RIPARIAN.

streamside management zone (SMZ) An area adjacent to a *stream in which vegetation is maintained or managed in order to protect water quality, provide diversity of habitat, and create wildlife *travel corridors.

street Any public thoroughfare or residential road that leads off a main road.

stress 1. The force per unit area that is applied to a body, and produces *strain on that material.
2. A physical or chemical process that causes a response within an *organism, *population, or *ecosystem.

stressor *See* AGENT.

striae *See* STRIATION.

striation A scratch mark that is caused by *abrasion of the surface of a rock by a *glacier, is aligned in the direction of ice movement, and is very useful in reconstructing directions of ice movement in the past. Also known as **striae**.

Strict Nature Reserve *See* SCIENTIFIC RESERVE/STRICT NATURE RESERVE.

strike The direction or trend of a *bedding plane or *fault, relative to the horizontal.

strike-slip fault A horizontal displacement *fault (such as the *San Andreas fault in California), in which movement occurs along the fault line as a result of *shearing. Also known as **transcurrent fault**.

strip cropping *See* STRIP FARMING.

strip cutting In *forest management, a system of *clearcutting in narrow strips in order minimize edge effects and allow natural *regeneration of the forest.

strip development A sprawling form of commercial land use in which each establishment is given direct access to a major thoroughfare, which usually has many large signs designed to attract passers-by. Visitors drive to and from the strip, which therefore requires extensive parking areas.

strip farming Planting different *crops in alternating strips along land contours, so that when one is harvested the other remains to protect the soil and prevent erosion. Also known as **strip cropping**.

strip mall North American term for a shopping centre (*shopping mall) that consists of a continuous line of single-storey shops, stores, businesses, and restaurants along a road or busy street, usually opening onto a car park. *Contrast* SHOPPING MALL.

strip mining The mining of *ore or *coal from an *open mine, which usually involves the use of large machines (bulldozers, power shovels, or stripping wheels) that scoop out the material in long, narrow strips. The two main types are *area mining and *contour mining.

Strong, Maurice A Canadian industrialist and public servant (born 1929) who was the first Executive Director of the *United Nations Environment Programme. He served as Secretary-General of the 1992 *United Nations Conference on Environment and Development and as a senior advisor to UN Secretary-General Kofi Annan.

strophic balance The state of balance between the *pressure gradient (which forces air to move from high to low pressure) and the *Coriolis force (which acts in the opposite direction), and causes *wind generally to blow at right angles to the *pressure gradient (deflected to the right in the *northern hemisphere and to the left in the *southern hemisphere).

structure 1. The way in which something is organized, such as the pattern of plant species within an *ecosystem.

2. Anything that is built for the support, shelter, or enclosure of people, animals, goods, or property.

Structure Plan A statutory *land use plan agreed by a county planning authority in the UK, which sets out strategic policies for town and country planning within that county. *Contrast* LOCAL PLAN.

stubble The short stalks that are left behind in a *field after the *crop has been harvested. ·

stubble mulch A soil covering that is composed of the unused stalks of crop plants (*stubble).

stump The lower part of a tree that remains standing after the tree has been felled or *coppiced.

stumpage The commercial value of standing trees.

stylet A long mouth part of an *insect, which is normally used to extract fluids.

stylops A genus of minute insects (of the group Strepsiptera), which in their larval state are *parasitic on bees and wasps.

subaerial Located on or near the ground surface.

subarctic The high latitude area in the northern hemisphere which has a *subarctic climate.

subarctic climate A *climate zone that is found across much of North America and most of northern Eurasia, north of the *humid continental climate and south of the *polar climate in the *northern hemisphere, and is characterized by bitterly cold winters and short cool summers.

subatomic Smaller than an *atom, or relating to the constituents of an atom or the forces within an atom.

sub-bituminous coal A rank of *coal that is intermediate between *lignite and *bituminous coal. Also known as **black lignite**.

subclimax A form of *arrested development of vegetation, where a *succession has been retarded or inhibited by natural arresting factors such as natural fires, storms, or lack of long-term environmental stability.

subcommunity A subdivision of a *community.

subdivision The act of subdividing land (for example into sites, blocks, or lots with streets or roads and open spaces), or the area of land that has been divided up.

subduction The process that takes place within *plate tectonics when two *crustal plates converge and one plate is forced below the other into the *mantle.

subduction zone A zone in the Earth's crust where one *crustal plate is pushed downwards into the *mantle beneath an adjacent one, driven by *sea-floor spreading.

subgenus One of two or more groups of closely related *species that constitute a *genus.

subglacial Beneath a *glacier or *ice cap. *Contrast* ENGLACIAL, SUPRAGLACIAL.

subglacial stream A *meltwater *stream that flows underneath a *glacier or *ice cap.

subhumid climate A warm, moist *climate which supports *grassland or *prairie vegetation.

sublimation The direct alteration (*phase change) of *ice (solid) to *water vapour (gas).

sublittoral Of or relating to the region of the *continental shelf, lying the sea *shore and the edge of the continental shelf.

sublittoral fringe The upper part of the *sublittoral zone, which is uncovered at *low tide.

sublittoral zone A zone within the *ocean that extends from the *low tide mark to the edge of the *continental shelf, with depths of up to about 300 metres.

submarine Beneath the surface of the *sea.

submarine canyon An underwater *canyon in the *continental shelf.

submerged aquatic vegetation *Vegetation that lives at or below the surface of a *waterbody.

subpolar and arctic climate zone The coldest of the world's *climate zones that is found at high *latitudes, and contains *continental subarctic, *marine subarctic, and *arctic and *ice cap climates.

subpolar climate A *climate zone that is found in the *northern hemisphere, south of the *polar climate, and is characterized by severely cold winters and short, cool summers. Also known as **boreal climate** or **taiga climate**.

subpolar low A major *wind belt that circles the Earth, in that easterly winds generated by the *polar high converge with westerly winds from the *subtropical high, at the subpolar low pressure belt, at about 60° latitude. Weather here is often very unsettled, with extensive cloud cover.

subpopulation A subset of a natural or artificial breeding *population. *See also* VARIETY.

subsidence 1. In meteorology, the slow sinking of air that is usually associated with *high pressure areas.
2. In geology, a settling of the ground surface as a result of the collapse of *porous formations from which large amounts of *groundwater, *oil, or other underground materials have been withdrawn.

subsidence inversion A temperature *inversion that develops in the atmosphere as a result of air gradually sinking over a wide area and being warmed by *adiabatic compression, often associated with *subtropical high pressure areas.

subsidy A grant that is paid by government to the supplier of a good or service that benefits the public.

subsist To support oneself.

subsistence The means of obtaining food and other items that are essential for basic survival.

subsistence economy An *economy in which production meets the minimum needs of the population, but produces no surplus.

subsistence farming Cultivation of *crops and *livestock in order to provide for the basic needs of the farmer and his/her family, without producing surpluses that can be sold at market. *Contrast* CASH CROP.

subsoil The lower part of a *soil profile, beneath the *topsoil, which generally contains little *organic matter.

subspecies A geographically isolated or physiologically distinct group within a *species, which is capable of interbreeding with other members of the subspecies but rarely does so. Also known as a **race**. *See also* CULTIVAR, VARIETY.

substance A material of a particular kind or composition.

substitutability The degree to which one material can be replaced or exchanged for another.

substrate The surface on which a plant or animal lives and grows; for example a rocky or sandy substrate. Also known as **substratum**.

substratum *See* SUBSTRATE.

subsurface drainage The removal of excess water from within a soil or deposit, often by means of *tile drains.

subsurface mining The extraction of *metal ore or *fossil fuel resources from beneath the surface of the ground.

subtropical The region between the *tropical and *temperate regions, which extends between about latitudes 35° and 40° North and South.

subtropical front A zone of temperature transition in the upper *troposphere over *subtropical latitudes, where warm air that is carried towards the poles by the *Hadley cell meets the cooler air of the middle latitudes.

subtropical high A semipermanent zone of *high pressure that is found in *subtropical latitudes.

subtropical jet A strong, generally westerly wind in the upper *troposphere that blows over *subtropical latitudes. *See also* JET STREAM.

suburb The outer part of a *city or *urban area, close to the urban–*rural boundary, with limited commercial land use and low-density residential development. Also known as **suburbia**. *See also* EXURBIA.

suburban Relating to, characteristic of, or situated in *suburbs. *See also* URBAN.

suburban sprawl The growth of a metropolitan area, particularly the *suburbs, over a large area. Also known as **sprawl**, **urban sprawl**, or **Los Angelization**.

suburbia *See* SUBURB.

succession 1. In general, a sequence or the action of following in order.
2. In ecology, the gradual and orderly sequence of different plant *communities that develop over time in the same area, as one community replaces another through a predictable series of stages (*seres), until a stable climax is established. The sequence normally involves five stages, beginning with creation of an area of bare land (*nudation), followed by the arrival of the first plant seeds (*migration), establishment of the seeds and growth of the plants (*ecesis), competition between the established plants, and their effects on the local environment (*reaction), until the populations of species eventually reach an equilibrium condition, in balance with the surrounding environment, which is sustainable (*stabilization). A typical succession in a temperate area might take up to 150 years from open ground to mature woodland, and each sere would have its characteristic plants and animals. *See also* PRIMARY SUCCESSION, SECONDARY SUCCESSION.

successional stage *See* SERE.

succulent A plant (such as a desert cactus) that is adapted to life in very dry conditions, often by having a soft fleshy body which can store water for use during times of drought.

sucker A *shoot that grows from the *stem of the plant below ground level.

suffocate To die due to a lack of *oxygen.

suitable forest land A North American term for land that is managed for timber production. Also known as **suitable timber land**.

suitable range A North American term for land that produces or is capable of producing enough *forage for *grazing by *livestock on a *sustainable basis, under intensive management, without causing environmental damage.

suitable timber land *See* SUITABLE FOREST LAND.

sulphate (SO_4) An *ion that contains *sulphur and *oxygen (SO_4), which react easily with *hydrogen to become *sulphuric acid (H_2SO_4). It is one of seven major ions that are found naturally in most *waterbodies; and is a good *indicator of *water pollution.

sulphate aerosol An *aerosol that consists of compounds of *sulphur formed by the interaction of *sulphur dioxide and sulphur trioxide with other compounds in the *atmosphere. These come from natural sources (particularly *volcanic eruptions) and from the combustion of *fossil fuels and the eruption of volcanoes like Mount Pinatubo.

sulphate reducer A *micro-organism that exists in an *anaerobic environment and reduces *sulphate to *hydrogen sulphide.

sulphur (S) An abundant, tasteless, odourless, yellowish, non-metallic solid element that is found in *hydrocarbons, which release *sulphur dioxide when burned. It is one of nine *macronutrients which plants require for healthy growth, and is an important but small constituent of some *proteins.

sulphur content The amount of *sulphur that is found in a particular coal, which strongly influences the amount of *sulphur dioxide that is released on combustion and thus affects *acid rain.

sulphur cycle The *biogeochemical cycle by which *sulphur circulates within the *biosphere. The sedimentary and atmospheric phases are both important in the sulphur cycle (unlike the *nitrogen and *carbon cycles), and they are of comparable size. One source of sulphur is volcanoes. Although major volcanic eruptions are relatively infrequent, and so release relatively small amounts of sulphur overall, when they do occur they produce sudden increases in atmospheric sulphur. Sulphur compounds are also released by *weathering of rocks. Roughly similar amounts of sulphur reach lakes and rivers from the atmosphere and from rock weathering. The oceans also play an important role in the sulphur cycle. Several species of marine phytoplankton produce *dimethyl sulphide (DMS), much of which breaks down in water. But some of the DMS enters the air where it is oxidized to *sulphur dioxide and *sulphate aerosol. The oxidation of DMS sulphate has implications for the control of *acid rain in many coastal areas, because it can be as important a source of sulphur as air pollution. Sulphate aerosols can act as *cloud condensation nuclei, further complicating the situation. In terrestrial ecosystems *bacteria break down sulphate and release gases, mainly *hydrogen sulphide. Human activities have doubled the amount of sulphur in the atmosphere since the Industrial Revolution in the

18th and 19th centuries. By far the most significant human impact is the burning of fossil fuels, particularly *coal (which contains on average 1–5% sulphur) and *oil (containing 2–3% on average).

sulphur dioxide (SO$_2$) A colourless, poisonous gas that is formed by burning *sulphur to create an *oxide, most commonly by the burning of *fossil fuels that contain sulphur in order to generate electricity in power stations. The sulphur dioxide is released into the atmosphere as an *air pollutant, and can cause *acid rain. It is a conventional or *criteria pollutant, and can give rise to *trans-frontier pollution. The amount of sulphur that enters the atmosphere from the burning of fossil fuel is similar to that produced by natural processes, but is not so widely distributed. Combustion emissions are spatially concentrated in major industrial regions and so they often overload regional *biogeochemical cycles. Much effort has been invested in devising ways of reducing sulphur dioxide emissions at source, particularly from the burning of fossil fuels in power stations. Strategies include burning less fuel, switching to fuel with a lower sulphur content, reducing the *sulphur content of the fuel before burning it, and reducing the sulphur content of the fuel at combustion (by *fluidized bed technology) or the sulphur content of the waste gases (by *flue gas desulphurization).

sulphur hexafluoride (SF$_6$) A colourless gas that is soluble in alcohol and ether. It is a powerful *greenhouse gas that is widely used in the electrical utility industry to insulate high-voltage equipment. It has a global warming potential of 24 900 and persists in the atmosphere for up to 3200 years. It is one of the six greenhouse gases to be limited under the *Kyoto Protocol.

sulphuric acid (H$_2$SO$_4$) A *toxic, *corrosive, strongly acid, colourless compound that is formed when *sulphur oxides combine with atmospheric

S

*moisture, and is a major ingredient of *acid rain.

summer The warmest *season of the year, between *spring *and autumn/ fall, when the Sun is highest in the sky. Astronomically this is the period between the *summer solstice and the *autumnal equinox, which covers the months of June, July, and August in the northern hemisphere, and December, January, and February in the southern hemisphere.

summer solstice The longest day of the year, when the Sun is highest in the sky (because of the Earth's tilt), which marks the start of the *summer season. This occurs in the northern hemisphere on or near 21 June, and in the southern hemisphere on 21 or 22 December. *See also* SOLSTICE, WINTER SOLSTICE.

sump A pit or tank that is used to catch liquid *runoff for drainage or disposal.

Sun The star at the centre of our *solar system, around which the planets (including the Earth) orbit. It radiates energy (particularly heat and light) derived from thermonuclear reactions in its interior. The Earth is on average about 150 million kilometres away from the Sun, but this varies between 147 million and 152 million during the elliptical orbit. In galactic terms our Sun is quite insignificant, one of perhaps 100 000 million stars, but it is the most important member of the solar system and dominates the solar system in two senses—its size and its energy output. With a diameter of nearly 1.4 million kilometres, the Sun is more than 100 times bigger than the Earth and has a volume equivalent to that of a million Earths. Almost 99% of the mass of the solar system is contained within this one star. The Sun is composed mostly of *hydrogen (about 70%) and *helium (30%); other elements make up less than 1%. It generates energy by nuclear fusion reactions that burn hydrogen into helium in its interior. These reactions produce immense heat— temperatures at the surface of the Sun

are about 5530°C and at the centre about 15 000 000°C. The reactions also produce vast amounts of energy, and cause *sunspots and solar flares that are huge relative to the size of the Earth.

suncup A small, bowl-shaped depression in ice that is melted by *insolation.

sundown *See* SUNSET.

sunlight The light or rays from the *Sun. *See also* SOLAR HEATING.

sunlight zone *See* PHOTIC ZONE.

sunrise The first light of day, as the *Sun rises above the horizon in the morning.

sunset The time in the evening at which the *Sun begins to fall below the horizon. Also known as **sundown**.

sunspot A dark-coloured region on the surface of the *Sun that represents an area of cooler temperatures and intense magnetic fields. The number of sunspots is greater when the Sun is more active. Sunspots vary in size from about 1500 kilometres to groups up to 200 000 kilometres in diameter. They constantly change size and appearance. Most spots appear and disappear within a day or so, but some last longer than a month. Quite why sunspots vary so much is not known, but they might be caused by strong magnetic fields that block the outward flow of heat to the Sun's surface.

sunspot cycle *See* SOLAR CYCLE.

supercell A severe *thunderstorm that is characterized by a rotating, long-lived, intense updraft (*mesocycle), which produces severe *weather, including extremely large hailstones, damaging *winds, and violent *tornadoes. Also known as a **mesocyclone**.

supercity *See* MEGALOPOLIS.

supercooled Cooled below *freezing point.

supercooled water Liquid water droplets at temperatures well below 0°C that would freeze immediately if particles were present to start the freezing process.

supercritical water Water which has undergone a thermal treatment in which moderate temperatures and high pressures are used to increase its ability to break down large organic molecules into smaller, less toxic ones by combining *oxygen with simple organic compounds to form *carbon dioxide and water.

Superfund A fund established by Congress in the USA under the 1980 *Comprehensive Environmental Response, Compensation, and Liability Act, financed by fees paid by toxic waste generators and by cost recovery from cleanup projects, in order to pay for the cleanup of inactive and abandoned *hazardous waste sites or *hazardous materials that have been accidentally spilled or illegally dumped. Also known as the **Superfund Trust Fund**.

Superfund Amendments and Reauthorization Act (SARA) (1986) US legislation that amended the 1986 *Comprehensive Environmental Response, Compensation, and Liability Act to provide additional funding for environmental cleanup of hazardous waste sites.

Superfund Innovative Technology Evaluation (SITE) In the USA, a programme of the *Environmental Protection Agency designed to promote the development and use of innovative treatment technologies in the cleanup of *Superfund sites.

Superfund Program See COMPREHENSIVE ENVIRONMENTAL RESPONSE, COMPENSATION, AND LIABILITY ACT.

Superfund Trust Fund See SUPERFUND.

superimposed river A *river network in which an initial pattern, determined by original rock structures and types, has been preserved as the river has cut down through underlying rocks, superimposing the original pattern on underlying structures and rock types.

superparasitism The infestation of *parasites by other parasites.

superposition, principle of The basis for interpreting the *stratigraphic sequences within *sedimentary rocks, based on the assumption that older rocks are overlain by younger rocks so long as the rocks have not been disturbed by *folding, *faulting, or other geological processes. This helps geologists to establish a *relative chronology of deposition, and to make geological associations from place to place.

supersaturated air A condition that occurs when the *relative humidity of air in the *atmosphere is greater that 100%.

supersaturation A *relative humidity of more than 100% in the *atmosphere.

supersonic Faster than the speed of sound in the surrounding air. See also MACH.

supersonic transport A jet aeroplane that flies faster than the speed of sound (*Mach > 1).

supplementarity A requirement within the *Kyoto Protocol that adequate domestic energy and other policies exist to ensure the achievement of long-term *emissions reduction goals. See also FLEXIBILITY MECHANISM.

supplementary feed Feed (such as hay or *silage, feed blocks or concentrates) that is used to supplement the dietary requirements of *livestock, usually during the winter.

supply The quantity of a good or service that producers are willing to provide at the specified price, time period, and condition of sale. Contrast DEMAND. See also MARKET EQUILIBRIUM, SUPPLY SCHEDULE.

supply chain A series of business transactions that starts with raw material and ends with the sale of the finished product or service.

S

supply schedule The relationship between price and the quantity of a good or service that producers are willing to provide.

suppressed One of the four major *crown classes in trees, in which trees which have crowns below the general level of the crown cover receive no direct light either from above or from the sides. Also known as **overtopped**.

suppression In *fire management, all of the activities involved in extinguishing or containing a fire, starting with its discovery.

supraglacial On top of a *glacier or *ice cap. Contrast ENGLACIAL, SUBGLACIAL.

supraglacial stream A *meltwater *river that flows over the surface of the ice in a *glacier or *ice cap.

surf *Waves breaking on the *shore.

surf zone The zone of *wave action that extends from the *shore out to the most seaward point of the zone (the *breaker zone) at which waves approaching the coastline start to break.

surface air temperature The *temperature of the air near the surface of the Earth, the global average of which is 15°C.

surface current The horizontal movement of seawater at the surface of a *sea or *ocean.

surface drainage The removal of excess water from the surface of the ground, either naturally by a *stream, or artificially by means of a *drainage ditch.

surface fire A *forest fire of moderate intensity in which low vegetation (such as *shrubs) and some of the surface (*bark) of trees are burned, but not the *crowns of trees, so the trees survive.

surface hoar A deposit of *ice crystals that builds up on a surface when the temperature of the surface is colder than the air above and colder than the frost point of that air. See also HOAR.

surface impoundment Any natural or constructed facility that is used to store liquids on the ground surface, including *water storage ponds and *lagoons.

surface inversion See RADIATION INVERSION.

surface layer The lowest few metres of the *atmosphere, or the upper few metres of the *ocean, within which the air or water is mixed by the *wind.

surface mining See OPEN CAST MINING.

Surface Mining Control and Reclamation Act (1977) US legislation that established a programme for the regulation of surface mining activities and the reclamation of coal-mined lands, under the administration of the Office of Surface Mining, Reclamation and Enforcement, in the Department of the Interior.

surface resource Any *renewable resource (such as *timber) that is found on the surface of the Earth, in contrast to resources such as *groundwater and *minerals which are found beneath the surface.

surface rights Ownership of or rights to use the surface of the land. Subsurface rights (including *mineral rights) are often treated separately.

surface runoff *Runoff that flows overland to a *stream channel.

surface storage All forms of natural storage of water (including *ponds and depressions) on the ground surface.

surface tension The tension of the surface film of a liquid, which controls the shape of the liquid surface.

surface water All water that is naturally open to the atmosphere, including *rivers, *lakes, *reservoirs, *ponds, *streams, *impoundments, *seas, *estuaries, and *wetlands.

surfactant A *soluble chemical compound that reduces the surface tension between two liquids, as used in many *detergents.

surge 1. A sudden forceful flow, for example in the flow of a *glacier.
2. A large, destructive *ocean wave that is created by very low *atmospheric pressure and *strong winds, for example associated with a *hurricane.

surplus Excess; a quantity that is much larger than is needed.

surrogate A substitute. *See also* PROXY.

surveillance The ongoing systematic collection, collation, and analysis of data and the timely distribution of information to those who need to know so that action can be taken.

survey 1. A drawing showing the legal boundaries of a property.
2. A count or *census of the number of individuals in an area, from which inferences about the *population can be made.

survival The act of living or continuing to exist.

survival of the fittest *See* NATURAL SELECTION.

survivorship The condition of outliving others; the likelihood of a representative newly born individual surviving to various ages. *Contrast* LONGEVITY.

survivorship curve The curve that describes changes in *mortality rate as a function of age.

susceptibility The degree to which an organism is sensitive to a particular factor, such as *infection by a *pathogen. *See also* SENSITIVITY.

suspended load *See* SUSPENDED SEDIMENT.

suspended sediment The fine particles of *sediment that are carried in *suspension by a *river, when turbulent mixing exceeds gravitational sinking. Also known as **suspended load**.

suspended solids Solids (*colloidal and particulate matter such as *clay,

*sand, and finely divided *organic material) that either float on the surface of water or are suspended in it. Suspended solids can usually be removed by filtering.

suspension A mixture in which fine particles are suspended in a fluid, supported by buoyancy.

suspension feeder Any *organism (such as sponges and corals) that feeds on particulate *organic matter (including *plankton) that is suspended in a water column. Also known as a **filter feeder** or **suspensivore**.

suspensivore *See* SUSPENSION FEEDER.

sustainability A concept that is used to describe community and *economic development in terms of meeting the needs of the present without compromising the ability of future generations to meet their needs. *See also* ECOLOGICAL SUSTAINABILITY, SUSTAINABLE DEVELOPMENT.

sustainability indicator A measure (*indicator) of *sustainable development.

sustainable Capable of being sustained or continued over the long term, without adverse effects.

sustainable agriculture An *agricultural system that is ecologically sound, economically viable, and socially just. Sustainable agriculture uses techniques to grow crops and raise livestock that conserve soil and water, use organic *fertilizers, practice biological control of *pests, and minimize the use of non-renewable *fossil fuel energy.

sustainable community A *community of people that is capable of maintaining its present level of growth without damaging effects.

Sustainable Development Commission *See* UNITED NATIONS COMMISSION ON SUSTAINABLE DEVELOPMENT.

sustainable livelihood approach An approach to *sustainable development which involves public participation and integrates environmental, social, and

SUSTAINABLE DEVELOPMENT

According to the *World Commission on Environment and Development (WCED), sustainable development is 'development that meets the needs of the present without compromising the ability of future generations to meet their own needs', which includes *economic growth together with protection of the quality of the environment, each reinforcing the other. Many of the symptoms of the *environmental crisis illustrate what can go wrong if environmental resources are over-used or used in ways that create environmental damage or instability. This is particularly true in *developing countries, where developed-world notions of development have often been transplanted with dire consequences. A central objective of the 1992 *United Nations Conference on Environment and Development was to establish the need to replace existing exploitative and environmentally damaging forms of *economic development with more sustainable and environmentally friendly forms of development. *Sustainability is an in-built feature of all natural environmental systems provided that human interference is absent or minimized. It relates to the capacity of a system to maintain a continuous flow of whatever each part of that system needs for a healthy existence. Human use of environmental resources and interference with environmental systems disturbs this in-built capacity, which can make a system unsustainable. Economists argue that resource use and depletion can stimulate research and development, substitution of new materials, and the effective creation of new resources, but there are limits to what is possible. Conservationists and ecologists have long been aware of the significance of sustainability within natural environmental systems. However, it was not until the late 1980s that the broader concept of sustainable development was first introduced by the WCED. The search for a single definition of 'sustainability' seems elusive, partly because it embodies a number of ideas imported from different disciplines—including *economics (no growth or slow growth), *ecology (integrity of the *biosphere, *carrying capacity), *sociology (critique of *technology), and *environmental studies (*ecodevelopment, resource–environment links). Whilst the WCED definition has been widely accepted, there has none the less been widespread debate about what sustainability might actually mean in practice, and about how it might best be applied to different cultures and economies. It is generally accepted that development can be sustainable only if it is based on sound *ecological principles and practices, but beyond that there is little real consensus. Some *environmentalists argue that too much attention has been focused on defining the meaning of sustainability, and not enough on exploring the implications of sustainability as it is likely to affect the status quo within and between countries. One significant effect of the sustainable development debate has been to raise awareness of the need to adopt wider perspectives when making decisions that affect the environment. Moving towards a more sustainable way of living will inevitably require some radical changes in attitudes, *values, and behaviour. There is no clear answer to the question of how we can create a vibrant global economy that does not destroy the *ecosystems on which it is based. The environmentally sustainable 'brave new world' will have to be less polluting, probably heavily reliant on solar energy, make extensive use of new ways of using and reusing materials, adopt less resource-intensive means of growing food, and develop effective strategies for preserving forests. There would also need to be radical changes in energy systems, tax systems, international economic structures, and provision of international aid. A strategy for building

SUSTAINABLE DEVELOPMENT (*cont.*)

a sustainable society was proposed in *Caring for the Earth*, which suggested a set of principles for sustainable development which include respect and care for the community of life, improving the quality of human life, conserving the Earth's vitality and *diversity, minimizing the depletion of *non-renewable resources, keeping within the Earth's *carrying capacity, changing personal attitudes and practices, enabling communities to care for their own environments, providing a national framework for integrating development and conservation, and creating a global alliance.

economic issues into a holistic framework for analysis and management.

sustainable management Managing the use, development, and protection of *natural resources in a way or at a rate which enables people and communities to provide for their social, economic, and cultural well-being and for their health and safety. *See also* SUSTAINABLE DEVELOPMENT.

sustainable use The use of an *organism, *ecosystem, or other *renewable resource at a rate that is within its capacity for renewal.

sustainable yield The rate at which a *renewable resource may be used in a *sustainable way. Traditional ways of harvesting natural renewable resources —such as fish from the *oceans, wood from the *forests, and plants and products from natural *ecosystems—have usually been sustainable so long as the quantities extracted were not greater than natural processes were able to replace. Many ways in which *natural resources are used are now intensive and unsustainable, without deliberate management intervention to limit the quantities removed to below *threshold sustainable yields. Also known as **sustained production, sustained yield**.

sustained production *See* SUSTAINABLE YIELD.

sustained yield *See* SUSTAINABLE YIELD.

SUV *See* SPORTS UTILITY VEHICLE.

SVOC *See* SEMI-VOLATILE ORGANIC COMPOUND.

swale A low-lying depression on the ground surface, which is often marshy or swampy, or may contain small lakes.

swamp A low-lying area of *wetland that is usually at least partially flooded, is covered with grasses and trees, has better drainage than a *bog, has more woody plants than a *marsh, and does not accumulate deposits of *peat. Swamps may be *freshwater or *saltwater, and *tidal or non-tidal.

swamp gas *See* METHANE.

sward The above-ground component of *grassland vegetation, which comprises *grasses and *herbs.

swash A thin sheet of water that moves up a beach after a wave breaks on the *shore. *See also* SURF.

S-wave A type of seismic wave, created by an *earthquake, that causes *shear stress in the material it moves through. Also known as a **secondary wave** or **shear wave**. *Contrast* P-WAVE.

swell A relatively smooth ocean *wave (up to two metres high and 30 metres long) that may travel great distances from its source.

swidden agriculture *See* SLASH AND BURN.

symbiont One of two or more *organisms that are engaged in *symbiosis. *See also* FACULTATIVE MUTUALIST.

S

symbiosis A situation in which two different species live together and interact, with one benefiting and the other either benefiting (*mutualism), not being significantly affected (*commensalism), or being harmed (*parasitism).

symbiotic A relationship between members of two different *species which results in a mutual benefit (*symbiosis).

sympatric Describing closely related species that have naturally overlapping *ranges but do not interbreed. *Contrast* ALLOPATRIC.

synchronous Coinciding in time. *Contrast* ASYNCHRONOUS.

syncline A concave *fold of rock *strata, with the youngest rocks at the centre of the downfold. *Contrast* ANTICLINE.

synclinorium A complex type of *fold in rock, in which there are many small folds superimposed on the major *syncline.

syndrome A collection of characteristics that occur together, such as the symptoms that are characteristic of a particular *disease.

synecology The branch of *ecology that deals with whole *communities of plants and animals, and the interactions of the *organisms within them. *Contrast* AUTECOLOGY.

synergetic Working together in *synergy.

synergism The combined effect of cooperative interaction between two or more changes that is greater than the sum of their separate effects. *Contrast* ADDITIVE EFFECT.

synergy The combination of two. or more changes which is mutually reinforcing, making an overall impact greater than the sum of their separate effects.

synfuel *See* SYNTHETIC FUEL.

synoptic Overall, or relating to a large-scale or broad view.

synoptic chart *See* WEATHER MAP.

synoptic meteorology The branch of *meteorology that deals with regional-scale weather phenomena. *Contrast* MESOSCALE METEOROLOGY.

synthesis The ability to put things together to form a new whole, such as the development of new ideas from existing ones, or the combination of separate elements to form a new complex product, synthetic chemical compound, or material.

synthetic Produced artificially. *Contrast* NATURAL.

synthetic fuel Liquid or gaseous *fossil fuels that are produced from *coal, *lignite, or other solid *carbon sources. Also known as **synfuel**.

synthetic organic chemicals (SOCs) Artificial (*anthropogenic) *organic chemicals, some of which are *volatile and others of which tend to stay dissolved in water without undergoing *evaporation.

system Any collection of components that work together to perform a function, such as the hardware and software that comprise a computer system, or the *biotic and *abiotic elements that make up an *ecosystem. *See also* CLOSED SYSTEM, OPEN SYSTEM.

systematics The study of *species, which includes their description (identification) and naming (*taxonomy), and the study of relationships among and between *taxa (*phylogenetics).

Système Internationale (SI) The international system of units; the metric system of measurements used in science. Some SI units and conversion are given in Appendix 10.

Système Probatoire d'Observation de la Terre *See* SPOT.

systemic Affecting or spread through the whole body (for example of a plant or animal), not localized.

systemic herbicide A *herbicide that kills the whole of a plant; not just the part it is sprayed on. *Contrast* CONTACT HERBICIDE.

systems analysis The study of how the component parts of a *system interact and contribute to the functioning of the system.

tableland A large, relatively flat area of high ground.

tagmata A specialized group of segments in the body of an *arthropod.

tagmosis The fusion of individual segments to form a body wall, which occurs in some *insects.

taiga See CONIFEROUS FOREST.

taiga climate See SUBPOLAR CLIMATE.

tail water The *surface water that drains from the lower end of an irrigated field.

tailings Rock and other waste materials that are separated from crushed *ore in the *mining process.

tailpipe The exhaust pipe through which gases from an engine (for example in a motor vehicle) are released into the air.

tailpipe standard *Emission gas limitations that are applicable to the exhausts of mobile source engines.

take 1. The *abstraction of water from surface water or groundwater.
2. A North American term meaning to harass, harm, pursue, hunt, shoot, wound, kill, trap, capture, or collect any wild species.

taking A North American term for government appropriation of private property or property rights, without compensation.

taku The name of a *katabatic wind in southeastern Alaska.

tall stacks policy A North American approach to the reduction of *air pollution, by using tall *stacks (chimneys) which release gas emissions into the *atmosphere, from which they are dispersed by wind.

tallgrass A North American *prairie *grassland vegetation dominated by grasses (such as big bluestem, *Andropogon gerardii*) that are tall and that flourish with abundant moisture. This type of grassland originally covered portions of 14 states from Texas to Minnesota, but more than 90% of it has been lost to urban sprawl and converted to cropland.

tallgrass prairie A type of *prairie that grows in areas that receive more than 50 centimetres of precipitation per year and have *loamy or *clay-based soils.

talus Angular rock debris (*scree) at the base of a *cliff.

talus cone See SCREE.

tannin A naturally occurring substance that is found in the skins, seeds, and stems of grapes, or sometimes in oak barrels, that gives red wine its dry, slightly bitter taste.

tap root The main *root of a plant that grows vertically downwards into the *soil until it reaches either an impenetrable layer or a layer that lacks oxygen or moisture.

tap water Drinking water that comes directly from a tap, usually in a home. See also POTABLE WATER.

tar sand A *sedimentary deposit that consists of a mixture of *clay, *sand, *water, and a tar-like heavy oil (*bitumen), which can be extracted by heating, and can then be refined to produce synthetic *crude oil.

target area The area at which an application of *agrochemicals (such as an *insecticide or *pesticide) is targeted.

target level The desired or threshold value of an *environmental indicator, defined in terms of human health, safety, and environmental quality.

target species The intended catch of a *fishery. *Contrast* BYCATCH.

tarn A *freshwater *lake in a *cirque.

taxon A group of *organisms that are regarded as distinct enough to be treated as a separate unit. The plural of taxon is taxa.

taxonomic Of or relating to *tax-·onomy.

taxonomy The science of naming and classifying organisms into systematic groups (*taxa) based on shared character-istics and natural relationships. There are seven levels of classification in a hier-archy, with *kingdom at the top, fol-lowed by *phylum, *class, *order, *family, *genus, and *species. A branch of *systematics. *See also* LINNAEAN CLAS-SIFICATION.

Taylor Grazing Act (1934) US legisla-tion that authorized the Secretary of the Interior to establish grazing districts from unreserved public domain lands, and to make rules and regulations for their occupancy and use.

TBT *See* TRIBUTYLTIN.

TDS *See* TOTAL DISSOLVED SOLIDS.

TEA-21 *See* TRANSPORTATION EQUITY ACT FOR THE 21ST CENTURY.

tebufenozide A type of *insecticide that works by accelerating the *moult-ing process, and is used to control larval insects (butterflies and moths) in apple orchards.

technocentric Based on *technology and science, or reflecting trust in science. *Contrast* ANTHROPOCENTRIC, ECOCENTRIC.

technocracy The control or strong in-fluence of society and government by people with well-developed technical skills, particularly scientists and engin-eers.

techno-fix Reliance on engineering and *technology to find solutions to major human problems, including envir-onmental problems.

technological disaster A disaster that is caused by non-natural causes, in-cluding biological, chemical, *nuclear and transport disasters, and terrorism. Also known as **man-made disaster**. *Con-trast* NATURAL DISASTER.

technological hazard A hazard cre-ated by people, as opposed to a *natural hazard. Examples include the release of *air pollutants such as *CFCs, serious *industrial accidents such as *oil spills at sea and explosions at nuclear power stations and toxic chemical plants, and the creation of waste materials (such as *nuclear wastes) that are *toxic and *persistent and which natural *environ-mental systems are incapable of break-ing down.

technological imperative The view that if we *can* do something, we *should* do it, such as generating energy from *nuclear power, even if it has many risks and unknowns associated with it

technological optimist *See* OPTIMIST, ENVIRONMENTAL.

technology The application of scien-tific processes, methods, or knowledge to produce goods and services that people consider useful.

technology transfer The process of transferring scientific findings from re-search laboratories into useful products by the commercial sector.

tectonic Relating to the deformation of rocks in the Earth's *crust.

tectonic force Geological forces that buckle, distort, and fracture the Earth's *crust, including *warping, *folding, *faulting, and *volcanic activity.

tectonic plate *See* CRUSTAL PLATE.

tectonics The study of the large-scale movements of rocks in the Earth's *crust, including *diastrophism, *oro-genesis, and *plate tectonics.

teleconnection A linkage between changes in ocean and atmosphere circulations in widely separated parts of the globe, through which weather patterns in one region influence the weather patterns in a distant place. The best known example is the *El Niño/Southern Oscillation. *See also* BUTTERFLY EFFECT.

telemetry Sending and receiving data over long distance communication links, such as *satellite or telephone.

temperate Relating to the *mid-latitude area of the Earth's surface, between the *tropics (23.5°) and the *polar circles (66.5°) in both *hemispheres.

temperate climate A *mid-latitude *climate with mild temperatures and moderate levels of rainfall, both of which vary from *season to season.

temperate desert A *mid-latitude *desert *biome that is hot in the summer and cool in the winter, such as the Mojave Desert in the USA.

temperate forest A *biome that is characterized by *deciduous *broad-leaved trees, which includes *humid continental mixed forest and *marine west coast forest.

temperate grassland A *grassland *biome that is found in the warm climates of the continental interiors, such as the *prairie of North America, *pampas of South America, *veld of South Africa, and *steppe of Eurasia.

temperate rainforest A type of *rainforest *biome that is found in the mid-latitude zone, such as the cool, dense, rainy forest of the northern Pacific coast of North America, which is dominated by large *coniferous trees.

temperature A measure of the heat content of an object or substance, described using a number of different scales, particularly the commonly used *Fahrenheit and *Celsius scales, and the *Kelvin scale (which is used by scientists).

temperature inversion A layer of the *atmosphere in which air temperature increases with height, a reversal of the normal pattern.

tempered air *See* CONDITIONED AIR.

Tennessee Valley Authority (TVA) An agency established by the US Congress in 1933 to assist in the development of the Tennessee River and adjacent areas.

tension A force or stress that causes stretching, for example of *glacier ice as it passes over a rock step, or of crustal rocks stretched by *tectonic activity. *Contrast* COMPRESSION, SHEAR.

tephra *Volcanic ash and dust that is thrown out into the air during an explosive *volcanic eruption.

tephrochronology The study of layers of *volcanic ash (*tephra) to determine their *relative ages or *absolute ages.

teratogen A chemical substance or other factor (such as *radiation) that alters the formation of cells, tissues, and organs, which causes abnormal development of an embryo and thus birth defects. *See also* CARCINOGEN, MUTAGEN.

term 1. The period of time in which a *mammal completes gestation; from conception to birth.
2. The period of time during which a contract (such as the *Kyoto Protocol) is in force.

terminal moraine *See* END MORAINE.

terminator technology The *genetic engineering of plants to produce *sterile seeds, which makes it impossible for farmers to save and replant seeds from their harvest and makes them reliant on the commercial seed market. *See also* TRAITOR TECHNOLOGY.

terminus The end or snout of a *glacier.

TERRA A *satellite *remote sensing system that was launched by *NASA in 1999, which circled the Earth 16 times a day for the following six years in a polar orbit, and sent back information on how the oceans, continents, and atmosphere interact. Its sensors could scan the entire planet every one to two days.

terrace 1. A relatively flat, natural surface along a *river valley, above the level of the *floodplain.
   2. An artificial flat surface constructed on sloping ground in order to produce land for growing *crops and to reduce *soil erosion.

terracing Construction of a series of horizontal levels built on a hillside, in order to retain water and reduce soil *erosion.

terrain The natural surface features of an area of land, particularly the *landforms.

terrain analysis The collection, analysis, evaluation, and interpretation of information on the features of the terrain in a particular place, such as *slope, *aspect, and *relief.

terrain evaluation *See* LAND EVALUATION.

terrane A region of the Earth's *crust, usually defined by major *fault systems, in which the *geology differs significantly from that of surrounding regions.

terrarium A box, usually made of glass, which is used for the keeping and observation of small animals or plants.

terrestrial Related to, or living or growing in or on, land.

terrestrial ecosystem Any land-based *ecosystem, such as a *forest, *desert, *grassland, or *cropland.

terrestrial radiation Longwave *electromagnetic *radiation that is emitted by the Earth back into the *atmosphere, in contrast to the shortwave radiation that is emitted by the Sun.

territorial The behaviour of an animal defending its *territory against others of its own kind.

territorial sea A zone of the *high seas adjacent to each coastal state, that is generally either 5.6 or 22 kilometres wide, was established by the *Convention on the Territorial Sea and Contiguous Zone, and within which the state has sovereignty over air space, water, and the *seabed.

territoriality The defence of a given area (*territory), in order to partially or totally exclude others from it.

territory 1. A specific area of the Earth's surface over which a state exercises sovereignty.
   2. The area in which an animal *lives and which it defends against intruders, which is usually smaller than its *home range.

terrorism Using acts or threats of violence to civilians, for example as a political weapon to draw attention to a group's goals or to gain those goals through fear and intimidation.

tertiary consumer The top *carnivore, which occupies the top of a *food chain because it eats *herbivores and *carnivores.

tertiary economic activities Service activities, such as trade and transport. *Contrast* PRIMARY ECONOMIC ACTIVITIES, SECONDARY ECONOMIC ACTIVITIES.

Tertiary period In *geological history, the first period of the *Cenozoic era, which began 65 million years ago and ended 2.5 million years ago. It is subdivided into the *Palaeocene, *Eocene, *Oligocene, *Miocene, and *Pliocene. During the Tertiary period the continents, climatic zones, and vegetation zones assumed their present positions and animal life began to assume forms (particularly *mammals) that were similar to those we see today.

tertiary productivity The amount of new biomass (weight) that is produced by meat-eating animals (*carnivores) in a given period of time. *Contrast* PRIMARY PRODUCTIVITY, SECONDARY PRODUCTIVITY.

tertiary rock *See* METAMORPHIC ROCK.

tertiary sector That part of the *economy that produces services and information.

tertiary treatment *See* ADVANCED WASTEWATER TREATMENT.

Tethys Sea The sea that geologists believe lay between *Laurasia and *Gondwanaland in the original landmass *Pangaea, probably between 200 and 250 million years ago. *See also* CONTINENTAL DRIFT.

tetrachloromethane *See* CARBON TETRACHLORIDE.

texture, rock The characteristics of the particles of an individual *rock, including grain size, size variability, degree of rounding or angularity, and preferred *orientation.

thalweg The line of deepest water in a *stream channel as seen from above, where the water usually flows fastest.

thatch Plant stalks (straw, reeds, or even living grass) that are used as roofing material.

thematic mapper (TM) A *remote sensing device that is found on *Landsat *satellites, which scans images of the Earth's surface in seven spectral bands from visible to thermal infrared, with a spatial resolution of about 30 metres.

theoretical Based on theory or speculation, rather than on observation, experiment, or *empirical testing.

theory A coherent explanation or description, reasoned from known facts. *See also* HYPOTHESIS.

thermal 1. Relating to or caused by heat.
2. A small rising column of air occurring as a result of surface heating, which allows a glider to rise and fly in the sky.

thermal conductivity A measure of the capacity of a material (such as a rock) to transfer heat by direct contact.

thermal cover *Cover (such as shade beneath trees) that is used by animals to lessen the adverse effects of *weather.

thermal delay In climate change, the potential long *time lag before ocean warming affects *global warming.

thermal enrichment *See* THERMAL POLLUTION.

thermal equilibrium A condition that exists between two objects when they have reached the same *temperature, so that heat energy ceases to flow between them.

thermal expansion Expansion (for example of the volume of air) caused by an increase in temperature.

thermal flux *See* HEAT FLUX.

thermal plume A *plume of hot water that is discharged into a *stream or *lake by a source of heat, such as a power plant.

thermal pollution Artificially raising or lowering the temperature of a *waterbody in a way that damages *aquatic organisms or *water quality. Also known as **thermal enrichment**.

thermal radiation *See* INFRARED.

thermal reactor A *nuclear reactor in which the *fission *chain reaction is sustained mainly by slow (thermal) *neutrons.

thermal stratification The *stratification of layers of water of different temperature and density within a deep *waterbody.

thermal treatment The use of heat to treat *hazardous waste by changing its chemical, physical, or biological character or composition, for example by *incineration.

thermistor A *resistor with a *resistance that changes with temperature, and which can be used as a temperature sensor.

thermocline *See* METALIMNION.

thermodynamics The study of heat and energy flow in chemical reactions.

thermodynamics, first law of A physical law that states that energy is conserved; it is neither created nor destroyed under normal conditions. Also known as the **law of conservation of energy**. *See also* THERMODYNAMICS, SECOND LAW OF.

thermodynamics, second law of A physical law that states that less *energy is available to do work with each successive energy transfer or transformation in a system. *See also* THERMODYNAMICS, FIRST LAW.

thermograph An instrument that measures and records variations in air *temperature through time. *See also* THERMOMETER.

thermohaline Relating to density currents in the *ocean, associated with the combined effects of *temperature and *salinity.

thermohaline circulation Large-scale vertical motions of water in the *ocean which are driven by density differences caused by variations in *temperature and *salinity.

thermokarst A low-relief landscape dominated by irregular shallow depressions caused by selective thawing of ground ice (*permafrost).

thermometer An instrument that is used to measure *temperature, using either the *Fahrenheit or *Celsius scale. *See also* THERMOGRAPH.

thermonuclear reaction A *nuclear reaction that is self-sustaining and is triggered by the *fusion of light nuclei under very high temperatures, which releases a large amount of heat. The basis of a nuclear bomb.

thermopause The top of the *atmosphere; a transitional layer between the *thermosphere (where temperature increases with height) and the *exosphere. Its height above the ground varies between about 200 and 500 kilometres depending on *solar activity.

thermosphere The top layer within the upper *atmosphere that lies above the *mesosphere and extends to about 350 kilometres above the ground, where the air is very thin, temperature increases with altitude, and temperature may be as high as 1100°C.

thicket A dense growth of *bushes or *trees. *See also* BRUSH.

think-tank A group of individuals who seek collectively to generate new ideas or approaches to solving particular problems. Examples of US environmental think-tanks include the *Cato Institute, the *Heartland Institute, and the *Heritage Foundation.

thinning The act of removing some of the plants from a *crop, or immature trees from a *forest, in order to improve the growth of the remainder. Also known as **thinning out**. *See also* LINE THINNING, MECHANICAL THINNING, SELECTIVE THINNING.

thinning out *See* THINNING.

Third World *See* DEVELOPING COUNTRY.

Thirty Percent Club A group of countries that committed themselves to reduce their emissions of *sulphur dioxide by 30% between 1980 and 1993 in order to reduce the problem of *acid deposition.

Thoreau, Henry David US writer and social critic (1817–62) who is best known for the book *Walden* (1854), an account of his experiment in simple living, and for the essay *Civil Disobedience* (1849), which outlined a doctrine of passive resistance that was subsequently to influence the views of Mahatma Gandhi and Martin Luther King Jr.

thorium (Th) A naturally occurring *radioactive metal that is found at very low levels in soil, rocks, and water. It is a soft, *ductile, silver-grey, heavy metallic element which has a *radioactive *isotope, thorium-232, with a *half-life of 14 million years.

thorn forest A *deciduous forest of small thorny trees which grows in a *tropical *semi-arid climate.

thorn shrub A dry, open *woodland or *shrubland that is dominated by sparse, spiny shrubs.

threatened A *species or *community that is likely to become *endangered if limiting factors are not reversed. *Contrast* ENDANGERED, EXTINCT, EXTIRPATION, VULNERABLE SPECIES.

threatened species A category of species under threat, developed as *IUCN (*Red Data Book) Conservation Categories, and widely used around the world. The categories within this group are *extinct, *endangered, *vulnerable, *rare, and *indeterminate.

Three Gorges project A major engineering project along the Yangtze River in China to build the world's largest *hydroelectric *dam. The dam was designed as part of a large multipurpose water resource project, designed to achieve four main goals, namely flood control (to prevent flooding along the densely populated middle and lower reaches of the Yangtze), energy generation (to generate the large amounts of electricity needed for industrial expansion in central-south and east China), water supply (to significantly increase the availability of water for agriculture, industry, and domestic users in the region), and navigation improvement (to improve the navigation of modern boats between Chongqing and Yichang, which in the past has been severely restricted by a series of rapids, shoals, and shifting shallows). The scheme has already displaced up to 1.2 million people from that portion of the Yangtze valley, from more than 320 villages and 140 towns.

Three Mile Island A narrow island in the middle of the Susquehanna River near Harrisburg, Pennsylvania, USA, which is also the site of a *nuclear accident at the Three Mile Island Nuclear Generating Station on 28 March 1979, when one of the *nuclear reactors suffered a partial core *meltdown.

three Rs Common term for three methods of *waste management, which are *reduce, *reuse, and *recycle.

threshold A fixed value (such as the concentration of a particular *pollutant) at which an abrupt change in the behaviour of a *system is observed.

threshold dose The minimum *dose of a substance that will produce a measurable effect.

threshold level The concentration or amount of a particular substance or condition below which it cannot be detected, or below which a significant *adverse effect is not expected.

threshold limit value (TLV) The *concentration of a particular substance (such as an air *pollutant) to which most workers can be exposed without *adverse health effects.

threshold velocity The *flow velocity (of wind or water) that is needed to cause sediment to start to move.

throughfall The *precipitation that falls directly through vegetation to the ground surface below, and that which falls off leaves after *interception.

throughflow Water (*soil moisture) that moves downslope within a soil, along *impermeable *soil horizons. *Contrast* INTERFLOW.

throughput Output relative to input; the amount that passes through a *system from input to output.

throughput tax *See* DISPOSAL CHARGE.

thrust fault A vertical displacement *fault associated with low-angle *fault planes.

thunder The loud cracking or deep rumbling noise that is usually heard a short time after a *lightning strike within a *thunderstorm, and is generated by a shock wave caused by the rapid and violent expansion of atmospheric gases when they are suddenly heated (to perhaps 15 000°C) by *lightning. The delay between seeing the lightning and hearing the thunder increases with distance, generally by about one second per 350 metres.

thundercloud The dark, towering, thick, flat-topped *cumulonimbus clouds associated with *thunderstorms, caused by strong updraft.

thundershower A brief *rainstorm with *thunder and *lightning.

thunderstorm A *storm that results from strong rising air currents, which brings heavy *rain or *hail, and *thunder and *lightning. *See also* AIR MASS THUNDERSTORM, FRONTAL THUNDERSTORM.

tidal bore *See* BORE.

tidal current The cyclical rise and fall of the water level in confined coastal areas (such as *estuaries and inlets) that is caused by the *tide. Each tidal current lasts for about six hours 12 minutes, which corresponds to the two phases of high water and low water each *lunar day. Also known as **tidal stream**. *See also* FLOOD CURRENT, EBB CURRENT, SLACK WATER.

tidal cycle The 14 day period from one *spring tide to the next spring tide, or from one *neap tide to the next neap tide. Also known as the **tidal period**.

tidal flat A broad, flat *coastal wetland of muddy, marshy, or sandy sediment, which is covered and uncovered in each *tidal cycle. Also known as **tidal marsh**.

tidal marsh *See* TIDAL FLAT.

tidal period *See* TIDAL CYCLE.

tidal power Energy that is generated by using the motion of the *tides to drive water turbines that run electricity generators. Most tidal power schemes are based on barrages, usually across natural inlets. The reservoir is filled with water as the tide rises and the sluice gates are shut at high tide. As the tide falls water drains out from the reservoir. In traditional systems as the water rushed out it turned a water wheel to provide direct mechanical energy for grindstones and pulleys. In modern systems the water turns turbines to generate electricity. Some tidal energy schemes use only the out-flowing tide to produce energy and they operate for about 30% of the time. Others use both in-flowing and out-flowing tides, to generate energy for about 60% of the time. This is a traditional form of power generation, used in ancient Egypt and throughout Europe until the middle of the 20th century. It is sustainable and non-polluting, and in theory it could provide around 20% of present global energy production. Large tidal range schemes have been operating in France, China, and the former Soviet Union for some years, and they have proved that this approach is both sustainable and cost-effective.

tidal power station A *hydroelectric plant that exploits the rise and fall of *tides to drive *turbines to generate *electricity using *tidal power. Also known as **tidal station**.

tidal range The difference in height between successive high and low *tides at a particular place, which varies through time because of the *solar tides.

tidal station *See* TIDAL POWER STATION.

tidal stream *See* TIDAL CURRENT.

tidal wave An unusually high destructive water level along the shore, usually caused by a *storm surge. The term is often, but incorrectly, applied to a *tsunami.

tidal wetland Any *saltwater or *freshwater *wetland that is influenced by the cyclical movements of tidal waters.

tide The cyclical rising and falling of the water of the *oceans that occurs twice each day and results from the gravitational attraction of the Moon and Sun acting on the rotating Earth. *See also* TIDAL CURRENT, TIDAL CYCLE, TIDAL RANGE.

tile drain A perforated pipe (often made of plastic, concrete, or pottery) that is buried in *soil in order to drain off excess water.

till *See* BOULDER CLAY.

till fabric The alignment of elongated pebbles within *boulder clay, which gives an indication of the direction in which the *ice was moving when the sediment was deposited.

till plain An extensive, relatively flat or gently undulating land surface covered by *boulder clay, which was deposited by a retreating *glacier.

tillage The cultivation of land, particularly through ploughing, preparing the soil for sowing and raising crops, and weeding. *See also* CONSERVATION TILLAGE.

tilth The physical condition of a soil that determines its suitability for plant growth, and depends upon texture, structure, consistency, and the amount of *pore space.

timber 1. A general term for forests or other clusters of trees.
2. Wood or wood products that are used for construction. Also known as **lumber**.

timber belt *See* WINDBREAK.

timber line *See* TREE LINE.

timber stand A community (*stand) of trees of similar species, size, and age.

timberland *Forestland that produces or is capable of producing crops of industrial wood. Previously known as commercial forestland.

time lag *See* LAG.

time of travel The time it takes for *groundwater to move a given horizontal distance.

time scale The division of geological history into eras, periods, and epochs, usually on the basis of *stratigraphy and *palaeontology.

time zone A region across which clocks are set to the same time. The 360° circumference of the Earth is divided into 24 time zones on the basis of *longitude, each one roughly 15° wide and defined as one hour different from the adjacent time zone. The *International Date Line marks the cross-over from one day to the next.

time zoning A particular form of land use *zoning that allowing a particular activity in a particular place only for a defined period of time, after which it becomes illegal.

timeline The sequence of events over time.

tissue culture A technique that is used to grow body tissue of a plant or animal outside the body (*in vitro*) in a culture medium, for example in test tubes.

titanium (Ti) A lightweight non-ferrous metal element that is very strong, non-reactive and does not cause allergic reactions in humans.

TLV *See* THRESHOLD LIMIT VALUE.

TM *See* THEMATIC MAPPER.

TMDL *See* TOTAL MAXIMUM DAILY LOAD.

TNC *See* NATURE CONSERVANCY, THE.

TNT *See* TRINITROTOLUENE.

toad Any of various semiaquatic and terrestrial *amphibians that have stout bodies, dry skin, no tail, and long back limbs for leaping. *Contrast* FROG.

tolerable soil loss The amount of *soil that can be lost by *accelerated erosion over 100 years without significant or permanent reduction of soil productivity.

tolerance The ability of an *organism to survive exposure to potentially harmful amounts of a substance without showing an adverse effect.

tolerance limit The limit to which an *organism can withstand changes in the environment (such as exposure to a potentially harmful *pollutant). *See also* LIMITING FACTOR.

tolerant Not sensitive to adverse environmental conditions (such as shade or drought) and thus able to withstand them. *Contrast* INTOLERANT.

toluene A colourless, flammable liquid that is derived from *petroleum or *coal tar, and is used as a solvent for gums and lacquers and in high-octane fuels. *See also* TRINITROTOLUENE.

tombolo A coastal depositional landform (*bar) that grows by *longshore drift to join a small island with another island, or with the mainland.

TOMS *See* TOTAL OZONE MAPPING SPECTROMETER.

tonne *See* METRIC TON.

top carnivore The creature occupying the highest *trophic level in a *food web (such as a lion). It is generally not consumed by other organisms whilst it is alive.

topographic map A map that shows the *topography of an area, usually by means of *contour lines.

topography The character of the surface of an area, particularly its *relief and the position of its features.

toposequence *See* CATENA.

topsoil The upper layer within a *soil profile (comprising the *A-horizon and *B-horizon), which normally contains *organic matter, in which plants grow.

tor A rocky outcrop or hill. The distinctive granite tors on Dartmoor in Devon, England are believed to have been formed by the *chemical decomposition of *granite under *tropical climates during the *Tertiary period.

Torino scale A ten-point scale for assessing the risk of an *asteroid or *comet hitting the Earth; the higher the Torino value the greater the risk. The scale is given in Appendix 6.

tornado A violent storm with extremely high wind speeds in the form of a rapidly rotating column of air. Tornadoes can form when warm, humid air is sucked into a low pressure cell, where it comes into contact with a cold front moving towards it from the opposite direction. Also known as **cyclone** or **twister**.

tornado scale *See* FUJITA SCALE.

tornado warning A news bulletin (*warning) that is issued to warn the public and emergency and other agencies when a *tornado is forecast or is occurring.

tornado watch A news bulletin (*watch) announcing that atmospheric conditions are favourable for producing *tornadoes. *Contrast* TORNADO WARNING.

torpor A dormant state.

torrent 1. A very fast-moving stream of water.
2. A heavy downpour of rain that may cause flooding.

Torrey Canyon A large *oil tanker that sank and broke apart off southwest England in 1967 and caused a large *oil spill that polluted the *sea, *coastline, and marine *food chain.

total dissolved solids (TDS) The amount of dissolved substances (such as *salts or *minerals) in a sample of water, which reflects *salinity and is measured by *electrical conductivity. Also known as **dissolved solids**.

total economic value The overall economic *value of a particular *natural resource, taking into account both use and non-use values.

total exchange capacity *See* CATION EXCHANGE CAPACITY.

total fertility rate A measure of *fertility, based on the average number of children in a family. *See also* REPLACEMENT-LEVEL FERTILITY.

total growth rate The net rate of growth of a particular population, which results from births, deaths, *immigration, and *emigration.

total maximum daily load (TMDL) A North American term for the amount of a particular *pollutant that a *waterbody can receive and still meet *water quality standards.

Total Ozone Mapping Spectrometer (TOMS) A small *NASA spacecraft that makes global measurements of atmospheric *ozone on a daily basis.

total quality management (TQM) An approach to *management that focuses on achieving a high quality product and customer satisfaction.

total resource management An approach to *natural resource management in which all resources are managed as a whole, responsibly, and on a *sustainable basis. Also known as **stewardship management**.

total solids All of the *organic and *inorganic matter that is suspended or dissolved in water.

total suspended particles The particulate matter that is present in the air at a given place and time.

total suspended solids (TSS) A measure of the amount of *suspended solids that is present in *wastewater, *effluent, or a *waterbody, which can usually be removed by *filtration.

totally effluent free See ZERO DISCHARGE.

towering cumulus A tall *cumulus cloud that extends through low and middle cloud levels, but lacks the characteristic anvil-shaped top of a *cumulonimbus.

town An *urban settlement with a fixed boundary that is smaller than a *city. The threshold size for distinguishing between a town and a city varies from country to country.

townscape The *landscape of a *town or built-up area.

toxaphene An *insecticide that was developed as a substitute for *DDT. It causes adverse health effects their presenting in domestic water supplies, and is *toxic to freshwater and marine aquatic life. It is a *critical pollutant whose use is now restricted in North America.

toxic Poisonous; harmful to living organisms.

toxic chemical A chemical that can cause severe illness, poisoning, birth defects, disease, or death when ingested, inhaled, or absorbed by living organisms, even at low doses.

toxic cloud A *plume of gases, *vapours, *fumes, or *aerosols in the air, which contains *toxic materials.

toxic colonialism General name for the export, usually by shipping, of *toxic wastes to a weaker or poorer nation.

toxic concentration The *concentration at which a particular substance produces a *toxic effect.

toxic dose The *dose level at which a particular substance produces a *toxic effect.

toxic emission The release of *toxic materials into the air or water or onto the land.

toxic pollutant Any *pollutant that is listed as *toxic by a national agency, usually because it can cause death, disease, or birth defects in organisms that ingest or absorb it.

Toxic Release Inventory (TRI) A database of *toxic releases in the USA which is compiled from *SARA reports.

toxic site A site where *toxic wastes have been dumped or released, which has been designated for cleanup.

toxic substance Any material that can cause illness, death, disease, or birth defects when organisms are exposed to it, even in very low quantities.

Toxic Substance Control Act (TSCA) (1976) US legislation that seeks to ensure that the impacts of chemical substances on human health and the *environment are identified (by the *Environmental Protection Agency) and properly controlled before those materials are made available in the market, in order to protect people and the environment.

toxic waste Any waste that is poisonous to humans, animals, or plants. Contrast HAZARDOUS WASTE.

toxicant Any *poison or *toxic substance that may injure an *organism exposed to it. See also TOXIN.

toxicity The degree to which a particular substance is harmful or poisonous. *See also* ACUTE TOXICITY, CHRONIC TOXICITY.

toxicity assessment The process of defining the nature of injuries that may be caused to an *organism by *exposure to a given *toxic substance, and how these are affected by the time and *concentration of exposure.

toxicity test A controlled laboratory test to determine the *toxicity of a substance, using living organisms.

toxicological disaster An environmental *disaster caused by the large-scale, accidental release of a *toxic substance into the air, soil, or water, which damages organisms and *environmental systems.

toxicological profile The detailed analysis and description of a *toxic substance in order to determine safe levels of *exposure.

toxicology The study of the effects of poisons or *toxic substances on living organisms.

toxification Poisoning.

toxin A complex, highly *toxic *organic substance that is produced by certain animals, plants, or *bacteria and is harmful to other organisms. *See also* TOXICANT.

TQM *See* TOTAL QUALITY MANAGEMENT.

trace A very small amount of a material, such as a concentration of less than one to ten parts per million.

trace element *See* MICRONUTRIENT.

trace gas A *gas (such as *water vapour, *ozone, *nitrogen oxides, *sulphur dioxide, and *carbon dioxide) that is present in the *atmosphere in only small amounts, either because of its very *reactive nature or because it is produced or emitted at a very low rate. Many are *greenhouse gases.

trace metal A *metal (such as *cadmium, *copper, *chromium, *mercury, *nickel, *lead, and *zinc) that is naturally found in extremely small quantities in animal and plant cells and tissue but is necessary for growth.

tracer Any substance (such as *radioactive isotopes and certain dyes) that is used to find out how something else moves through an *organism or an *environmental system.

traction A process of river *bedload movement that involves the rolling or sliding of particles (coarse *sand and *gravel) along the bed of the river.

tractor logging A method of *logging in which *timber is dragged or carried by tractor from where it is cut to a collecting point.

tradable emission permit In *greenhouse gas *emissions trading, an authorization that allows an emitter to emit a specified number of tonnes of emission within a given period of time. This allow emitters to determine the most economic way of covering their emissions, including buying additional permits, selling excess permits, and/or taking actions to reduce emissions. Also known as **tradable permit**.

tradable permit *See* TRADABLE EMISSION PERMIT.

trade The commercial exchange (buying and selling) of goods and services.

trade wind A major *wind belt that circles the Earth, beyond the *equatorial low pressure zone up to about 30° latitude in each *hemisphere, in which the prevailing winds blow towards the equator, from the northeast in the northern hemisphere and from the southeast in the southern hemisphere. *See also* INTERTROPICAL CONVERGENCE ZONE.

trade-off An exchange that occurs as a compromise.

traditional agriculture *Agriculture that is based on traditional farming practices such as *crop rotation, use of animal *manures instead of chemical *fertilizers, and use of animal power rather than machines.

traditional knowledge The body of knowledge and beliefs that is handed down from generation to generation, within communities. Also known as **indigenous knowledge**.

traditional resource rights The bundle of rights associated with long-term *natural resource use within traditional communities, which includes *intellectual property rights, human rights, land rights, and religious rights.

Tragedy of the Commons The title of a paper published in 1968 in *Science* magazine (13 December, Vol. 162, pp. 1243–8) by American biologist Garrett Hardin, in which he discusses present-day attitudes towards the environment as being similar to how medieval villagers viewed common grazing land: all of the herdsmen had a right of access to the common pasture that (in those pre-enclosure times) had no field boundaries and was open to all. Each herdsman would be inclined to graze as many cows as possible on the common land, because that way he would get the maximum return. Such an arrangement—of use without restriction—would work well so long as the number of people and cattle remained below the *carrying capacity of the common land (perhaps because of disease and poaching). Once the carrying capacity is exceeded, each cow gets relatively less grass to eat so its yield declines, and *over-grazing decreases *environmental quality and leads to *soil erosion. Because each person is trying to maximize their income, they are all trapped in a system that can tolerate only certain levels of use, competing against each other for a larger share of the available resources (space and grass in this case). As Hardin concludes, 'each man is locked into a system that compels him to increase his herd without limit—in a world that is limited. Ruin is the destination toward which all men rush, each pursuing his own best interest in a society that believes in the freedom of the commons. Freedom in a commons brings ruin to all.' The 'Tragedy of the Commons' analogy applies just as well to common property resources, including the *atmosphere, the *oceans, and *wilderness. *See also* SUSTAINABLE DEVELOPMENT.

trail A North American term for a relatively narrow path that is designed for use by pedestrians, horse riders, and/or off-road vehicles; a designated corridor that is not classified as a highway, road, or street.

trait A genetically inherited feature of an organism. *See also* GENETIC ANALYSIS, PHENOTYPE.

traitor technology In *genetic engineering, the use of an external chemical to switch a plant's genetic traits on or off. Also known as **genetic use restriction technology**. *See also* TERMINATOR TECHNOLOGY.

trample To walk on and flatten.

trampling Physical damage to plants and/or soil that is caused by many people or animals walking over it.

Trans-Alaska Pipeline An *oil pipeline that runs 1200 kilometres from *oil wells at Prudhoe Bay in the north to the port of Valdez in the south, crossing large areas of *permafrost by being raised above the ground.

trans-basin diversion *See* INTERBASIN TRANSFER.

trans-boundary pollution *See* TRANSFRONTIER POLLUTION.

Transcendentalism A literary and philosophical movement that was begun during the early 19th century in New England by Ralph Waldo *Emerson and others. It was a reaction against the rationality of the Enlightenment movement, was strongly influenced by the Romantic movement, emphasized the essential unity of all creation and the inherent goodness of humans. Transcendentalists believed that insight would reveal more about fundamental truths than would logic and experience, and emphasized nature as a source of human inspiration. Transcendentalism played an important part in the emergence of the *environmental movement in the USA during the early 20th century, through such writers as Emerson and *Thoreau. *See also* ROMANTICISM.

transcurrent fault *See* STRIKE-SLIP FAULT.

transect A (usually straight) line on the ground along which observations are made at some (usually regular) interval.

transform fault A special type of *strike-slip fault which forms the boundary between two moving *crustal plates, usually along an offset segment of a *mid-oceanic ridge.

transform plate boundary A junction between adjacent *crustal plates which are moving alongside and parallel to one another.

transformation Progressive change that removes evidence of an earlier state, such as the conversion of land from one use to another (such as *deforestation).

transformer A device that is used to raise or lower the voltage of an electric current in distribution or transmission lines.

trans-frontier pollution A pollutant (such as *acid rain) that is carried by natural processes (the movement of air or water) across national or state borders. Also known as **international pollution** or **trans-boundary pollution**. *Contrast* INTERSTATE POLLUTION.

transgenic Containing *genes from another *species, which have probably been introduced using *recombinant DNA technology. *See also* GENE SPLICING.

transgression A rise in *sea level relative to the land. *See also* SEA LEVEL CHANGE.

transhumance A type of *pastoralism in which people and animals move sea-sonally in search of *pasture, with winters usually spent in snow-free lowlands and summers spent in the cooler uplands. *See also* NOMADISM.

transient Lasting only for a short period of time.

transition The process of changing from one state to another.

transition zone The outer area of a *biosphere reserve where reserve management is broadened to make it compatible with local socio-economic development.

transitional nation A country whose *economy is changing to advanced industrial and agricultural practices. *Contrast* DEVELOPED COUNTRY, DEVELOPING COUNTRY.

transitional zone A zone in which populations from two or more adjacent *plant communities meet and overlap.

translocated herbicide *See* SYSTEMIC HERBICIDE.

translocated species A *native species that has been introduced into suitable habitats within its own country, having previously been excluded by natural barriers. *See also* INTRODUCED SPECIES.

translocation 1. In *genetics, the switching of a segment of a *chromosome to another chromosome. **2.** In biology, the movement of water, nutrients, or chemicals within a plant. **3.** In ecology, the deliberate movement of whole habitats or populations of species, by people, from one place to another.

transmissible disease A *disease that is transmitted from one individual to another by *infection, through physical contact.

transmission line A pipeline that moves a liquid or gas, or wires or cables that move electric power from place to place.

transmissivity The rate at which *groundwater can flow through an *aquifer.

transmutation The conversion of one *element into another by a process that takes place in the *nucleus, such as the change from one *isotope of an element into another one by irradiating or bombarding it with *radioactive particles.

transnational corporation A company that is controlled from its home country but has large operations in many different countries. *Contrast* MULTI-NATIONAL CORPORATION.

transpiration The *evaporation of water from plants into the air through tiny holes (*stomata) in their leaves.

transplant To transfer a plant from one location to another, or transfer an organ from one individual to another.

transport To move or carry.

transportation The activity of moving things from one location to another, for example using vehicle to *transport people or goods.

Transportation Equity Act for the 21st Century (TEA-21) (1998) US legislation that replaced the *Intermodal Surface Transportation Efficiency Act, and authorized and funded highway, highway safety, transit, and other surface transportation programmes from 1998 to 2003.

transportation sector A North American term for all of the private and public vehicles that move people and commodities, including cars, trucks, buses, motorcycles, railways (including trains), aircraft, ships, barges, and natural gas pipelines. *Contrast* COMMERCIAL SECTOR, INDUSTRIAL SECTOR, RESIDENTIAL SECTOR.

transuranic waste (TRU) *Waste that arises mainly from weapons production, and consists of clothing, tools, rags, and similar items that are contaminated with small amounts of *radioactive elements (mostly *plutonium) with *half-lives greater than 20 years.

transverse dune A relatively straight, long *dune that is oriented at right angles to the prevailing wind or current.

trash Common term for *refuse.

trash-to-energy plan Burning *solid waste in order to produce *energy.

travel corridor The strip of land beside a *road or *trail that is normally visible from it.

travel cost method In *economic evaluation, a method for determining the value of things (such as recreational resources) that are generally not bought and sold, based on the assumption that the costs met by individuals who travel to a particular site can be used as surrogate prices.

travertine *See* TUFA.

trawl A type of *fishing net that is towed in the water or on the sea-bed behind a vessel. It has a wide mouth and tapers to a small, pointed end.

treated wastewater *Wastewater that has been processed (physically, chemically, or biologically) in order to reduce its potential to cause health problems.

treatment Any physical, chemical, or biological process that makes a hazardous substance less hazardous.

Treatment, Storage, and Disposal Facility (TSDF) A site in the USA that is regulated by the *Environmental Protection Agency, where a *hazardous substance is treated, stored, or disposed.

treaty A formal, binding, written agreement between two or more countries, arrived at by negotiation.

tree A large, *perennial, woody plant that usually has one main trunk, a number of branches, and a *crown of foliage.

tree cavity A hollow in a tree that provides a resting or nesting place for *wildlife.

tree dieback The large-scale death of trees that is caused by, or associated with, *air pollution. Also known as *forest death or, in Germany, **Waldsterben**.

tree farm A *stand of planted trees that are of the same age and species.

When mature they will usually be harvested by *clearcutting, followed by replanting.

tree farming Any *agroforestry practice that includes management of *trees within the *farmland. Also known as **farm forestry**.

tree line The upper limit of normal tree growth in mountains or high *latitudes, above which the climate is too cold or windy for trees to grow. Also known as **timber line**.

tree ring A *growth ring in the cross section of a tree trunk, which indicates growth during one year. The number of rings corresponds to the age of the tree, and variations in ring width reflect the pattern of weather during the life of the tree. See also DENDROCHRONOLOGY, DENDROCLIMATOLOGY.

trellis drainage pattern A *drainage pattern in which tributaries join at high angles, often approaching right angles, which is common in areas with rocks of different strengths (thus resistance to erosion) and in areas with regular series of folds (*anticlines and *synclines).

trench 1. A long, narrow ditch, often dug either to improve drainage or to install a pipe.
    2. See OCEAN TRENCH.

TRI See TOXIC RELEASE INVENTORY.

Triassic The first period of *geological time in the *Mesozoic era, between about 245 and 208 million years ago, when large predatory *reptiles (*dinosaurs) evolved.

tribal peoples See INDIGENOUS PEOPLES.

tributary A *stream or *river than flows into another river or a *lake.

tributyltin (TBT) A *synthetic *organic compound based on tin, which is sprayed on the hulls of ships to control the growth of barnacles and other organisms and is extremely *toxic to *marine life. Also known as **organotin**.

trickle irrigation See DRIP IRRIGATION.

trickling filter A *wastewater treatment system in which the water is trickled over a bed of stones that are covered with *bacteria. The bacteria break down the *organic waste and produce clean water.

trilobite A class of marine *arthropod with three segments to its body. Trilobites were abundant in the *Palaeozoic era but became extinct during the *Permian.

trinitrotoluene (TNT) An explosive consisting of a pale yellow crystalline compound ($C_7H_5N_3O_6$) that is a *flammable and *toxic derivative of *toluene.

triphenyltin See TRIBUTYLTIN.

troglobion See TROGLOBITE.

troglobite An animal that lives its entire life within a cave and is specifically adapted to life in total darkness. Also known as a **troglobion**.

troglodyte A cave-dweller.

trophic Relating to food or nutrition, or to levels in a *food chain.

trophic level A level in a *food chain that is defined by the method of obtaining food and defines the number of steps (energy transfers) up from the original source of energy, from *primary producer, to *primary consumer, *secondary consumer, and *tertiary consumer.

trophic state/status The degree of biological *productivity of a *waterbody, which largely reflects the availability of *nutrients within it. It may be *oligotrophic (nutrient poor), *mesotrophic (moderately productive), or *eutrophic (very productive and fertile).

trophic structure The organization of feeding relationships in an *ecosystem into the different levels in a *food chain, such as *producers (plants), *primary consumers (herbivores), and *secondary consumers (carnivores).

t

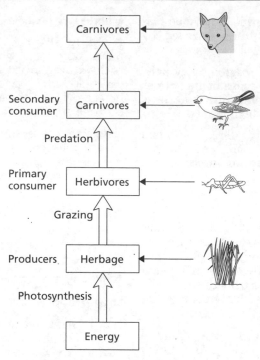

Trophic levels and energy flow

Fig 20 Trophic level

Tropic of Cancer The line of *latitude that runs 23°2′ north of the *equator, which is the farthest north that the *Sun can appear directly overhead, and marks the northern limit of the *tropics.

Tropic of Capricorn The line of *latitude that runs 23°27′ south of the *equator, which is the farthest south that the *Sun can appear directly overhead, and marks the southern limit of the *tropics.

tropical Relating to, situated in, or characteristic of the *tropics, the region on either side of the *equator, between the *Tropic of Cancer and the *Tropic of Capricorn, in which climate varies relatively little from season to season.

tropical air mass A warm/hot *air mass that forms in the *tropics.

tropical climate A *climate zone which covers much of the *equatorial and *tropical regions, in which temperatures remain high throughout the year and precipitation can be very high at least during part of the year. There are five different climate zones within the tropics—*equatorial wet, *tropical wet-and-dry, *monsoon, *tropical steppe and desert, and *west coast desert.

tropical continental air mass An *air mass that develops over large land masses in *subtropical latitudes, contains hot dry air, and brings clear skies and low *relative humidity.

tropical cyclone A general term for all *cyclonic circulations originating over the warm waters of the *tropics, which includes tropical depressions, tropical storms, and *hurricanes.

tropical depression A *tropical cyclone with wind speeds of up to 60 kilometres per hour.

tropical desert A *desert *biome that is hot throughout the year, such as the Sahara.

tropical disturbance A moving mass of *thunderstorms in the *tropics that persists for at least 24 hours and has relatively light winds.

tropical forest Forest that grows in *tropical climates with high temperatures and generally high annual rainfall. The two main types are *tropical rainforest and *monsoon forest.

tropical maritime air mass An *air mass that develops over *tropical or *subtropical oceans, contains hot moist air, and often brings convectional rain in summer and *drizzle in winter.

tropical rainforest clearance *Deforestation of *tropical rainforests that is caused by many things, including over-intensive *shifting cultivation, the collection of *fuelwood for cooking and heating and for making charcoal, encroachment and clearance by landless peasant farmers, clearance for *pasture or *crops, commercial *logging, *grazing, road construction, ranching, mining, and fire. Ecologists are concerned at the pace and pattern of the clearance of tropical rainforest, which once covered about a tenth of the Earth's surface. More than 40% of the rainforest has been cleared since the 1940s, and clearance continues at a rate of about 200 000 square kilometres a year. Less than 1% the Brazilian Amazon had been cleared before 1975, but between 1975 and 1987 the rate of clearance increased exponentially. Overall, about a quarter of the loss is due to forestry, but the balance of factors varies from place to place. Much of the wood felled from rainforests is destined for the international timber market, where tropical hardwood commands a high price. Ecologists argue that if clearance continues at recent rates all of the world's primary (undisturbed) rainforest is likely to disappear or be damaged by 2020. This would mean the loss of an irreplaceable biological asset—rainforests contain about half of all the wood growing on the Earth and at least 40% of all known species of plants and animals. They are amongst the most diverse and complex *ecosystems on the planet. As well as a significant loss of *biodiversity, clearance of rainforests causes the loss of valuable natural resources including hardwoods (such as mahogany, rosewood, and teak) and tree products (such as quinine, vegetable gums, and rubber). Tropical rainforest clearance may also contribute to the *greenhouse effect and *global warming, by removing an important *carbon sink. Various attempts have been made to conserve the remaining rainforests. For example, some governments (including those of Costa Rica, Panama, Brazil, and the Democratic Republic of the Congo) have designated forest reserves to conserve representative tracts of this important biome, and some endangered tropical hardwood species have been listed in the *Convention on International Trade in Endangered Species (CITES).

tropical seasonal forest A *tropical forest *biome comprising a mixture of forest, open woodland, and grassy *savanna that grows in a hot climate that has distinct wet and dry seasons.

tropical steppe and desert climate zone A very hot, dry *tropical climate zone beyond the *tropical wet-and-dry climate zone, at higher *latitudes, dominated by relatively stable subtropical high pressure cells throughout the year. Many of the world's great deserts (including the Sahara, the Sonoran, and the Great Australian) are found in this climate zone.

tropical storm A *tropical cyclone with maximum winds of between 64.4 and 120.75 kilometres per hour.

tropical wave A long area of low pressure in the *tropics, accompanied by a cluster of *thunderstorms, that has the potential to develop into a *tropical depression.

tropical wet climate A *tropical climate with enough rainfall to allow a dense *tropical rainforest to grow.

TROPICAL RAINFOREST

A *rainforest *biome which is found in the hot, wet *tropical climate zone, mostly between latitudes 5° and 10°. In this humid zone temperatures remain uniformly high throughout the year. There are high *diurnal variations in temperature, each month receives at least 60 millimetres of rainfall (so there is no dry season), annual rainfall usually exceeds 1500 millimetres, and there are periodic torrential downpours accompanied by thunder and lightning. The rainforest environment is hot, wet, and humid. Ecologists also believe that it has been remarkably stable over exceedingly long periods of time. This has encouraged plants and animals to adapt to their environment, which in turn has promoted *speciation and the evolution of unparalleled *biodiversity. More than half of the remaining rainforests are in Central and South America (particularly in Amazonia), and the rest are in South East Asia (particularly Sumatra, Borneo, and Papua New Guinea) and in Africa (particularly the Congo Basin). The forest is dominated by tall *evergreen trees and dense *undergrowth and it usually has very distinct *stratification. The vegetation is dominated by tall broadleaved *evergreen trees, *epiphytes (plants such as mosses which grow on other plants but are not parasitic on them), and lianas (climbing vines) and giant ferns. Rainforest trees are closely spaced, and they are often tied together by lianas and epiphytes. Most are tall and straight, have very shallow root systems, and are anchored to the ground by a series of buttresses. Many different tree species are present in even quite small areas of rainforest. Animal life within the rainforest is rich and varied. It includes forest elephants (whose ivory tusks are highly prized by poachers), many different reptiles (including snakes, alligators and crocodiles, turtles, and lizards), a range of primates (including monkeys and apes), some predatory big cats and a huge variety of birds and insects. Contrary to popular belief (because they look so abundant and productive) rainforests grow on very poor soils. Most of the *nutrients within the rainforest ecosystem are stored within the *biomass rather than in the soils, and the decomposer part of the *food web (which includes fungi, bacteria, and insects) rapidly breaks down organic material including dead trees, broken limbs and branches, leaf litter, and dead animals. The nutrients are quickly recycled through the forest soil into tree roots, so that forest soils have little organic matter and only a very thin cover of leaf litter. Because the forest soils contain so few nutrients, they are not suitable for continuous cultivation after the forest has been cleared. Rapid *leaching caused by heavy tropical rainfall makes the problem even worse. Soil productivity quickly declines (after between three and six years) and cultivated areas are then abandoned. *Secondary succession leads to the growth of scrubby vegetation, and then to *savanna (if the cleared area is large) or secondary rainforest (if it is a relatively small patch which can revegetate and self-repair). Also known as **equatorial forest** or **selva**. *See also* TROPICAL RAINFOREST CLEARANCE, TROPICAL FOREST.

tropical wet-and-dry climate A *tropical climate zone that lies outside the *tropical wet climate and has a summer wet season and a winter dry season. This promotes the growth of *savanna landscapes, dominated by scattered trees within extensive *grassland.

tropics The part of the Earth's surface that lies between the *Tropic of Cancer and the *Tropic of Capricorn, which has a hot climate and no cold season. In America this zone is known as the *neotropics and in Africa and South East Asia as the *palaeotropics.

tropopause The boundary in the *atmosphere between the *troposphere and the *stratosphere, which is located at an altitude of about 20 kilometres in the *tropics and falls with *latitude to 10 kilometres at the poles. *Jet streams are found in the tropopause.

troposphere The portion of the *atmosphere closest to the Earth's surface, which stretches up to the lower boundary of the *stratosphere (the *tropopause), and through which both pressure and temperature usually decrease with increasing height. It extends up to about eight kilometres at the poles and 16 kilometres at the *equator, although the height varies through the seasons. Most of the troposphere lies within 10.5 kilometres of the Earth's surface. *Gravity pulls the atmosphere towards the Earth, and about 75 % of the total weight of the atmosphere is squashed into this lowest layer (which is why we have *atmospheric pressure). As a result, air pressure is greatest closest to the ground, and it decreases with height. The troposphere also contains about 90 % of the moisture and *dust within the atmosphere. This is the most important part of the atmosphere for life on Earth because weather patterns and weather changes originate in this zone along with the wind systems and air currents that distribute heat, moisture, and air pollutants from place to place. It is also the warmest part of the atmosphere, and the only part where temperatures are usually above 0°C. The troposphere is warmed by the Earth, which absorbs *insolation and heats the overlying air. Except in local areas of *temperature inversion, temperature decreases with height in the troposphere at an average rate of about 6.5°C per kilometres (the *normal lapse rate). At the top of the troposphere, marked by the *tropopause, temperatures are down to around −60°C. The troposphere interacts with the *stratosphere above, and with the land and oceans below, playing a vital role in linking together many important environmental flows and systems. It ex-changes heat, energy, water, chemicals, and *aerosols with the Earth's surface. Atmospheric processes within the troposphere strongly influence movements and patterns of these important environmental factors across the Earth's surface. Weather changes are largely confined to the troposphere.

tropospheric ozone *Ozone that is located in the *troposphere. It is created by air pollution through the interaction of *nitrogen oxides, *volatile organic compounds, and sunlight, is a significant *greenhouse gas, and causes widespread human health problems via *photochemical smog. *Contrast* STRATOSPHERIC OZONE.

trough The lowest point in a wave of air or water; the opposite of *ridge.

troughing The deepening of a centre of low pressure in the atmosphere.

Trout Unlimited A US-based non-profit conservation organization that is committed to conserving, restoring, and managing the country's trout and salmon fisheries and their *watersheds.

TRU *See* TRANSURANIC WASTE.

true cost The sum of the *market and non-market (*social) costs associated with the production and use of goods and services.

true north The direction of the Earth's actual *North Pole, as opposed to *magnetic north.

true south The direction of the Earth's actual *South Pole, as opposed to *magnetic south.

truncated spur A former headland in a *river valley which has been sliced off by an advancing valley *glacier.

trunk The main stem of a tree.

TSCA *See* TOXIC SUBSTANCE CONTROL ACT.

TSS *See* TOTAL SUSPENDED SOLIDS.

tsunami A giant *wave (a seismic wave not a tidal wave) that is triggered by a submarine *earthquake or *volcanic eruption and moves across the ocean at a speed of more than 700 kilometres per hour. The word tsunami is Japanese for 'harbour wave', and tsunamis cause widespread damage when they reach the coast. In the open ocean these giant waves have long wavelengths (up to several hundred kilometres) and small heights (less than a metre). When they reach shallow water at the *coast they slow down and build up, which creates large, powerful waves that can sweep inland and cause widespread damage and destruction, with great loss of life.

tufa A light-coloured *limestone that is formed in caves and around hot springs where *groundwater saturated in *carbonate is exposed to the air. Also known as **travertine**.

tuff A general term for an *extrusive *volcanic rock formed of compacted *volcanic ash and dust. See also PUMICE.

tumour Any abnormal mass of cells that results from uncontrolled cell division and multiplication, and can be *benign (not cancerous) or *malignant (cancerous).

tundra A treeless *biome found in at high latitudes in *arctic and *subarctic regions that have long freezing winters and brief summers and where grasses, mosses, lichen, low shrubs, and a few flowering plants survive. The land is usually a flat or gently rolling plain, with black soil and permanently frozen subsoil (*permafrost), and is usually very wet.

tunnel In *mining, a lateral or horizontal underground passage that is dug to reach a *vein or mineral deposit.

turbidimeter A device that measures the density of *suspended solids in a liquid, based on the cloudiness of the mixture.

turbidite A deposit of sediment from a *turbidity current.

turbidity The degree of cloudiness in water (or air) that is caused by the presence of *suspended solids.

turbidity current A dense, flowing mass of water and *suspended sediment, that is heavier than clear water and thus flows downhill along the bottom of the sea or a lake, often at high speed. See also DENSITY CURRENT.

turbine A device in which blades on a wheel are turned by the force of moving water or steam. The blades are connected by a shaft to a generator in order to produce electricity.

turbulence Unstable flow of a gas or liquid, or instability in the *atmosphere that produces *gusts and eddies.

turbulent flow A *flow type that involves mixing within moving air or water, caused by irregular *eddies. Contrast LAMINAR FLOW.

turbulent mixing The physical mixing of air or water by *turbulent flow.

turf The surface layer of *soil that includes a matted layer of *grass and grass roots. See also SWARD.

turnover The process by which the water in a large body of *freshwater is completely mixed by circulation during *spring and *autumn. Also known as **overturn**. See also MIXING CYCLE.

turnover rate See RETENTION TIME.

tussock A clump of *grasses or *sedges that are bound together by their roots.

TVOC Total volatile organic compound. See also VOLATILE ORGANIC COMPOUND.

twilight See DUSK.

twilight zone See MESOPELAGIC.

twiner A *root climber plant that raises itself up by twisting itself around another upright plant. Contrast SCRAMBLER.

twister The popular name in the USA for a *tornado.

type locality The location of the first described specimen of a *species or *subspecies, or where a particular rock type, *stratigraphic unit, *fossil, or *mineral, was first identified.

typhoid fever An acute, highly infectious and often fatal disease caused by the typhoid bacillus (*Salmonella typhosa*) that is transmitted by contaminated food or water.

typhoon The name for a *severe storm or *hurricane in the western Pacific.

typicalness *See* REPRESENTATIVENESS.

ubiquitous Widely present, being or seeming to be everywhere.

Udall Scholarship The most prestigious scholarship awarded to undergraduate students in the USA, given to those 'who have demonstrated outstanding potential and a commitment to pursuing careers related to the environment'.

UK *See* UNITED KINGDOM.

UK Biodiversity Action Plan *See* BIODIVERSITY ACTION PLAN.

UKMET The medium-range numerical *weather prediction model that is used in the United Kingdom.

ultimate disposal The process of returning *waste materials to the environment in a form which will have the minimal *environmental impacts.

ultimate factor Aspects of the environment (such as food) that are directly important to the well-being and fitness of an *organism. *Contrast* PROXIMATE FACTOR.

Ultimate Resource A controversial book by Julian *Simon (1996) that attacks *neo-Malthusian thinking and supports continued population and economic growth, arguing that there are few physical limits to the availability of *natural resources or the ability of environmental systems to assimilate wastes.

ultisol A type (*order) of soil that develops in the *tropics and *subtropics. It is created largely by *chemical weathering, has more *clay in the *B-horizon than the *A-horizon, and has an *acid subsoil.

ultrabasic An *igneous rock that has a relatively high content of *magnesium and *iron, but very little *silica. Also known as **ultramafic**.

ultraclean coal Coal that has been washed, ground into fine particles, then chemically treated to remove *sulphur, ash, and other substances.

ultramafic *See* ULTRABASIC.

ultraplankton Any *plankton organism that is less than two micrometres in size.

ultrasheltered In terms of *wave exposure, a fully enclosed *coast with a *fetch of up to a few hundred metres.

ultraviolet (UV) *Radiation from the Sun that is within the non-visible portion of the *electromagnetic spectrum with a *wavelength (0.39 micrometres to 0.01 micrometres) shorter than violet light, which is so blue that humans cannot see it. It can be useful or potentially harmful, depending on its wavelength. Type A (UV-A) (longer wavelength) is generally beneficial to plants but type B (UV-B) (shorter wavelength) can cause a variety of skin problems (including sunburn, skin cancer, and premature aging) and cataracts in humans, and is directly affected by levels of *ozone in the atmosphere; a decrease in ozone levels leads to increased UV-B radiation on the ground.

umbrella species A species of plant or animal that has a large home *range and broad *habitat requirements, both of which overlap with other species, so that if it is given a large enough area for its own protection the other species will also benefit.

UN *See* UNITED NATIONS.

unacceptable risk The level of *risk above which specific action by a manager or government is regarded as necessary in order to protect life and property, and to achieve stated objectives. *Contrast* ACCEPTABLE RISK, AVOIDABLE RISK.

unavailable water Water that is held so tightly in the *pores of the soil that it cannot be taken up by plant roots. *Contrast* AVAILABLE WATER.

unbuffered Lacking a resistance to change in *pH, having a low *buffering capacity.

UNCBD *See* UNITED NATIONS FRAMEWORK CONVENTION ON BIOLOGICAL DIVERSITY.

UNCCC *See* UNITED NATIONS FRAMEWORK CONVENTION ON CLIMATE CHANGE.

UNCCD *See* UNITED NATIONS CONVENTION TO COMBAT DESERTIFICATION.

UNCED *See* UNITED NATIONS CONFERENCE ON ENVIRONMENT AND DEVELOPMENT.

UNCHE *See* UNITED NATIONS CONFERENCE ON THE HUMAN ENVIRONMENT.

UNCLOS *See* UNITED NATIONS CONVENTION ON THE LAW OF THE SEA.

UNCOD *See* UNITED NATIONS CONFERENCE ON DESERTIFICATION.

unconfined aquifer An *aquifer that contains water which is not under pressure, so the water level in a *well is the same as in the water table outside the well.

unconformity A gap in the rock record (particularly within a *stratigraphic sequence) at a particular place, caused by the *erosion of existing rocks before younger rocks are subsequently deposited on top.

unconsolidated Loose, separate, or unattached, for example the individual particles in a sediment such as *alluvium.

uncontrolled fire Any fire that threatens to destroy life, property, or *natural resources.

unconventional oil *Natural resources such as *shale oil and *tar sands that can be liquefied and used like *oil.

unconventional pollutant *See* NON-CRITERIA POLLUTANT.

UNCSD *See* UNITED NATIONS COMMISSION ON SUSTAINABLE DEVELOPMENT.

UNCTAD *See* UNITED NATIONS CONFERENCE ON TRADE AND DEVELOPMENT.

underburn A fire that can consume ground vegetation but not *trees or *shrubs.

underfit stream *See also* MISFIT STREAM.

underground mining A technique for extracting *minerals that consists of subsurface excavation with minimal disturbance of the ground surface.

underground storage tank A large tank buried underground, used for storing hazardous substances or petroleum products such as *gasoline or heating oil. *See also* LEAKING UNDERGROUND STORAGE TANK.

undergrowth *See* BRUSH.

undergrowth layer *See* UNDERSTOREY.

underplant To plant young trees or sow seed under an existing *stand of vegetation.

underprivileged group *See* DISADVANTAGED GROUP.

understorey The plants beneath the *canopy in a stand of *stratified vegetation. Also known as **undergrowth layer**. *Contrast* OVERSTOREY.

underwood *See* BRUSH.

undesirable species Any species of plant that is not desirable for a particular reason, for example because it is unpalatable, poisonous, or can injure animals.

undeveloped land *See* RAW LAND.

undiscovered resource A *natural resource that is assumed to exist, or has yet to be thought about. *Contrast* UNIDENTIFIED RESOURCE.

UNDP *See* UNITED NATIONS DEVELOPMENT PROGRAMME.

u

unenclosed land Land that has not been enclosed with a fence or wall.

UNEP *See* UNITED NATIONS ENVIRONMENT PROGRAMME.

UNESCO *See* UNITED NATIONS EDUCATIONAL, SCIENTIFIC AND CULTURAL ORGANIZATION.

uneven-aged stand A forest stand composed of trees of different ages and sizes, all growing together. *Contrast* ALL-AGED STAND, EVEN-AGED STAND.

ungrazed Land or plants that have not been grazed by animals.

ungulate A *grazing animal (such as a cow, deer, bison, or sheep) that has hooves.

unidentified resource A *mineral resource that is assumed to be present within known geological areas, but has not yet been specifically located or described in detail. *Contrast* UNDISCOVERED RESOURCE.

UNIDO *See* UNITED NATIONS INDUSTRIAL DEVELOPMENT ORGANIZATION.

uniform effluent standard A pollution control *standard that is designed to regulate the discharge of *point sources of pollution by enforcing compliance with *effluent quality standards.

uniformitarianism The belief that the geological processes that have shaped the Earth through geological time are the same as those operating and observable today, that is they are uniform through time. *Contrast* CATASTROPHISM.

unimproved grassland *Grassland that has not been agriculturally improved by being treated with chemical *fertilizers or *herbicides, so it is usually rich in *wildlife.

unintended release Any non-deliberate release into the environment of *genetically modified organisms.

unique species A *native species that is found in only one place.

unit 1. A single undivided whole (thing or person).

2. A fixed quantity (for example of length, time, or value) that is used as a standard of measurement.

United Kingdom (UK) The nation state that consists of England, Scotland, Wales and Northern Ireland.

United Nations (UN) An international organization of countries that was established in 1945 with 51 members, which by July 2000 had expanded to 188 members. Its purpose is 'to preserve peace through international co-operation and collective security'.

United Nations Commission on Sustainable Development (UNCSD) A commission of the *United Nations that was created in December 1992, as a result of the *United Nations Conference on Environment and Development, in order to monitor and report on implementation of the UNCED agreements at the local, national, regional, and international levels.

United Nations Conference on Desertification (UNCOD) A major international conference which met in Nairobi in August 1977, and agreed a Plan of Action to Combat Desertification which was designed to prevent and stop the advance of *desertification and, where possible, to reclaim desertified land for productive use.

United Nations Conference on the Human Environment (UNCHE) The first major international conference to address the relationships between human activities and the *environment, held in Stockholm, Sweden, in June 1972. It produced a set of principles in the Stockholm Declaration, led to the founding of the *United Nations Environment Programme, and was a catalyst for development of the modern *environmental movement. It was followed up in 1992 by the *United Nations Conference on Environment and Development.

United Nations Conference on Trade and Development (UNCTAD) An organization set up by the United Nations

UNITED NATIONS CONFERENCE ON ENVIRONMENT AND DEVELOPMENT (UNCED)

A major international conference that was held in Rio de Janeiro, Brazil, in June 1992 and is popularly known as the Earth Summit. The conference was the culmination of more than a decade of preparatory work, and it established the tone, pace, and direction of the international environmental agenda for the foreseeable future. The seeds for the Earth Summit were sown in Stockholm in 1972, where the first international *United Nations Conference on the Human Environment was held. A series of preparatory meetings were held between 1990 and 1992 in various countries, at which governments, *non-governmental organizations (NGOs), and expert scientists discussed and largely reached agreement on a series of basic documents which would be formally debated at Rio. Most (178) countries were represented and more than 100 heads of state attended the Rio summit. Five important agreements were reached—the *Rio Declaration on Environment and Development, *Agenda 21, a *United Nations Framework Convention on Biological Diversity, a *United Nations Framework Convention on Climate Change, and a *Statement of Forest Principles. Despite the huge amount of media coverage it attracted, and the promising tone of the conventions and other items which were agreed, the Rio Earth Summit has had a mixed reaction. *Environmentalists from *developed countries hailed the conference as the 'last chance to save the planet', whilst delegates from *developing countries saw it as an opportunity to redress long-standing economic grievances. Supporters argue that although the conference attracted a great deal of bad press coverage, and was widely criticized by environmental *pressure groups, it did mark a welcome and substantial change in international political attitudes towards the environment. They also applaud the recognition formally given at Rio of the need to tackle over-consumption in industrialized states, and *poverty and *resource scarcity in the developing world. Agenda 21 recognized that growth, the alleviation of poverty, *population policy, and *environmental protection were mutually reinforcing and supporting. Critics argue that whilst UNCED delivered an impressive number of international agreements on a variety of topics, it hardly began to address the fundamental driving forces of global environmental change—such as trade, population growth, and institutional change. They also point out that the legal agreements signed at Rio are relatively weak and lack binding commitments and timetables. On balance, even though some aims were not achieved at Rio, it was nonetheless an important step on the long-term path towards environmentally *sustainable development.

in 1964, which 'promotes the development-friendly integration of developing countries into the world economy ... with a particular focus on ensuring that domestic policies and international action are mutually supportive in bringing about sustainable development'.

United Nations Convention Concerning the Protection of the World Cultural and Natural Heritage An international convention that

was adopted in November 1972 and came into force in 1975, under which each state that signed it 'recognizes [its] duty of ensuring the identification, protection, conservation, presentation and transmission to future generations of the cultural and natural heritage'.

United Nations Convention on the Law of the Sea (UNCLOS) (1982) An international convention that came into force in 1994 and was designed 'to regulate all aspects of the resources of the sea and uses of the ocean'. Amongst other things, it introduced the concept of the 200 mile (322 kilometre) *Exclusive Economic Zones.

United Nations Convention to Combat Desertification (UNCCD) (1992) An international convention that is designed to promote a new, integrated approach to the problem of land degradation in arid, semi-arid, and dry sub-humid areas, emphasizing action to promote *sustainable development at the community level.

United Nations Development Program (UNDP) The global development network set up by the United Nations in 1965, which operates by 'advocating for change and connecting countries to knowledge, experience and resources to help people build a better life', and provides technical support for *development.

United Nations Educational, Scientific and Cultural Organization (UNESCO) An agency of the United Nations that was created in 1946 to promote collaboration between countries through education, science, and culture. It has headquarters in Paris, France, and national commissions in many countries.

United Nations Environment Programme (UNEP) A major programme that was established in 1972 at the *United Nations Conference on the Human Environment, to stimulate and coordinate *sustainable environmental practices worldwide. *See also* EARTHWATCH.

United Nations Food and Agriculture Organization (FAO) The first specialized United Nations agency that was created in 1945 to combat hunger and poverty, with a mission to raise living standards, agricultural productivity, nutrition levels, and rural living conditions.

United Nations Framework Convention on Biological Diversity (UNCBD) A framework for international action in protecting *biodiversity that

was agreed at the *United Nations Conference on Environment and Development in 1992. It calls on countries to identify *endangered species and conserve the places where they live, and under it each nation agreed to prepare a National Action Plan or Biodiversity Strategy.

United Nations Framework Convention on Climate Change (UNCCC) A framework for international action in tackling the prospect of *global warming associated with emission of *greenhouse gases which was agreed at the *United Nations Conference on Environment and Development in 1992. It called for the stabilization of greenhouse gas emissions by the year 2000 to prevent serious induced global warming, and led to the *Kyoto Protocol. Oil-producing countries (particularly the USA) were concerned that limitations on the burning of *fossil fuels would be a direct threat to their economies, so there was heated debate and a great deal of behind-the-scenes negotiation was required to get the convention accepted.

United Nations Industrial Development Organization (UNIDO) A part of the *United Nations whose mission is to improve people's living conditions and promote global prosperity through offering tailor-made solutions for the sustainable industrial development of developing countries and countries with economies in transition.

United Nations System-wide Earthwatch *See* EARTHWATCH.

universal soil loss equation A formula that is used to estimate rates of *soil erosion by considering climate, soils, and topographic conditions at a site, as well as the extent to which the use and management of the soil reduce erosion. It has the general form $A = R \times K \times L \times S \times C \times P$ in which A is the amount of soil lost from that field each year, R is the *erosivity of the rainfall (based on amount and type of rainfall), K is erosivity of the soil (liability of the particular soil to erosion), L is the length factor of the field (ratio of the length of the field to a standard field of 22.6 metres), S is the slope factor (ratio of the soil, lost to the amount lost from a field with a 9% gradient), C is the crop management factor (ratio of soil loss to that from a field under cultivated bare fallow), and P is the soil conservation practice factor (ratio of soil loss to that from a field where no care is taken to prevent erosion). Also known as **soil loss equation**. *See also* WIND EROSION EQUATION.

universal solvent A *solvent (such as water) that can dissolve almost anything, including both *bases and *acids.

univoltine Having one brood and generation in a year or season. *See also* VOLTINISM. *Contrast* BIVOLTINE, MULTIVOLTINE.

unknown (threat) category of species A category for species under threat that is defined by the *IUCN *Red Data Book as including *status unknown and *insufficiently known.

unleaded fuel *Petrol that has not been treated with *lead. *Contrast* LEADED FUEL.

unpalatable species Species that are not readily eaten by animals because they taste unpleasant.

unplanned ignition A *fire that has started at random by either natural or human causes, or on purpose by an arsonist.

unregulated harvest Tree *harvest that is not part of the *allowable annual harvest, which can include the removal of dead material or non-commercial species.

unsaturated flow The movement of water in a *soil that is not completely filled with water. *Contrast* SATURATED FLOW.

unsaturated zone That portion of a soil or an *aquifer in which not all of the pore space is filled with water, although some water may be present. Also known as **vadose zone**. *Contrast* SATURATED *See also* AERATION ZONE.

unstable air Air that rises easily and can form clouds and rain. *See also* INSTABILITY.

unsuitable forestland *Forestland that is not managed primarily for *timber production.

updraft A small-scale current of rising air, often within a *cloud.

upland High, hilly, or mountainous country. Also known as **highland**. *Contrast* LOWLAND.

uplift The lifting up of a region of the Earth's *crust, for example during *mountain building.

upslope fog A *fog that forms when moist, stable air is cooled when it is lifted up a mountain slope.

upslope precipitation *Precipitation that forms when moist, stable air is forced to rise along an elevated plain. *See also* RAIN SHADOW.

upslope wind *See* ANABATIC WIND.

upstream In the direction from which the flow is coming; against the current. *Contrast* DOWNSTREAM.

upwarping The *warping or bending upwards of the Earth's *crust.

upwelling The rising to the surface of *nutrient-rich waters from the bottom of the *ocean.

upwelling current A type of *ocean current that lifts cold, nutrient-rich water from the ocean floor.

upwind In the direction from which the wind is blowing; against the wind. *Contrast* DOWNWIND.

uraninite A strongly *radioactive, metallic mineral that is the main *ore of *uranium. It is found in *veins with the minerals of *lead, *tin, and *copper, and in *sandstone deposits.

uranium (U) A silver-coloured, heavy, *radioactive metal that is found especially in *pitchblende and *uranite,

u

exists naturally as a mixture of three *isotopes (U-234, U-235, and U-238), and is used in *nuclear reactors and the production of *nuclear weapons.

urban Relating to, characteristic of, or situated in a *city. *See also* SUBURBAN.

urban and built-up area/land A land use category in the USA that consists of residential, industrial, commercial, and institutional land. This includes construction sites, railway yards, cemeteries, airports, golf courses, *sanitary landfills, *sewage treatment plants, water control structures and spillways, small parks, and highways.

urban area A geographical area that constitutes a *town or *city.

urban climate A local climate that is affected by the presence of a *town or *city, which can include lower *relative humidity, lower *wind speeds, and higher *rainfall. *See also* URBAN HEAT ISLAND.

urban development The development of fabric and infrastructure in an *urban area.

urban district A subdivision of a county that covers an *urban area in the United Kingdom.

urban forestry The management of naturally occurring and planted trees in *urban areas.

urban heat island A dome of raised air temperatures that lies over an *urban area and is caused by the heat absorbed by buildings and structures. Also known as **heat island**.

urban planning The design and organization of *urban areas and activities. Also known as **city planning**.

urban renewal The redevelopment or rehabilitation of property in an *urban area.

urban runoff *Runoff that is generated from *urban areas.

urban settlement A densely populated settlement, which can take the form of a *town, *city, *metropolis, *conurbation, *megalopolis, or *world city. *See also* RURAL SETTLEMENT.

urban sprawl *See* SUBURBAN SPRAWL.

urbanization The social, *demographic, and economic processes involved in the growth of *towns and *cities.

urtication A burning or itching sensation which is usually caused by an *allergic response to insect bites or food or drugs.

US United States.

US Army Corps of Engineers (ACE) A US federal agency whose mission is to provide engineering services, including planning, designing, building, and operating dams and other civil engineering projects, designing and managing the construction of military facilities for the Army and Air Force, and providing design and construction management support for other defence and federal agencies. The Corps has played a major role in altering the US landscape.

US Bureau of Land Management (BLM) A federal agency of the *US Department of the Interior that administers large areas of public land which is located mostly in 12 western states. Its mission is to sustain the health, diversity, and productivity of the public lands for the use and enjoyment of present and future generations.

US Bureau of Reclamation (BR) A federal agency of the *US Department of the Interior that was established in 1902 to provide public funds for irrigation projects in arid regions. The BR played a major role in the construction of *dams, *reservoirs, and *irrigation systems designed to reclaim arid and semi-arid lands in western states, especially since in the 1930s.

US Department of Agriculture (USDA) The federal department of the US government that develops and executes policy on farming, agriculture, and food. Also known as the **Agriculture Department**.

US Department of the Interior (DOI) A Cabinet department of the US Government that manages and conserves most federally owned land. It is not responsible for local government or for civil administration, except in the cases of Indian reservations. Also known as the **Interior Department**.

US Fish and Wildlife Service (FWS) The federal agency of the *US Department of the Interior that is responsible for managing and preserving the country's *wildlife. Also known as the **Fish and Wildlife Service**.

US Forest Service The federal agency of the *US Department of Agriculture that is responsible for, among other things, managing the country's *National Forests. Also known as **Forest Service**.

USA United States of America.

USDA See US Department of Agriculture.

use district An area within a city that is covered by a *zoning ordinance that prescribes what land use is permitted, and what type of structures may be placed on it.

use value The economic value of a particular *resource for a specific use. Contrast NON-CONSUMPTIVE VALUE.

U-shaped valley A deep *valley with steep sides and a flat floor, usually eroded by a *glacier. Also known as **glacial trough**. Contrast V-SHAPED VALLEY.

utilitarian Having a useful function.

utilitarianism A philosophy that judges everything in terms of its utility or usefulness, and which argues that all action should be directed towards achieving the greatest happiness for the greatest number of people.

utility 1. The quality of being of practical use. See also UTILITARIAN.
2. A company that performs a public service (such as delivering water or energy to a region), subject to government regulation.

utility corridor A North American term for a strip of land that forms a passageway through which commodities such as oil, gas, and electricity are transported.

UV See ULTRAVIOLET.

u

vaccination Taking a *vaccine as a precaution against contracting an *infectious disease, through a needle injection, by mouth, or by aerosol. Also known as **innoculation**.

vaccine A product that produces immunity from an *infectious organism and therefore protects the body from the *disease.

vadose zone *See* AERATION ZONE, UNSATURATED ZONE.

vagile Wandering; freely motile, mobile. *Contrast* SESSILE.

vagrant An individual of a *species that naturally wanders, moving from place to place outside their usual range or away from the usual migratory routes, without being part of a self-regenerating population.

valence The number of hydrogen atoms that typically bond to an atom of an *element, which indicates the charge of the atom. For example, oxygen has a valence of two in water (H_2O).

valence electron The outermost electrons in an atom, which are particularly active in bonding.

validation The process of establishing the integrity and correctness of data. *See also* GROUND TRUTH.

valley Any long area of low-lying land that has higher ground on either side, and usually contains a *river.

valley breeze A daytime local *katabatic wind in a *mountain environment that blows down the *pressure gradient up-slope and up-valley. It is caused by warm air that rises up the slopes under stable, calm weather conditions.

valley fog *See* RADIATION FOG.

valley glacier *See* ALPINE GLACIER.

valley train A glacial *outwash plain that is confined within a narrow valley.

valuation The act of determining or estimating the *value or worth of an object.

value An expression of monetary worth of a particular object, or its power to command other things in exchange.

value judgement A judgement of the rightness or wrongness of something, based on a particular set of *values.

value system *See* VALUES.

values A set of ethical beliefs and preferences that determine our sense of right and wrong. Also known as **value system**. *See* VALUE JUDGEMENT.

vanadium (V) A naturally occurring, white, soft, ductile, toxic *heavy metal that is *carcinogenic and is used to make *alloys in iron and steel making.

vantage point *See* VIEWPOINT.

vaporization The conversion of a *liquid or *solid to a *vapour.

vapour The gaseous form of a substance that is normally in the solid or liquid state at room temperature and pressure, such as *water vapour.

vapour density The density of a *gas relative to the density of *hydrogen at the same temperature and pressure, which determines its *buoyancy in air.

vapour pressure The pressure exerted by the molecules of a *vapour, which is affected by *humidity and is a measure of the tendency of a material to *evaporate. *See also* SATURATION VAPOUR PRESSURE.

variability Absence of uniformity.

variable Liable to, or capable of, change.

variety In *botany, a genetically distinct division of a *species, next below the rank of *subspecies. *See also* CULTIVAR.

varve A thin layer of *sediment that is deposited in a lake within one year by *glacial outwash.

vascular plant Any plant (including *ferns, *conifers, and *flowering plants) that contains a system of vessels (*phloem and *xylem) which transport water and nutrients between different parts of the plant, for example from the roots to the leaves. *Contrast* NON-VASCULAR PLANT.

VC *See* VINYL CHLORIDE.

vector A carrier. **1.** Any organism that is capable of transmitting *disease. Examples include flies, mites, fleas, ticks, rats, and dogs.
2. In *genetic manipulation, the agent by which *DNA is transferred from one cell to another.

veering winds Wind that changes direction in a clockwise manner.

vegan A *vegetarian who also excludes from their diet all other animal products, such as milk and eggs.

vegetarian A person who only eats plants, not meat. A *herbivore.

vegetation All of the plants and trees that grow in a particular region or area.

vegetation cover type *See* COVER TYPE.

vegetation succession *See* SUCCESSION.

vegetative Plant parts or plant growth that is not involved in the production of seed, such as the roots, stems, and leaves.

vehicle Any non-living means of transport (usually made by humans), including cars, motorcycles, trains, ships, and aircraft.

vehicle emission The emission of *pollution (including *carbon monoxide, *hydrocarbons, and *nitrogen oxides) by a motor vehicle.

vein 1. A layer of *ore between layers of rock.
2. A structural part of a *leaf.
3. In the human body, a blood vessel that carries blood from the capillaries towards the heart.

veld *Temperate grassland in South Africa.

velocity Speed, or distance travelled per unit time. Also known as **flow rate, flow velocity**.

vent A vertical pipe in a *volcano, up which the molten *magma moves to the surface. *See also* HYDROTHERMAL VENT, VOLCANIC PLUG.

ventifact A stone that has flat surfaces that are often highly polished and have been formed by *sandblasting in strong winds; often found across a *desert floor.

ventilation The movement of air, particularly the process of supplying fresh air to, or removing contaminated air from, a building or structure. *See also* AIR CONDITIONING.

ventilation index A measure of the potential of the atmosphere to disperse ^airborne pollutants from a stationary source, based on the product of the *mixing depth and transport wind speed. Also known as a **clearing index**.

venturi effect The speeding up of the flow of air or water as it passes through a constriction (such as a pipe or narrow channel), caused by the pressure rise on the upflow side of the constriction and the pressure drop on the downflow side as it diverges.

venturi scrubber An *air pollution control device that use water to remove fine particles from volatile, hazardous, or corrosive gas streams, or from gas streams containing solid materials that are difficult to handle.

vermiculite The mineral *mica that has been processed and expanded by heat. It is used as a medium for starting seedlings and root cuttings.

V

vermiculture A *composting process that uses worms and *micro-organisms to convert *organic wastes into nutrient-rich *humus.

vermin The collective name for various small animals (including mice, rats, and birds) and insects (including cockroaches and flies) that are regarded as *pests.

vernal Of or relating to *spring.

vernal equinox The *equinox at which the sun approaches the *northern hemisphere and passes directly over the *equator, which occurs around 20 March. *Contrast* AUTUMNAL EQUINOX.

vertebrate An animal (such as an *amphibian, *bird, fish, *mammal, or *reptile) that has a backbone and a spinal column. *Contrast* INVERTEBRATE.

vertical Up and down, as opposed to *lateral or sideways.

vertical stratification The vertical distribution of *subcommunities within a *community of plants and animals. *See also* STRATIFICATION.

vertisol A type (*order) of clay soil in the *Comprehensive Soil Classification System that develops in climates with clear wet and dry seasons. Vertisols expand when wet and crack when dry.

very exposed In terms of *wave exposure, an open *coast that faces into prevailing winds and receives wind-driven waves and *swell with a *fetch of at least several hundred kilometres, but where deep water (50 metres) is no closer to the shore than about 300 metres.

very sheltered In terms of *wave exposure, a coast with a *fetch of less than about three kilometres where it faces prevailing winds, or about 20 kilometres where it faces away from prevailing winds, or which has offshore obstructions (such as *reefs).

vesicle A small air pocket that forms in *volcanic rock as it solidifies.

vessel 1. General name for a boat or ship.

2. Any tube that carries fluids within a body, such as *veins in humans or the *xylem tissue in plants.

vestigial A structure that no longer serves a useful purpose, such as an organ whose original function has been lost during *evolution of the organism (for example, the appendix).

viability The ability to survive, or to live and develop normally.

viable Capable of living, developing, or germinating normally.

viable population A self-sustaining *population with a high probability of survival because it has sufficient numbers and *reproductive potential.

vicariance *Speciation that occurs as a result of the separation and subsequent isolation of portions of an original population by a geographical barrier, such as a mountain or a *waterbody.

vicariant Species that occupy similar *niches but in geographical isolation from each other, as a result of *vicariance.

Vienna Convention for the Protection of the Ozone Layer (1985) An international agreement designed to minimize human destruction of the *ozone layer, mainly by reducing the production and emission of *CFCs into the atmosphere. The *Montreal Protocol fits within the framework of this convention.

viewpoint A place from which the surrounding *landscape or *scenery can be viewed or observed. Also known as a **vantage point**.

viewscape The visible *landscape.

viewshed Those parts of a *landscape that can be seen from a particular point. *Contrast* AIRSHED, WATERSHED.

village A *rural settlement that is much smaller than a *town. Sometimes used to refer to local centres within a *city, that were previously separate villages.

vineyard A place where grapes are grown for making wine.

v

vinyl chloride (VC) A *halogenated, *hydrocarbon compound that is used in producing some plastics (including *PVC), and is believed to be *carcinogenic.

viral agent An *agent that is know to cause particular diseases, which may be used for example as a *biological warfare agent. *See also* DISEASE AGENT.

virga Precipitation that falls from a cloud but evaporates before reaching the ground. Also known as **fallstreak**.

virgin forest An area of *old growth forest in its natural state, untouched by humans.

virgin material 1. Any undeveloped *resource that is or could become a source of raw materials.
   2. Previously unused raw material, such as *copper, *aluminium, *lead, *zinc, *iron, or other *metal or metal ore.

virulence The ability of an organism to cause *disease.

virus An ultra-microscopic non-cellular infectious agent (the smallest known type of *organism) that contains genetic information but replicates itself only within the cells of living *hosts. Many are *pathogens and can cause disease (such as *AIDS) in humans.

viscosity A measure of the resistance of a liquid to flow.

viscous Possessing a high *viscosity.

visibility Ability to be seen or the distance a person can see with normal vision.

Visibility Protection Program The programme specified by the Clean Air Act in the USA, which is designed to restore visibility caused by air pollution and prevent future reduction of visibility.

visible light The portion of the *electromagnetic spectrum that includes light that can be seen by humans, which we see as colours ranging from red (longer *wavelengths, around 0.77 micrometres) to violet (shorter wavelengths, around 0.4 micrometres), as in a *rainbow.

visitor day A way of measuring the use of a particular site or facility by visitors, defined as the use of an area for a total of 12 person-hours by one or more people, either continuously or spread over several visits. Also known as **recreation visitor day**.

vista A point or area along a route which offers a panoramic, unusual, or very attractive view of a *landscape or *scenery.

visual blight Any change to a *landscape that adversely affects *visual quality.

visual impact Any positive or negative change in the appearance of the *landscape as a result of development.

visual quality The nature and quality of a *visual resource, and how it is perceived and valued by the general public.

visual resource The visible physical features of a *landscape.

vitamin Organic *micronutrients that are present in minute quantities in natural foodstuffs, that the body needs in small amounts to grow and stay strong but is unable to produce in sufficient quantities for itself.

vitrification A process for treating *radioactive waste that immobilizes it by mixing it with molten glass.

viviparous Animals (including most *mammals, some *reptiles, and some *fish) that are born live and do not hatch from eggs.

VOC *See* VOLATILE ORGANIC COMPOUND.

void A space within something solid, for example within a *soil or a *sedimentary rock.

volatile A substance that evaporates readily at normal temperatures and pressures.

volatile organic compound (VOC) Compounds (such as *gasoline, *alcohol, and the solvents used in paints) that contain carbon and evaporate readily into the air. They contribute to the formation of smog and may be toxic.

volcanic Relating to, produced by, or consisting of *volcanoes.

volcanic activity Any activity of a *volcano, which includes both underground activity (*intrusive volcanic activity) and activity on the Earth's surface (*extrusive volcanic activity). Also known as **vulcanism**.

volcanic aerosol The cloud of particles which is thrown out into the *stratosphere by explosive *volcanic eruptions. See also VOLCANIC RADIATIVE FORCING.

volcanic ash See TEPHRA.

volcanic bomb A small lump of *magma that is thrown up into the air during an explosive *volcanic eruption as molten rock and then cools into a solid fragment before it reaches the ground.

volcanic breccia A *pyroclastic rock formed from large pieces of *volcanic rock, including *volcanic bombs. Contrast BRECCIA.

volcanic cone A conical mountain that is built up around a central *vent by *lava and *pyroclastic materials released when a volcano erupts. Also known as **volcanic dome**.

volcanic dome See VOLCANIC CONE.

volcanic dust See TEPHRA.

volcanic eruption The eruption of a *volcano as a result of *extrusive volcanic activity, which causes molten *magma to reach and spill out over the Earth's surface, and large amounts of *volcanic gas and steam to be released. Products include *volcanic bombs, *tephra, *lava flows, and the build-up of *volcanic cones.

volcanic gas Dissolved *gases contained in *magma that are released into the atmosphere during *volcanic eruptions.

volcanic glass See OBSIDIAN.

volcanic plug A volcanic landform that is formed when *lava hardens within a *vent in an active *volcano. A volcanic plug can cause a buildup of pressure in the magma trapped behind it which may be followed by a violent explosive eruption.

volcanic radiative forcing The effect of *volcanic aerosols on radiation transfers within the atmosphere. Regional air temperature is directly affected via changes in *albedo.

volcanic rock See IGNEOUS ROCK.

volcanic steam Hot steam that is released into the air in *volcanic eruptions, and from *fumaroles and *solfataras.

volcanic zone An area in which there are active or dormant *volcanoes.

volcano A mountain formed by *volcanic activity and material.

voltinism The number of generations or broods that an individual has each year. See also BIVOLTINE, MULTIVOLTINE, UNIVOLTINE.

volume A measure of how much space an object occupies, expressed in *SI units as cubic metres (m³).

volume reduction An approach to *waste management that involves reducing the volume of *waste, usually by compacting, shredding, *incineration, or *composting.

voluntary Without being required to; by free will.

voluntary commitment The formal name, under the *Kyoto Protocol, for any actions that a government or organization takes to reduce *pollution emissions on a voluntary basis, outside what is required by regulations.

voluntary simplicity Deliberately choosing a *lifestyle that is based on a lower level of *consumption, for environmental or ethical reasons.

vortex A whirling mass of air in the form of a vertical column or spiral.

vorticity A measure of the amount of local rotation in an airflow.

voucher specimen A preserved and archived specimen of a plant or animal that documents the use of a particular name, or its presence in a particular place.

V-shaped valley A river *valley that has relatively uniform gently sloping walls and has been eroded by *fluvial processes. *Contrast* U-SHAPED VALLEY.

vulcanism *See* VOLCANIC ACTIVITY.

vulnerability The *sensitivity, *resilience, and capacity of a system to adapt to stress or perturbation. Also known as **fragility**.

vulnerability analysis The process of estimating the *vulnerability to potential *natural hazards or *environmental change (such as the release of a particular pollutant) of specified elements at risk (such as elderly people or rare species).

vulnerable species A category of species defined by the *IUCN *Red Data Book as 'taxa believed likely to move into the *endangered category in the near future if the causal factors continue operating'. *Contrast* ENDANGERED, EXTINCT, EXTIRPATED, THREATENED.

vulnerable zone An area over which the concentration of a chemical that has been accidentally released into the air could reach the *level of concern.

V

wader A general term for long-legged birds that wade in water in search of food and *overwinter in the *intertidal zone in *estuaries. Also known as a **wading bird**.

wadi An *ephemeral river channel in a *desert area, that contains water only immediately after a storm.

wading bird *See* WADER.

wait-and-see principle A conservative approach to *environmental management, in which action is postponed until complete, correct, and detailed predictions of the likely impacts of different courses of action are available. *Contrast* PRECAUTIONARY PRINCIPLE.

wake The region of turbulence immediately behind a solid body caused by the flow of air over or around the body.

Walden A book by Henry David *Thoreau (1854), in which he describes his retreat from civilization between 1845 and 1847 to live in a small wooden cabin by Walden Pond in Massachusetts. The book also outlines Thoreau's philosophy of self-reliance and his *transcendentalist beliefs, both of which were heavily influenced by the writings of Ralph Waldo *Emerson.

Waldsterben *See* TREE DIEBACK.

warm front The boundary between a warm *air mass and the cooler air mass it replaces as it advances and flows over it. *Contrast* COLD FRONT. *See also* FRONT, FRONTAL LIFTING.

warm occlusion A *frontal zone formed when a *cold front overtakes a *warm front and rises up and over the denser air of the *warm sector. *Contrast* COLD OCCLUSION.

warm sector The region of warm air between a *warm front and the following *cold front. *Contrast* COLD SECTOR.

warm-blooded An animal whose body temperature is internally regulated. Also known as an **endotherm**. *Contrast* COLD-BLOODED.

warning A message to inform about impending danger (such as a *tornado warning), particularly to alert emergency services and the general public about a hazard which has already been detected or observed. *Contrast* WATCH.

warping The gentle deformation of the Earth's crust over a large area, without *folding or *faulting, which can be upwards (*upwarping) or downwards (*downwarping), as a response to great pressure.

wash 1. A common form of *slope erosion process, involving *overland flow across a sloping surface.
2. The usually dry channel of an *intermittent or *ephemeral stream in a *semi-arid area.

wash zone The depth zone in which sediments are disturbed by *wave action near the *shoreline.

washout The removal of an *air pollutant by *precipitation below clouds.

wastage zone The area on a *glacier where there is a net loss of snow and ice. Also known as **zone of wastage**. *Contrast* ACCUMULATION ZONE.

waste Any material that is unused and rejected as worthless or unwanted. Also known as **garbage, refuse**, or **trash**.

waste characterization The process of identifying the chemical and microbiological constituents of a particular *waste material.

waste disposal Getting rid (*disposal) of *toxic, *radioactive, or other *waste materials, most commonly by *landfill or *incineration, rather than by *reuse, *recovery, or *recycling.

waste disposal system A system for the *disposal of *wastes, including *sewer systems, *treatment works, and *disposal wells.

waste exchange An arrangement in which companies exchange their *waste materials, for the benefit of both parties.

waste feed The flow of *waste materials into an *incinerator, which can be either continuous or intermittent.

waste generation The amount of *waste material generated by a particular source, or that enters the *waste stream before *recycling, *composting, *landfilling, or *incineration takes place.

waste heat Heat that is left after useful energy generation, such as the warm water discharged into a storage pond after cooling a commercial *nuclear reactor.

waste heat recovery Using *waste heat that would otherwise be released into the environment to provide space heating or water heating in buildings.

waste load allocation The maximum amount (relative or absolute) of *pollutants that an individual discharger of wastes is allowed to release into a particular waterway.

waste management The management of *waste materials, usually based on the management of wastes at all stages (production, handling, storage transport, processing, and ultimate disposal) in such a way as to minimize the risks to human health, wildlife, and environmental systems. Traditional approaches to the management of nontoxic solid waste rely heavily on *sanitary landfill. Disposing of *toxic wastes in this cavalier manner is unsuitable,

largely because of the risk of contaminating *groundwater. The nuclear industry produces sizeable amounts of waste, much of which is safe and can be disposed of in traditional ways. But some *nuclear waste is highly radioactive, poses serious health risks to people and wildlife, and remains dangerously radioactive for long periods of time (hundreds of thousands of years or longer for many isotopes).

waste minimization *See* WASTE REDUCTION.

waste reclamation The reclamation of solid wastes by converting them into useful products. Examples include composting *organic waste into soil and separating *aluminium and other metals for *recycling.

waste reduction The process of reducing the amount of *waste that is disposed of, for example by avoiding and/or reducing the generation of waste in the first place, or by reusing, recycling, and recovering waste through *resource recovery. Also known as **waste minimization**.

waste site A dumping ground.

waste stream The output of wastes from a particular area, which includes *domestic waste, *yard wastes, and industrial, commercial, and construction *refuse.

waste treatment The process by which *hazardous waste materials are stored or treated in order to minimize their impacts on the environment.

waste treatment lagoon A storage pond in which *wastewater is stored for biological treatment.

waste-to-energy facility *See* MUNICIPAL WASTE COMBUSTOR.

wastewater Water that carries *wastes from homes, businesses, and industries and usually contains *dissolved solids and/or *suspended solids.

wastewater reclamation Any process which involves the reuse of *wastewater (for example for cooling or *irrigation), with or without treatment.

wastewater treatment Chemical, biological, and mechanical procedures that are applied to *wastewater in order to remove, reduce, or neutralize *contaminants before the water is discharged into a *waterbody. *See also* ADVANCED WASTEWATER TREATMENT, PRIMARY TREATMENT, SECONDARY TREATMENT.

wastewater treatment plant A *facility that contains a series of tanks, screens, filters, and other means by which *pollutants are removed from *wastewater.

watch An announcement that conditions are favourable for the development of a particular natural hazard (usually weather-related, such as a *tornado watch) in and close to the watch area. Watches are usually in effect for a specified time period, usually six hours. *Contrast* WARNING.

water (H$_2$O) An odourless, tasteless, colourless liquid formed by a combination of hydrogen and oxygen which freezes at 0°C (32°F) and boils at 100°C (212°F) and is a major constituent of all living matter.

water balance *See* WATER BUDGET.

water budget An assessment of all of the inputs, outputs, and net changes to a particular *water resource system (such as a *drainage basin) over a given period of time, usually a year. Also known as **water balance**.

water catchment *See* DRAINAGE BASIN.

water chemistry The chemical composition and characteristics of *water. *See also* WATER QUALITY.

water column The area between the surface and the bottom of a *waterbody.

water conservation Any practice that promotes the efficient use of *water, such as minimizing losses, reducing wasteful use, and protecting availability for future use.

water consumption The use of water.

water cycle The continuous natural cycling of water from the *atmosphere to the land and oceans and back. The basic structure of the global water cycle is quite simple. Water is evaporated from the oceans, seas, lakes, rivers, and vegetated land areas, and it becomes part of the atmospheric store of water vapour. Global and regional wind systems redistribute the water vapour across the Earth's surface. Condensation creates *clouds and *precipitation, and the latter brings water back to the surface where it enters the soil or *groundwater, or flows directly into rivers or lakes. Rivers transport water (as *runoff) from land surfaces into the oceans. Naturally, much of the precipitation falls directly over seas and oceans, effectively short-circuiting the land phase of the water cycle. The total volume of water in the global cycle is estimated at about 1384 million cubic kilometres. Just over 2% of the total water in the global cycle is freshwater, and most of that is locked up as polar ice caps and in glaciers. If all of the ice were to melt it would release enough water to keep the world's rivers flowing at their normal rates for up to 1000 years. The major stores are the *oceans (97.41%), *ice caps and *glaciers (1.9%), *groundwater (0.5%), *soil moisture (0.01), *lakes and *rivers (0.009%), and the *atmosphere (0.0001%). Also known as the **hydrogeological cycle** or **hydrological cycle**.

water equivalent The depth of *water that would result from the melting of a sample of *snow or *ice.

water filter A filter (such as charcoal or a fine *membrane) that is used to remove impurities from water.

Water Framework Directive A directive of the *European Union that was introduced in October 2000 and is designed to promote a *sustainable water policy within the European Union.

water harvesting The process of capturing and saving rainwater for later use

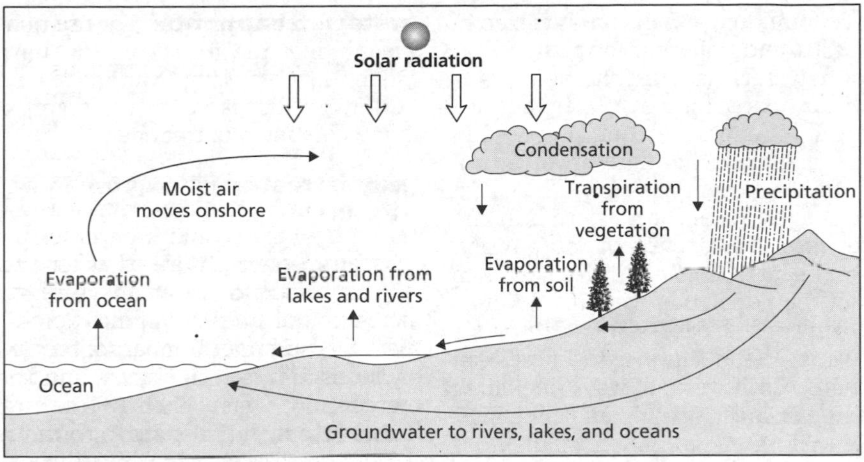

Fig 21 Water cycle

(for example storing water in barrels for use in watering a garden).

water management The planned development, distribution, and use of *water resources.

water mass A discrete body of *seawater in which physical properties (such as *temperature and *salinity) are uniform, but different from surrounding water masses with which it will mix only very slowly. *Contrast* AIR MASS.

water mill A mill that is powered by a water wheel, a traditional form of *hydropower. *Contrast* WINDMILL.

water pollutant Any material (usually a chemical) that *pollutes freshwater.

water pollution The *pollution of fresh water (for example through *eutrophication, *acidification, and *groundwater pollution), which decreases its purity and often makes it unsuitable for use as a *water resource if not dangerous to human health.

water power *See* HYDROPOWER.

water purification The process of removing *contaminants from a water supply in order to make it safe and *palatable for human use.

water quality The chemical, physical, and biological characteristics of a particular *waterbody, usually in relation to its suitability for a particular use. *See also* WATER CHEMISTRY.

water quality criteria Standards that are used to protect *water quality for designated uses such as drinking, swimming, raising fish, farming, or industrial use.

water quality objective A statutory target for *water quality that is used to provide a common framework for dischargers and regulators.

water quality standard A *standard that defines the management goals for a particular *waterbody, by designating the beneficial uses to be made of the water and by setting *water quality criteria appropriate for protecting those uses.

water quality testing Monitoring *water quality to ensure that it is safe for particular uses (such as protection of fish, drinking, and swimming).

water resource Any source of water that is useful to people (for example, for drinking, recreation, irrigation, livestock production, industry). Only 3% of the water within the *water cycle is *freshwater, and 99% of that is frozen in *ice caps and *glaciers, or is stored as *groundwater. As a result, only 1% is readily available for use. Water shortage is a major and recurrent problem in many regions, including most of Africa, the Middle East, much of South Asia, a large proportion of the western USA and northwest Mexico, parts of South America, and nearly all of Australia. Less than 20% of the population in many developing countries has access to clean drinking water. Water supplies to an area can be increased in various ways (including building *dams, *canals and pipelines, and building *desalination plants), but these engineering solutions are expensive. Access to and shared use of water are major issues for many neighbouring countries, and *hydropolitics can led to major confrontation between countries. World-wide, irrigation agriculture consumes over 70% of all the water used by people, industry accounts for a further 20%, and domestic and municipal uses account for most of the remaining 8%.

water resource management The deliberate *management (protection, allocation, and development) of *water resources in order to benefit people and/or the environment.

water softening The reduction or removal of *calcium and *magnesium ions from water in order to reduce its *hardness.

water storage pond An artificial pond or holding tank in which liquid *wastes are stored before treatment. Also known as a **water treatment lagoon**.

water supply The total amount of *water that is available within a particular area for human and other uses.

water supply system The collection, treatment, storage, and distribution of *potable water from source to use.

water table The level beneath the ground below which *permeable *rock is saturated with *groundwater. *See also* AQUIFER, PERCHED AQUIFER.

water treatment Any method of cleaning water for a specific purpose (such as drinking water, irrigation water, or discharge to a stream). Also known as **sewage treatment**.

water treatment lagoon *See* WATER STORAGE POND.

water vapour *Water that has evaporated into the air as an invisible gas (it accounts for up to 5% of air by volume), which is an important part of the *water cycle and also the most important *greenhouse gas. When it condenses it turns into a *liquid state to form *mist, *fog, *rain, *sleet, *hail, or *snow.

water well A vertical shaft that is sunk into the ground, usually to a depth of less than 100 metres, in order to access an underground *water supply (*aquifer).

water year A 12-month period, usually selected to begin and end during a relatively dry season, which is used a basis for consistent processing and reporting of hydrological data.

water yield The total amount of water (*streamflow and *groundwater) that drains from a *drainage basin over a given period of time (usually a year), usually expressed as equivalent depth of water spread evenly across the basin area.

waterbar A diversion ditch or hump that is used on a road or trail in order to divert water off and prevent *erosion.

waterbody Any body of surface water, such as a *lake, *river, or *pond.

waterborne Floating on, supported by, or transmitted in *water.

waterborne disease A *disease that is spread by *contaminated water.

watercourse *See* CHANNEL.

waterfall A cascade of water over resistant rocks or a cliff. *See also* LONG PROFILE.

waterfowl Any bird that spends much of its life in *wetland, particularly lakes, rivers, or streams

water-holding capacity The amount of water that a particular soil can hold.

waterhole A natural depression in the ground, in which *rainwater collects.

waterlogged Saturated with *water.

waterlogging The *saturation of soil with water, often because of *impeded drainage.

watershed *See* DRAINAGE DIVIDE.

watershed approach *See* WATERSHED MANAGEMENT.

watershed area All of the land and water within a *drainage divide.

watershed management The planned use of all of the *natural resources of *drainage basins in order to meet agreed objectives. Also known as **watershed approach**.

waterspout A column of rotating air over a body of water; in effect a *tornado over the water. *Contrast* LANDSPOUT.

waterway A navigable *watercourse, such as a *river or *canal.

watt(W) The *SI unit of power that is equal to one *joule per second ($J\ s^{-1}$).

wave A movement up and down or back and forth, such as a rhythmic movement on the surface of the *ocean, produced by wind as it blows over the water, or the intersection of a *warm front and a *cold front.

wave amplitude *See* WAVE HEIGHT.

wave crest The highest part of a *wave. *Contrast* WAVE TROUGH.

wave cyclone A low-pressure centre (*cyclone) that forms and moves along a *front. Also known as **wave depression**.

wave depression *See* WAVE CYCLONE.

wave energy *See* WAVE POWER.

wave frequency The number of *waves per second to pass a given point, which reflects *wavelength.

wave height The vertical distance between the *wave trough and *wave crest. Also known as **wave amplitude**. *Contrast* WAVELENGTH.

wave period The time that elapses between the passage of two successive *wave crests or *wave troughs past a fixed point.

wave power Capturing the energy that is available in the motion of ocean *waves and converting it into usable *energy (*electricity). Records of wave size and frequency, collected using a wave recorder, can be used to compute the potential for wave power in a given area. Various approaches have been tested since the early 1970s, including transforming the motion of the water surface into a flow of air, which can then drive a turbine, or transforming the motion of the water surface using floating devices which rise and fall at different rates. Wave power currently contributes very little to the global energy supply, but it has great potential—up to 1 million MW (1 TW) according to some estimates. It also has minimal *environmental impacts. Also known as **wave energy**.

wave refraction The bending of *wave crests to parallel the topography of the sea floor as they arrive ^onshore. *See also* LONGSHORE DRIFT.

wave steepness The ratio of *wave height to *wavelength, which affects the stability of ships and marine structures.

wave train A series of *waves from the same direction.

wave trough The lowest part of a wave, between two *wave crests.

wavelength 1. The horizontal distance between two successive *wave crests or *wave troughs. *Contrast* WAVE HEIGHT.
2. The wavelength of *electromagnetic radiation is often measured in *angstroms (Å), although scientists now prefer to express wavelengths in micrometres (μm). There are 10 000 Å in

w

one micrometre. Very long wavelengths are usually expressed in terms of frequency, as *hertz (Hz).

WBCSD *See* WORLD BUSINESS COUNCIL FOR SUSTAINABLE DEVELOPMENT.

WCED *See* WORLD COMMISSION ON ENVIRONMENT AND DEVELOPMENT.

WCMC *See* WORLD CONSERVATION MONITORING CENTRE.

WCS *See* WILDLIFE CONSERVATION SOCIETY, WORLD CONSERVATION STRATEGY.

weak acid An *acid that does not ionize completely in water. *Contrast* STRONG ACID.

wealth The state of being rich and affluent, or having an abundance of material possessions and resources.

weapons of mass destruction (WMD) A weapon that kills or injures large numbers of civilians as well as military personnel. This includes *biological weapons, *chemical weapons, and *nuclear weapons.

weather The physical condition of the *atmosphere at a particular place and time, particularly in terms of moisture, temperature, pressure, and wind.

weather balloon A large balloon filled with helium or hydrogen, which lifts a *radiosonde aloft to measure temperature, pressure, and humidity as it rises through the air.

weather forecast The *prediction of future *weather for a place and over a specified time period, based on knowledge of meteorological principles and processes, and information about current and recent weather conditions.

weather front *See* FRONT.

weather map A map or chart that shows the *meteorological conditions over a particular geographical area at a given time. Also known as a **synoptic chart**.

weather modification The deliberate management of weather systems to benefit humans, for example through *cloud seeding to encourage rainfall over dry places.

weather side *See* WINDWARD SIDE.

weathering The chemical *decomposition and physical *disintegration of rocks *in situ* by the action of external factors, such as rain, as a form of *erosion. *See also* CHEMICAL WEATHERING, MECHANICAL WEATHERING.

WEDOS *See* WORLD ENVIRONMENT AND DISASTER OBSERVATION SYSTEM.

weed A plant that is growing where it is not wanted, or where is does not normally grow, and which is considered undesirable or unattractive. *Compare* PEST.

weed control The process of reducing the growth of *weeds to an acceptable level, for aesthetic, economic, public health, or other benefits.

weeding Removing unwanted plants (*weeds), particularly at the *seedling stage.

weedkiller *See* HERBICIDE.

Weeks Law (1911) US legislation that authorized purchase and addition to *National Forest System lands within the *watersheds of navigable streams, in order to regulate the flow of navigable streams or to produce timber.

weir A low *dam that is built across a *stream in order to raise the water level or divert the flow.

welfare An index of well-being.

well A deep hole or shaft that has been dug or drilled in order to access underground resources (particularly water, *oil, *gas, or *brine).

well field An area that contains one or more productive *wells.

well injection Putting fluids underground by means of a *well.

well monitoring Measurement of *water quality in a *well, using on-site instruments or laboratory methods.

well plug A watertight and gastight seal that is installed in a *well in order to prevent the movement of fluids.

wellhead The top of a *well, at ground level.

wellhead protection area A designated area around a *public water supply *well that is protected from *contaminants that may damage human health.

west The compass point that is at 270° from north.

west-coast desert climate zone A *tropical climate zone of dry *desert climates that is confined to narrow bands along the west coasts of all continents except Eurasia and Antarctica.

westerlies The dominant winds that blow from the west in middle latitudes on the *poleward side of the *subtropical high-pressure areas. Also known as **prevailing westerlies**.

wet deposition The deposition of *pollutants from the atmosphere by rain or snow. *Contrast* DRY DEPOSITION. *See also* ACID RAIN.

wet meadow A *grassland with *waterlogged soil.

wet prairie A type of *prairie that grows in sites that have *waterlogged soils throughout much of the *growing season.

wetland Land that is covered with water for at least part of each year, and is thus transitional between *terrestrial and *aquatic ecosystems. It supports aquatic vegetation that is specifically adapted for *waterlogged soil conditions, as found for example in *swamps, *bogs, *fens, *marshes, and *estuaries.

wetland soil *See* HYDRIC SOIL.

whale Any of the larger *cetacean mammals that have a streamlined body and breathe through a blowhole on the head.

whaling The intentional hunting and killing of *whales for their meat, blubber, and other products.

whirlwind *See* DUST DEVIL.

white frost *See* HOAR.

white water Water that moves across *rapids or a *waterfall, and has white foam on the surface. It is created by breaking waves.

white-out Reduced daylight *visibility caused by heavy *fog, *snow, or *rain.

WHO *See* WORLD HEALTH ORGANIZATION.

whole-tree removal In *forest management, felling and transporting the whole tree with its crown and sometimes even its roots, for chipping or trimming and cutting at a mill.

wild 1. In a natural state, not tamed, *domesticated, or *cultivated.
  2. The wild—an uninhabited area in a primitive state

Wild and Scenic River Corridor The area on both sides of a *wild or scenic river.

Wild and Scenic Rivers Act (1968) US legislation that provided for designation of *wild and scenic rivers and for comprehensive studies of other rivers designated as potential additions to the *National Wild and Scenic Rivers System.

wild animal *See* WILDLIFE.

wild river A river or section of a river designated under the *Wild and Scenic Rivers Act, which is unpolluted, free of impoundments, generally inaccessible except by trail, and flows through primitive countryside.

wild species An organism that naturally lives in the *wild and has not been subject to breeding to alter it from its *native state.

wild type The normal form of an *organism, which occurs most frequently in nature.

wilderness Defined by the US *Wilderness Act (1964) as 'an area where the earth and its community of life are untrammelled by man, where man himself is a visitor who does not remain'. In this formal sense, wilderness is a place where vehicles are not allowed, where no permanent camps or structures can be made, and where *wildlife and its *habitat would be kept in as primitive a condition as possible.

Wilderness Act (1964) US legislation that defined *wilderness, established the *National Wilderness Preservation System, and created a process by which Congress could in future add areas to the national system of national forests, parks, and wildlife refuges.

Wilderness Area A large area of *wilderness in the USA that has been set aside for preservation by the US Congress by including it in the *National Wilderness Preservation System. Previously known as **Primitive Area**.

Wilderness Lands Sagebrush Rebellion See SAGEBUSH REBELLION.

wildfire An unplanned *fire, often in *wildland, that is burning out of control. See also PRESCRIBED FIRE.

wildflower A wild or uncultivated flowering plant.

wildfowl Any *game bird, such as ducks, swans, and geese.

wildland Land that is covered mainly by natural vegetation.

wildland fire Any fire that occurs on undeveloped land.

wildlife Any *animal which is now or historically has been found in the *wild, or in the wild state, within a particular area (such as a country). This includes *mammals, *birds, *reptiles and *amphibians, and including both *game species and non-game species and *vermin. Also known as **wild animal**. See also CONSUMPTIVE WILDLIFE, NON-CONSUMPTIVE WILDLIFE.

Wildlife Conservation Society (WCS) One of the first *conservation organizations in the USA, established in

1895, whose mission is to advance wildlife conservation, promote the study of zoology, and create and manage urban wildlife parks.

wildlife corridor A linear *habitat or feature (such as a *canal or *hedgerow) that allows animals and plants to move between isolated sites. Also known as **wildlife travel corridor**.

wildlife habitat An area of land and water that supports particular *wildlife or groups of wildlife, such as *woodland, *cropland, *rangeland, and *wetland.

wildlife refuge See WILDLIFE SANCTUARY.

wildlife sanctuary An area that is designated for the protection of wild animals (*refuge), where animals can breed without interference or danger and *hunting and *fishing are either prohibited or strictly controlled. Also known as **wildlife refuge**. See also MANAGED NATURE RESERVE/WILDLIFE SANCTUARY.

wildlife travel corridor See WILDLIFE CORRIDOR.

wildlife tree A standing live or dead tree that provides a *habitat for the *conservation or enhancement of *wildlife.

wildscape A rugged or *wilderness *landscape.

wildwood Original *woodland cover, unaffected by human activity.

willingness to accept The amount of compensation that an individual is willing to accept in exchange for giving up a particular good or service.

willingness to pay The amount that an individual is willing to pay to acquire a particular good or service.

willow Any of the large number of *deciduous trees and shrubs of the *genus Salix.

willy-willy The name given to *severe storms in some South Pacific islands.

wilt To lose strength; a plant wilts if it has insufficient water.

wilting point The water content of a soil below which plants are no longer able to draw water from it, so they *wilt.

wind The horizontal movement of air, that is caused by variations in *atmospheric pressure.

wind belt A zone of the Earth's surface over which particular *wind systems flow.

wind direction The direction from which the wind is blowing, reported with reference to *true north.

wind energy Energy from moving air, usually in the form of *electrical energy that is produced by wind-driven *turbines. Wind energy has been traditionally exploited to drive *windmills, and more recently by *wind turbines and *wind farms. The use of wind power around the world has risen rapidly since the early 1980s. Electricity generation from wind power looks likely to increase further in the future, because it is cheap, environmentally clean, and eases distribution problems in remote areas. Many countries have great untapped wind energy potential. Also known as **wind power**. *See also* WIND FARM, WINDMILL.

wind erosion The process by which soil is detached, transported, and deposited by *wind. Also known as **aeolian erosion**.

wind erosion equation A formula that is used to estimate rates of *soil erosion from a field that has particular characteristics, by considering *erodibility, *roughness, climate, shelter, and vegetation.

wind farm A group of *wind turbines that generate electricity, usually owned and operated by one company.

wind field The three-dimensional spatial pattern of *winds at a particular place and time.

wind power *See* WIND ENERGY.

wind power plant *See* WIND FARM.

wind rose A circular diagram that shows, for a given place or area, the frequency and strength of the *wind that blows from various directions.

wind shear A sudden change in *wind speed or direction.

wind sock/cone An open-ended fabric cone that is used to indicate the direction and relative speed of the *wind, often at airfields. *Compare* WIND VANE.

wind speed The average speed at which *wind blows in a particular place, which is usually measured in kilometres per hour (km h^{-1}) and is described using the *Beaufort scale.

wind turbine A mechanical device that converts the energy of *wind into *electricity; a modern form of *windmill.

wind vane An instrument that is attached to a high structure, and rotates freely to show *wind direction. *Compare* WIND SOCK.

wind wave A wave in a large *waterbody (such as an ocean) that is generated by the flow of air over the water surface.

windbreak *See* SHELTERBELT.

windchill The cooling effect of strong wind and low temperature, which can cause *hypothermia in people.

windfall 1. Fruit that has fallen from a tree.
   2. A tree that is blown down by the wind.
   3. An unexpected financial gain.

windmill A mill that is powered by the wind, traditionally used to grind grain for flour. *Contrast* WATER MILL.

windstrip *See* SHELTERBELT.

windthrow Trees that have been uprooted by strong *wind.

windward The side toward the *wind. *Contrast* LEEWARD.

winter The coldest *season of the year between *autumn/fall and *spring, when the Sun is mainly in the opposite hemisphere. Astronomically this is the

W

period between the *winter solstice and the *vernal equinox, which covers the months of December, January, and February in the northern hemisphere, and June, July, and August in the southern hemisphere.

winter cover *Cover that is required for *overwintering, such as *den trees for squirrels or dense evergreen *thickets for deer.

winter range *Rangeland at low altitude that is used by *migratory animals (such as deer, elk, caribou, and moose) during the winter months.

winter solstice The shortest day of the year, when the Sun is lowest in the sky (because of the Earth's tilt), which marks the start of the *winter season. This occurs in the northern hemisphere on 21 or 22 December, and in the southern hemisphere on or near 21 June. *See also* SOLSTICE, SUMMER SOLSTICE.

winterkill The death of *aquatic organisms during the winter that is caused by a lack of *dissolved oxygen in the frozen water.

Wisconsinan The North American name for the fourth and most recent period of ice advance (*glaciation) of the *Pleistocene *ice age, starting about 75 000 years ago and ending with the start of the Holocene around 10 000 years ago. Known in Europe as the **Würm**. *See also* DEVENSIAN.

Wise Use Movement A US non-profit *anti-environmental *front group comprising a loosely affiliated network of people and organizations (including ranchers, loggers, miners, industrialists, hunters, off-road vehicle users, and land developers) who favour widespread privatization and oppose environmental regulation, particularly any which restricts access to *natural resources and public lands. *Compare* SMART GROWTH NETWORK.

withdrawal The removal of water from *surface water or *groundwater.

WMD *See* WEAPONS OF MASS DESTRUCTION.

WMO *See* WORLD METEOROLOGICAL ORGANIZATION.

wolf tree A large older tree (such as an oak or beech) that has a broad *crown and little or no *timber value, but is valuable for *wildlife. Often a remnant from a previous *stand.

wood 1. The structural parts of woody *perennial plants, especially *trees. *See also* TIMBER.
  2. A small *grove of trees of mixed species, including *undergrowth.

wood energy Wood products that are used as fuel, which includes *industrial timber, wood chips, *bark, sawdust, forest residues, *charcoal, and pulp waste.

woodland A *forest in which the *canopy is not closed and there are many large open (treeless) spaces. Also known as **open canopy forest** or **open forest**.

woodlot A small plot of land on which trees are grown and managed for *timber.

woodscape A wooded or *woodland *landscape.

woody perennial A plant with thick, tough, *lignified stems which usually lives longer than three years and increases in size each year.

work The energy that is transferred by a force to an object as the object moves.

World Bank The popular name for the International Bank for Reconstruction and Development, an international financial institution that provides funds for development, which was established in 1947.

World Business Council for Sustainable Development (WBCSD) A coalition of 175 international companies drawn from more than 30 countries and 20 major industrial sectors, who share a commitment to *sustainable development through *economic growth, ecological balance, and social progress.

World city A *city which is important nationally and globally as a centre of trade, banking, finance, industry, and markets.

World Commission on Environment and Development (WCED) An international committee that was set up by the United Nations General Assembly in 1983 to examine international and global environmental problems and to propose strategies for *sustainable development. It was chaired by the then Norwegian Prime Minister Gro Harlem Brundtland, and is also known as the Brundtland Commission. Its report, *Our Common Future* (1987), defined and popularized the notion of sustainable development through global cooperation, and was a major catalyst for the 1992 *United Nations Conference on Environment and Development.

World Conservation Monitoring Centre (WCMC) A research centre for monitoring *endangered species that was set up in 1979 by the *World Conservation Union, and established in 2000 as the world *biodiversity information and assessment centre of the United Nations Environment Programme, which seeks 'to promote wiser decision-making and a sustainable future by providing information on the conservation and sustainable management of the living world'.

World Conservation Strategy (WCS) A major international *nature conservation initiative that was launched in 1980 and presented a single integrated approach to global problems based on the concepts of helping species and populations, conserving planetary life support systems, and maintaining genetic diversity.

World Conservation Union (IUCN) A *non-governmental organization that was founded in 1948. It promotes scientifically based action for the conservation of wild living resources and publishes the *Red List and the *Red Data Book. Previously called the International Union for Conservation of Nature and Natural Resources.

World Environment and Disaster Observation System (WEDOS) A proposed environmental *remote sensing system based on launching 26 Earth observation *satellites at low altitude, which would allow any area of the world to be observed at least once a day with a 20 metre resolution.

World Environment Day Occurring on 5 June each year, World Environment Day was designated in 1972 by the *United Nations Conference on the Human Environment as a day to focus on environmental problems.

World Health Organization (WHO) An agency of the *United Nations that was founded in 1948, is based in Geneva, and serves to promote cooperation among nations in controlling *disease.

world heritage Natural and cultural *heritage sites that belong to all the peoples of the world, irrespective of the territory in which they are located.

World Heritage Convention See CONVENTION CONCERNING THE PROTECTION OF THE WORLD CULTURAL AND NATURAL HERITAGE.

World Heritage Fund A trust fund that was established by *UNESCO, under the 1972 *Convention Concerning the Protection of the World Cultural and Natural Heritage, for 'the protection of the world cultural and natural heritage of outstanding universal value'. It funds the protection of *World Heritage Sites. Contrast GLOBAL HERITAGE FUND.

World Heritage List The formal list of *World Heritage Sites administered by *UNESCO.

World Heritage Site An *IUCN Management Category (X) for protected areas, a specific site (such as a forest, mountain range, lake, desert, building, complex, or city) that has been nominated for the international World Heritage programme administered by UNESCO, under the *United Nations Convention Concerning the Protection of the World Cultural and Natural Heritage.

World Meteorological Organization (WMO) An intergovernmental organization that was established by the *United Nations in 1947 to monitor and standardize collection of global weather data. It is based in Geneva, Switzerland.

World Park An international version of a *national park, which belongs to no single nation but is the responsibility of all nations. None have yet been declared, although the *Antarctic Treaty envisaged it as a means of protecting the unique landscape of Antarctica.

World Population Conference, Bucharest (1974) An international conference organized by the *United Nations which brought together representatives from more developed and less developed countries for the first time. The conference agreed that 'development is the best contraceptive', and proposed the *World Population Plan of Action. It was followed up by the *International Conference on Population which was held in Mexico City in 1984.

World Population Plan of Action A plan of action that was developed at the 1974 *World Population Conference, which recognized that population growth and economic development are mutually related. Amongst other things, the plan was designed to encourage countries to develop their own population policies, reduce mortality and infant mortality, make family planning education and services available for all individuals, promote regional and rural development policies in order to reduce urban pressure, improve national population and development planning, and promote legal, educational, and employment equality for women.

world problematique A concept created by the *Club of Rome to describe the complex set of the crucial problems

(political, social, economic, technological, environmental, psychological, and cultural) that humanity faces, and that need solving.

World Resources Institute (WRI) An environmental *think-tank and independent research organization based in Washington, DC, that was set up in 1982 to address global environmental issues.

World Summit on Sustainable Development (WSSD) An international gathering of governments, *UN agencies and *non-governmental organizations that met in Johannesburg, South Africa, in August to September 2002, in order to assess progress in dealing with *sustainable development since the 1992 *United Nations Conference on Environment and Development.

World Trade Organization (WTO) An international organization with 136 member countries that was set up in 1994 to replace the *General Agreement on Tariffs and Trade, in order to regulate international trade.

World Wide Fund for Nature (WWF) The world's largest and most experienced independent *conservation organization, founded in 1961 as the World Wildlife Fund, whose mission is to protect nature and biological diversity around the world. It has 4.7 million supporters and a global network active in around 100 countries.

World Wildlife Fund (WWF) *See* WORLDWIDE FUND FOR NATURE.

worldview A largely unconscious but generally coherent set of presuppositions and beliefs that every person has which shape how we make sense of the world and everything in it. This in turn influences such things as how we see ourselves as individuals, how we interpret our role in society, how we deal with social issues, and what we regard as truth. *See also* ANTHROPOCENTRISM, ECOCENTRISM.

Worldwatch Institute An independent research organization that was founded in 1974. It is based in the USA and 'works for an environmentally sustainable and socially-just society, in which the needs of all people are met without threatening the health of the natural environment or the well-being of future generations'.

worst case scenario The *scenario that produces the worst outcome imaginable for a given set of assumptions and conditions.

wrack Seaweed, other organic matter, and drift material (including solid wastes and other pollutants) that accumulates on a *beach, usually at the *high tide mark.

wrack zone The collection of material (*wrack) at the *high tide mark on a beach.

WRI *See* WORLD RESOURCES INSTITUTE.

WSSD *See* WORLD SUMMIT ON SUSTAINABLE DEVELOPMENT.

WTO *See* WORLD TRADE ORGANIZATION.

Würm The European name for the *Wisconsinan *glaciation, the fourth and final phase of glaciation during the *Pleistocene.

WWF *See* WORLD WIDE FUND FOR NATURE.

W

xenobiosis A form of *symbiosis among ants in which two colonies of different species live together on friendly terms without rearing their broods in common.

xenobiota A plant that is displaced from its normal *habitat, or a chemical that is not normally found in a particular biological system.

xenolith A fragment of pre-existing rock that is embedded inside an *igneous rock. Also known as **inclusion**.

xenon (Xe) A heavy, colourless, relatively inert, natural *gas that is a trace element in the Earth's *atmosphere.

xeric Having very little moisture, tolerant of or adapted to dry conditions. *Contrast* MESIC, HYDRIC.

xerophilous Able to withstand dry ground and *drought.

xerophyte A plant that is adapted to grow in very dry conditions. *Contrast* HYDROPHYTE, MESOPHYTE.

xerosere The vegetation *succession that begins in a dry *habitat where little moisture is freely available, such as solid bedrock, broken rocks, or fine rock and sand particles.

X-ray diffraction In *mineralogy, a method for studying microscopic crystal form and structure, using *X-rays.

X-rays *Electromagnetic radiation of very short *wavelength and very high energy, which can penetrate many substances (including skin tissue), hence X-rays are used for photographing internal tissues, bones, and organs in humans, and for detecting flaws in metal objects.

xylem The layer of cells in the trunk of a *tree that carries water and nutrients, which are absorbed from the soil by the roots, to the leaves. *See also* GRAIN.

xylem sap *See* SAP.

xylophagous Feeding on wood, which certain insects or insect larvae do.

yard A North American term for the enclosed land around a house or building.

yard waste A North American term for *solid waste in the form of grass clippings, leaves, twigs, branches, and other garden refuse.

yarding In forest management, moving logs from where the tree fell to a central place (*landing) for hauling away from the *stand.

year The period of time that it takes for a *planet (such as the *Earth or Mars) to make a complete revolution around the *Sun. An Earth year is 365 days, but a Pluto year is almost 250 Earth years.

yellowcake An impure mixture of *uranium oxides (U_3O_8) that is extracted from crushed uranium *ore during processing.

yellow-fever mosquito The mosquito (*Aedes aegypti*) that transmits yellow fever and dengue.

Yellowstone National Park The world's first national park, established in 1872, located in the US states of Idaho, Montana, and Wyoming. It is famous for its *geysers, *hot springs, and other *geothermal features, and is home to grizzly bears and wolves and free-ranging herds of bison and elk.

yield The amount of a *natural resource (such as freshwater, fish, or crop growth) that is harvested over a given period of time, usually a year.

Younger Dryas A *stadial or relatively cold period from about 12 900 to 11 600 years ago, which interrupted the warming of the Earth after the last *ice age.

youth 1. The first part of the *life cycle of an organism, before it reaches *maturity.

2. An early stage in the *cycle of erosion in which a landscape has been uplifted and is beginning to be dissected by canyons cut by young streams. *Contrast* MATURITY, OLD AGE.

Yucca Mountain The site in southern Nevada, USA, of a controversial development of an underground *repository for long-term storage of *high-level radioactive waste.

zero discharge A condition in which no *pollutants are released into the environment. Also known as **totally effluent free**.

zero population growth (ZPG) **1.** A stable *population, with a *growth rate of zero, which is achieved when births plus *immigration equal deaths plus *emigration. Also known as **replacement-level fertility**.
  2. The name of an environmental *pressure group that was founded by Paul *Ehrlich in 1968 and which in 2002 changed its name to *population connection.

zero till *See* NO TILL.

zero-emission vehicle A vehicle (usually powered by *electricity, *fuel cells or *hydrogen) that theoretically produces no *pollution when stationary or operating.

zero-grazing Animal *husbandry that is based on bringing *fodder to the animals rather than letting them *graze freely.

zinc (Zn) A blue-white, lustrous, *metallic element, one of the eight *micronutrients that are essential for plant health. It is soluble and found naturally in air, soil, water, and foods. It is an essential nutrient for humans but is harmful in large amounts. It is *brittle at ordinary temperatures but *malleable when heated, and is used in a wide variety of *alloys and in galvanizing *iron.

zircon A hard *mineral that is common in all types of *igneous rock and many *metamorphic rocks, that yields clear gemstones called cubic zircons which are a cheap alternative to diamond.

zonal Along a line of *latitude. *Contrast* MERIDIONAL.

zonal soil A mature soil that has well-developed characteristics typical of the *climatic zone in which it occurs. *Contrast* AZONAL SOIL, INTRAZONAL SOIL.

zonal wind The large-scale movement of air in a dominantly west–east or east–west direction, with little movement north–south or south–north.

zonation The distribution of plants and animals into different geographical zones, such as the variations in water depth in the *intertidal zone.

zone fossil *See* INDEX FOSSIL.

zone of ablation *See* ABLATION ZONE.

zone of accumulation *See* ACCUMULATION ZONE.

zone of aeration *See* AERATION ZONE.

zone of leaching The layer of soil, just beneath the *topsoil, from which soluble *nutrients are removed (leached) by water.

zone of saturation *See* SATURATED ZONE.

zone of wastage *See* WASTAGE ZONE.

zoning In North America, the rules and regulations that are established by local government to govern the use of land, particularly the size, type, structure, nature, and use of buildings. *See also* AESTHETIC ZONING, AGRICULTURAL ZONING, CLUSTER ZONING, CONSERVATION ZONING, EXCLUSIVE USE ZONING, FLOODPLAIN ZONING, FOREST USE ZONING, LARGE-LOT ZONING, NATURAL RESOURCE ZONING, OPEN SPACE ZONING, PERFORMANCE STANDARD ZONING, PROHIBITION ZONING, SPOT ZONING, TIME ZONING, ZONING ORDINANCE, ZONING PERMIT.

zoning ordinance In North America, a local law or regulation that is established by a city council or similar body to govern land use within a particular designated area (zone) and the nature of the buildings within each zone. *See also* ZONING, ZONING PERMIT.

zoning permit In North America, an official finding that a planned use or structure complies with existing *zoning ordinances, or is allowed by the granting of a special exception.

zoning variance In land use *zoning, a licence to act contrary to the usual rule.

zoo Short for zoological garden. A place (usually a garden or park) where live animals are kept for exhibition or for *captive breeding programmes for *endangered species. *See also* BOTANICAL GARDEN.

zooecology The study of the relationship between animals and their environment. *Contrast* ZOOLOGY.

zoological area An area in the USA that has been designated by the US Forest Service or others as containing *animals, animal groups, or animal communities that are natural and important in some particular way.

zoological garden *See* ZOO.

zoological kingdom *See* ANIMAL KINGDOM.

zoology The study of animals (including *mammals, *birds, *reptiles, *amphibians, *fish, *insects, spiders, and *molluscs) and their structures and functions. *Contrast* ZOOECOLOGY. *See also* BOTANY.

zoonosis An *infection or infectious *disease that occurs mainly in wild and domestic animals but can be transmitted to humans.

zooplankton The animal component of the *plankton community, which feed on *phytoplankton and other zooplankton, and are eaten by tiny fish.

zooxanthellae Brownish *algae that live in a *symbiotic relationship with certain *corals, clams, and some sponges and are responsible for the brilliant green, yellow, and blue colours in corals and clams.

ZPG *See* ZERO POPULATION GROWTH.

zygote A fertilized egg. *See also* FERTILIZATION, GAMETE.

Z

Appendix 1. Sources of online data on the environment

The most useful and accessible general source of up-to-date information are the *GEO: Global Environment Outlook* reports, published by the United Nations Environment Programme, both in print and online (the reports can be downloaded in full as PDF files from http://www.unep.org/geo). The GEO Indicators section contains data on atmosphere, disasters caused by natural hazards, biodiversity, coastal and marine areas, freshwater, urban areas, and global environmental issues.

Useful online sources of information and data on particular environmental issues include-

Biodiversity Global Biodiversity Information Facility (GBIF); http://www.gbif.net/portal/index.jsp

Biodiversity Species 2000; http://www.species2000.org/

Carbon dioxide Carbon Dioxide Information Analysis Center (CDIAC); http://cdiac.esd.ornl.gov/trends/emis/em_cont.htm

Climate change Intergovernmental Panel on Climate Change (IPCC); http://www.ipcc.ch/

Conservation World Conservation Monitoring Center (WCMC); http://www.unep-wcmc.org/

Conservation World Conservation Union (International Union for the Conservation of Nature - IUCN); http://www.iucn.org/

Conservation World Wildlife Fund (WWF); http://www.wwf.org/

Development Organization for Economic Co-operation and Development (OECD); http://www.oecd.org

Development United Nations Development Programme (UNDP); http://www.undp.org/

Development United Nations Educational, Scientific and Cultural Organization (UNESCO); http://www.unesco.org

Development World Bank; http://www.worldbank.org/data/wdi2000/index.htm

Energy International Energy Agency (IEA); http://www.iea.org/

Energy World Energy Council (WEC); http://www.worldenergy.org/wec-geis/

Environment Environmental Protection Agency Envirofacts Data Warehouse; http://www.epa.gov/enviro/

Environment United Nations Environment Programme GEO Data Portal; http://geodata.-grid.unep.ch/

Environment World Resources Institute (WRI); http://earthtrends.wri.org/

Environment Worldwatch Institute; http://www.worldwatch.org/pubs/

Environmental organizations World Directory of Environmental Organisations online; http://www.interenvironment.org/wd/

Food and agriculture United Nations Food and Agriculture Organization of the United Nations (FAO); http://faostat.fao.org/

Forests Forest Stewardship Council (FSC); http://www.fsc.org/en/

Greenhouse gases United Nations Framework Convention on Climate Change (UNFCCC); http://ghg.unfccc.int/index.html

Health United Nations World Health Organization (WHO); http://www.who.int/whr/en/

Heritage UNESCO World Heritage Centre; http://whc.unesco.org/

Human settlements United Nations Human Settlements Programme (UN-HABITAT); http://www.unchs.org/habrdd/statprog.htm

National and global statistics United Nations Statistics Division (UN-STAT); http://unstats.un.org/unsd/

Natural disasters International Disasters Database; http://www.em-dat.net/

Nuclear energy International Atomic Energy Agency (IAEA); http://www.iaea.org/

Ozone United Nations Environment Programme Ozone Secretariat; http://www.unep.ch/ozone/DataReport99.shtml

Population United Nations Population Division; http://www.un.org/esa/population/unpop.htm

Resources Global Resource Information Database, Sioux Falls (GRID-Sioux Falls); http://grid2.cr.usgs.gov/

Soils International Soil Reference and Information Centre (ISRIC); http://www.isric.org/

Sustainable development International Institute for Sustainable Development (IISD); http://www.iisd.org/

Wetlands Ramsar Convention on Wetlands; http://www.ramsar.org/

Appendix 2. International environmental treaties

Up-to-date information on international environmental treaties, and on which countries have signed up to which treaties, can be found online at the FBI *World Factbook* web site (http://www.cia.gov/cia/publications/factbook/appendix/appendix-c.html), on which this appendix is based.

a. Subjects

Air Pollution *See* CONVENTION ON LONG-RANGE TRANSBOUNDARY AIR POLLUTION

Air Pollution-Nitrogen Oxides *See* PROTOCOL TO THE 1979 CONVENTION ON LONG-RANGE TRANSBOUNDARY AIR POLLUTION CONCERNING THE CONTROL OF EMISSIONS OF NITROGEN OXIDES OR THEIR TRANSBOUNDARY FLUXES

Air Pollution-Persistent Organic Pollutants *See* PROTOCOL TO THE 1979 CONVENTION ON LONG-RANGE TRANSBOUNDARY AIR POLLUTION ON PERSISTENT ORGANIC POLLUTANTS

Air Pollution-Sulphur 85 *See* PROTOCOL TO THE 1979 CONVENTION ON LONG-RANGE TRANSBOUNDARY AIR POLLUTION ON THE REDUCTION OF SULPHUR EMISSIONS OR THEIR TRANSBOUNDARY FLUXES BY AT LEAST 30%

Air Pollution-Sulphur 94 *See* PROTOCOL TO THE 1979 CONVENTION ON LONG-RANGE TRANSBOUNDARY AIR POLLUTION ON FURTHER REDUCTION OF SULPHUR EMISSIONS

Air Pollution-Volatile Organic Compounds *See* PROTOCOL TO THE 1979 CONVENTION ON LONG-RANGE TRANSBOUNDARY AIR POLLUTION CONCERNING THE CONTROL OF EMISSIONS OF VOLATILE ORGANIC COMPOUNDS OR THEIR TRANSBOUNDARY FLUXES

Antarctic - Environmental Protocol *See* PROTOCOL ON ENVIRONMENTAL PROTECTION TO THE ANTARCTIC TREATY

Biodiversity *See* CONVENTION ON BIOLOGICAL DIVERSITY

Climate Change *See* UNITED NATIONS FRAMEWORK CONVENTION ON CLIMATE CHANGE

Climate Change-Kyoto Protocol *See* KYOTO PROTOCOL TO THE UNITED NATIONS FRAMEWORK CONVENTION ON CLIMATE CHANGE

Desertification *See* UNITED NATIONS CONVENTION TO COMBAT DESERTIFICATION IN THOSE COUNTRIES EXPERIENCING SERIOUS DROUGHT AND/OR DESERTIFICATION, PARTICULARLY IN AFRICA

Endangered Species *See* CONVENTION ON THE INTERNATIONAL TRADE IN ENDANGERED SPECIES OF WILD FLORA AND FAUNA (CITES)

Environmental Modification *See* CONVENTION ON THE PROHIBITION OF MILITARY OR ANY OTHER HOSTILE USE OF ENVIRONMENTAL MODIFICATION TECHNIQUES

Hazardous Wastes *See* BASEL CONVENTION ON THE CONTROL OF TRANSBOUNDARY MOVEMENTS OF HAZARDOUS WASTES AND THEIR DISPOSAL

Law of the Sea *See* UNITED NATIONS CONVENTION ON THE LAW OF THE SEA (LOS)

Marine Dumping *See* CONVENTION ON THE PREVENTION OF MARINE POLLUTION BY DUMPING WASTES AND OTHER MATTER (LONDON CONVENTION)

Marine Life Conservation *See* CONVENTION ON FISHING AND CONSERVATION OF LIVING RESOURCES OF THE HIGH SEAS

Nuclear Test Ban *See* TREATY BANNING NUCLEAR WEAPONS TESTS IN THE ATMOSPHERE, IN OUTER SPACE, AND UNDER WATER

Ozone Layer Protection See MONTREAL PROTOCOL ON SUBSTANCES THAT DEPLETE THE OZONE LAYER

Ship Pollution See PROTOCOL OF 1978 RELATING TO THE INTERNATIONAL CONVENTION FOR THE PREVENTION OF POLLUTION FROM SHIPS, 1973 (MARPOL)

Tropical Timber 83 See INTERNATIONAL TROPICAL TIMBER AGREEMENT, 1983

Tropical Timber 94 See INTERNATIONAL TROPICAL TIMBER AGREEMENT, 1994

Wetlands See CONVENTION ON WETLANDS OF INTERNATIONAL IMPORTANCE ESPECIALLY AS WATERFOWL HABITAT (RAMSAR)

Whaling See INTERNATIONAL CONVENTION FOR THE REGULATION OF WHALING

b. Treaties

Antarctic Treaty
opened for signature 1 December 1959
entered into force 23 June 1961
objective to ensure that Antarctica is used for peaceful purposes only (such as international cooperation in scientific research); to defer the question of territorial claims asserted by some nations and not recognized by others; to provide an international forum for management of the region; applies to land and ice shelves south of 60°S latitude
parties 45
further details http://www.scar.org/treaty/at_text.html

Basel Convention on the Control of Transboundary Movements of Hazardous Wastes and Their Disposal
shown in the table below as Hazardous Wastes
opened for signature 22 March 1989
entered into force 5 May 1992
objective to reduce transboundary movements of wastes subject to the Convention to a minimum consistent with the environmentally sound and efficient management of such wastes; to minimize the amount and toxicity of wastes generated and ensure their environmentally sound management as close as possible to the source of generation; and to assist less developed countries in environmentally sound management of the hazardous and other wastes they generate
parties 149
further details http://www.basel.int/about.html

Convention for the Conservation of Antarctic Seals
shown in the table below as Antarctic Seals
opened for signature 1 June 1972
entered into force 11 March 1978
objective to promote and achieve the protection, scientific study, and rational use of Antarctic seals, and to maintain a satisfactory balance within the ecological system of Antarctica
parties 16
further details http://untreaty.un.org/English/UNEP/antarticseals_english.pdf

Convention on Biological Diversity
shown in the table below as Biodiversity
opened for signature 5 June 1992
entered into force 29 December 1993
objective to develop national strategies for the conservation and sustainable use of biological diversity
parties 182
further details http://www.biodiv.org/convention/articles.asp

Convention on Fishing and Conservation of Living Resources of the High Seas
Shown in the table below as Marine Life Conservation

opened for signature 29 April 1958
entered into force 20 March 1966
objective to solve through international cooperation the problems involved in the conservation of living resources of the high seas, considering that because of the development of modern technology some of these resources are in danger of being overexploited
parties 37
further details http://www.oceanlaw.net/texts/genevafish.htm

Convention on Long-Range Transboundary Air Pollution
shown in the table below as Air Pollution
opened for signature 13 November 1979
entered into force 16 March 1983
objective to protect the human environment against air pollution and to gradually reduce and prevent air pollution, including long-range transboundary air pollution
parties 48
further details http://www.unece.org/env/lrtap/lrtap_h1.htm

Convention on Wetlands of International Importance Especially as Waterfowl Habitat (Ramsar)
shown in the table below as Wetlands
opened for signature 2 February 1971
entered into force 21 December 1975
objective to stem the progressive encroachment on and loss of wetlands now and in the future, recognizing the fundamental ecological functions of wetlands and their economic, cultural, scientific, and recreational value
parties 125
further details http://www.ramsar.org/key_conv_e.htm

Convention on the Conservation of Antarctic Marine Living Resources
shown in the table below as Antarctic-Marine Living Resources
opened for signature 5 May 1980
entered into force 7 April 1982
objective to safeguard the environment and protect the integrity of the ecosystem of the seas surrounding Antarctica, and to conserve Antarctic marine living resources
parties 31
further details http://www.ccamlr.org/pu/e/e_pubs/bd/pt1.pdf

Convention on the International Trade in Endangered Species of Wild Flora and Fauna (CITES)
shown in the table below as Endangered Species
opened for signature 3 March 1973
entered into force 1 July 1975
objective to protect certain endangered species from overexploitation by means of a system of import/export permits
parties 156
further details http://www.cites.org/eng/disc/text.shtml

Convention on the Prevention of Marine Pollution by Dumping Wastes and Other Matter (London Convention)
shown in the table below as Marine Dumping
opened for signature 29 December 1972
entered into force 30 August 1975
objective to control pollution of the sea by dumping and to encourage regional agreements supplementary to the Convention
parties 78
further details http://www.londonconvention.org

Convention on the Prohibition of Military or Any Other Hostile Use of Environmental Modification Techniques
shown in the table below as Environmental Modification
opened for signature 10 December 1976
entered into force 5 October 1978
objective to prohibit the military or other hostile use of environmental modification techniques in order to further world peace and trust among nations
parties 66
further details http·//www.opcw.org/html/db/cwc/more/enmod.html

International Convention for the Regulation of Whaling
shown in the table below as Whaling
opened for signature 2 December 1946
entered into force 10 November 1948
objective to protect all species of whales from overhunting; to establish a system of international regulation for the whale fisheries to ensure proper conservation and development of whale stocks; and to safeguard for future generations the great natural resources represented by whale stocks
parties 42
further details http://www.iwcoffice.org/commission/convention.htm

International Tropical Timber Agreement, 1983
shown in the table below as Tropical Timber 83
opened for signature 18 November 1983
entered into force 1 April 1985; this agreement expired when the International Tropical Timber Agreement, 1994, went into force
objective to provide an effective framework for cooperation between tropical timber producers and consumers and to encourage the development of national policies aimed at sustainable utilization and conservation of tropical forests and their genetic resources
parties 54
further details http://eelink.net/~asilwildlife/ITTA83.html

International Tropical Timber Agreement, 1994
shown in the table below as Tropical Timber 94
opened for signature 26 January 1994
entered into force 1 January 1997
objective to ensure that by the year 2000 exports of tropical timber originate from sustainably managed sources; to establish a fund to assist tropical timber producers in obtaining the resources necessary to reach this objective
parties 58
further details http://www.itto.or.jp/live/Live_Server/144/ITTA1994e.doc

Kyoto Protocol to the United Nations Framework Convention on Climate Change
shown in the table below as Climate Change-Kyoto Protocol
opened for signature 16 March 1998
entered into force 23 February 2005
objective to further reduce greenhouse gas emissions by enhancing the national programmes of developed countries aimed at this goal and by establishing percentage reduction targets for the developed countries
parties 144
further details http://unfccc.int/resource/docs/convkp/kpeng.html

Montreal Protocol on Substances That Deplete the Ozone Layer
shown in the table below as Ozone Layer Protection
opened for signature 16 September 1987
entered into force 1 January 1989
objective to protect the ozone layer by controlling emissions of substances that deplete it

parties 183
further details http://www.unep.org/ozone/Montreal-Protocol/Montreal-Protocol2000.shtml

Protocol of 1978 Relating to the International Convention for the Prevention of Pollution From Ships, 1973 (MARPOL)

shown in the table below as Ship Pollution
opened for signature 17 February 1978
entered into force 2 October 1983
objective to preserve the marine environment through the complete elimination of pollution by oil and other harmful substances and the minimization of accidental discharge of such substances
parties 119
further details http://www.imo.org/Conventions/contents.asp?doc_id=678&topic_id=258

Protocol on Environmental Protection to the Antarctic Treaty

shown in the table below as Antarctic-Environmental Protocol
opened for signature 4 October 1991
entered into force 14 January 1998
objective to provide for comprehensive protection of the Antarctic environment and dependent and associated ecosystems; applies to the area covered by the Antarctic Treaty
consultative parties 27
further details http://www.antarctica.ac.uk/About_Antarctica/Treaty/protocol.html

Protocol to the 1979 Convention on Long-Range Transboundary Air Pollution Concerning the Control of Emissions of Nitrogen Oxides or Their Transboundary Fluxes

shown in the table below as Air Pollution-Nitrogen Oxides
opened for signature 31 October 1988
entered into force 14 February 1991
objective to provide for the control or reduction of nitrogen oxides and their transboundary fluxes
parties 28
further details http://www.unece.org/env/lrtap/full%20text/1988.NOX.e.pdf

Protocol to the 1979 Convention on Long-Range Transboundary Air Pollution Concerning the Control of Emissions of Volatile Organic Compounds or Their Transboundary Fluxes

shown in the table below as Air Pollution-Volatile Organic Compounds
opened for signature 18 November 1991
entered into force 29 September 1997
objective to provide for the control and reduction of emissions of volatile organic compounds in order to reduce their transboundary fluxes so as to protect human health and the environment from adverse effects
parties 21
further details http://www.unece.org/env/lrtap/full%20text/1991.VOC.e.pdf

Protocol to the 1979 Convention on Long-Range Transboundary Air Pollution on Further Reduction of Sulphur Emissions

shown in the table below as Air Pollution-Sulphur 94
opened for signature 14 June 1994
entered into force 5 August 1998
objective to provide for a further reduction in sulphur emissions or transboundary fluxes
parties 23
further details http://www.unece.org/env/lrtap/full%20text/1994.Sulphur.e.pdf

Protocol to the 1979 Convention on Long-Range Transboundary Air Pollution on Persistent Organic Pollutants

shown in the table below as Air Pollution-Persistent Organic Pollutants
opened for signature 24 June 1998

entered into force 23 October 2003
objective to provide for the control and reduction of emissions of persistent organic pollutants in order to reduce their transboundary fluxes so as to protect human health and the environment from adverse effects
parties 22
further details http://www.unece.org/env/lrtap/full%20text/1998.POPs.e.pdf

Protocol to the 1979 Convention on Long-Range Transboundary Air Pollution on the Reduction of Sulphur Emissions or Their Transboundary Fluxes by at Least 30%
shown in the table below as Air Pollution-Sulphur 85
opened for signature 8 July 1985
entered into force 2 September 1987
objective to provide for a 30% reduction in sulphur emissions or transboundary fluxes by 1993
parties 22
further details http://www.unece.org/env/lrtap/full%20text/1985.Sulphur.e.pdf

Treaty Banning Nuclear Weapon Tests in the Atmosphere, in Outer Space, and Under Water
shown in the table below as Nuclear Test Ban
opened for signature 5 August 1963
entered into force 10 October 1963
objective to obtain an agreement on general and complete disarmament under strict international control in accordance with the objectives of the United Nations; to put an end to the armaments race and eliminate incentives for the production and testing of all kinds of weapons, including nuclear weapons
parties 113
further details http://pws.ctbto.org/treaty/treaty_text.pdf

United Nations Convention on the Law of the Sea (LOS)
shown in the table below as Law of the Sea
opened for signature 10 December 1982
entered into force 16 November 1994
objective to set up a comprehensive new legal regime for the sea and oceans; to include rules concerning environmental standards as well as enforcement provisions dealing with pollution of the marine environment
parties 148
further details http://www.un.org/Depts/los/convention_agreements/texts/unclos/unclos_e.pdf

United Nations Convention to Combat Desertification in Those Countries Experiencing Serious Drought and/or Desertification, Particularly in Africa
shown in the table below as Desertification
opened for signature 14 October 1994
entered into force 26 December 1996
objective to combat desertification and mitigate the effects of drought through national action programmes that incorporate long-term strategies supported by international cooperation and partnership arrangements
parties 178
further details http://www.unccd.int/

United Nations Framework Convention on Climate Change
shown in the table below as Climate Change
opened for signature 9 May 1992
entered into force 21 March 1994
objective to achieve stabilization of greenhouse gas concentrations in the atmosphere at a low enough level to prevent dangerous anthropogenic interference with the climate system
parties 189
further details http://unfccc.int/resource/docs/convkp/conveng.pdf

C. Countries that had ratified by June 2005

Country	Air Pollution	Air Pollution-Nitrogen Oxides	Air Pollution-Persistent Organic Pollutants	Air Pollution-Sulphur 85	Air Pollution-Sulphur 94	Air Pollution-Volatile Organic Compounds	Antarctic-Environmental Protocol	Antarctic-Marine Living Resources	Antarctic Seals	Antarctic Treaty	Biodiversity	Climate Change	Climate Change-Kyoto Protocol	Desertification	Endangered Species	Environmental Modification	Hazardous Wastes	Law of the Sea	Marine Dumping	Marine Life Conservation	Nuclear Test Ban	Ozone Layer Protection	Ship Pollution	Tropical Timber 83	Tropical Timber 94	Wetlands	Whaling
Afghanistan											X	X		X	X	X		X	X		X	X					
Albania											X	X		X	X		X	X			X	X				X	
Algeria											X	X		X	X		X	X			X	X	X			X	
Andorra																											
Angola											X	X		X	X	X	X	X	X			X	X			X	X
Antigua & Barbuda											X	X		X	X		X	X	X		X	X	X			X	
Argentina	X						X	X	X	X	X	X	X	X	X	X	X	X	X		X	X	X	X	X	X	X
Armenia	X										X	X		X	X		X	X			X	X				X	
Australia	X	X			X	X	X	X	X	X	X	X		X	X		X	X	X		X	X	X			X	X
Austria	X	X		X	X	X					X	X	X	X	X	X	X	X	X		X	X	X	X	X	X	X
Azerbaijan				X	X						X	X		X	X		X	X			X	X				X	
Bahamas											X	X		X	X		X	X	X			X	X			X	
Bahrain											X	X		X	X		X	X			X	X	X			X	
Bangladesh											X	X		X	X		X	X	X			X				X	
Barbados											X	X		X	X		X	X	X			X	X			X	
Belarus	X	X		X							X	X	X	X	X		X	X			X	X				X	
Belgium	X	X			X	X	X	X	X	X	X	X	X	X	X	X	X	X	X		X	X	X	X	X	X	X
Belize											X	X	X	X	X		X	X			X	X				X	
Benin											X	X		X	X		X	X			X	X	X			X	
Bhutan											X	X		X	X		X	X				X				X	
Bolivia											X	X	X	X	X	X	X	X			X	X	X			X	X
Bosnia & Herzegovina	X										X	X		X	X		X	X		X	X	X	X			X	
Botswana							X				X	X		X	X		X	X	X			X	X	X		X	
Brazil							X	X	X	X	X	X	X	X	X	X	X	X	X		X	X	X	X	X	X	X
Brunei											X	X		X	X							X				X	
Bulgaria	X	X		X	X	X	X	X	X	X	X	X	X	X	X	X	X	X	X		X	X	X	X	X	X	X
Burkino Faso	X	X									X	X		X	X		X	X		X	X	X				X	
Burma											X	X		X	X		X	X			X	X	X	X		X	
Burundi											X	X		X	X		X	X			X	X				X	
Cambodia											X	X		X	X		X	X		X	X	X	X	X	X	X	

Cameroon	X	X					X	X		X	X	X		X		X		X	X			X
Canada	X	X		X			X	X	X	X	X	X	X	X		X	X	X	X			X
Cape Verde			X		X	X		X	X	X	X	X		X		X			X			X
Central African Republic	X									X	X	X	X	X		X	X	X	X			X
Chad					X		X	X		X	X	X		X	X	X	X		X		X	X
Chile		X			X		X	X	X	X	X	X	X	X	X	X	X		X		X	X
China						X	X	X	X	X	X	X	X	X		X	X		X		X	X
Colombia							X	X		X	X	X	X	X		X	X	X	X		X	X
Comoros								X		X	X	X	X	X		X		X	X		X	X
Congo, Democratic Republic								X		X	X	X		X	X	X	X		X		X	X
Congo, Republic								X		X	X	X	X	X		X		X	X		X	X
Cook Islands								X		X	X	X		X		X			X			
Costa Rica								X		X	X	X		X	X	X	X	X	X		X	X
Cote D'Ivoire								X		X	X	X	X	X		X		X	X		X	X
Croatia	X				X			X		X	X	X	X	X		X		X	X		X	X
Cuba	X	X						X	X	X	X	X	X	X	X	X			X		X	X
Cyprus	X	X					X	X	X	X	X	X	X	X		X	X	X	X		X	X
Czech Republic	X	X	X	X	X		X	X	X	X	X	X	X	X		X	X	X	X		X	X
Denmark	X	X	X	X	X		X	X	X	X	X	X	X	X		X	X	X	X		X	X
Djibouti								X		X	X	X		X		X			X		X	X
Dominica								X	X	X	X	X	X	X	X	X			X			X
Dominican Republic					X			X		X	X	X	X	X		X	X		X		X	X
Ecuador					X	X		X		X	X	X	X	X	X	X	X		X		X	X
Egypt								X		X	X	X	X	X		X	X	X	X		X	X
El Salvador								X	X	X	X	X	X	X	X	X	X		X		X	X
Equatorial Guinea								X			X	X		X		X			X			
Eritrea	X	X		X			X			X	X	X		X		X			X			
Estonia	X	X						X	X	X	X	X	X	X	X	X		X	X		X	X
Ethiopia								X		X	X	X	X	X		X	X	X	X		X	X
Fiji	X	X			X	X		X	X	X	X	X	X	X	X	X	X	X	X		X	X
Finland	X	X		X	X	X		X	X	X	X	X	X	X	X	X	X	X	X		X	X
France	X	X			X	X		X	X	X	X	X	X	X	X	X	X	X	X		X	X
Cameroon								X	X	X	X	X	X	X	X	X			X			X
Gabon								X		X	X	X		X	X	X			X		X	X

(cont.)

C. Countries that had ratified by June 2005 (contd.)

Country	Air Pollution	Air Pollution-Nitrogen Oxides	Air Pollution-Persistent Organic Pollutants	Air Pollution-Sulphur 85	Air Pollution-Sulphur 94	Air Pollution-Volatile Organic Compounds	Antarctic-Environmental Protocol	Antarctic-Marine Living Resources	Antarctic Seals	Antarctic Treaty	Biodiversity	Climate Change	Climate Change-Kyoto Protocol	Desertification	Endangered Species	Environmental Modification	Hazardous Wastes	Law of the Sea	Marine Dumping	Marine Life Conservation	Nuclear Test Ban	Ozone Layer Protection	Ship Pollution	Tropical Timber 83	Tropical Timber 94	Wetlands	Whaling
Gambia											X	X	X	X	X		X	X			X	X	X			X	
Georgia	X	X									X	X	X	X	X		X	X				X	X			X	
Germany	X	X		X	X	X	X	X	X	X	X	X		X	X	X	X	X	X		X	X	X	X	X	X	X
Ghana							X	X			X	X		X	X	X		X			X	X	X	X	X	X	X
Greece	X	X		X	X	X	X		X	X	X	X		X	X	X	X	X	X		X	X	X	X	X	X	X
Grenada											X	X		X	X			X				X				X	X
Guatemala										X	X	X	X	X	X		X	X	X		X	X	X	X	X	X	X
Guinea											X	X	X	X	X		X	X				X	X			X	X
Guinea-Bissau											X	X		X	X			X				X				X	X
Guyana											X	X		X	X		X	X				X	X			X	
Haiti																		X		X							
Holy See (Vatican City)																					X	X					
Honduras											X	X	X	X	X		X	X	X		X	X	X	X	X	X	X
Hong Kong																						X	X				
Hungary	X	X		X						X	X	X	X	X	X	X	X	X	X		X	X	X			X	
Iceland	X	X	X							X	X	X		X	X		X	X	X		X	X	X			X	X
India							X	X		X	X	X	X	X	X	X	X	X			X	X	X	X	X	X	
Indonesia											X	X		X	X		X	X				X	X	X	X	X	
Iran											X	X	X	X	X		X	X				X	X			X	
Iraq																		X				X					
Ireland	X	X		X	X	X				X	X	X	X	X	X	X	X	X	X		X	X	X		X	X	X
Israel									X		X	X	X	X	X		X				X	X	X			X	X
Italy	X	X		X	X	X	X	X	X	X	X	X	X	X	X	X	X	X	X		X	X	X	X	X	X	X
Jamaica											X	X	X	X	X		X	X		X	X	X	X			X	X
Japan							X	X	X	X	X	X	X	X	X		X	X	X		X	X	X	X	X	X	X
Jordan											X	X	X	X	X		X	X	X		X	X	X			X	
Kazakhstan	X										X	X		X	X	X	X				X	X					
Kenya											X	X		X	X		X	X		X	X	X	X			X	X
Kiribati											X	X	X	X	X		X	X			X	X	X			X	X

Korea, North
Korea, South
Kuwait
Kyrgyzstan
Laos
Latvia
Lebanon
Lesotho
Liberia
Libya
Liechtenstein
Lithuania
Luxembourg
Macedonia
Madagascar
Malawi
Malaysia
Maldives
Mali
Malta
Marshall Islands
Mauritania
Mauritius
Mexico
Micronesia, Federated States
Moldova
Monaco
Mongolia
Morocco
Mozambique
Namibia
Nauru
Nepal

(cont.)

C. Countries that had ratified by June 2005 (contd.)

Country	Air Pollution	Air Pollution-Nitrogen Oxides	Air Pollution-Persistent Organic Pollutants	Air Pollution-Sulphur 85	Air Pollution-Sulphur 94	Air Pollution-Volatile Organic Compounds	Antarctic-Environmental Protocol	Antarctic-Marine Living Resources	Antarctic Seals	Antarctic Treaty	Biodiversity	Climate Change	Climate Change-Kyoto Protocol	Desertification	Endangered Species	Environmental Modification	Hazardous Wastes	Law of the Sea	Marine Dumping	Marine Life Conservation	Nuclear Test Ban	Ozone Layer Protection	Ship Pollution	Tropical Timber 83	Tropical Timber 94	Wetlands	Whaling
Netherlands	X	X	X	X	X	X	X	X		X	X	X	X	X	X	X	X	X	X			X	X	X	X	X	X
New Zealand	X	X	X	X	X	X	X	X	X	X	X	X		X	X	X	X	X	X			X	X	X	X	X	X
Nicaragua											X	X	X	X	X		X	X			X	X	X		X	X	
Niger											X	X	X	X	X	X	X	X			X	X	X		X	X	
Nigeria											X	X		X	X		X	X		X	X	X	X			X	
Niue											X	X	X	X	X			X		X		X			X		
Norway	X	X		X	X	X	X	X	X	X	X	X	X	X	X	X	X	X	X		X	X	X	X	X	X	X
Oman											X	X	X	X	X		X	X				X	X			X	X
Pakistan											X	X		X	X	X	X	X	X			X	X			X	
Palau											X	X	X	X	X			X				X				X	
Panama											X	X	X	X	X		X	X	X		X	X	X	X	X	X	X
Papua New Guinea										X	X	X	X	X	X	X		X			X	X	X	X	X	X	
Paraguay											X	X	X	X	X		X	X	X		X	X				X	
Peru							X	X		X	X	X	X	X	X		X	X	X		X	X	X	X	X	X	X
Philippines	X						X			X	X	X		X	X		X	X	X		X	X	X	X	X	X	X
Poland	X	X					X	X	X	X	X	X	X	X	X	X	X	X	X		X	X	X			X	
Portugal	X										X	X	X	X	X		X	X		X		X	X		X	X	
Qatar											X	X	X	X	X		X	X				X					
Romania	X	X	X	X			X			X	X	X	X		X	X	X	X	X		X	X	X	X		X	X
Russia	X	X					X	X	X	X	X	X		X	X	X	X	X	X		X	X	X			X	X
Rwanda											X	X		X	X		X				X	X	X			X	
Saint Kitts & Nevis											X	X			X			X				X	X			X	
Saint Lucia											X	X		X	X		X	X				X	X	X		X	X
Saint Vincent & the Grenadines											X	X		X	X	X	X					X	X			X	X
Samoa											X	X	X	X				X			X	X					

Country																				
San Marino							X	X								X				X
Sao Tome & Principe					X		X	X	X		X	X	X	X		X		X		X
Saudi Arabia						X	X	X	X	X	X	X	X			X		X		X
Senegal	X	X	X	X		X	X	X	X	X	X	X	X			X	X	X		X
Serbia & Montenegro	X								X											
Seychelles						X	X	X	X	X	X	X	X			X		X		X
Sierra Leone						X	X	X	X	X	X	X	X			X		X	X	X
Singapore	X					X	X	X	X	X	X	X	X			X	X	X		X
Slovakia	X	X	X	X		X	X	X	X	X	X	X				X	X	X		X
Slovenia	X	X	X	X		X	X	X	X	X	X	X				X		X	X	X
Solomon Islands																X	X			
Somalia						X	X	X	X	X	X	X	X			X	X	X		X
South Africa	X	X	X	X		X	X	X	X	X	X	X	X			X	X	X	X	X
Spain	X	X				X	X	X	X	X	X	X	X			X	X	X	X	X
Sri Lanka	X	X				X	X	X	X	X	X	X	X			X	X	X		X
Sudan						X	X	X	X	X	X	X	X			X	X	X		X
Suriname	X	X	X	X		X	X	X	X	X	X	X				X	X	X	X	X
Swaziland	X	X	X	X	X	X	X	X	X	X	X	X				X		X		X
Sweden	X	X	X	X	X	X	X	X	X	X	X	X	X			X	X	X	X	X
Switzerland	X	X	X	X		X	X	X	X	X	X	X	X			X	X	X	X	X
Syria						X	X	X	X	X	X	X				X		X		X
Taiwan																				
Tajikistan						X	X	X	X	X	X	X	X			X	X	X	X	X
Tanzania						X	X	X	X	X	X	X	X			X	X	X	X	X
Thailand							X	X	X		X	X	X			X	X	X		X
Togo						X	X	X	X	X	X	X	X			X	X	X	X	X
Tonga			X	X	X	X	X	X	X	X	X	X								
Trinidad & Tobago			X	X	X	X	X	X	X	X	X	X	X			X	X	X	X	X
Tunisia	X	X				X	X	X	X	X	X	X	X			X	X	X	X	X
Turkey	X					X	X	X	X	X	X	X	X			X	X	X	X	X
Turkmenistan						X	X	X	X		X	X				X		X	X	X
Tuvalu						X	X	X	X											

(cont.)

C. Countries that had ratified by June 2005 (contd.)

Country	Air Pollution	Air Pollution-Nitrogen Oxides	Air Pollution-Persistent Organic Pollutants	Air Pollution-Sulphur 85	Air Pollution-Sulphur 94	Air Pollution-Volatile Organic Compounds	Antarctic-Environmental Protocol	Antarctic-Marine Living Resources	Antarctic Seals	Antarctic Treaty	Biodiversity	Climate Change	Climate Change-Kyoto Protocol	Desertification	Endangered Species	Environmental Modification	Hazardous Wastes	Law of the Sea	Marine Dumping	Marine Life Conservation	Nuclear Test Ban	Ozone Layer Protection	Ship Pollution	Tropical Timber 83	Tropical Timber 94	Wetlands	Whaling
Uganda	X										X	X		X	X	X	X	X		X	X	X					
Ukraine	X	X		X						X	X	X		X	X		X	X	X	X	X	X	X			X	
United Arab Emirates		X						X		X	X	X		X	X		X	X	X		X	X				X	
United Kingdom	X	X			X	X	X	X	X	X	X	X		X	X	X	X	X	X	X	X	X	X	X	X	X	X
United States	X	X							X	X	X	X		X	X	X			X	X	X	X	X			X	X
Uruguay							X	X	X	X	X	X	X	X	X	X	X	X			X	X	X	X	X	X	
Uzbekistan							X	X			X	X	X	X	X		X					X		X	X	X	
Vanuatu											X	X	X	X	X				X			X				X	
Venezuela								X		X	X	X		X	X		X	X		X	X	X	X	X		X	
Vietnam											X	X		X	X	X	X	X			X	X	X	X	X	X	
Western Sahara														X									X				
Yemen											X	X		X	X	X	X	X			X	X				X	
Zambia											X	X		X	X		X	X			X	X			X	X	
Zimbabwe											X	X		X	X		X	X				X				X	

Appendix 3. The Beaufort scale of wind strength

Force	Wind speed (kilometres per hour)	Description	Evidence of impact on land
0	1.5 or less	Calm	The air is still, smoke rises vertically
1	1.6–5	Light air	Wind vanes and flags do not move, but rising smoke drifts
2	6–12	Light breeze	Wind is felt on the face, leaves rustle, wind vanes move
3	13–20	Gentle breeze	Leaves rustle, small twigs move, and the wind extends light flags
4	21–29	Moderate breeze	Dust, loose leaves, and pieces of paper blow about
5	30–39	Fresh breeze	Small trees that are in full leaf wave in the wind
6	40–50	Strong breeze	Large branches move, telephone wires whistle, it becomes difficult to use an open umbrella
7	51–61	Moderate gale	Trees start to sway, people walking into the wind start to have difficulties
8	62–74	Gale	Small twigs are torn from trees
9	75–87	Severe gale	Slight structural damage (chimneys are blown down, slates and tiles are torn from roofs)
10	88–101	Storm	Seldom experienced inland, but causes considerable structural damage; trees are broken or uprooted
11	102–120	Violent storm	Very rarely experienced, but causes widespread damage; trees are uprooted and blown some distance, cars are overturned
12	Greater than 120	Hurricane	Widespread devastation, with buildings destroyed and many trees uprooted

Appendix 4. The Saffir–Simpson hurricane scale

Hurricane category	Wind speed (kilometres per hour)	Evidence of impact on land
Category 1 hurricane	118–152	No real damage to buildings, but damage to mobile homes and poorly constructed signs. Some coastal flooding and minor damage to piers
Category 2 hurricane	153–176	Some damage to building roofs, doors, and windows, considerable damage to mobile homes, and some trees blown down. Piers damaged by flooding, and small boats in unprotected moorings may break loose
Category 3 hurricane	177–208	Some structural damage to small residences and utility buildings, large trees blown down, mobile homes and poorly built signs destroyed. Coastal flooding destroys smaller structures, with larger structures damaged by floating debris; flooding may extend well inland
Category 4 hurricane	209–248	More extensive structural damage, including collapse of roofs on small residences. Major erosion of beach areas, and flooding well inland
Category 5 hurricane	More than 248	Complete roof failure on many residences and industrial buildings, some complete building failures, small utility buildings blown over or away. Flooding causes major damage to lower floors of all structures near the shoreline. Large-scale evacuation of low-lying residential areas may be required

Appendix 5. The Fujita scale of tornado intensity

F-scale number	Description	Wind speed (kilometres per hour)	Type of damage
F0	Gale tornado	64–115	Some damage to chimneys, branches broken off trees, shallow-rooted trees pushed over, sign boards damaged
F1	Moderate tornado	116–179	Surface peeled off roofs, mobile homes moved or overturned, moving vehicles pushed off road, attached garages may be destroyed
F2	Significant tornado	180–251	Considerable damage; roofs torn off houses, mobile homes destroyed, train carriages blown over, large trees snapped or uprooted, light objects blown around
F3	Severe tornado	252–330	Roof and some walls blown off well-constructed buildings, trains overturned, most trees uprooted
F4	Devastating tornado	331–416	Well-constructed buildings are flattened, structures with weak foundations are blown some distance, vehicles blown over, large objects blown around
F5	Incredible tornado	417–509	Strong buildings lifted off foundations, blown around and disintegrate, vehicles fly through the air, reinforced concrete structures badly damaged
F6	Inconceivable tornado	510–606	Very unlikely, and very localized (surrounded by F4 and F5 winds); serious damage caused by large objects, including vehicles, blowing around.

Appendix 6. Torino asteroid and comet impact hazard scale

Torino level	Category	Description
0	No hazard (white zone)	Zero likelihood of a collision
1	Normal (green zone)	Routine discovery in which a pass near the Earth is predicted, but the chance of collision is extremely unlikely
2	Meriting attention by astronomers (yellow zone)	Discovery of an object making a close but not highly unusual pass near the Earth, with an actual collision very unlikely
3		Close encounter, with a 1% or greater chance of collision capable of localized destruction
4		Close encounter, with a 1% or greater chance of collision capable of regional devastation
5	Threatening (orange zone)	Close encounter that poses a serious but still uncertain threat of regional devastation
6		Close encounter by a large object which poses a serious but still uncertain threat of global catastrophe
7		Very close encounter by a large object, which (if occurring within 100 years) poses an unprecedented but still uncertain threat of a global catastrophe
8	Certain collisions (red zone)	Collision is certain, capable of causing localized destruction
9		Collision is certain, capable of causing unprecedented regional devastation
10		Collision is certain, capable of causing global climatic catastrophe

Appendix 7. Richter earthquake magnitude and modified Mercalli earthquake intensity scales

Richter magnitude	Mercalli intensity (at epicentre)	Experience of observers
1–2	I	Felt by very few people, barely noticeable
2–3	II	Felt by a few people, especially on upper floors
3–4	III	Noticeable indoors, especially on upper floors, but may not be recognizable as an earthquake
4	IV	Felt by many indoors, few outdoors
4–5	V	Felt by almost everyone, some people awakened, small objects moved, trees and poles may shake
5–6	VI	Felt by everyone, difficult to stand, some heavy furniture is moved, some plaster falls, chimneys may be slightly damaged
6	VII	Well-built structures may have slight to moderate damage, poorly built structures may be badly damaged, some walls may collapse
6–7	VIII	Little damage in specially built structures, considerable damage to ordinary buildings, severe damage to poorly built structures, some walls collapse
7	IX	Considerable damage to specially built structures, buildings moved off foundations, large ground cracks, large-scale destruction, landslides
7–8	X	Destruction of most structures and foundations, severe ground cracking, landslides, large-scale destruction
8	XI	Total damage, few structures left standing, bridges destroyed, wide cracks in the ground, waves seen on the ground
8 or greater	XII	Total damage, waves seen on the ground, objects thrown up in the air.

Appendix 8.   The geological time-scale

Eon	Era	Sub-era	Period	Epoch	Starting (millions of years ago)
Phanerozoic	Cenozoic	Quaternary	Pleistogene	Holocene	0.01
				Pleistocene	1.64
			Neogene	Pliocene	5.2
		Tertiary		Miocene	23.3
			Palaeogene	Oligocene	35.4
				Eocene	56.5
				Palaeocene	65
	Mesozoic		Cretaceous		145.6
			Jurassic		208
			Triassic		245
	Palaeozoic	Upper Palaeozoic	Permian		290
			Carboniferous		362.5
			Devonian		408.5
		Lower Palaeozoic	Silurian		439
			Ordovician		510
			Cambrian		570
Precambrian Proterozoic Archaean Priscoan					2500 4000 4600

Appendix 9.   Chemical elements

Details of the properties of each element can be found online at http://www.lenntech.com/Periodic-chart-elements/alphabetic.htm

Chemical element	Symbol	Atomic number
Actinium	Ac	89
Aluminium	Al	13
Americium	Am	95
Antimony	Sb	51
Argon	Ar	18
Arsenic	As	33
Astatine	At	85
Barium	Ba	56
Berkelium	Bk	97
Beryllium	Be	4
Bismuth	Bi	83
Bohrium	Bh	107
Boron	B	5
Bromine	Br	35
Cadmium	Cd	48
Calcium	Ca	20
Californium	Cf	98
Carbon	C	6
Cerium	Ce	58
Cacsium	Cs	55
Chlorine	Cl	17
Chromium	Cr	24
Cobalt	Co	27
Copper	Cu	29
Curium	Cm	96
Darmstadtium	Ds	110
Dubnium	Db	105
Dysprosium	Dy	66
Einsteinium	Es	99
Erbium	Er	68
Europium	Eu	63
Fermium	Fm	100
Fluorine	F	9
Francium	Fr	87
Gadolinium	Gd	64
Gallium	Ga	31
Germanium	Ge	32
Gold	Au	79
Hafnium	Hf	72
Hassium	Hs	108
Helium	He	2
Holmium	Ho	67
Hydrogen	H	1
Indium	In	49
Iodine	I	53
Iridium	Ir	77
Iron	Fe	26
Krypton	Kr	36
Lanthanum	La	57
Lawrencium	Lr	103
Lead	Pb	82
Lithium	Li	3
Lutetium	Lu	71
Magnesium	Mg	12
Manganese	Mn	25
Meitnerium	Mt	109
Mendelevium	Md	101
Mercury	Hg	80
Molybdenum	Mo	42
Neodymium	Nd	60
Neon	Ne	10
Neptunium	Np	93
Nickel	Ni	28
Niobium	Nb	41
Nitrogen	N	7
Nobelium	No	102
Osmium	Os	76
Oxygen	O	8
Palladium	Pd	46
Phosphorus	P	15
Platinum	Pt	78
Plutonium	Pu	94
Polonium	Po	84
Potassium	K	19
Praseodymium	Pr	59
Promethium	Pm	61
Protactinium	Pa	91
Radium	Ra	88
Radon	Rn	86
Rhenium	Re	75
Rhodium	Rh	45
Rubidium	Rb	37
Ruthenium	Ru	44
Rutherfordium	Rf	104
Samarium	Sm	62
Scandium	Sc	21
Seaborgium	Sg	106
Selenium	Se	34
Silicon	Si	14
Silver	Ag	47
Sodium	Na	11
Strontium	Sr	38

Sulphur	S	16	Ununpentium	Uup	115
Tantalum	Ta	73	Ununquadium	Uuq	114
Technetium	Tc	43	Ununseptium	Uus	117
Tellurium	Te	52	Ununtrium	Uut	113
Terbium	Tb	65	Ununium	Uuu	111
Thallium	Tl	81	Uranium	U	92
Thorium	Th	90	Vanadium	V	23
Thulium	Tm	69	Xenon	Xe	54
Tin	Sn	50	Ytterbium	Yb	70
Titanium	Ti	22	Yttrium	Y	39
Tungsten	W	74	Zinc	Zn	30
Ununbium	Uub	112	Zirconium	Zr	40
Ununhexium	Uuh	116			
Ununoctium	Uuo	118			

Appendix 10. SI units and conversions

LENGTH

Imperial		Metric
1 mile	=	1.61 kilometres
1 foot	=	30.5 centimetres
1 inch	=	2.54 centimetres
0.621 mile	=	1 kilometre
39.4 inches	=	1 metre

Unit		Number of metres
kilometres (km)	=	1,000
metre (m)	=	1
centimetre (cm)	=	0.01
millimetre (mm)	=	0.001
micrometre (micron)	=	0.000001

MASS AND WEIGHT

Imperial		Metric
1 pound	=	0.454 kilograms
1 ounce	=	28.3 grams
2.20 pounds	=	1 kilogram
0.0353 ounce	=	1 gram

Unit		Number of grams
kilogram (kg)	=	1000
hectogram	=	100
dekagram	=	10
gram (g)	–	1
milligram (mg)	=	0.001
microgram (μg)	=	0.000001
nanogram	=	0.000000001
picogram (pg)	=	0.000000000001

CONCENTRATION

		Weight/weight	Weight/volume
1 part per million	=	1 milligram per kilogram	1 milligram per litre
	=	1 microgram per gram	1 microgram per millilitre
1 part per billion	=	1 microgram per kilogram	1 microgram per litre
1 part per trillion	=	1 nanogram per kilogram	1 nanogram per litre

VOLUME (LIQUIDS)

Imperial		Metric
1 gallon	=	3.79 litres
1 quart	=	0.946 litres
1.06 quarts	=	1 litre
0.264 gallons	=	1 litre

Unit		Number of litres
kilolitre	=	1000
hectolitre	=	100

dekalitre	=	10
litre	=	1
decilitre	=	0.1
centilitre	=	0.01
millilitre	=	0.001
cubic centimetre	=	0.001
microlitre	=	0.000001

VOLUME (GASES)

Imperial		**Metric**
1 cubic foot	=	0.0283 cubic metre
35.3 cubic feet	=	1 cubic metre

Unit		**Number of cubic metre**
1 cubic metre	=	1
1 litre	=	0.001
1 cubic centimetre	=	0.000001